AF828983

Neuroscience and Social Science

Agustín Ibáñez • Lucas Sedeño
Adolfo M. García
Editors

Neuroscience and Social Science

The Missing Link

Springer

Editors

Agustín Ibáñez
Laboratory of Experimental Psychology and Neuroscience (LPEN)
Institute of Cognitive and Translational Neuroscience (INCYT) INECO Foundation, Favaloro University
Buenos Aires, Argentina

National Scientific and Technical Research Council (CONICET)
Buenos Aires, Argentina

Universidad Autónoma del Caribe
Barranquilla, Colombia

Center for Social and Cognitive Neuroscience (CSCN)
School of Psychology
Universidad Adolfo Ibáñez
Santiago, Chile

Centre of Excellence in Cognition and its Disorders
Australian Research Council (ACR)
Sydney, NSW, Australia

Lucas Sedeño
Laboratory of Experimental Psychology and Neuroscience (LPEN)
Institute of Cognitive and Translational Neuroscience (INCYT) INECO Foundation, Favaloro University
Buenos Aires, Argentina

National Scientific and Technical Research Council (CONICET)
Buenos Aires, Argentina

Adolfo M. García
Laboratory of Experimental Psychology and Neuroscience (LPEN)
Institute of Cognitive and Translational Neuroscience (INCYT) INECO Foundation, Favaloro University
Buenos Aires, Argentina

National Scientific and Technical Research Council (CONICET)
Buenos Aires, Argentina

Faculty of Education
National University of Cuyo (UNCuyo)
Mendoza, Argentina

ISBN 978-3-319-68420-8 ISBN 978-3-319-68421-5 (eBook)
DOI 10.1007/978-3-319-68421-5

Library of Congress Control Number: 2017957844

© Springer International Publishing Switzerland 2017
This work is subject to copyright. All rights are reserved by the Publisher, whether the whole or part of the material is concerned, specifically the rights of translation, reprinting, reuse of illustrations, recitation, broadcasting, reproduction on microfilms or in any other physical way, and transmission or information storage and retrieval, electronic adaptation, computer software, or by similar or dissimilar methodology now known or hereafter developed.
The use of general descriptive names, registered names, trademarks, service marks, etc. in this publication does not imply, even in the absence of a specific statement, that such names are exempt from the relevant protective laws and regulations and therefore free for general use.
The publisher, the authors and the editors are safe to assume that the advice and information in this book are believed to be true and accurate at the date of publication. Neither the publisher nor the authors or the editors give a warranty, express or implied, with respect to the material contained herein or for any errors or omissions that may have been made. The publisher remains neutral with regard to jurisdictional claims in published maps and institutional affiliations.

Printed on acid-free paper

This Springer imprint is published by Springer Nature
The registered company is Springer International Publishing AG
The registered company address is: Gewerbestrasse 11, 6330 Cham, Switzerland

Preface

This book results from discontent: although, in the last two decades, social cognitive affective neuroscience has been recognized as a strong field with potentially huge societal impact, the translation of findings from the laboratory to society remains markedly limited, if not altogether null. To face this scenario, here we aimed to provide a novel reconsideration of the borderlands of neuroscience and the social sciences, offering diverse, multidimensional perspectives about their current and potential interactions.

The volume comprises four sections. In Part I, we bring together neuroscientific perspectives on hot topics within social cognition, such as emotions, morality, and different forms of interpersonal dynamics. The works in Part II examine specific translational outlets of social neuroscience, including clinical settings and mass communication. Societally relevant implications of the field are further expounded in Part III, which focuses on poverty, social equality, and public health. To conclude, Part IV contains provocative reflections on conceptual, methodological, and translational issues which pervade the dialogue between neuroscience and the social sciences.

Such a vast array of topics come from the hand of renowned international experts operating in neuroscience, psychology, psychiatry, neurology, journalism, philosophy, biology, sociology, and therapy, among other fields. Together, their contributions provide a multidisciplinary and multi-domain view of the most recent interactions between social cognitive affective neuroscience and several social sciences. Each part offers a comprehensive vision about both the state-of-the-art and future trends in relevant areas, as well as an intrinsic discussion regarding the intertwine of neuroscience with other social sciences.

We would like to note that this is not a handbook, given that we are not aiming for exhaustiveness; rather, we are targeting selected prototypical interactions of neuroscience and social sciences in terms of complementarity, tensions, and fertile bidirectional critiques, as well as empirical and theoretical reconsiderations. By presenting contributions from diverse scientific and disciplinary domains, this book offers a comprehensive description of the present and future of neuroscience in different fields of society. Thus, we hope this endeavor will come to inform a necessary milestone for a more organic and active dialogue between multiple disciplines that are typically separated by individual approaches. After a long period of passionate work from the authors and ourselves, we believe that the result not only proves appealing to a wide audience but that it also overcomes classical discussions between neuroscience and varied humanistic fields, presenting the current and future developments which are critical for our society.

This book would not have been possible without the active participation of several actors. The authors have generously provided their outstanding knowledge to discuss the relevance of the interactions between neuroscience and social sciences. Expert reviewers have selflessly contributed with rigorous reviews, ensuring that only submissions of the highest quality made it to print. Our gratitude goes out to all these remarkable scholars who anonymously helped during the revision process. Also, we want to acknowledge the support from our network of institutions, which provided the time and support to develop this project: the Institute of Cognitive and Translational Neuroscience (INCYT); the INECO Foundation; Favaloro University; the National Scientific and Technical Research Council (CONICET); the National University of Cuyo (UNCuyo); Universidad Autónoma del Caribe; the Center for Social and Cognitive Neuroscience (CSCN), Universidad Adolfo Ibáñez; and the Centre of Excellence in Cognition and Its Disorders, Australian Research Council (ACR). We would also like to thank the different sources of funding directly or indirectly related to this work, namely, CONICET, CONICYT/FONDECYT (Regular 1170010), FONCyT-PICT (2012-0412 and 2012-1309), the Initiative for the Promotion of Regional Public Goods of the Inter-American Development Bank (IDB), and the INECO Foundation. We also extend our gratitude to the editorial board of Springer, especially to Bruno Fiuza: The idea of this book was conceived in a cafe in *Puerto Madero*, Buenos Aires, close to the *Río de la Plata*, one afternoon in which Bruno gently proposed Springer as a platform for our current hopes and concerns regarding the limits and possibilities of neuroscience. Finally, we express our deep and warmest regards to our life companions, Margherita, Pamela, and María, for their patience and understanding during the elaboration of this book. Their support is perhaps the best illustration of how crucial social factors are in the daily workings of neuroscience.

Buenos Aires, Argentina

Agustín Ibáñez
Lucas Sedeño
Adolfo M. García

Contents

Exploring the Borderlands of Neuroscience and Social Science 1
Agustín Ibáñez, Lucas Sedeño, and Adolfo M. García

Part I Neuroscientific Research on Social Cognition

Valuing Others: Evidence from Economics, Developmental Psychology, and Neurobiology 21
Pablo Billeke, Patricia Soto-Icaza, Mauricio Aspé-Sánchez, Verónica Villarroel, and Carlos Rodríguez-Sickert

Bias and Control in Social Decision-Making 47
Paloma Díaz-Gutiérrez, Sonia Alguacil, and María Ruz

Neurobiological Approaches to Interpersonal Coordination: Achievements and Pitfalls 69
Carlos Cornejo, Zamara Cuadros, and Ricardo Morales

The Social Neuroscience of Attachment 95
Pascal Vrtička

Mind-Reading in Altruists and Psychopaths 121
Fatima Maria Felisberti and Robert King

From Primary Emotions to the Spectrum of Affect: An Evolutionary Neurosociology of the Emotions 141
Warren D. TenHouten

Moral Cognition and Moral Emotions 169
Sandra Baez, Adolfo M. García, and Hernando Santamaría-García

On the Cognitive (Neuro)science of Moral Cognition: Utilitarianism, Deontology, and the "Fragmentation of Value" 199
Alejandro Rosas

The Social/Neuroscience: Bridging or Polarizing Culture and Biology? 217
Andrés Haye, Ricardo Morales, and Sebastián Niño

Part II Impact of Social Neuroscience on Social Spheres

Dementia and Social Neuroscience: Historical and Cultural Perspectives 243
Olivier Piguet

Clinical Studies of Social Neuroscience: A Lesion Model Approach 255
Fiona Kumfor, Jessica L. Hazelton, François-Laurent De Winter, Laurent Cleret de Langavant, and Jan Van den Stock

Psychotherapy and Social Neuroscience: Forging Links Together 297
Andrés Roussos, Malena Braun, Saskia Aufenacker, and Julieta Olivera

The Brain in the Public Space: Social Neuroscience and the Media 329
María Jimena Mantilla, Martín H. Di Marco, and Diego A. Golombek

Part III Integration of Social and Neuroscientific Insights

Electrophysiological Approaches in the Study of the Influence of Childhood Poverty on Cognition 349
Marcos Luis Pietto, Juan E. Kamienkowski, and Sebastián J. Lipina

The Cultural Neuroscience of Socioeconomic Status 383
Jung Yul Kwon, Ryan S. Hampton, and Michael E.W. Varnum

Social Ties, Health and Wellbeing: A Literature Review and Model 397
Andrew Haddon Kemp, Juan Antonio Arias, and Zoe Fisher

Part IV Philosophical Contributions to Theoretical, Methodological, and Ethical Questions

The Self-Domesticated Animal and Its Study 431
Mario Bunge

How Is Our Self Related to Its Brain? Neurophilosophical Concepts 443
Georg Northoff

Enaction and Neurophenomenology in Language 471
Roberto Arístegui

A Pluralist Framework for the Philosophy of Social Neuroscience 501
Sergio Daniel Barberis, M. Itatí Branca, and A. Nicolás Venturelli

Social Neuroscience and Neuroethics: A Fruitful Synergy 531
Arleen Salles and Kathinka Evers

Contributors

Sonia Alguacil Mind, Brain and Behavior Research Center, University of Granada, Granada, Spain

Department of Experimental Psychology, University of Granada, Granada, Spain

Juan Antonio Arias Department of Psychology, Swansea University, Swansea, UK

Roberto Arístegui Escuela de Psicología, Universidad Adolfo Ibáñez, Santiago, Chile

Mauricio Aspé-Sánchez División de Neurociencias (NeuroCICS), Centro de Investigación en Compleijidad Social, Facultad de Gobierno, Universidad del Desarrollo, Santiago, Chile

Saskia Aufenacker Universidad del Salvador, Buenos Aires, Argentina

Sandra Baez Grupo de Investigación Cerebro y Cognición Social, Bogotá, Colombia

Universidad de los Andes, Bogotá, Colombia

Laboratory of Experimental Psychology & Neuroscience (LPEN), Institute of Cognitive and Translational Neuroscience (INCYT), Institute of Cognitive Neurology (INECO) & CONICET, Favaloro University, Pacheco de Melo 1860, Buenos Aires, Argentina

National Scientific and Technical Research Council (CONICET), Buenos Aires, Argentina

Sergio Daniel Barberis Universidad de Buenos Aires (UBA), Buenos Aires, Argentina

Agencia Nacional de Promoción Científica y Tecnológica (ANPCyT), Buenos Aires, Argentina

Pablo Billeke División de Neurociencias (NeuroCICS), Centro de Investigación en Compleijidad Social, Facultad de Gobierno, Universidad del Desarrollo, Santiago, Chile

Malena Braun Equipo de Investigación en Psicología Clínica, Universidad de Belgrano, Buenos Aires, Argentina

Mario Bunge Department of Philosophy, McGill University, West Montreal, Quebec, Canada

Carlos Cornejo Laboratorio de Lenguaje, Interacción y Fenomenología (LIF), Escuela de Psicología, Pontificia Universidad Católica de Chile, Santiago, Chile

Zamara Cuadros Laboratorio de Lenguaje, Interacción y Fenomenología (LIF), Escuela de Psicología, Pontificia Universidad Católica de Chile, Santiago, Chile

Paloma Díaz-Gutiérrez Mind, Brain and Behavior Research Center, University of Granada, Granada, Spain

Department of Experimental Psychology, University of Granada, Granada, Spain

Kathinka Evers Centre for Research Ethics and Bioethics, Uppsala University, Uppsala, Sweden

Fatima Maria Felisberti Psychology Department, Kingston University London, London, UK

Zoe Fisher Traumatic Brain Injury Service, Morriston Hospital, Swansea, UK

Adolfo M. García Laboratory of Experimental Psychology and Neuroscience (LPEN), Institute of Cognitive and Translational Neuroscience (INCYT), INECO Foundation, Favaloro University, Buenos Aires, Argentina

National Scientific and Technical Research Council (CONICET), Buenos Aires, Argentina

Faculty of Education, National University of Cuyo (UNCuyo), Mendoza, Argentina

Diego A. Golombek Departamento de Ciencia y Tecnología, Universidad Nacional de Quilmes, Buenos Aires, Argentina

Consejo Nacional de Investigaciones Científicas y Técnicas (CONICET), Buenos Aires, Argentina

Ryan S. Hampton Department of Psychology, Arizona State University, Tempe, AZ, USA

Andrés Haye Pontificia Universidad Católica de Chile, Santiago, Chile

Jessica L. Hazelton The University of Sydney, School of Psychology, Sydney, NSW, Australia

The University of Sydney, Brain and Mind Centre, Sydney, NSW, Australia

Agustín Ibáñez Laboratory of Experimental Psychology and Neuroscience (LPEN), Institute of Cognitive and Translational Neuroscience (INCYT), INECO Foundation, Favaloro University, Buenos Aires, Argentina

National Scientific and Technical Research Council (CONICET), Buenos Aires, Argentina

Universidad Autónoma del Caribe, Barranquilla, Colombia

Center for Social and Cognitive Neuroscience (CSCN), School of Psychology, Universidad Adolfo Ibáñez, Santiago, Chile

Centre of Excellence in Cognition and its Disorders, Australian Research Council (ACR), Sydney, NSW, Australia

M. Itatí Branca Universidad Nacional de Córdoba (UNC), Córdoba, Argentina

Consejo Nacional de Investigaciones Científicas y Técnicas (CONICET), Buenos Aires, Argentina

Juan E. Kamienkowski Laboratorio de Inteligencia Artificial Aplicada (Departamento de Computación, FCEyN-UBA, CONICET), Buenos Aires, Argentina

Departamento de Física (FCEyN-UBA, CONICET), Buenos Aires, Argentina

Andrew Haddon Kemp Department of Psychology and the Health and Wellbeing Academy, College of Human and Health Sciences, Swansea University, Swansea, UK

Robert King Applied Psychology, University College Cork, Cork, Ireland

Fiona Kumfor The University of Sydney, School of Psychology, Sydney, NSW, Australia

The University of Sydney, Brain and Mind Centre, Sydney, NSW, Australia

ARC Centre of Excellence in Cognition and its Disorders, University of Sydney, Sydney, NSW, Australia

Jung Yul Kwon Department of Psychology, Arizona State University, Tempe, AZ, USA

Laurent Cleret de Langavant Faculté de Médecine, Université Paris Est, Créteil, France

Centre de référence maladie de Huntington, Hôpital Henri Mondor, AP-HP, Créteil, France

Laboratoire de NeuroPsychologie Interventionnelle, Institut National de la Santé et Recherche Médical (INSERM) U955, Equipe 01, Créteil, France

Département d'Etudes Cognitives, Ecole Normale Supérieure – PSL* Research University, Paris, France

Sebastián J. Lipina Unidad de Neurobiología Aplicada (UNA, CEMIC-CONICET), Buenos Aires, Argentina

María Jimena Mantilla Instituto de Investigaciones Gino Germani, Buenos Aires, Argentina

Consejo Nacional de Investigaciones Científicas y Técnicas (CONICET), Buenos Aires, Argentina

Martín H. Di Marco Consejo Nacional de Investigaciones Científicas y Técnicas (CONICET), Buenos Aires, Argentina

Instituto de Investigaciones Gino Germani, Buenos Aires, Argentina

Ricardo Morales Laboratorio de Lenguaje, Interacción y Fenomenología (LIF), Escuela de Psicología, Pontificia Universidad Católica de Chile, Santiago, Chile

A. Nicolás Venturelli Instituto de Humanidades (UNC/CONICET), Buenos Aires, Argentina

Sebastián Niño Pontificia Universidad Católica de Chile, Santiago, Chile

Georg Northoff Mind, Brain Imaging and Neuroethics Research Unit, Institute of Mental Health Research, Royal Ottawa Mental Health Centre, Ottawa, ON, Canada

Julieta Olivera Equipo de Investigación en Psicología Clínica, Universidad de Buenos Aires, Buenos Aires, Argentina

Consejo Nacional de Investigaciones Científicas y Técnicas (CONICET), Buenos Aires, Argentina

Marcos Luis Pietto Unidad de Neurobiología Aplicada (UNA, CEMIC-CONICET), Buenos Aires, Argentina

Laboratorio de Inteligencia Artificial Aplicada (Departamento de Computación, FCEyN-UBA, CONICET), Buenos Aires, Argentina

Olivier Piguet The University of Sydney, School of Psychology and Brain & Mind Centre, Sydney, NSW, Australia

ARC Centre of Excellence in Cognition and its Disorder, Sydney, NSW, Australia

Carlos Rodríguez-Sickert División de Neurociencias (NeuroCICS), Centro de Investigación en Compleijidad Social, Facultad de Gobierno, Universidad del Desarrollo, Santiago, Chile

Alejandro Rosas Philosophy Department, National University of Colombia, Bogotá, Colombia

Andrés Roussos Equipo de Investigación en Psicología Clínica, Universidad de Buenos Aires, Buenos Aires, Argentina

Consejo Nacional de Investigaciones Científicas y Técnicas (CONICET), Buenos Aires, Argentina

María Ruz Mind, Brain and Behavior Research Center, University of Granada, Granada, Spain

Department of Experimental Psychology, University of Granada, Granada, Spain

Arleen Salles Centre for Research Ethics and Bioethics, Uppsala University, Uppsala, Sweden

Centro de Investigaciones Filosóficas, Buenos Aires, Argentina

Hernando Santamaría-García Universidad Javeriana, Bogotá, Colombia

Centro de Memoria y Cognición Intellectus, Hospital Universitario San Ignacio, Bogotá, Colombia

Laboratory of Experimental Psychology & Neuroscience (LPEN), Institute of Cognitive and Translational Neuroscience (INCYT), Institute of Cognitive Neurology (INECO) & CONICET, Favaloro University, Buenos Aires, Argentina

Grupo de Investigación Cerebro y Cognición Social, Bogotá, Colombia

National Scientific and Technical Research Council (CONICET), Buenos Aires, Argentina

Lucas Sedeño Laboratory of Experimental Psychology and Neuroscience (LPEN), Institute of Cognitive and Translational Neuroscience (INCYT), INECO Foundation, Favaloro University, Buenos Aires, Argentina

National Scientific and Technical Research Council (CONICET), Buenos Aires, Argentina

Patricia Soto-Icaza Interdisciplinary Center of Neuroscience, Pontificia Universidad Católica de Chile, Santiago, Chile

Jan Van den Stock Laboratory for Translational Neuropsychiatry, Department of Neurosciences, KU Leuven, Leuven, Belgium

Department of Old Age Psychiatry, University Psychiatric Center KU Leuven, Leuven, Belgium

Warren D. TenHouten Department of Sociology, University of California at Los Angeles, Los Angeles, CA, USA

Michael E.W. Varnum Department of Psychology, Arizona State University, Tempe, AZ, USA

Verónica Villarroel Centro de Investigación y Mejoramiento de la Educación (CIME), Facultad de Psicología, Universidad del Desarrollo, Concepción, Chile

Pascal Vrtička Max Planck Institute for Human Cognitive and Brain Sciences, Leipzig, Germany

François-Laurent De Winter Laboratory for Translational Neuropsychiatry, Department of Neurosciences, KU Leuven, Leuven, Belgium

Department of Old Age Psychiatry, University Psychiatric Center KU Leuven, Leuven, Belgium

Abbreviations

ABRs	Auditory brainstem responses
ACC	Anterior cingulate cortex
ACG	Active control group
AD	Alzheimer's disease
AI	Anterior insula
AIC	Anterior insula cortex
aPFC	Anterior prefrontal cortex
AR	Authority ranking
ASD	Autism spectrum disorder
aSTG	Anterior superior temporal gyrus
ATP	Anterior temporal pole
AV	Attachment avoidance
AWM(s)	Attachment working model(s)
AX	Attachment anxiety
BA	Brodmann area
BDI	Beck depression inventory
BDNF	Brain-derived neurotrophic factor
BEAST	Bodily expressive action stimulus test
BEES	Balanced emotional empathy scale
BES	Basic empathy scale
BLRI	Barret–Lennard relationship inventory
BMI	Body Mass Index
BOLD	Blood-oxygen-level dependent
BPD	Borderline personality disorder
BT	Behavior therapy
bvFTD	Behavioral-variant frontotemporal dementia
CBT	Cognitive behavioral therapy
CCRT	Core conflictual relationship theme
CE	Corrective experience
CeA	Central nucleus of the amygdala
CHD	Coronary heart disease

CMS	Cortical midline structures
CS	Communal sharing
CTRA	Conserved transcriptional response to adversity
dACC	Dorsal anterior cingulate cortex
DG	Dictator game
dlPFC	Dorsolateral prefrontal cortex
DMN	Default mode network
dmPFC	Dorsomedial prefrontal cortex
DNA	Deoxyribonucleic acid
DSM	Diagnostic and statistical manual
EEG	Electroencephalography
EM	Equality matching
EP	Explanatory pluralism
EPT	Empathy-for-pain task
EQ	Empathy quotient
ERN	Error-related negativity
ERP	Event-related potential
ERP	Event-related potentials
ESUP	Expressive suppression
EVC	Expected value of control theory
FBT	False belief task
FEAST	Facial expressive action stimulus test
FFA	Fusiform face area
FG	Fusiform gyrus
fMRI	Functional magnetic resonance imaging
fNIRS	Functional near-infrared spectroscopy
FOEs	Fortune-of-other emotions
FTLD	Frontotemporal lobar degeneration
FVT	Foraging value theory
GAD	General anxiety disorder
GAF	Global assessment of functioning
GENIAL model	*G*enomics-*e*nvironment-vagus *n*erve-social *i*nteraction-*a*llostatic regulation-*l*ongevity
HD	Huntington's disease
HME	Higher maternal education
HNPS	Hanse neuropsychoanalysis study
HPA	Hypothalamic-pituitary-adrenal
HR	Hazard ratio
HRV	Heart rate variability
HS	Head start
iCBT	Internet-based cognitive behavioral therapy
ICD	International classification of disease
IFG	Inferior frontal gyrus
IOS	Inclusion of the other in the self

IPP	Interpersonal psychotherapy
IQ	Intelligence quotient
IRI	Interpersonal reactivity index
LDL	Low-density lipoprotein
LG	Licking and grooming
LME	Lower maternal education
LPP	Late positive potential
mACC	Middle anterior cingulate cortex
MASC	Movie for the assessment of social cognition
MBCT	Mindfulness-based cognitive therapy
MDD	Major depressive disorder
MET	Multifaceted empathy test
MFN	Medial frontal negativity
Mini-SEA	Mini-social cognition and emotional assessment
MNS	Mirror neuron system
MOR	μ-opioid receptor
MP	Market pricing
mPFC	Medial prefrontal cortex
MRI	Magnetic resonance imaging
MRS	Modified ranking scale
MVPA	Multi-voxel pattern analyses
NAT	Natural viewing
NCS	Neural correlates
NES	Neural enabling condition of self
NIMH	National institute of mental health
NPS	Neural predisposition of self
OCD	Obsessive compulsive disorder
OFC	Orbitofrontal cortex
OR	Odds ratio
OXT	Oxytocin
OXTR	Oxytocin receptor
PACC	Perigenual anterior cingulate cortex
PCC	Posterior cingulate cortex
PCC	Precuneus cingulate
PET	Positron emission tomography
PFC	Prefrontal cortex
POR	Practice-oriented research
PPDT	Psychodynamic therapy
PreHD	Presymptomatic HD carriers
pSTS	Posterior STS
PT	Perspective taking
PTSD	Post-traumatic stress disorder
RCT	Randomized clinical trial
RDoC	Research domain criteria
REAP	(Cognitive) reappraisal

RMET	Reading the mind in the eyes test
RO	Response of the others
RS	Response of the self
RT	reaction time(s)
SACC	Supragenual anterior cingulate cortex
SAD	Social anxiety disorder
SApNS	Social approach neural system
SAvNS	Social aversion neural system
SCE	Self-conscious emotions
SCL-90	Symptom check list 90
SCNM	social context network model
SD	Semantic dementia
SES	Socioeconomic status
SN	Social neuroscience
SN	Substantia nigra
SPECT	Single-photon-emission computed tomography
SRE	Self-reference effect
ST	Simulation theory
STS	Superior temporal sulcus
SVO	Social value orientation
TASIT	The awareness of social inference test
TC	Temporal cortex
TG	Trust game
TMS	Transcranial magnetic stimulation
ToM	Theory of mind
TP	Temporal pole
TPr	Training program
TPJ	Temporoparietal junction
TT	Theory theory
UR	Utilitarian response(s)
vlPFC	Ventrolateral prefrontal cortex
vmOFC	Ventromedial orbitofrontal cortex
vmPFC	Ventromedial prefrontal cortex
VNS	Vagal nerve stimulation
VPT	Visual perspective taking
VS	Ventral striatum
VTA	Ventral tegmental area
W	Wish
WAIS	Wechsler adult intelligence scale
WEIRD	Western, educated, industrialized, rich, and democratic

Exploring the Borderlands of Neuroscience and Social Science

Agustín Ibáñez, Lucas Sedeño, and Adolfo M. García

Abstract The field of social cognitive affective neuroscience seems to overcome long-standing problems undermining old-fashioned cognitive neuroscience, such as its reductionist approach; its exclusion of affect, body, and culture in the comprehension of mental phenomena; and its propensity toward isolationist models over integrative or multilevel theories. Moreover, in this developing field, centuries-old arguments of incommensurability between natural and human sciences can be reframed as little more than pseudoproblems. The apparent paradigm shift inherent in social cognitive neuroscience entails new conceptual, methodological, metatheoretical, and aesthetic questions. Also, it gives rise to novel problems as

A. Ibáñez (✉)
Laboratory of Experimental Psychology and Neuroscience (LPEN), Institute of Cognitive and Translational Neuroscience (INCYT), INECO Foundation, Favaloro University, Buenos Aires, Argentina

National Scientific and Technical Research Council (CONICET), Buenos Aires, Argentina

Universidad Autónoma del Caribe, Barranquilla, Colombia

Center for Social and Cognitive Neuroscience (CSCN), School of Psychology, Universidad Adolfo Ibáñez, Santiago, Chile

Centre of Excellence in Cognition and its Disorders, Australian Research Council (ACR), Sydney, NSW, Australia
e-mail: aibanez@ineco.org.ar

L. Sedeño
Laboratory of Experimental Psychology and Neuroscience (LPEN), Institute of Cognitive and Translational Neuroscience (INCYT), INECO Foundation, Favaloro University, Buenos Aires, Argentina

National Scientific and Technical Research Council (CONICET), Buenos Aires, Argentina
e-mail: lucas.sedeno@gmail.com

A.M. García
Laboratory of Experimental Psychology and Neuroscience (LPEN), Institute of Cognitive and Translational Neuroscience (INCYT), INECO Foundation, Favaloro University, Buenos Aires, Argentina

National Scientific and Technical Research Council (CONICET), Buenos Aires, Argentina

Faculty of Education, National University of Cuyo (UNCuyo), Mendoza, Argentina
e-mail: adolfomartingarcia@gmail.com

© Springer International Publishing Switzerland 2017

A. Ibáñez et al. (eds.), *Neuroscience and Social Science*,
DOI 10.1007/978-3-319-68421-5_1

it taxes the boundaries with other disciplines. Many of these dynamical tensions among related fields of knowledge, which are often left implicit, continue to change across domains and periods. Here we chart such new borderlands and summarize the contributions comprised in the present book. *Neuroscience and Social Science: The Missing Link* engages empirical researchers and theorists around the world in an attempt to integrate perspectives from many relevant disciplines, separating real from spurious divides between them and delineating new challenges for future investigation. The volume is organized in four sections. Section A is devoted to neuroscientific research on specific domains of social cognition, ranging from social emotions, negotiation, cooperation, and interpersonal coordination to empathy and morality. Section B focuses on the impact of social neuroscience in specific social spheres, namely, the clinical field, psychotherapeutic settings, and the mass media. Section C encompasses works on the integration of social and neuroscientific insights to approach matters as pressing as poverty, socioeconomic inequality, health, and well-being. Finally, Section D offers philosophical contributions on theoretical, methodological, and even ethical questions arising from such promising interdisciplinary encounter. Through this wide-ranging proposal, the volume promotes novel reflections on a much-needed marriage while opening opportunities for social neuroscience to plunge from the laboratory into the core of social life.

Keywords Social neuroscience • Social sciences • Multilevel approaches • Explanatory pluralism • Translational neuroscience

1 New Vistas for the Dialogue Between Neuroscience and the Social Sciences

Old-fashioned cognitive neuroscience (as practiced from the 1970s to the 1990s) has been plagued with empirical, metatheoretical, and transdisciplinary obstacles, such as a reductionist approach; the exclusion of affect, body, and culture in the comprehension of mental phenomena; and a propensity toward isolationist models over integrative theories. Many of the ensuing caveats have been circumvented by the new field of social and cognitive neuroscience, which has fully reconsidered the mind from a pluralistic, situated agenda.

Since its inception in the early 2000s, multilevel social neuroscience [1–8] (or social cognitive affective neuroscience) has dramatically shaped our understanding of the affective and cultural dimensions of neurocognition. Thanks to its explanatory pluralism, this field has moved beyond long-standing dichotomies and reductionisms, offering a neurobiological perspective on topics classically monopolized by nonscientific traditions or purely social-oriented sciences, such as consciousness, subjectivity, and free will. Moreover, it has forged new avenues for dialogue with disciplines which directly address societal dynamics, such as economics, law, education, public policy making, and sociology. This interaction has even given rise to new specialties, such as neurosociology [9] or cultural neuroscience [10, 11]. Consequently, unprecedented opportunities have emerged to explore the intimate links between individual biological processes and interpersonal sociocultural phenomena.

In particular, social neuroscience has emphasized the need to conceptualize human cognition as a situated, context-dependent, embodied phenomenon which is variously modulated by affect, action, society, and culture [12–14]. Under this light, centuries-old arguments of incommensurability between natural and human sciences can be reframed as little more than pseudoproblems.

2 New Approaches, New Problems, New Opportunities

This book was conceived in our everyday research on social and cognitive neuroscience. The group we three lead together has always operated in interdisciplinary frontiers. Even in the projects which target specialized domains (such as social, affective, and cognitive aspects of psychiatric and neurological conditions), our team has integrated neuroscientific approaches with eclectic contributions from various fields.

In particular, we have aimed to overcome the psychiatry/neurology divide by proposing dimensional and transnosological accounts of social cognition across multiple disorders. We have thus sought to escape from an old dualistic approach which still impacts current theoretical and applied frameworks in clinical neuroscience, as seen in persistent oppositions such as brain structure vs. brain functions, neurological diseases vs. mental diseases, neural causes vs. psychological causes, and so on [15, 16]. We also have proposed a multilevel, partially emergentist, and explanatory pluralism based on metatheoretical roots for a new social cognitive neuroscience [13, 14, 17–19].

These theoretical changes promote a reconsideration of the cognitive research agenda in terms of *intercognition* [15, 20]. In this sense, through our laboratory work, we have striven to assess (1) contextual modulations of social phenomena [12, 21, 22]; (2) the neural basis of social prejudice [23–25]; (3) cross-domain synergies between language and actions [26–32] and between dance, anticipatory processes, and plastic changes induced by expertise [33, 34]; (4) multidimensional signatures of specific neurocognitive processes [35, 36]; and (5) automated analyses of spontaneous verbal behavior in noncontrolled and naturalistic settings [37]. We have also conducted work outside the laboratory and applied behavioral insights inspired by social neuroscience to uncover extreme cognitive conditions in real life, such as the mindset and moral cognition of extremist terrorists [38]. We also have relied on insights from social neuroscience to better understand the dynamics of economic cooperation and betrayal beyond the limits of classical behavioral economics paradigms [35, 39, 40]. Finally, we have pursued alternatives to compartmentalized conceptions of cognitive phenomena, moving toward more ecological tasks which better reflect the situated impact of mental disorders [21, 35, 41–51].

Yet, despite our efforts to explore the borderlands of social cognitive neuroscience and other social sciences, we still feel constrained by disciplinary isolation, building models of neurocognitive processes that are still largely confined to the limits of the laboratory.

Despite its limitations, this apparent paradigm shift inspired by the new arena of social and cognitive neuroscience lays the ground for novel challenges at multiple levels. Most of them can be captured by three outstanding questions: (a) how does social neuroscience appraise current definitions of moral cognition, socialization, and cooperation?; (b) how can we reconceptualize mental (psychiatric or neurological) diseases by integrating social, psychological, and biological perspectives?; and (c) how can researchers cope with the tension between experimental control and ecological validity in trying to investigate social cognition as a situated phenomenon?

At the same time, beyond internal changes in the field of neuroscience, new problems emerge in the dialogue with other disciplines. Even when the need for explanatory pluralism is explicitly acknowledged, experimental research about cognition typically targets a set of compartmentalized, universal, context-free operations via atomistic spectatorial paradigms. More particularly, a theoretical abyss still exists between neuroscience and the social sciences, even though terminological overlaps may create the illusion of a vast shared ground.

Finally, although it has the potential to do so, social cognitive neuroscience does not provide comprehensive theories to broadly understand the biological, psychological, and communal emergence of the mind, limiting the potential growth of theories from multiple fields. Most efforts in social cognitive neuroscience come from *models of* specific domains (e.g., empathy, theory of mind, cooperation, moral judgment, moral emotion, moral reasoning, basic and social emotions, social perception, etc.). These *models of* lack theoretically integrated roots and, above all, common ideas for interdisciplinary metatheorization. The absence of an integrated theoretical corpus underlying disparate microprocess was early noted by journalist John Horgan in his book *The Undiscovered Mind.* As it happens, there are no such things as fully isolated cognitive domains in everyday cognition [20, 52]. On the contrary, there is a natural blending of processes influenced by internal and external milieus, which jointly confer our ongoing experience with a *sensus communis* [20]. This emergent gestalt (beyond perception) is not reducible to the sum of their (hypothesized) components, and it is continually signified as phenomenological and sociocultural constraints unfold through time. Thus, a Babel Tower of models without theoretical integration can act as a demagnetized compass when interdisciplinary convergence is of the essence.

So far, the efforts of neuroscience to come close to social science rest on strong but poorly recognized barriers. Of course, some relevant maneuvers must be highlighted, such as the use of ecological tasks resembling everyday cognition, the emergence of "second-person" or "two-person" neuroscience [53, 54] and hyperscanning techniques [55], the avoidance of excessively artificial settings and stimuli, the implementation of multisource recording technologies, the pursuit of more direct links with new disciplines within the social sciences, and the call for greater transdisciplinary discussion. Notwithstanding, a critical and more direct interaction between neuroscience and social sciences is urgently required.

Like many other neuroscientists, we acknowledge the pressing need to forge the abovementioned dialogue. This book is part of our attempt to explore, discuss, promote, criticize, and reconsider the limits between social neuroscience and social science. We believe that many of these dynamical tensions among related fields of knowledge are often left implicit and that they continue to change across domains and periods. In the *neuroboom* era, it seems that a critical and polyphonic approach is required to discuss the limits and possibilities of cross-fertilization between both disciplines. Thus, an explicit assessment of these borderlands, even if incomplete, biased, or subsampled, could represent a promissory starting point to explore covert and overt rapprochements across relevant subfields.

A deeper dialogue calls for new insights into the limits of interdisciplinary theory construction, the guises of shared terminology, and the translational possibilities of brain-based research. More generally, these imperatives prompt consideration of an overarching issue: to what extent can knowledge from social neuroscience and social science foster concrete progress in the other fields? If the answer proves elusive, it is because the synergy between both fields has been preliminary at best. Most research groups still work in isolation or in sporadic collaboration, without developing pluralistic studies or active cooperation networks. Wide-ranging breakthroughs could be achieved by combining the strengths of both sides, enhancing global networking, and instilling a joint translational philosophy. Importantly, collaborative developments may promote new cross-disciplinary approaches to specific problems of social life.

The incipient field of social neuroscience has begun to break the traditional and artificial separations between biology and social realms, thus giving rise to new challenges. These involve potential differences in the definition of social phenomena across disciplines, the limits of multilevel and transdisciplinary co-construction, the methodological tension between experimental frameworks and daily life phenomena, and the current gaps between the promises and achievements of translational neuroscience. Aimed to explore those new borderlands, the present book engages empirical researchers and theorists operating in neuroscience and social science around the world. Specifically, it intends to integrate perspectives from both disciplines, separating real from spurious divides between them and delineating new challenges for future investigation.

3 The Contributions of This Book

The volume includes contributions by experts interested in the convergences, divergences, and controversies across these fields, including studies on the interplay among relevant levels of inquiry (e.g., neural, psychological, and social dimensions), articles rooted in specific scholarly traditions (e.g., neuroscience, sociology, philosophy of science), and essays on new theoretical foundations to enhance the rapprochement in question.

The volume is organized in four sections. Section A is devoted to neuroscientific research on specific domains of social cognition, ranging from social emotions, moral cognition, negotiation, cooperation, and interpersonal coordination to empathy and morality. Section B focuses on the impact of social neuroscience in specific social spheres, namely, the clinical field, psychotherapeutic settings, and the mass media. Section C encompasses works on the integration of social and neuroscientific insights to contribute to matters as pressing as poverty, socioeconomic inequality, health, and well-being. Finally, Section D offers philosophical contributions on theoretical, methodological, and even ethical questions arising from this promising interdisciplinary encounter.

Part I opens with the chapter "Valuing Others: Evidence from Economics, Developmental Psychology, and Neurobiology", in which Billeke et al. [56] examine how different disciplines (such as economic research, social psychology, and neuroscience) address the relation between decision-making and others' preferences and behaviors. According to the authors, we weigh others' preferences to adapt our own behavior and achieve adequate social interactions. In the first part of the chapter, they discuss how the economic perspective has started to include social preferences as a key influence on social decision-making, as well as how neuroscience can contribute to improve behavioral models. Then, they review research from developmental psychology related to the way human beings understand and integrate others' perspectives into their own behavior and decisions. Finally, findings from social neuroscience and neuroeconomics are presented, with emphasis on the neurobiological mechanisms underlying social decision-making. With this chapter, the authors propose a starting point for building a multilevel approach to study complex social behavior in humans.

The topic of social decision-making is also targeted by Díaz-Gutiérrez et al. [57]. Their chapter, titled "Bias and Control in Social Decision-Making", focuses on the neurocognitive bases of relevant biases and control mechanisms. First, an overview is offered of the main tasks (e.g., behavioral economics games, implicit association tasks), methods (e.g., electroencephalography, functional magnetic resonance imaging), and analytical tools (e.g., univariate and multivariate approaches) used to assess the neural basis of social biases. Second, biases in social decision-making are empirically shown to depend on individual factors, including gender, age, prosociality, and emotional states. Likewise, this domain seems permeable to several stimulus-related variables, such as facial cues driving social categorization, the appraisal of emotional expressions and trustworthiness, and the presence of personal information about the agents involved. Third, several studies are reviewed which show how economic and moral decisions, despite the role of automatic and implicit biases, can be regulated via control mechanisms. Overall, the cognitive systems mediating biases and adaptive control are proposed to depend on widespread brain networks spanning several cortical (mainly parietal, temporal, and prefrontal) and subcortical (e.g., insula, anterior cingulate, amygdala) regions.

Further insights into the neurobiology of social phenomena are offered by Cornejo et al. in "Neurobiological Approaches to Interpersonal Coordination: Achievements and Pitfalls" [58], which addresses the study of interpersonal coordination. Through a wide-ranging review of experimental research (based on neurophysiological, hemodynamic, behavioral, and peripheral measures), the authors address the breakthroughs and shortcomings of two leading approaches: social neuroscience and the dynamical systems theory. Both trends are shown to have fostered greater understanding of intersubjective couplings in highly constrained scenarios. However, they are argued to be limited in their capacity to characterize situated, real-life interactions among individuals. First, the review shows that most studies in social neuroscience measure dyads' actions (movements or neural activity) in structured tasks, focusing on individual brains over genuinely bipersonal phenomena. Second, while ensuing solipsist biases are apparently circumvented by dynamical systems theory, this framework reduces social life to the same set of principles governing any other complex system (e.g., traffic or weather changes), thus proving blind to the key defining features of human interaction. In light of such shortcomings, Cornejo et al. close their chapter by advocating an emergentist, holistic, context-sensitive, and meaning-driven conceptualization of interpersonal coordination.

Next, in "The Social Neuroscience of Attachment", Pascal Vrtička [59] vouches for attachment theory as a good example of a profitable interaction between neuroscience and social science in the quest to understand human development. First, the author presents the fundamental assumptions of attachment theory and their implications from an evolutionary and a sociocultural perspective. Then, he shows how this theory has motivated experimental studies in social neuroscience and how their findings may give rise to potential prevention and intervention strategies in the fields of mental health, physical health, and policy making. Finally, the author discusses future avenues for the "social neuroscience of attachment" as a field for promising interdisciplinary breakthroughs.

Then, in the chapter titled "Mindreading in Altruists and Psychopaths", Felisberti and King [60] discuss mind reading as a neurocognitive capacity which allows navigating through the intricacies of social interactions. In particular, the authors focus on altruists and psychopaths as relevant models to illuminate this domain. First, the very notion of mind is introduced and framed as a (problematic) conceptual cornerstone to examine our ability to attribute cognitive and affective states to others. Second, the notion of mind reading is defined and analyzed in terms of current relevant theories. A distinction is then made between cognitive and affective mind reading, with emphasis on the domains of empathy and mimicry, and critical neurological correlates of such faculties are identified. Third, the authors review classical paradigms in the study of mind reading, namely, the Reading the Mind in the Eyes Test, the false-belief task, the Sally-Anne task, the Yoni test, the Heider-Simmel illusion, and the Movie for the Assessment of Social Cognition. Fourth, a review is offered of empirical results illuminating mind reading tendencies in altruists, psychopathic individuals, and other expert manipulators, such as con-artists and magicians. In sum, this chapter offers a concise overview of the neurocognitive basis of a crucial form of social interaction.

The following chapter, "From Primary Emotions to the Spectrum of Affect: An Evolutionary Neurosociology of the Emotions", is a contribution from Warren TenHouten [61], who presents an evolutionary neurosociology of emotions acknowledging the links among mind, brain, and society as three interactive levels of analysis. TenHouten proposes that Fiske's fourfold social relations model and the evolutionary neuroethology of Maclean nurture Plutchik's fourfold model of life problems, which is the basis of a promising psychoevolutionary framework to conceptualize emotions. The four relational models consist in eight fundamental socio-relational situations (based on both positive and negative valences) and the prototypical adaptive reactions to them, which has led to the evolutionary development of primary emotions. The four pairs of opposite emotions from this conception are described as natural kinds, whose combination gives rise to complex emotions. Then, the author presents a classification of secondary and tertiary emotions. Finally, he shows results from an empirical analysis to evaluate the relation between two opposite secondary emotions and their opposite primary components, as well as their valenced social relations.

Also in the line of socio-affective processes, the chapter titled "Moral Cognition and Moral Emotions", by Baez et al. [62], offers a comprehensive assessment of the neurocognitive mechanisms subserving moral cognition and moral emotions. First, this contribution summarizes the history of insights into moral cognition, characterizing three of its key subdomains (moral sensitivity, moral reasoning, and moral judgment). Next, based on neuroimaging evidence from clinical and neurotypical samples as well as behavioral studies on neuropsychiatric populations, the authors trace neurobiological and phenomenological links between moral cognition and three socio-cognitive domains: theory of mind, empathy, and moral emotions. Thereupon, they focus on the latter to specify the neurocognitive underpinnings of two broad emotion types, namely, fortune-of-other emotions (in particular, envy and *Schadenfreude*) and self-conscious emotions (including shame and guilt). The chapter then introduces current context-sensitive models of moral cognition rooted in brain network approaches. Finally, from a translational perspective, several real-life scenarios are considered which could profit from ongoing and future rapprochements between social sciences and cognitive neuroscience. In sum, Baez et al. propose a wide-ranging characterization of multiple cognitive and affective processes which influence our morality on a daily basis and contribute to our uniqueness as a species.

The theme of moral cognition is recapped by Alejandro Rosas [63]. His chapter, titled "On the Cognitibe (Neuro)science of Moral Cognition: Utilitarianism, Deontology and the 'Fragmentation of Value'", shows that normative conclusions about moral judgment based on neuroscientific research crucially depend on theoretical commitments. First, the author briefly describes the dual-process theory of cognition, its impact in moral cognition, and the assumptions that support normative conclusions. Then, new data from tasks based on this theory are presented alongside findings from cognitive load studies. Together, these are shown to support an alternative version of the dual-process model, more akin to the philosophical-normative notion of the "fragmentation of value." This alternative version of the dual-process model aims to overcome the neat overlap between the deontological/utilitarian and the intuitive/reflective divides. Thereon, the author proposes that both utilitarian and deontological intuitions are equally fundamental and partially in tension. In addition, the concept of variable utilitarian and deontological sensitivities among individuals is introduced. Finally, the author presents the normative conclusion of this alternative dual-process theory.

This section ends with "The Social/Neuro Science: Bridging or Polarizing Culture and Biology?", a work in which Haye et al. [64] examine the conceptual foundations of self-regulation and discuss the biological assumptions of this phenomenon at the crossing of psychology and neuroscience. To this end, they review contemporary research on self-regulation in experimental psychology and social neuroscience, focusing on theoretical models of self-regulation, emotion regulation, and attentional regulation. Thereon, the authors address the caveats framing current conceptions of the biological dimension of self-regulation, considering dualistic, individualistic, aprioristic, adaptationistic, and anthropocentric limitations. In their view, these features represent a theoretical shortcoming that compromises the interplay between culture and biology and which may eventually increase the gap that social neuroscience aims to bridge.

Part II begins with "Dementia and Social Neuroscience: Historical and Cultural Perspectives". In this contribution, Olivier Piguet [65] offers a chronicle of how neuroscientists have changed their perception of social and emotional behaviors from nuisance variables of "higher" cognitive functions to integral aspects of human cognition. Focusing on neurodegenerative diseases, he shows that diagnosis of dementia has evolved from a primary classical cognitive approach (emphasizing alterations of memory, language, and attention) to the inclusion of social/emotional impairments as key signs of the disease. Moreover, the author highlights that the inclusion of social neuroscience methods in the clinical evaluation of neurodegenerative disease can increase the diagnostic accuracy and specificity and that it may also contribute to predicting the rate of disease progression and underlying neuropathological patterns. In short, this historical overview maps the evolution of our contemporary conceptions of dementia while foregrounding the critical role that social neuroscience has played in such a process.

On a similar note, Kumfor et al. [66] devote their chapter, "Clinical Studies of Social Neuroscience: A Lesion Model Approach", to examining how clinically based research in social neuroscience provides key evidence to understand the neurobiological basis of complex human behaviors. Emphasis is placed on the lesion model approach (now extended to neurodegenerative disorders, thanks to advances in structural and functional neuroimaging techniques), which allows identifying potential brain networks underpinning social behaviors, such as face processing, emotion recognition, theory of mind, and empathy. Specifically, the authors review research that combines behavioral and neuroimaging approaches in four progressive neurodegenerative disorders: behavioral variant frontotemporal dementia, semantic dementia, Huntington's disease, and Alzheimer's disease. In addition, the chapter presents the paradigms that have been employed to evaluate social behavior in these conditions and the current patterns of findings which enlighten our understanding of the "social brain." Finally, the authors discuss the importance of including social cognition assessments in current diagnostic criteria for neurodegenerative diseases to enhance comprehension of these conditions.

For their own part, in the chapter titled "Psychotherapy and Social Neuroscience: Forging Links Together", Roussos et al. [67] discuss the need to forge new links between psychotherapy, as an essentially interpersonal practice, and social neuroscience, as a potentially useful translational arena, to refine theories relevant for both fields. To this end, they review studies integrating elements from both fields, describing their approaches, contributions, obstacles, and methodological challenges. Emphasis is placed on the notions of empathy and interpersonal relationships (including therapeutic alliance and attachment), as key constructs on which to anchor this interdisciplinary pursuit. Finally, the authors reflect on how ensuing practice-oriented research could empower psychotherapists and other mental health professionals in clinical practice.

Then, in "The Brain in the Public Space: Social Neuroscience and the Media" Mantilla et al. [68] examine how research in neuroscience (and, in particular, social neuroscience) is disseminated by the mass media. The authors state that in the last decades, knowledge about the brain has increased significantly and that it has started to circulate outside traditional academic spheres. They then discuss the particular features of this phenomenon in the field of neuroscience. By way of illustration, they present an analysis of how neuroscience has been covered in a national newspaper from Argentina. According to their findings, social neuroscience research does not represent a significant proportion of the reports disseminated in the press, although their prevalence is on the increase. This is supported by the growth of press publications related to interpersonal ties and emotional mechanisms. To conclude, the authors propose recommendations to overcome the gap between scientific research and its popularization in mass media.

Section C comprises three contributions examining how research at the interface of neuroscience and social science can illuminate major problems affecting multiple communities. In "Electrophysiological Approaches in the Study of the Influence of Childhood Poverty on Cognition", Pietto et al. [69] discuss how joint insights from neuroscience and social sciences might contribute to establishing indicators of the effects of childhood poverty. Specifically, the authors offer a systematic review of electrophysiological studies addressing the influence of childhood poverty on cognitive development. Most of the paradigms implemented in these studies measure neural activity during inhibitory control, selective attention, and resting-state designs, comparing children with low and high socioeconomic status. The findings show between-group differences in neural markers of interference control and auditory sensory processing, together with differential patterns of oscillatory activity in frontal regions. Based on this mainly correlational and preliminary evidence, the authors propose that electrophysiological markers might be used to evaluate interventions in children who live in adverse social conditions due to poverty. Furthermore, they recognize the relevance of employing electrophysiological methods to assess these interventions outside the laboratory setting, so as to incorporate more ecological measures in the study of childhood poverty.

The following chapter, "The Cultural Neuroscience of Socioeconomic Status", comes courtesy of Kwon et al. [70] who venture into cultural neuroscience by addressing the impact of socioeconomic status (SES) on brain functioning. The authors propose that individuals with low and high SES are inserted in different ecologies (marked by distinctive norms, values, resources, and threats) which significantly modulate neurocognition and outward behavior. After highlighting the benefits of electrophysiological and neuroimaging methods to explore such phenomena, they review critical research on several relevant topics. The evidence suggests that low SES involves higher attunement to others, increased sensitivity to potential threats, lower trait inference skills, and greater attentional capacities. Moreover, the construct of SES is discussed in terms of cross-cultural differences, life history theory, and its role in shaping adaptive responses to particular environments. To conclude, Kwon et al. outline future directions to broaden the scope of SES research from the perspective of cultural neuroscience.

Next, the chapter titled "Social Ties, Health and Wellbeing: A Literature Review and Model", by Kemp et al. [71], touches on the relation between social ties and health outcomes. First, through a review of epidemiological evidence, the authors assess the possible association between social ties, health, and well-being. Then, they characterize the potential mechanisms that might mediate or moderate such an association. Finally, based on this background, they introduce the GENIAL (Genomics, Environment, vagus Nerve, social Interaction, Allostatic regulation, Longevity) model, which integrates behavioral, psychological, and physiological mechanisms driving the health of individuals, together with sociostructural factors that may either facilitate or hinder the desired health outcomes in a community. This model ascribes a major regulatory role to the vagus nerve (as indexed by heart rate variability), due to its relation with psychological and physiological processes that influence social ties, health, and well-being. Finally, the authors suggest that future health studies should continue to focus on the value of interpersonal relations while adopting a multidisciplinary approach to research.

Finally, Section D brings together various philosophical reflections on the interdisciplinary synergies underlying the previous contributions. First, "The Self-Domesticated Animal and Its Study", by Mario Bunge [72], sets forth a body of philosophical reflections on how social neuroscientists study human interactions. Bunge departs from the assumptions that all mental processes are brain processes and that human mental life can only be properly understood from a sociopsychological perspective. The core of the chapter is devoted to the presentation and illustration of a formula for various types of mental activity, aimed to formalize the links between their intensity, automaticity, and controllability. In particular, a distinction is proposed between "exo-endo" processes (environmentally biased mental constructions or moral deliberations) and "endo-exo" processes (in which action is biased by intellectual or moral processes). Thereupon, the author relies on specific findings from diverse disciplines to discuss six mechanisms operative during social cognition: spontaneous processes (those that occur without external inputs), automatic processes (such as raw perceptions, feelings, and conditioned reflexes), controlled processes (considering their role in imitation, theory of mind, and empathy), exo-endo processes (with emphasis on religiosity), endo-exo processes (such as free will), and exo-endo-exo processes (a loop proposed to underlie specific emotions and behaviors driven by false beliefs). The chapter concludes with a call to merge biopsychology and social sciences in the pursuit of a better understanding of the interactive dealings of "self-domesticated animals."

Second, in "How is Our Self Related to Its Brain? Neurophilosophical Concepts", Georg Northoff [73] discusses links between philosophical concepts of the self and neuroscientific findings on self-reference processes. First, the author addresses notions such as the mental substance (originally introduced by Descartes), the representational self (the self as a cognitive function), the phenomenological self (characterized in terms of self-consciousness), the minimal self (the self as implicitly, tacitly, and immediately experienced in consciousness), and the social self (described as the linkage and integration of the self into the social context). Then, he reviews current neuroscientific research regarding the self-reference effect, which implies distinct links between the activity of middle regions of the brain and stimuli that are closely related to the self. The specificity of such findings is discussed given that the same regions are also associated with several other cognitive functions. This also raises the issue of how to evaluate the psychological and experimental specificity of the self via experimental paradigms. Finally, the author discusses conceptual frameworks that may allow neuroscience and philosophy to link their respective domains to develop an empirical plausible concept of self.

Third, the chapter titled "Enaction and Neurophenomenology in Language", by Roberto Arístegui [74], addresses a species-specific semiotics involved in multiple forms of human interaction: language. The author frames language from an enactivist perspective, highlighting the limitations of reductionist autopoietic conceptions and arguing for a neurophenomenological account which incorporates notions from pragmatics, including expressive speech acts. More particularly, this work seems to establish links between the notion of enaction, the expressive dimension of language, and holistic approaches to meaning and social semiotics.

Fourth, in "A Pluralist Framework for the Philosophy of Social Neuroscience", Barberis et al. [75] offer a philosophical perspective on theoretical and modeling issues in social neuroscience. To this end, they outline a pluralistic framework addressing the proliferation of modeling approaches, explanatory styles, and integrative trends in the literature. Drawing on current examples based on multiple methods and theoretical inclinations, the authors illustrate the particularities of mechanistic, dynamical, computational, and optimality models in the field. Moreover, they consider the role of causal/compositional or noncausal/structural information in the development of models of the social brain, while assessing the impact of precision, generality, simplicity, and other representational ideals. Finally, integrative trends are discussed considering their prospects for inter-theoretical reduction, mechanistic mosaic unity, and multilevel integrative analysis. In short, this chapter discusses varied epistemological tensions and possibilities in social neuroscience, foregrounding a pluralistic view for the critical assessment of extant and prospective models.

The book closes with a contribution titled "Social Neuroscience and Neuroethics: A Fruitful Synergy", wherein Salles and Evers [76] reflect on how social neuroscience can tackle philosophical and social issues by an association with neuroethics, given the complementarity between their respective areas and explanatory methods. The authors explain that neuroethics focuses on the analysis of the several findings about the brain that have an impact on philosophical analysis, medical and legal practice, and health and social policy. They then propose that, to succeed, this alliance should be based on a deeper understanding of neuroethics, in comparison to the common descriptions from social neuroscience. Critical reflection and conceptual examination are highlighted as the basis to fully address the ontological, epistemological, and ethical impact of social neuroscience. To illustrate this approach, the authors examine the "nature-nurture" distinction in the light of social neuroscience contributions. In short, this chapter emphasizes that social neuroscience might benefit from partnering with neuroethics.

As shown by the vast repertoire of topics and approaches listed above, this book caters to a wide audience of readers interested in the social dimension of the human mind from a transdisciplinary perspective. Although the combination of experimental, theoretical, and metatheoretical elements from neuroscience and the social sciences may at times prove complex material, the volume is explicit and transparent enough for beginners operating in relevant fields, including not only social neuroscience *per se* but also cognitive science, psychology, behavioral science, linguistics, philosophy, sociology, cultural psychology, economics, and policy making. Through this wide-ranging proposal, *Neuroscience and Social Science: The Missing Link* promotes novel reflections on this much-needed marriage while opening opportunities for social neuroscience to plunge from the laboratory into the core of social life.

Acknowledgments This work was partially supported by CONICET, CONICYT/FONDECYT (Regular 1170010), FONCyT-PICT (2012-0412 and 2012-1309), FONDAP (15150012), and the INECO Foundation.

References

1. Stanley DA, Adolphs R. Toward a neural basis for social behavior. Neuron. 2013;80(3):816–26.
2. Eisenberger NI, Cole SW. Social neuroscience and health: neurophysiological mechanisms linking social ties with physical health. Nat Neurosci. 2012;15(5):669–74.
3. Adolphs R. Conceptual challenges and directions for social neuroscience. Neuron. 2010;65(6):752–67.
4. Singer T. The past, present and future of social neuroscience: a European perspective. NeuroImage. 2012;61(2):437–49.
5. Todorov A, Harris LT, Fiske ST. Toward socially inspired social neuroscience. Brain Res. 2006;1079(1):76–85.

6. Cacioppo JT, Decety J. Social neuroscience: challenges and opportunities in the study of complex behavior. Ann N Y Acad Sci. 2011;1224:162–73.
7. Bernhardt BC, Singer T. The neural basis of empathy. Annu Rev Neurosci. 2012;35:1–23.
8. Parkinson C, Wheatley T. The repurposed social brain. Trends Cogn Sci. 2015;19(3):133–41.
9. Franks DD, Turner JH. Handbook of neurosociology. Dordrecht: Springer; 2012.
10. Rule NO, Freeman JB, Ambady N. Culture in social neuroscience: a review. Soc Neurosci. 2013;8(1):3–10.
11. Kim HS, Sasaki JY. Cultural neuroscience: biology of the mind in cultural contexts. Annu Rev Psychol. 2014;65:487–514.
12. Ibanez A, Manes F. Contextual social cognition and the behavioral variant of frontotemporal dementia. Neurology. 2012;78(17):1354–62.
13. Barutta J, Gleichgerrcht E, Cornejo C, Ibanez A. Neurodynamics of mind: the arrow illusion of conscious intentionality as downward causation. Integr Psychol Behav Sci. 2010;44(2):127–43.
14. Cosmelli D, Ibanez A. Human cognition in context: on the biologic, cognitive and social reconsideration of meaning as making sense of action. Integr Psychol Behav Sci. 2008;42(2):233–44.
15. Ibanez A, García AM, Esteves S, Yoris A, Muñoz E, Reynaldo L, et al. Social neuroscience: undoing the schism between neurology and psychiatry. Soc Neurosci. 2016:1–39. http://dx.doi.org/10.1080/17470919.2016.1245214.
16. Ibanez A, Kuljis RO, Matallana D, Manes F. Bridging psychiatry and neurology through social neuroscience. World Psychiatry. 2014;13(2):148–9.
17. Barutta J, Cornejo C, Ibanez A. Theories and theorizers: a contextual approach to theories of cognition. Integr Psychol Behav Sci. 2011;45(2):223–46.
18. Barutta J, Aravena P, Ibáñez A. The machine paradigm and alternative approaches in cognitive science. Integr Psychol Behav Sci. 2010;44(2):176–83.
19. Ibanez A. Dinámica de la cognición. J Saez Editor: Santiago; 2008.
20. Ibanez A, Garcia AM. Contextual cognition. The invisible sensus communis of a situated mind. Heidelberg: Springer; 2017. In press.
21. Baez S, Garcia AM, Ibanez A. The social context network model in psychiatric and neurological diseases. Curr Top Behav Neurosci. 2017;30:379–96.
22. Ibáñez A, Riveros R, Aravena P, Vergara V, Cardona JF, García L, et al. When context is difficult to integrate: cortical measures of congruency in schizophrenics and healthy relatives from multiplex families. Schizophr Res. 2011;126(1):303–5.
23. Ibanez A, Gleichgerrcht E, Hurtado E, Gonzalez R, Haye A, Manes FF. Early neural markers of implicit attitudes: N170 modulated by intergroup and evaluative contexts in IAT. Front Hum Neurosci. 2010;4:188.
24. IbÁÑEz A, Haye A, GonzÁLez R, Hurtado E, HenrÍQuez R. Multi-level analysis of cultural phenomena: the role of ERPs approach to prejudice. J Theory Soc Behav. 2009;39(1):81–110.
25. Hurtado E, Haye A, Gonzalez R, Manes F, Ibanez A. Contextual blending of ingroup/outgroup face stimuli and word valence: LPP modulation and convergence of measures. BMC Neurosci. 2009;10(1):69.
26. Cardona J, Kargieman L, Sinay V, Gershanik O, Gelormini C, Amoruso L, et al. How embodied is action language? Neurological evidence from motor diseases. Cognition. 2014;131:311–22.
27. García A, Ibáñez A. Words in motion: motor-language coupling in Parkinson's disease. Transl Neurosci. 2014;5(2):152–9.
28. Kargieman L, Herrera E, Baez S, Garcia AM, Dottori M, Gelormini C, et al. Motor-language coupling in Huntington's disease families. Front Aging Neurosci. 2014;6:122.
29. Melloni M, Sedeno L, Hesse E, Garcia-Cordero I, Mikulan E, Plastino A, et al. Cortical dynamics and subcortical signatures of motor-language coupling in Parkinson's disease. Sci Rep. 2015;5:11899.
30. Garcia AM, Ibanez A. A touch with words: dynamic synergies between manual actions and language. Neurosci Biobehav Rev. 2016;68:59–95.
31. García AM, Ibáñez A. Hands typing what hands do: action-semantic integration dynamics throughout written verb production. Cognition. 2016;149:56–66.

32. Birba A, Garcia-Cordero I, Kozono G, Legaz A, Ibanez A, Sedeno L, et al. Losing ground: frontostriatal atrophy disrupts language embodiment in Parkinson's and Huntington's disease. Neurosci Biobehav Rev. 2017;80:673–87.
33. Amoruso L, Sedeno L, Huepe D, Tomio A, Kamienkowski J, Hurtado E, et al. Time to tango: expertise and contextual anticipation during action observation. NeuroImage. 2014;98:366–85.
34. Amoruso L, Ibanez A, Fonseca B, Gadea S, Sedeno L, Sigman M, et al. Variability in functional brain networks predicts expertise during action observation. NeuroImage. 2017;146:690–700.
35. Melloni M, Billeke P, Baez S, Hesse E, de la Fuente L, Forno G, et al. Your perspective and my benefit: multiple lesion models of self-other integration strategies during social bargaining. Brain. 2016;139(Pt 11):3022–3040. https://doi.org/10.1093/brain/aww231.
36. García-Cordero I, Sedeño L, de la Fuente L, Slachevsky A, Forno G, Klein F, et al. Feeling, learning from, and being aware of inner states: interoceptive dimensions in neurodegeneration and stroke. Philos Trans R Soc Lond Ser B Biol Sci. 2016. https://doi.org/10.1098/rstb.2016-0006.
37. Garcia AM, Carrillo F, Orozco-Arroyave JR, Trujillo N, Vargas Bonilla JF, Fittipaldi S, et al. How language flows when movements don't: an automated analysis of spontaneous discourse in Parkinson's disease. Brain Lang. 2016;162:19–28.
38. Baez S, Herrera E, Garcia A, Manes F, Young L, Ibanez A. Outcome-oriented moral evaluation in terrorists. Nat Hum Behav. 2017;1:0118.
39. Ibanez A, Billeke P, de la Fuente L, Salamone P, Garcia AM, Melloni M. Reply: Towards a neurocomputational account of social dysfunction in neurodegenerative disease. Brain. 2016;140:e15.
40. Gonzalez-Gadea ML, Sigman M, Rattazzi A, Lavin C, Rivera-Rei A, Marino J, et al. Neural markers of social and monetary rewards in children with attention-deficit/hyperactivity disorder and autism spectrum disorder. Sci Rep. 2016;6:30588.
41. Hesse E, Mikulan E, Decety J, Sigman M, Garcia Mdel C, Silva W, et al. Early detection of intentional harm in the human amygdala. Brain. 2016;139(Pt 1):54–61.
42. Baez S, Pino M, Berrío M, Santamaría-García H, Sedeño L, García A, et al. Corticostriatal signatures of Schadenfreude: evidence from Huntington's disease. J Neurol Neurosurg Psychiatry. 2017. https://doi.org/10.1136/jnnp-2017-316055.
43. Baez S, Santamaría-García H, Orozco J, Fittipaldi S, García A, Pino M, et al. Your misery is no longer my pleasure: reduced schadenfreude in Huntington's disease families. Cortex. 2016;83:78–85.
44. Baez S, Morales JP, Slachevsky A, Torralva T, Matus C, Manes F, et al. Orbitofrontal and limbic signatures of empathic concern and intentional harm in the behavioral variant frontotemporal dementia. Cortex. 2015;75:20–32.
45. Escobar MJ, Huepe D, Decety J, Sedeno L, Messow MK, Baez S, et al. Brain signatures of moral sensitivity in adolescents with early social deprivation. Sci Rep. 2014;4:5354.
46. Baez S, Manes F, Huepe D, Torralva T, Fiorentino N, Richter F, et al. Primary empathy deficits in frontotemporal dementia. Front Aging Neurosci. 2014;6:262.
47. Baez S, Ibanez A. The effects of context processing on social cognition impairments in adults with Asperger's syndrome. Front Neurosci. 2014;8:270.
48. Baez S, Couto B, Torralva T, Sposato LA, Huepe D, Montañes P, et al. Comparing moral judgments of patients with frontotemporal dementia and frontal stroke. JAMA Neurol. 2014;71(9):1172–6.
49. Ibanez A, Aguado J, Baez S, Huepe D, Lopez V, Ortega R, et al. From neural signatures of emotional modulation to social cognition: individual differences in healthy volunteers and psychiatric participants. Soc Cogn Affect Neurosci. 2013;9:939–50.
50. Baez S, Herrera E, Villarin L, Theil D, Gonzalez-Gadea ML, Gomez P, et al. Contextual social cognition impairments in schizophrenia and bipolar disorder. PLoS One. 2013;8(3):e57664.
51. Garcia AM, Bocanegra Y, Herrera E, Moreno L, Carmona J, Baena A, et al. Parkinson's disease compromises the appraisal of action meanings evoked by naturalistic texts. Cortex. 2017. https://doi.org/10.1016/j.cortex.2017.07.003.

52. Spunt RP, Adolphs R. A new look at domain specificity: insights from social neuroscience. Nat Rev. 2017;18:559–67.
53. Schilbach L. Towards a second-person neuropsychiatry. Philos Trans R Soc Lond Ser B Biol Sci. 2016;371(1686):20150081.
54. Garcia AM, Ibanez A. Two-person neuroscience and naturalistic social communication: the role of language and linguistic variables in brain-coupling research. Front Psychiatry. 2014;5:124.
55. Babiloni F, Astolfi L. Social neuroscience and hyperscanning techniques: past, present and future. Neurosci Biobehav Rev. 2014;44:76–93.
56. Billeke P, Soto-Icaza P. Valuing others: evidence from economics, developmental psychology, and neurobiology. In: Ibáñez A, Sedeño L, García AM, editors. Neuroscience and social science. Cham: Springer; 2017.
57. Díaz-Gutiérrez P, Alguacil S, Ruz M. Bias and control in social decision-making. In: Ibáñez A, Sedeño L, García AM, editors. Neuroscience and social science. Cham: Springer; 2017.
58. Cornejo C, Cuadros Z, Morales R. Neurobiological approaches to interpersonal coordination: achievements and pitfalls. In: Ibáñez A, Sedeño L, García AM, editors. Neuroscience and social science. Cham: Springer; 2017.
59. Vrtička P. The social neuroscience of attachment. In: Ibáñez A, Sedeño L, García AM, editors. Neuroscience and social science. Cham: Springer; 2017.
60. Felisberti FM, King R. Mindreading in altruists and psychopaths. In: Ibáñez A, Sedeño L, García AM, editors. Neuroscience and social science. Cham: Springer; 2017.
61. TenHouten WD. From primary emotions to the spectrum of affect: an evolutionary neurosociology of the emotions. In: Ibáñez A, Sedeño L, García AM, editors. Neuroscience and social science. Cham: Springer; 2017.
62. Baez S, García AM, Santamaría-García H. Moral cognition and moral emotions. In: Ibáñez A, Sedeño L, García AM, editors. Neuroscience and social science. Cham: Springer; 2017.
63. Rosas A. On the cognitive (neuro)science of moral cognition: utilitarianism, deontology and the 'fragmentation of value'. In: Ibáñez A, Sedeño L, García AM, editors. Neuroscience and social science. Cham: Springer; 2017.
64. Haye A, Morales R, Niño S. The social/neuro science: bridging or polarizing culture and biology? In: Ibáñez A, Sedeño L, García AM, editors. Neuroscience and social science. Cham: Springer; 2017.
65. Piguet O. Dementia and social neuroscience: historical and cultural perspectives. In: Ibáñez A, Sedeño L, García AM, editors. Neuroscience and social science. Cham: Springer; 2017.
66. Kumfor F, Hazelton JL, De Winter F-L, de Langavant LC, Van den Stock J. Clinical studies of social neuroscience: a lesion model approach. In: Ibáñez A, Sedeño L, García AM, editors. Neuroscience and social science. Cham: Springer; 2017.
67. Roussos A, Braun M, Aufenacker S, Olivera J. Psychotherapy and social neuroscience: forging links together. In: Ibáñez A, Sedeño L, García AM, editors. Neuroscience and social science. Cham: Springer; 2017.
68. Mantilla MJ, Di Marco MH, Golombek DA. The brain in the public space: social neuroscience and the media. In: Ibáñez A, Sedeño L, García AM, editors. Neuroscience and social science. Cham: Springer; 2017.
69. Pietto ML, Kamienkowski JE, Lipina SJ. Electrophysiological approaches in the study of the influence of childhood poverty on cognition. In: Ibáñez A, Sedeño L, García AM, editors. Neuroscience and social science. Cham: Springer; 2017.
70. Kwon JY, Hampton RS, Varnum MEW. The cultural neuroscience of socioeconomic status. In: Ibáñez A, Sedeño L, García AM, editors. Neuroscience and social science. Cham: Springer; 2017.
71. Kemp AH, Arias JA, Fisher Z. Social ties, health and wellbeing: a literature review and model. In: Ibáñez A, Sedeño L, García AM, editors. Neuroscience and social science. Cham: Springer; 2017.
72. Bunge M. The self-domesticated animal and its study. In: Ibáñez A, Sedeño L, García AM, editors. Neuroscience and social science. Cham: Springer; 2017.

73. Northoff G. How is our self related to its brain? Neurophilosophical concepts. In: Ibáñez A, Sedeño L, García AM, editors. Neuroscience and social science. Cham: Springer; 2017.
74. Arístegui R. Enaction and neurophenomenology in language. In: Ibáñez A, Sedeño L, García AM, editors. Neuroscience and social science. Cham: Springer; 2017.
75. Sergio Daniel Barberis M, Itatí Branca A, Venturelli N. A pluralist framework for the philosophy of social neuroscience. In: Ibáñez A, Sedeño L, García AM, editors. Neuroscience and social science. Cham: Springer; 2017.
76. Salles A, Evers K. Social neuroscience and neuroethics: a fruitful synergy. In: Ibáñez A, Sedeño L, García AM, editors. Neuroscience and social science. Cham: Springer; 2017.

Part I
Neuroscientific Research on Social Cognition

Valuing Others: Evidence from Economics, Developmental Psychology, and Neurobiology

Pablo Billeke, Patricia Soto-Icaza, Mauricio Aspé-Sánchez, Verónica Villarroel, and Carlos Rodríguez-Sickert

Abstract Human social skills are widely studied among very different disciplines. In this chapter, we review, discuss, and relate evidence concerning the process of valuing others' perspectives, preferences, and behaviors from an economic, psychological, and neurobiological viewpoint. This process of valuing others (or other-regarding preferences) can be understood as weighing others' preferences to adapt our own behavior and achieve adequate social interaction. We first review economic research related to decision-making in social contexts, with emphasis on how decision-making has integrated other-regarding preferences into the decision-making algorithm. By means of social and developmental psychology research, we then review how social skills develop from identification to understanding others. Finally, we discuss the neurobiological mechanisms underlying social skills and social decision-making, focusing on those systems that can participate in processes of valuing others preferences. As a conclusion, we highlight five points that we believe an interdisciplinary approach should take into account. We thus intend to generate a starting point for building a more extensive explicatory bridge among the different disciplines that study complex human social behavior.

Keywords Neuroeconomics • Decision-making • Other-regarding preferences • Mentalization • Theory of mind • Social cognition • Interdisciplinary approach • Game theory

---

P. Billeke (✉) • M. Aspé-Sánchez • C. Rodríguez-Sickert
División de Neurociencias (NeuroCICS), Centro de Investigación en Compleijidad Social, Facultad de Gobierno, Universidad del Desarrollo,
Av. Las Condes 12461, Las Condes, Santiago 7590943, Chile
e-mail: pbilleke@udd.cl; maspes@udd.cl; carlosrodriguez@udd.cl

P. Soto-Icaza
Interdisciplinary Center of Neuroscience, Pontificia Universidad Católica de Chile, Santiago, Chile
e-mail: pasoto@uc.cl

V. Villarroel
Centro de Investigación y Mejoramiento de la Educación (CIME), Facultad de Psicología, Universidad del Desarrollo, Concepción, Chile
e-mail: vvillarroel@udd.cl

© Springer International Publishing Switzerland 2017
A. Ibáñez et al. (eds.), *Neuroscience and Social Science*,
DOI 10.1007/978-3-319-68421-5_2

1 Introduction

We are an extremely social species; almost all of our behavior is related to other human beings. Currently, various disciplines deal with the problem of understanding human social behavior. However, few proposals that combine different approaches and findings have been elaborated. In this chapter, we discuss the evidence and research approaches from an array of disciplines related to the idea of how humans consider other preferences and behaviors during this decision-making process. We shall use the term "valuing others" to refer to the processes by which humans weigh the preferences and behaviors of others as to adapt or guide their behavior during social interactions. In the following pages, our endeavor will be to present and discuss the evidence from three research programs, namely, (1) economics research related to decision-making in social contexts, (2) social psychology research related to the development of mentalizing and perspective-taking skills, and (3) neuroscience research related to neuronal mechanisms underlying vicarious human behaviors.

The fundamental aim of this chapter is to show some of the current efforts to build an interdisciplinary understanding of social behavior instead of giving a global integrative approach. In order to build a fully interdisciplinary research programming between social science and neuroscience, the authors have established some basic bridges which are necessary to discuss and begin to build this understanding. Therefore, with the purpose of contributing to this global aim, we have structured this chapter in three sections. In the first one, we discuss how the approach from economics toward the social decision-making process has started to incorporate social preferences and how neuroscience approaches can contribute to improving the predictive ability of the behavioral model. In the second section, we review evidence from developmental psychology related to how human beings begin to understand and integrate the perspective of others into their own behavior and decisions. Finally, we discuss findings from social neuroscience and neuroeconomics related to the neurobiological mechanisms that underlie social decision-making, in order to suggest possible interdisciplinary approaches, and their possible pitfalls.

2 Behavioral Models of Human Conduct and the Black Box

In recent years, the emergence of subfields such as neuroeconomics and social neuroscience has driven the dialogue between behavioral economics and natural science. Especially, behavioral economics has relied on game theory an experimental paradigm for neuroscientists when studying complex social behavior inside the controlled settings of a laboratory. Likewise, to concurrently record or modulate brain activity—by means of techniques such as electroencephalography (EEG), functional magnetic resonance imaging (fMRI), and transcranial magnetic stimulation

(TMS) (see below)—could shed a light on the cognitive mechanisms that underlie the behavior of experimental subjects and their reactions against the behavior of their fellow partners.

When there is a confluence of disciplines, the potential gains of combining both perspectives might be hampered by language barriers (e.g., jargon that is discipline specific) and incongruities between the widespread research practices within each discipline (e.g., the importance that is given to generality in contrast with parsimony or to prediction over explanation). In this section, we suggest three perspectives that can lead to a fruitful interdisciplinary interaction from the perspective of economics. We focus on (1) the neurophysiological foundations of behavioral models of social preferences, (2) general guidelines for modeling social behavior and social cognition, and (3) specific instantiations of neurophysiological variables within those behavioral models.

2.1 Homo Behavioralis *and the Influx of Ideas from Psychology and Other Disciplines*

When scholars from disciplines such as psychology or anthropology began to question the plausibility of the prevalent model of human agency in economics, the reply came from one of his most renowned representatives. Milton Friedman wrote his famous *Essays in Positive Economics* (1951), which strongly influenced future generations of economist researchers [1]. There he claimed that "the only relevant test of the validity of a hypothesis is the comparison of its predictions with experience." Furthermore, Friedman argued that even if assumptions appear false or implausible, their empirical weakness should be tolerated if they lead to accurate predictions. When Friedman adds this second statement, not only can one infer that he was oblivious to the advances of neuroscience but also that the aim of Friedman and his fellow custodians was to keep the black box closed and to keep the *homo economicus* locked inside [2].

It is not that the members of the congregation for the Doctrine of the Economic Faith denied the existence of other drivers of human behavior beyond self-interest—e.g., altruism. Nor did they believe we are perfect optimizers. Their stance relied on an argument of parsimony: the benefits of generalizing the utility function to account for possible anomalies and produce more accurate predictions would be negligible against the loss of parsimony and tractability of adding new parameters to the utility function. The overwhelming amount evidence from laboratory and field experiments showed that this view on the trade-off between prediction power and parsimony was not accurate. The effort to correct this mistake was assumed by a new breed of "behavioral" economists. Indeed, one can say that there is nothing new in this approach. They are just continuing the enterprise launched by Adam Smith himself, as a moral philosopher, in *The Theory of Moral Sentiments* (1759) [3].

The first task undertaken by the behavioral squad was to upgrade the utility function so that these “anomalies” could be captured within an augmented utility function. Around psychological constructs, such as loss aversion and reference dependence, Kahneman and Tversky developed prospect theory [4]. While Kahneman, Tversky, and their followers focused on decision under uncertainty, and later on issues such as intertemporal inconsistency [5, 6], a separate group of behavioral economists reacted to the strong evidence against the self-interest hypothesis provided by experimental studies. These studies showed that agents do cooperate in social dilemmas such as trust games [7–11], public good games [12–14], even when cooperating is against their (material) self-interest. And, within bargaining games such as the ultimatum game [15–17], agents are willing to incur in material costs to avoid unfair outcomes and sanction free riders in collective action problems [18].

Taking their insights from social psychology, sociology, and anthropology, a family of models was produced within behavioral economics. These models, referred to as models of social (or other-regarding) preferences, can be either outcome based, e.g., models of inequity aversion [19, 20], or intention based, e.g., models that capture norms of both positive and negative reciprocity [21–23]. Cooperation in trust games was initially understood as the result of positive reciprocity (intention-based social preferences). The trustee is willing to spend resources to reward trust placed in him. On the other hand, rejection on the ultimatum game was initially understood as the result of inequity aversion. However, later studies provided evidence for a more complex structure of moral response. Trustees in a trust game are also motivated by outcome-based preferences [8], and rejection in the ultimatum game also involves negative reciprocity [24]. Furthermore, current studies show that the research on social preferences can also be extracted by the research produced in other areas of behavioral economics. For instance, time inconsistency can also affect the nature of social preferences [25].

To the extent that neuroeconomic studies have provided neurophysiological mechanisms for experimental anomalies and, thus, biological foundations for social preferences models, neuroeconomists were welcomed as part of the new tribe of behavioral economists but were not so well received by old-school orthodox economists who were still concerned with keeping the black box closed even for the new model of human agency: the *homo behavioralis* and its representation in an augmented utility function. For instance, it has been argued that neuroscience could not transform economics because what goes on inside the brain is irrelevant to the discipline. As if nothing had changed since Friedman’s influential piece, they put forward the idea of a “mindless economics,” arguing that what matters are the decisions people make, not the process by which they reach them [25]. We will develop this idea in the opposite direction and claim that the major challenges posited by neuroeconomics precisely relate to our understanding of the neurocognitive processes that underlie social behavior and, furthermore, open the possibility to embed economics in the biological processes taking place in the brain.

2.2 *Impact of Neurosciences on Modeling Individual and Social Behavior*

In the same way that behavioral economics has used insights from psychology to develop more "realistic" models of individual decision-making, in which people often did things that were not in their best interests, the evidence coming from neurobiology presents an additional challenge to the standard economic assumptions. Thus, evidence from neuroeconomics indicates that decision-making is far from being a unitary process (a simple matter of integrated and coherent utility maximization), suggesting instead that it is driven by the interaction of multiple systems or processes [26]. This range from the more basic dual-process approach that has influenced our general comprehension of human cognition and behavior beyond Descartes' error (fast/hot module and the slow/cold, automatic vs. controlled processes [26–28]) to more complex multiple system approaches toward social behavior and social decision-making [29–31]. Steinbeis et al. [32], for instance, show that behavioral inhibition—modulated by the neuroanatomical development of the cognitive control systems—plays a crucial role in the implementation of fair behavior in bargaining games.

2.3 *Prediction Accuracy of Behavioral Models: Combining Psychological and Neurobiological Parameters*

A specific aspect of the relevance of the neuroeconomic program refers to its capacity to inform behavioral models in such a way that prediction accuracy can be improved. This point is very important, because if we do not build a bridge between neuroscience and algorithmic social decision theory, it will be very difficult for this program to reach the academic community of economists. To discuss the issues that could emerge from this challenge, we consider a distributional problem in the spirit of Andreoni and Miller [33], in which an agent i decides how to split an amount m between himself and another agent $-i$ for different budget constraints. For every monetary unit agent i sacrifices $(m - x_i)$, his partner will receive $(m - x_i)/p$ monetary units. Thus, p can be interpreted as the price of altruism and agent i's choice can be represented as the consumer's choice problem.

2.3.1 Neoclassical Model (*Homo Economicus*, Black Box)

$$\max_{x_i} \{U_i(x_i)\} \text{s.t} \, x_i + px_{-i} = m,$$
$$\text{which yields to} \, x_i(m,p) = x_i^* \text{ with } x_i^* = m \text{ and } x_{-i}^* = 0.$$

In the case above, the only relevant argument of $U_i(\bullet)$ is his own material self-interest x_i. If, alternatively, we consider that agent i's choice is also affected by the material welfare of his partner $-i$, we could represent his choice problem as follows.

2.3.2 Behavioral Model (Other-Regarding Preferences, Black Box)

$$\max_{x_i,x_{-i}} \{U_i(x_i, x_{-i}, \theta_i)\} \text{ s.t } x_i + px_{-i} = m,$$
$$\text{which yields to } x_i(m, p, \theta_i) = x_i^* \text{ with } x_i^* \langle m \text{ and } x_{-i}^* \rangle 0$$

where θ_i is a parameter that represents the intensity of the moral dispositions of the agent that counterbalances his self-interest.[1] Most models assume that θ_i is private. Now consider the possibility that θ_i can be estimated from the neurobiological activation n_i, $\hat{\theta}_i(n_i) = \theta_i + \varepsilon_i$. If this is the case, the lower the measurement error, the greater will be the predictive gains of opening the black box. The registered neurobiological activation n_i could give us information about θ_i through two channels: the individual's idiosyncratic characteristics and the dimensions of the stimuli not captured by the model. For the sake of simplicity, we will assume that n_i is simply a contextual modulator of θ_i. Thus, the structure of choice could be represented as follows.

2.3.3 Neurobiological Model (Other-Regarding Preferences, Neurobiological State)

$$\max_{x_i,x_{-i}} \{U_i(x_i, x_{-i}, \theta_i \mid n_i) s_i\} \text{ s.t } x_i + px_{-i} = m,$$
$$\text{which yields to } x_i(m, p, \theta_i \mid n_i) = x_i^{**}; x_i(m, p) = x_i^{**} > 0, x_{-i}^{**} > 0$$

The improvement in prediction accuracy of a model that incorporates n_i is an indicator of the incompleteness of the behavioral model. However, it is not only important to come up with a model that accurately predicts behavior in a particular context. Fehr and Camerer [34], among others, argue that prosocial behaviors occur in one-shot anonymous games as the result of a reflexive behavior that is highly adapted for repeated interactions where immediate prosocial behavior earns future benefits. Under this view, prosociality in one-shot games results from bounds on rationality in full response to changes in the economic structure. Alternatively, pro-

[1] A simple functional specification of the agent's social preferences could be expressed as $U_i(x_i, x_{-i}, \theta_i) = (1-\theta_i) \times u(x_{-i}) + \theta_i \times u(x_{-i}) = (1-\theta_i) \times \sqrt{x_i} + \theta_i \times \sqrt{x_{-i}}$ where θ_i represents the weight agent i attaches to his partner individual welfare. In some alternative functional specifications, both considerations to the efficiency and equity of the final distribution have been introduced (see [164]).

social behavior could reflect robust social preferences for treating others generously or reciprocally, and those preferences are similar to preferences for other kinds of primary and secondary rewards. Within this scheme, different arrangements of neurobiological activation $n_i^0 \uparrow n_i^1$ could lead to similar predictions in terms of cooperation that could indicate the motives underlying both cases. Such a case has been shown recently; see below [35]. Furthermore, these neural traits could provide crucial information to distinguish different types of individuals and, consequently, have more information about their behavior in the future or in different social contexts.

A crucial issue in this point is what are precisely these neurobiological traits and states and how these states weigh the parameters of self-interest and other-regarding preferences. Although neuroscientists are far from reaching consensus, there is accumulative evidence that can indicate some general structures of these traits and states. In the following section, we will review some critical evidence from developmental psychology and developmental neuroscience in order to give insight on how these neurobiological states mature and change during the ontogeny. Then, in the final section, we shall analyze how these neurobiological processes can be structured, with special focus on how the system weighs and values the regarding preferences of others.

3 Development of Social Preferences

One of the most relevant facts indicates that the neurobiological state has a decisive influence in the decision-making process is the human development. The maturity of different brain systems in different timelines generates several behavioral manifestations that are characteristic to a specific age [36, 37]. This is true not only during childhood and adolescence but also for older adults where pathological neuronal degeneration is expected [38, 39].

Regarding early human development research, one of the most intriguing human social phenomena is the ability to read the minds of others, known as "mentalization" or "theory of mind". This ability has been described as one of the major landmarks in social development, because it enables children to handle more complex social interactions. Indeed, the ability to figure out and finally to attribute and understand the other person's thoughts and feelings has been depicted as a distinctive human trait [40]. However, the mechanism by which this ability emerged has been the subject of drawn-out controversy [41–43]. The analysis of the development of human social functioning is a useful tool for understanding how social skills are structured. This analysis reveals that social ability development is not a unitary or an "all-or-nothing" type of outcome. Instead, it is an interactive specialization that entails both the association of an ability with a brain system and the specialization of this function in interaction with others [44]. In this context, one of the main drivers for this development is the necessity to anticipate and predict the behavior of others, which is crucial for both primate and human survival [45]. Certainly, the newborn ability to discriminate a relevant biological agent seems to be coordinated to, first, a guarantee that the partner is actually a living being and, second, that this living being is actually human. As human babies are born premature [43], their extreme dependency puts them at higher risk; hence, they must draw the interlocutor's attention directly to them in order to modify the performance of others to get what he or she needs to survive. It seems possible that the later human ability to "read minds" arises from all those previous early stages of social development as a guarantee for survival since it constitutes a specialized expertise of social prediction. This section is organized in three overlapping stages of development, starting with the early capacity to identify biological/social agents and ending with the explicit manifestation of mentalization skills.

3.1 Identification of Social Agents in Newborns and Infants

The early stages of social development are the building blocks in which further social skills have grown. Certainly, the only way that a human infant can survive is if there is another being that can provide food, water, etc. Evidence in newborns showed that toddlers as young as only few days of age are able to discriminate different perceptual signs that indicate the existence of a social agent [46–48]. For example, they can identify points that emulate a coherent biological motion [48], face-like patterns [49], and direct versus averted gaze in faces [50, 51], and they can even imitate basic movements from another human being [52]. Indeed, from 2 months old, infants show a preference for looking at eyes rather than mouths or bodies [53]. This preference also describes a specialization process in 3-month-old toddlers, who prefer eyes only when they are accurately located in the upper part of the face configuration rather than placed in another location of the face [54]. All these findings are showing that there is an ontogenetic orientation toward the social agents, which seems to be in a growing process of behavioral and neural specialization. Indeed, comparative studies between preterm and full-term infants and among subjects of different ages [55–58] emphasize the role of the experience in the cerebral functions refinement [44]. From biological motion detection to imitation and face-like stimuli and direct gaze preference in newborns, human social development seems to be organized to detect, understand, and finally predict and manipulate the social agent [59].

EEG findings in infants and children are in accordance with this developmental perspective. The EEG technique is a noninvasive measurement of the brain activity through scalp electrodes widely used in neuroscience [60]. The evidence has shown that the electrical brain activity phase related to stimulus presentation, called event-related potentials (ERPs) [61], follows a developmental trajectory. An illustrative example is the N170, that is a negative deflection occurring at 170 ms after presentation of a human [60, 62–69], whose likely source is the ventral visual stream near the fusiform face area. In adults, the N170 evidenced a higher amplitude and latency for inverted human faces, while in infants it did not show any modulation by stimuli orientation. In 6-month-old infants, there is an "infant N170" (called P400 component) characterized by higher amplitude in response to faces displaying direct gaze rather than an averted gaze [50], as well as to inverted faces only in the case of their mother [70], evidencing a specific selection process present in early life.

3.2 *Being Able to Interact with Other Humans*

It is important to note that these skills are present in a context of reciprocal interaction [71, 72]. While it is clear that infant behaviors like crying, screaming, gazing, and smiling are aimed to make the social partner answer their requirements, it is also clear that the partner cannot remain indifferent to those calls of attention. What actually happens when infants and their caregivers are coordinated or synchronized? It has been described that in mother-toddler relationships with infants from 3 to 6 months, the engagement periods came in a burst mode, with periods of asynchronous states [72]. Interestingly, these mismatch states are followed by repair sequences of the interactive errors by both the infant and the mother. These repair behaviors can have functionality in the interaction skills development. Indeed, the importance of stages as "reparation" contexts has been widely described in the attachment theory [71]. Precisely, these bonding-recovering stages emphasize the importance of the mutuality of the attachment between the caregiver and the infant which is crucial to underline [71, 73]. The higher social skills like mentalization abilities were the result of all these precursors or early stages of development, which are the building blocks in which further social skills are grown [59].

An important step in the development of the capacity to interact with other human beings is the joint attention (JA) skill. JA has been described as the capacity to share an interest with another person by alternating the gaze in order to coordinate the interest in an object with a social partner [74–80]. A key component of JA is the division and the alternation of the subject's attention between the object and the partner [77, 81]. Several studies agree that JA emerges around the age of 9 months [74, 76, 77, 82], when children learn to use eye contact to derive information about another person's goal-directed behavior [76]. Importantly, the ability to attend to an object jointly with another person has proved to be crucial for several capacities such as social synchronization, development of language [74, 76, 78, 79, 83, 84], and development of theory of mind [80]. The knowledge of the latter tends to be ambiguous to clarify if JA involves a level of "self-awareness" of the social agent [45]. Does the infant actually "know" the agent's state of mind when is engaged in a JA interaction? There is a line of studies that defines JA as the situation in which two subjects are looking at the same object but without the awareness that the focus of attention is a common interest. The real capacity to realize that the focus of attention is a common element between the infant and the agent is what is called "shared attention" [45]. Accordingly, what is clearly a higher development of social knowledge is the mentalization ability, which is the capacity to understand and predict the behavior of other people and their knowledge, intentions, emotions, and beliefs [85, 86]. Furthermore, JA and shared attention would be intermediate stages toward mentalization inasmuch as the theory of mind ability solely enables to notice and take into account the agent's mental state. Interestingly, the neuroimaging evidence revealed that JA and mentalization might be related. Specifically, fMRI is a method that measures changes in the hemodynamic brain response associated with neural activity—specifically, the blood-oxygen-level-dependent (BOLD) signal [87]. There is broad consensus about the brain network that is recruited when adult subjects participate in mentalizing tasks (see next section below). Interestingly, the same network is involved when participants show JA behavior in adulthood and later childhood. During early childhood, the EEG evidence shows that responses to JA are associated with the Nc component. This ERP refers to a negative deflection that occurs around 300–850 ms after stimulus onset [56, 66, 77, 82], and it is associated with attentional reorientation. In children during the age when they can achieve the false-belief mentalization, this component did not seem to present any differences. However, two neuronal measures seem to mark the mentalization achievement. One of these is the presence of a specific oscillatory activity in the temporoparietal areas of the mentalization network (see next section) and the maturation on neural fiber that connects the frontal and temporoparietal regions [88]. Thus, specific neuronal development seems to be a marker for more complex social skills achievement.

3.3 Knowing the Others' Mental States

What do infants know about the mental states of others? Do they actually try to modify the actions of others because they can infer what is in their minds? Premack and Woodruff [89] stated that the mentalization ability is a system of inference that enables us to attribute mental states both to oneself and to another—for instance, purposes, intentions, knowledge, belief, and thinking. Certainly, this system of inference is needed because such "mental states" are not directly observable, making it a "theory" of what are the others' mental states (i.e., theory of mind). The explicit skill to identify other people's false beliefs becomes evident not before 4 years of age [85, 90]. However, there is a line of research that describes how infants are able to do some kind of inferences about others' feelings and thoughts [91–94]. That line of studies appeared as alternative experimental paradigms to overcome the language-dependent bias which standard/classic false-belief tasks [86] have. Hence, the infants' difficulty both to inhibit their own knowledge about something that another person does not know and to think over different representations makes this task impossible to solve for children under 4 years old [95, 96]. Therefore, researchers use infants' longer looking time as a measure of children's anticipatory belief [94] or surprise as measure of a violation of the expectation paradigm [92, 93, 97] in nonverbal false-belief tasks. Thus, this line of research has shown that there is evidence of an "implicit" theory of mind [91]. However, there is another line of research that has been skeptical about this interpretation [41–43, 98, 99]. This evidence can be interpreted just as perceptual processes and competences rather than high-level cognitive processes. Furthermore, high-level constructs that come from this experimental paradigm might be revealing the researcher's over-interpretation instead of the ability for which it was created [43]. Indeed, the increase in looking times that these studies have shown might be revealing a visual perception process related with a new arrangement of the stimuli rather than an interpretation of the agent's belief [99].

At this point of the controversy, it is important to consider that the implicit mentalization ability, the JA ability, the different levels of visual perspective taking (mentioned below), and the explicit theory of mind itself could be understood as stages of complexity inside the development process of the same capacity. The visual perspective taking (VPT) is the capacity to know that an object can be seen from a certain point of view and that someone else could not see it because there is a physical barrier [100]. Research of VPT should also be considered to understand the mentalization development as a dynamic building block process. These studies provide interesting evidence to consider the existence of an intermediate level of mentalization [59]. The first level of VPT [101, 102] can be understood as a previous step toward a well-consolidated theory of mind, because, around the age of 2, the child is only able to identify whether another person can see an object or not, but it says nothing about a genuine capacity to attribute the mental state of the agent. Nevertheless, this VPT level becomes more complex a couple of years after when it allows the child to identify the others' references and perspectives [90, 98, 101, 102]. This higher VPT level, known as Level 2 VPT, allows the child to understand that objects can be seen in different ways, depending on the form of presentation and point of view [98, 101, 102]. There is evidence that correlates Level 2 VPT with the development of mentalization ability [101]. Although the first theories point out that the visual perspective taking is the basic process from which more complex (social) perspectives arise, recent evidence indicates an opposite ontological development [102]. Early infants can track others' experiential backgrounds. In fact, several studies have found that infants take what others have witnessed into account when acting and responding toward them. In other words, the infants revert to the background constituted by past experiences and use it to understand an agent's desires, goals, and intentions. This ability becomes evident before infants can solve complex visual perspective-taking tasks (Level 2) and even before they can solve explicit mentalizing problems, like the false-belief task [91]. This evidence indicates that the developmental processes that lead to the explicit mentalizing ability are related to the integration of others' preferences into our behaviors. This skill, as an integrative process, becomes more complex through aging, incorporating more sources of information, such as memories, social knowledge, and visual skills, among others. Thus, the development of this skill serves as the basis for more complex explicit mentalizing or the theory of mind skill. Following the deconstruction of the mentalizing concept proposed elsewhere [103], the skill of valuing others can help us gather not well matching evidence, which has come from cognitive neuroscience and neuroeconomics. In the next section, we will review neuroscience evidence related to brain components of the system of other-regarding preferences.

4 Neurobiological System Related to Other-Regarding Preferences

Our brain has evolved to solve complex cognitive demands required for living in social groups of increasing size [104]. Experimental evidence has established that, unlike other social species, humans display a large amount of cooperative behaviors, including altruism, trust, and reciprocity [105, 106]. These behaviors are observed even when individuals interact with strangers and with individuals they will never meet again [107]. Trust, altruism, and reciprocity are crucial to establish and maintain cooperative links between different individuals. Recent work using neuroscience techniques has begun to reveal the brain states related to these prosocial dispositions [108]. In the following subsection, we will review evidence from neuroeconomic studies using two game theory experimental paradigms, namely, trust and dictator games. Then, we shall discuss evidence from the two putative systems related to other-regarding preferences or "valuing others" processes that can underlie human prosocial behaviors.

4.1 Trust and Reciprocity

The most widely used experimental setting to study trust and reciprocity is the trust game (TG) or invested game. In this game, two players, who do not know each other, engage in an anonymous interaction. The experimenter gives the "investor" (or trustor) some amount T of money. The trustor then decides how much of T send (or "invest") in the other player, referred to as the trustee. The amount A_1 sent by the trustor is multiplied by an exchange factor r (typically 3). Thus, the trustee receives an amount of money three times the amount sent by the trustor (rA_1). Finally, the trustee decides how much of the money received (rA_1) is sent back to the trustor (A_2) [7]. The prediction from the self-interest hypothesis for TG is that the trustees will keep all the money. Assuming that the trustors have mentalizing capabilities (see above), they should anticipate this betrayal and send nothing. In the very first test of this game, 0.6% of the trustors sent nothing to the trustee, 66% sent half or more of their endowment, and about 50% ended the game with more money than their initial endowment (which implies, of course, that $A_2 > A_1$; in other words, trustees were trustworthy; [7]. These behaviors have been replicated in several studies. In a recent meta-analysis, Johnson and Mislim [109] collected the data from the 162 replications of the TG available at the time and found that, on average, trustors send 0.5 of his/her endowment to the trustee (n = 23,900; std = 0.12; min = 0.22; max = 0.89), while the trustee returns 0.37 of their total endowment (n = 21,529; std = 0.11; min = 0.11; max = 0.81 [109]. Repeated interactions of the TG show a similar pattern, indicating a high tendency toward trust and reciprocity by both players [107].

Trustee behavior is interesting. While, for trustors, there is an expected gain, this is not so clear for trustees. The trustee has the opportunity to break the trust, which is, as stated above, the classical self-interest prediction. This is particularly true for one-shot, anonymous interactions since there are no incentives to build reputation and create a greater amount of trust for future interactions. Classically, trustee's behavior has been considered just reciprocity, but this is only true if allocations made by trustees are different from allocations made by a subject in a context where his/her behavior is unrelated with the perceived intentions of cooperation from the other player [110]. There is a difference between intention-based behaviors, such as the behavior in the TG, where trustee's behavior depends on ascribing cooperative intentions to the trustor, and outcome-based behaviors, such as the behavior in the dictator game (DG, described below), where subject behavior depends only on the final share of the game and not on the others' intentions.

4.2 The Neural Dynamics of the TG

In a TG, the very first decision by the trustor involves deciding whether to trust the other player or not. From the trustors' perspective, this involves (1) knowing whether they are playing with another human or a non-intentional entity (generally a computer which makes random allocations) and (2) then deciding to send or not to send some amount of money to the trustee. Several reports have shown increased activity in the medial prefrontal cortex (mPFC, a structure involved in metallization processes; [111] when trustors decide to trust another human partner [112–115]. In addition, during the first stage, the trustor has not received any feedback on the trustworthiness of his/her partner; therefore, the reinforcement learning system must be engaged to adjust trustor behavior based on feedback reward. Delgado et al. [114] read the descriptions of the life events of different trustees to trustors, indicating praiseworthy, neutral, or suspicious moral characters for each of them. Not surprisingly, rates of cooperation were higher when playing with the praiseworthy partner. Interestingly, trustors showed different activation in the ventral striatum (VS) for positive and negative feedback but only when they were playing with the neutral trustee. The VS has been involved in processing feedback and prediction error [114, 116], suggesting that, in the neutral condition, trustors activate the reinforcement system to learn about the trustworthiness of their partners, while praiseworthy and suspicious moral characters bias the behavior of trustors [114]. Interestingly, the neuropeptide oxytocin (OXT) has been associated with trust behaviors in humans [117, 118]. Kosfeld et al. [119] used a TG experiment to show that intranasal infusions of OXT increase trust in humans (but not in other nonsocial interactions), do not increase risk-taking behavior, and did not change trustees' behavior. Although the mechanism of action of OXT is not clear, evidence suggests that OXT decreases stress responses and anxiety in social interactions, likely modulating the amygdala and anterior cingulate cortex (ACC) activity [117, 120].

Considering now the situation of the trustees, reports show that the mentalization system becomes active when they receive an allocation from trustors. Van den Bos et al. [121] has shown that the mPFC increases its activation when trustees defect.

On the other hand, when trustees reciprocate a high-risk allocation (i.e., the trustor could lose a large amount of money if the trustee chose to defect), there is greater activation of the temporoparietal junction, which is also a part of the mentalization system [122–124]. Moreover, trustees' reciprocity in low-risk allocations correlated with the activity in the anterior insula cortex (AIC), a structure involved in emotional and salience processing [113, 125]. Furthermore, trustees reciprocating low benefit allocations (i.e., when the monetary incentives to reciprocate are low) were associated with an increased activity in the ACC and the dorsolateral prefrontal cortex (dlPFC), which are structures involved in cognitive control and the inhibition of selfish impulses [126–129].

Another interesting finding is the effect of individual traits in reciprocal interaction [121, 130]. For example, people with more traits characterized by positive emotionality trust more in others, while people with less tendency to psychopathic traits show more reciprocate behaviors [130]. Other study shows that when a prosocial subject reciprocated, they showed an increased activation in VS, while defection increased the activity in ACC, AIC, and right TPJ. In contrast, pro-self individuals showed the opposite pattern, showing increasing ACC, AIC, and right TPJ activity after they reciprocated. This shows that these structures were more active when participants chose their less frequent behavior, considering their personal trait or past history [121].

Trustees' reciprocal behavior is also influenced by expectations [131]. Chang et al. [131] asked trustees about their second-order beliefs (i.e., how much money they think the trustor expects) and compared these second-order beliefs with the amount that trustees actually send. With this information, they could categorize the allocations made by trustees as "minimizing guilt" (when the amount sent was close to the trustees' second-order beliefs) or "maximizing outcome" (when trustees sent an amount significantly smaller than what they expected based on their second-order beliefs). When trustees minimized guilt, they exhibited higher activation in dlPFC, AIC, and dorsal ACC, which are structures reported to be activated by negative affective states [132–134]. On the other hand, when trustees maximized outcome, higher activation occurred in ventral mPFC, VS, and dorsal mPFC. The authors proposed a model where minimizing guilt increased AIC activation, which increased activation in dorsal mPFC, while maximizing outcome decreased AIC activation, which increased activation in the VS [131].

4.3 Altruism

Historically, altruism has been studied by means of the dictator game (DG). In this game, there are also two players involved in an anonymous one-shot interaction. The first player, called "dictator," receives an amount T of money and donates some a part of it ($A1 \in [0, T]$) to the second player, called the "recipient." This decision ends the game and the recipient has no participation in deciding about this distribution. Crucially, the recipient has no chances of punishing the dictator if the amount is not acceptable to him. Thus, there are not direct incentives for a strictly self-interested dictator to share any portion of the received money, and any donation is defined as an altruistic act [108, 135]. Behavioral evidence shows that even when participants play this game with unknown others, dictators tend to donate around 25% of their money to the recipient [136]. Interesting variants have been introduced to the game. Cherry et al. [137], for instance, made the dictators earn their own money, thereby giving subjects a sense of ownership. In this case, about 91% of the dictators don't send anything to the recipient. In addition, there have been recent efforts to include social knowledge about the recipient in the DG [138, 139]. Such experimental settings have shown that there are important variables which explain allocations, such as the knowledge about who the recipient is and how the game is explicitly described to the players [136]. Likewise, social distance is an important modulator of behavior in the DG. Hoffman et al. [140] showed that 64% of dictators kept all the money when social distance was maximized. In addition, some authors have shown that donations tend to be higher when people are informed that the recipient is a real charitable organization [138, 139].

4.4 Neuronal Dynamic of the DG

Despite its simplicity, and the fact that it has been used widely in behavioral economics, few neuroeconomics experiments have used the DG to assess the neural basis of altruism. In a recent article, Hutcherson et al. [141] made subjects participate in a DG where subjects had to choose between two options of allocation. By using this protocol, they induced choices between the default 50–50% split, generous (benefiting the other at a cost to oneself) or selfish behavior (benefiting oneself at a cost to another). The authors fitted a drift-diffusion model which assumes that choices are the output of a noisy process that weighs the linear sum of monetary outcomes for self and others. In this model, the choice is made when sufficient neural evidence has accumulated in favor of one of the options, and it assumes that the valuing of self and other outcomes is computed independently and then integrated in an overall value signal. At the neural level, the authors found that ventromedial prefrontal cortex (vmPFC) activity correlated positively with the value that subjects assigned to proposals, as measured by the Likert response scale. vmPFC has been reported to encode stimulus values at the time of decision in a wide range of tasks [142, 143]. Moreover, fitting general linear models (GLM), they found that valuations toward self-outcomes correlate with the activity in both vmPFC and VS, while valuations toward other outcomes correlate with the activity in the right TPJ, precuneus, and vmPFC. These results, further discussed below, show that the right TPJ is an area that becomes activated specifically when focusing on others, while vmPFC combines information about self and others.

In another experiment, Hein et al. [35] studied the role of empathy and reciprocity motives in human altruism. Using a DG, they investigated differences in altruistic behavior from experimental subjects when they observed recipients (1) receiving painful shocks (empathy partner) or (2) giving an amount of money to save some of those empathy partners from painful shocks (reciprocity partner), an action perceived as kind and, thus, one that should elicit reciprocity motives. A baseline partner neither received painful shock nor was instructed to give money for saving subjects from shock. Authors observed that subjects behave more altruistically toward the empathy and the reciprocity partners, noteworthy, without significant differences in allocations between the two motive inductions. At the neural level, a network consisting of AIC, VS, and ACC was activated in both motive-induction conditions. Moreover, individual pattern of brain connectivity in this network predicts subjects' altruist behavior. Interestingly, this prediction was particular for each treatment. Thus, a positive connectivity between ACC and AIC and a slightly negative connectivity between AIC and VS predict empathy-driven altruism, while a strong bidirectional projection between AIC and ACC and a positive connectivity between AIC and VS predict reciprocity-driven altruism. Additionally, the ACC connectivity to AIC correlates positively with baseline levels of altruism. Notice that, at the behavioral level, both motives were indistinguishable, because motives are a mental construct hidden to revealed preferences. A neuroeconomic approach is able to unravel both motives and their influence on altruistic behavior.

4.5 Two Putative Systems for Valuing Others' Outcomes

4.5.1 Anterior Cingulate Cortex and Vicarious Performance Monitoring

As seen above, a set of cognitive and affective functions determining the need for adaptive control prove central to economic decision-making [144]. A key neural structure that participates in these functions is the ACC, which is involved in interactions such as reciprocity, choosing the less common behavior [128, 145, 146], empathy and reciprocity-driven motives in human altruism [35], violations of social norms [147, 148], and mediating the effects of OXT in trust behavior [117].

The ACC is the frontal part of the cingulate cortex. Anatomically, the ACC has classically been subdivided in a rostral (rACC) and a dorsal part (dACC) [149]. The inputs to dACC include the amygdala, AI, orbitofrontal cortex, vmPFC and midbrain, and prominent ventral tegmental area. Its outputs target the lateral PFC, the motor cortex, striatum, subthalamic nucleus, and locus coeruleus [150]. The activity of the dACC has been correlated with almost the whole set of known psychological variables. Broadly speaking, dACC has been considered a key hub in a network of brain regions implicated in domain-general executive functions in humans [127], being important for cognitive control (i.e., our ability to flexibly adjust behavior according to internally maintained goals and away from behaviors that are more automatic but distracted from those goals [149]. Consequentially, there exists some agreement relating the involvement of the dACC in motivation and reward-based decision-making [127, 151].

However, there is no clear consensus on the function of dACC. Currently, two main proposals interpret its functioning: the expected value of control (EVC) theory and the foraging value theory (FVT). EVC [150] proposes that dACC plays a central role in decisions about the allocation of cognitive control based on a cost (for instance, the effort needed) and benefit (for instance, improved performance) analysis that identifies the highest EVC. The FVT theory, on the other hand, argues that difficulty or control allocation is insufficient to account for all dACC activity [152]. Instead, the dACC plays a key role in behavioral flexibility. Its activity reflects the history, weighted by time of occurrence, of previously chosen rewards, computing the value of persisting in the current environment versus the value of switching away from it [153].

Following the evidence review above and other experiments using economic social exchanges [115, 147, 154], some researchers argue that particular areas of ACC track, specifically, behavioral motivation and prediction errors not of self but specifically of others [149]. In this line, studies suggest that the gyral region of the ACC (ACCg) computes "other-oriented" information (i.e., information about other agents that might be animals or people, rather than ourselves). Apps et al. [155], for instance, examined the brain activity of human subjects when they received cues about the level of an economic reward and the cost incurred for receiving this reward, under conditions in which the costs and rewards pertained to the same experimental subjects or to a third person. In this experiment, ACCg activity correlated with the net value of rewards to be received by the third person when the third person incurred the cost of the effort. By contrast, the ACC sulcus signaled the effort level regardless of whether the effort was exerted by the subject or by a third person [149]. Authors found, "with a striking consistency," that the ACCg responds exclusively to other-oriented information.

4.5.2 Temporoparietal Regions and Valuing Others' Processes

As reviewed above, mentalization is our ability to represent and attribute others' mental or internal states, such as ideas, beliefs, desires, emotions, and motivations [31, 156] Similarly, perspective taking (PT) is the ability to comprehend that the same event or object can be seen or constructed in multiple ways, depending on each subject's point of view. Both processes enable humans to weight others' behaviors and preferences into the subjective valuations that underlie decision-making, a process that can be called "valuing others" [38, 39, 124]. At the neurobiological level, meta-analysis studies have shown that this area becomes active in all the tasks involving PT or mentalization [157]. Furthermore, some scholars have proposed that TPJ is a key neural structure underlying the distinction between self and others' perspectives [156, 158–160].

The involvement of the TPJ in general mentalizing functions can be linked to its anatomical characteristics. TPJ is constituted by the posterior part of the temporal lobe, the inferior part of parietal lobe, and the lateral part of occipital lobe [161]. This area is a heteromodal association cortex integrating multiple sources of sensory (and non-sensory) information. In addition, this region is located at a maximum synaptic/geodesic distance from sensory and motor areas. This seems to be useful for generating integrative computations addressing inner (abstract) and social processes [162].

There is plenty of evidence highlighting a consistent role of the TPJ in other preferences and how much these preferences affect personal decisions. TPJ is engaged, for instance, when subjects must anticipate others' decisions and behaviors [38, 39, 123, 141, 160, 163], when trustees reciprocate a high-risk allocation when pro-self individuals reciprocate [121], or when dictators evaluate the outcomes of others [141]. All these findings point to the existence of neuronal processes that compute others' preferences and behaviors, where TPJ is a key structure underlying the mechanism that allows us to integrate the others' preferences during a social interaction.

5 Conclusions

Currently there is a broad interest to combine evidence from different fields to better understand our complex social behavior. Our review suggests that, while the integration between social and natural sciences is still elusive, the evidence warrants five conclusions that may guide interdisciplinary discussion among behavioral economics, developmental psychology, and neuroscience. In particular, we believe that it is necessary to take care of the following observations:

1. The process of social decision-making can be understood as an algorithmic process that necessarily needs to be in contrast with real decision-making data.
2. In this algorithmic process, humans take into account multiple motivators (parameters), where self-interest (wellbeing/survival) and other-regarding preferences (valuing others' processing) are the most relevant.
3. The ways by which these motivators are finally integrated strongly depend on the neurobiological organization of multiple (not unitary) systems.
4. The neurobiological system (understood as neurophysiological states and traits) implicates both a general and a variable organization.
5. The variations of these neurobiological systems (not only one black box) depend at least on ontogenic (developmental) states, contextual constraints, and individual predispositions.

The social skills analyzed here are only an example of the areas where multiple disciplines have focused their efforts. Currently, it is extremely necessary to work on establishing common concepts in order to gather disperse perspectives. Through this chapter, we intend to generate a conceptual bridge among the knowledge input from psychology, neuroscience, and economics. This is certainly not a global theoretical framework but rather a starting point for building common conceptual framings in order to increase an interdisciplinary dialogue. In this way, we expect to be able to address difficult and unanswered questions about our amazing and, at the same time, conflictive social behavior.

Acknowledgments This work was supported by Comisión Nacional de Investigación Científica y Tecnológica CONICYT (Grant FONDECYT 11405268 to CR-S, Grant FONDECYT Inicio 11140535 to PB and Grant PCHA/DoctoradoNacional/2014-21140043 to PS-I).

References

1. Friedman M. Essays in positive economics. Chicago: University of Chicago Press; 1953.
2. Stigler G, Becker G. De Gustibus Non Est Disputandum. Am Econ Rev. 1977;67(2):76–90.
3. Ashraf N, Camerer CF, Loewenstein G. Adam Smith, behavioral economist. J Econ Perspect. 2005;19(3):131–45.
4. Kahneman D, Tversky A. Prospect theory: an analysis of decision under risk. Econometrica JSTOR. 1979;47(2):263–91.
5. O'Donoghue T, Rabin M. Doing it now or later. Am Econ Rev. 1999;89(1):103–24.
6. Laibson D. golden eggs and hyperbolic discounting. Q J Econ. 1997;112(2):443–78.
7. Berg J, Dickhaut J, McCabe K. Trust, reciprocity, and social history. Games Econ Behav. 1995;10(1):122–42.
8. Cox JC. How to identify trust and reciprocity. Games Econ Behav. 2004;46(2):260–81.
9. Falk A, Kosfeld M. The hidden costs of control. Am Econ Rev. 2006;96(5):1611–30.
10. Sheremeta RM, Zhang J. Three-player trust game with insider communication. Econ Inq. 2014;52(2):576–91.
11. Heyes A, List JA. Supply and demand for discrimination: strategic revelation of own characteristics in a trust game. Am Econ Rev. 2016;106(5):319–23.
12. Fehr E, Gächter S. Cooperation and punishment in public goods experiments. Am Econ Rev. 2000;90(4):980–94.
13. Reuben E, Riedl A. Enforcement of contribution norms in public good games with heterogeneous populations. Games Econ Behav. 2013;77(1):122–37.
14. Oprea R, Charness G, Friedman D. Continuous time and communication in a public-goods experiment. J Econ Behav Organ. 2014;108:212–23.
15. Brañas-Garza P, Espín AM, Exadaktylos F, Herrmann B. Fair and unfair punishers coexist in the ultimatum game. Sci Rep. 2015;4(1):6025.
16. Nowak MA. Fairness versus reason in the ultimatum game. Science. 2000;289(5485):1773–5.
17. Güth W, Kocher MG. More than thirty years of ultimatum bargaining experiments: motives, variations, and a survey of the recent literature. J Econ Behav Organ. 2014;108:396–409.
18. Fehr E, Gächter S. Altruistic punishment in humans. Nature. 2002;415(6868):137–40.
19. Bolton GE, Ockenfels A. ERC: a theory of equity, reciprocity, and competition. Am Econ Rev. 2000;90(1):166–93.
20. Fehr E, Schmidt K. A Theory of fairness, competition and cooperation. Q J Econ. 1999;114(August):817–68.
21. Rabin M. Incorporating fairness into game theory and economics. Am Econ Rev. 1993;83:1281–302.
22. Falk A, Fischbacher U. A theory of reciprocity. Games Econ Behav. 2006;54(2):293–315.
23. Dufwenberg M, Kirchsteiger G. A theory of sequential reciprocity. Games Econ Behav. 2004;47(2):268–98.
24. Falk A, Fehr E, Fischbacher U. On the nature of fair behavior. Econ Inq. 2003;41(1):20–6.
25. Andreoni J, Barton B, Bernheim BD, Aydin D, Naecker J. When fair isn't fair: sophisticated time inconsistency in social preferences. Work Pap. 2016;1996:58.
26. Loewenstein G, Rick S, Cohen JD. Neuroeconomics. Annu Rev Psychol. 2008;59:647–72.
27. Damasio A. Feelings of emotion and the self. Ann N Y Acad Sci. 2003 Oct;1001:253–61.
28. Kahneman D. Thinking, fast and slow. New York: Macmillan and Company; 2011.

29. King-Casas B, Chiu PH. Understanding interpersonal function in psychiatric illness through multiplayer economic games. Biol Psychiatry. 2012;72(2):119–25.
30. Declerck CH, Boone C, Emonds G. When do people cooperate? The neuroeconomics of prosocial decision making. Brain Cogn. 2013;81(1):95–117.
31. Billeke P, Aboitiz F. Social cognition in schizophrenia: from social stimuli processing to social engagement. Front Psychiatry. 2013;4(February):1–12.
32. Steinbeis N, Bernhardt BC, Singer T. Impulse control and underlying functions of the left DLPFC mediate age-related and age-independent individual differences in strategic social behavior. Neuron. 2012;73(5):1040–51.
33. Andreoni J, Miller J. Giving according to GARP: an experimental test of the consistency of preferences for altruism. Econometrica. 2002;70(2):737–53.
34. Fehr E, Camerer CF. Social neuroeconomics: the neural circuitry of social preferences. Trends Cogn Sci. 2007;11:419–27.
35. Hein G, Morishima Y, Leiberg S, Sul S, Fehr E. The brains functional network architecture reveals human motives. Science. 2016;351(6277):1074–8.
36. McAuliffe K, Blake PR, Steinbeis N, Warneken F. The developmental foundations of human fairness. Nat Hum Behav. 2017;1(2):42.
37. Dalgleish T, Walsh ND, Mobbs D, Schweizer S, van Harmelen A-L, Dunn B, et al. Social pain and social gain in the adolescent brain: a common neural circuitry underlying both positive and negative social evaluation. Sci Rep. 2017;7(February 2016):42010.
38. Ibáñez A, Billeke P, de la Fuente L, Salamone P, García AM, Melloni M. Reply: Towards a neurocomputational account of social dysfunction in neurodegenerative disease. Brain. 2017;140(3):e15.
39. Melloni M, Billeke P, Baez S, Hesse E, de la Fuente L, Forno G, et al. Your perspective and my benefit: multiple lesion models of self-other integration strategies during social bargaining. Brain. 2016;139(11):3022–40.
40. Baars B, Gage N. Social cognition: perceiving the mental states of others. In: Cognition, brain and consciousness: introduction to cognitive neuroscience. 2nd ed. San Diego, CA: Elsevier; 2010.
41. Penn DC, Povinelli DJ. On the lack of evidence that non-human animals possess anything remotely resembling a "theory of mind". Philos Trans R Soc Lond Ser B Biol Sci. 2007;362(January):731–44.
42. Povinelli DJ, Vonk J. Chimpanzee minds: suspiciously human? Trends Cogn Sci. 2003;7(4):157–60.
43. Aboitiz FA. Brain for speech. A view from evolutionary neuroanatomy. London: Palgrave Macmillan; 2017.
44. Johnson MH. Interactive specialization: a domain-general framework for human functional brain development? Dev Cogn Neurosci. 2011;1(1):7–21.
45. Emery NJ. The eyes have it: the neuroethology, function and evolution of social gaze. Neurosci Biobehav Rev. 2000;24(6):581–604.
46. Bertenthal BI, Proffitt DR, Cutting JE. Infant sensitivity to figural coherence in biomechanical motions. J Exp Child Psychol. 1984;37(2):213–30.
47. Pavlova M, Sokolov A. Orientation specificity in biological motion perception. Percept Psychophys. 2000;62(5):889–99.
48. Simion F, Regolin L, Bulf H. A predisposition for biological motion in the newborn baby. Proc Natl Acad Sci U S A. 2008;105(2):809–13.
49. Macchi Cassia V, Simion F, Umiltaa C. Face preference at birth: the role of an orienting mechanism. Dev Sci. 2001;4(1):101–8.
50. Farroni T, Csibra G, Simion F, Johnson MH. Eye contact detection in humans from birth. Proc Natl Acad Sci. 2002;99(14):9602–5.
51. Farroni T, Mansfield EM, Lai C, Johnson MH. Infants perceiving and acting on the eyes: tests of an evolutionary hypothesis. J Exp Child Psychol. 2003;85(3):199–212.

52. Meltzoff AN, Moore MK. Imitation of facial and manual gestures by human neonates. Published by: American Association for the Advancement of Science Stable. URL: http://www.jstor.org/stable/1744187. 1977;198(4312):75–8.
53. Jones W, Klin A. Attention to eyes is present but in decline in 2-6-month-old infants later diagnosed with autism. Nature. 2013;504(7480):427–31.
54. Turati C, Valenza E, Leo I, Simion F. Three-month-olds' visual preference for faces and its underlying visual processing mechanisms. J Exp Child Psychol. 2005;90(3):255–73.
55. Macchi Cassia V, Bulf H, Quadrelli E, Proietti V. Age-related face processing bias in infancy: evidence of perceptual narrowing for adult faces. Dev Psychobiol. 2014;56(2):238–48.
56. Luyster RJ, Powell C, Tager-Flusberg H, C a N. Neural measures of social attention across the first years of life: characterizing typical development and markers of autism risk. Dev Cogn Neurosci. 2014;8:131–43.
57. De Haan M, Johnson MH, Halit H. Development of face-sensitive event-related potentials during infancy. In: De Haan M, editor. Infant EEG and event-related potentials. 1st ed. New York: Psychology Press; 2007.
58. Pena M, Arias D, Dehaene-Lambertz G. Gaze following is accelerated in healthy preterm infants. Psychol Sci. 2014;25(10):1884–92.
59. Soto-Icaza P, Aboitiz F, Billeke P. Development of social skills in children: neural and behavioral evidence for the elaboration of cognitive models. Front Neurosci. 2015;9(September):1–16.
60. Haan M De. Introduction to infant EEG and event-related potentials. In: Haan M, editor. Infant EEG and event-related potentials. New York, USA: Psychology Press Ltd New York; 2002. p. 39–76.
61. Luck SJ. Ten simple rules for designing and interpreting ERP experiments University of Iowa. In: Handy TC, editor. Event related potentials: a methods handbook. Cambridge, MA: MIT Press; 2004.
62. Csibra G, Kushnerenko E, Grossmann T. Electrophysiological methods in studying infant cognitive development. In: Nelson CA, Luciana M, editors. Handbook of developmental cognitive neuroscience. Cambridge, MA: MIT Press; 2008. p. 1–50.
63. Hileman CM, Henderson H, Mundy P, Newell L, Jaime M. Developmental and individual differences on the P1 and N170 ERP components in children with and without autism. Dev Neuropsychol. 2013;36(2):214–36.
64. Itier RJ. N170 or N1? Spatiotemporal differences between object and face processing using ERPs. Cereb Cortex. 2004;14(2):132–42.
65. Courchesne E, Ganz L, Norcia a M. Event-related brain potentials to human faces in infants. Child Dev. 1981;52(3):804–11.
66. Dawson G, Webb SJ, McPartland J. Understanding the nature of face processing impairment in autism: insights from behavioral and electrophysiological studies. Dev Neuropsychol. 2005;27(3):403–24.
67. de Haan M, CA N. Brain activity differentiates face and object processing in 6-month-old infants. Dev Psychol. 1999;35(4):1113–21.
68. Elsabbagh M, Volein A, Csibra G, Holmboe K, Garwood H, Tucker L, et al. Neural correlates of eye gaze processing in the infant broader autism phenotype. Biol Psychiatry. 2009;65(1):31–8.
69. Johnson MH, Griffin R, Csibra G, Halit H, Farroni T, de Haan M, et al. The emergence of the social brain network: evidence from typical and atypical development. Dev Psychopathol. 2005;17(3):599–619.
70. Balas BJ, Nelson CA, Westerlund A, Vogel-Farley V, Riggins T, Kuefner D. Personal familiarity influences the processing of upright and inverted faces in infants. Front Hum Neurosci. 2010;4(February):1.
71. Bretherton I. The origins of attachment theory: John Bowlby and Mary Ainsworth. Dev Psychol. 1992;28(5):759–75.
72. Tronick EZ, Cohn JF. Infant-mother face-to-face interaction: age and gender differences in coordination and the occurrence of miscoordination. Child Dev. 1989;60(1):85.

73. Harlow HF, Zimmermann RR. Affectional response in the infant monkey: orphaned baby monkeys develop a strong and persistent attachment to inanimate surrogate mothers. Science. 1959;130(3373):421–32.
74. Mundy P, Card J, Fox N. EEG correlates of the development of infant joint attention skills. Dev Psychobiol. 2000;36:325–38.
75. Charman T. Why is joint attention a pivotal skill in autism? Philos Trans R Soc Lond Ser B Biol Sci. 2003;358(January):315–24.
76. Morgan B, Maybery M, Durkin K. Weak central coherence, poor joint attention, and low verbal ability: independent deficits in early autism. Dev Psychol. 2003;39(4):646–56.
77. Striano T, Reid VM, Hoehl S. Neural mechanisms of joint attention in infancy. Eur J Neurosci. 2006;23(10):2819–23.
78. Lachat F, Hugueville L, Lemaréchal J-D, Conty L, George N. Oscillatory brain correlates of live joint attention: a dual-EEG study. Front Hum Neurosci. 2012;6(June):156.
79. Hopkins WD, Taglialatela JP. Initiation of joint attention is associated with morphometric variation in the anterior cingulate cortex of chimpanzees (Pan troglodytes). Am J Primatol. 2013;75(5):441–9.
80. Charman T, Baron-Cohen S, Swettenham J, Baird G, Cox A, Drew A. Testing joint attention, imitation, and play as infancy precursors to language and theory of mind. Cogn Dev. 2000;15(4):481–98.
81. Bakeman R, Adamson LB. Coordinating attention to people and objects in mother-infant and peer-infant interaction. Child Dev. 1984;55(4):1278–89.
82. Kopp F, Lindenberger U. Effects of joint attention on long-term memory in 9-month-old infants: an event-related potentials study. Dev Sci. 2011;14(4):660–72.
83. Striano T, Reid VM. Social cognition in the first year. Trends Cogn Sci. 2006;10(10):471–6.
84. Hirotani M, Stets M, Striano T, Friederici AD. Joint attention helps infants learn new words: event-related potential evidence. Neuroreport. 2009;20(6):600–5.
85. Wimmer H, Perner J. Beliefs about beliefs: representation and constraining function of wrong beliefs in young children's understanding of deception. Cognition. 1983;13(1):103–28.
86. Baron-Cohen S, Leslie AM, Frith U. Does the autistic child have a "theory of mind"? Cognition. 1985;21(1):37–46.
87. Auer DP. Spontaneous low-frequency blood oxygenation level-dependent fluctuations and functional connectivity analysis of the "resting" brain. Magn Reson Imaging. 2008;26(7):1055–64.
88. Grosse Wiesmann C, Schreiber J, Singer T, Steinbeis N, Friederici AD. White matter maturation is associated with the emergence of theory of mind in early childhood. Nat Commun. 2017;8:14692.
89. Premack D, Woodruff G. Does the chimpanzee have a theory of mind. Behav Brain Sci. 1978;1:515–26.
90. Perner J, Roessler J. From infants' to children's appreciation of belief. Trends Cogn Sci. 2012;16:519–25.
91. Baillargeon R, Scott RM, He Z. False-belief understanding in infants. Trends Cogn Sci. 2010;14(3):110–8.
92. Choi YJ, Luo Y. 13-Month-olds' understanding of social interactions. Psychol Sci. 2015;26(3):274–83.
93. Kovács ÁM, Téglás E, Endress AD. The social sense: susceptibility to others' beliefs in human infants and adults. Science. 2010;330(6012):1830–4.
94. Southgate V, Senju a CG. Action anticipation through attribution of false belief by 2-year-olds. Psychol Sci. 2007;18(7):587–92.
95. SAJ B, Bernstein DM. What can children tell us about hindsight bias: a fundamental constraint on perspective–taking? Soc Cogn. 2007;25(1):98–113.
96. Bloom P, German TP. Two reasons to abandon the false belief task as a test of theory of mind. Cognition. 2000;77:25–31.

97. Surian L, Caldi S, Sperber D. Attribution of beliefs by 13-month-old infants. Psychol Sci. 2007;18(7):580–6.
98. Moll H, Meltzoff AN. How does it look? Level 2 perspective-taking at 36 months of age. Child Dev. 2011;82(2):661–73.
99. Aichhorn M, Perner J, Kronbichler M, Staffen W, Ladurner G. Do visual perspective tasks need theory of mind? NeuroImage. 2006;30(3):1059–68.
100. Moll H, Tomasello M. Level 1 perspective-taking at 24 months of age. Br J Dev Psychol. 2006;24(3):603–13.
101. Hamilton AF de C, Brindley R, Frith U. Visual perspective taking impairment in children with autistic spectrum disorder. Cognition. 2009;113(1):37–44.
102. Moll H, Kadipasaoglu D. The primacy of social over visual perspective-taking. Front Hum Neurosci. 2013;7(September):558.
103. Schaafsma SM, Pfaff DW, Spunt RP, Adolphs R. Deconstructing and reconstructing theory of mind. Trends Cogn Sci. 2015;19(2):65–72.
104. Dunbar RIM, Shultz S. Evolution in the social brain. Science. 2007;317(5843):1344–7.
105. Fehr E, Fischbancher U. Third-party punishment and social norms. Evol Hum Behav. 2004;25(2):63–87.
106. Camerer CF, Fehr E. When does "economic man" dominate social behavior? Science. 2006;311(5757):47–52.
107. Krueger F, Grafman J, McCabe K. Neural correlates of economic game playing. Philos Trans R Soc Lond Ser B Biol Sci. 2008;363(1511):3859–74.
108. Lee D. Game theory and neural basis of social decision making. Nat Neurosci. 2008;11(4):404–9.
109. Johnson ND, Mislin AA. Trust games: a meta-analysis. J Econ Psychol. 2011;32(5):865–89.
110. Camerer CF, Loewenstein G, Prelec D. Neuroeconomics: How neuroscience can inform economics. J Econ Lit. 2005;43(1):9–64.
111. Amodio DM, Frith CD. Meeting of minds: the medial frontal cortex and social cognition. Nat Rev Neurosci. 2006;7(4):268–77.
112. McCabe K, Houser D, Ryan L, Smith V, Trouard T. A functional imaging study of cooperation in two-person reciprocal exchange. Proc Natl Acad Sci U S A. 2001;98:11832–5.
113. Rilling JK, Sanfey AG, Aronson JA, Nystrom LE, Cohen JD. The neural correlates of theory of mind within interpersonal interactions. NeuroImage. 2004;22(4):1694–703.
114. Delgado MR, Frank RH, Phelps EA. Perceptions of moral character modulate the neural systems of reward during the trust game. Nat Neurosci. 2005;8:1611–8.
115. King-Casas B, Tomlin D, Anen C, Camerer CF, Quartz SR, Montague PR. Getting to know you: reputation and trust in a two-person economic exchange. Science. 2005;308:78–83.
116. Delgado MR, Li J, Schiller D, E a P. The role of the striatum in aversive learning and aversive prediction errors. Philos Trans R Soc Lond Ser B Biol Sci. 2008;363(1511):3787–800.
117. Baumgartner T, Heinrichs M, Vonlanthen A, Fischbacher U, Fehr E. Oxytocin shapes the neural circuitry of trust and trust adaptation in humans. Neuron. 2008;58(4):639–50.
118. Zak PJ, Kurzban R, Ahmadi S, Swerdloff RS, Park J, Efremidze L, et al. Testosterone administration decreases generosity in the ultimatum game. PLoS One. 2009;4(12):e8330.
119. Kosfeld M, Heinrichs M, Zak PJ, Fischbacher U, Fehr E. Oxytocin increases trust in humans. Nature. 2005;435(June):673–6.
120. Aspé-sánchez M, Moreno M, Rivera MI, Rossi A. Oxytocin and vasopressin receptor gene polymorphisms: role in social and psychiatric traits. Front Neurosci. 2016;9(January):510.
121. van den Bos W, Güroğlu B, van den Bulk BG, Rombouts SA, Crone E. Better than expected or as bad as you thought? The neurocognitive development of probabilistic feedback processing. Front Hum Neurosci. 2009;3(December):52.
122. Mitchell JP. Activity in right temporo-parietal junction is not selective for theory-of-mind. Cereb Cortex. 2008;18(2):262–71.
123. Billeke P, Boardman S, Doraiswamy PM. Social cognition in major depressive disorder: a new paradigm? Transl Neurosci. 2013;4(4):437–47.

124. Billeke P. The more I get to know you, the more I distrust you? Non-linear relationship between social skills and social behavior. Front Psychiatry. 2016;7:49.
125. de Vignemont F, Singer T. The empathic brain: how, when and why? Trends Cogn Sci. 2006;10:435–41.
126. Shenhav A, Botvinick MM, Cohen JD. The expected value of control: an integrative theory of anterior cingulate cortex function. Neuron. 2013;79(2):217–40.
127. Ebitz RB, Platt ML, Ebitz RB, Platt ML. Neuronal activity in primate dorsal anterior cingulate cortex signals task conflict and predicts adjustments in pupil-linked arousal Article Neuronal Activity in Primate Dorsal Anterior Cingulate Cortex Signals Task Conflict and Predicts Adjustments in Pu. Neuron. 2015;85(3):628–40.
128. Billeke P, Zamorano F, López T, Rodriguez C, Cosmelli D, Aboitiz F. Someone has to give in: theta oscillations correlate with adaptive behavior in social bargaining. Soc Cogn Affect Neurosci. 2014;9(12):2041–8.
129. Billeke P, Zamorano F, Cosmelli D, Aboitiz F. Oscillatory brain activity correlates with risk perception and predicts social decisions. Cereb Cortex. 2013;23(12):2872–83.
130. Ibáñez MI, Sabater-Grande G, Barreda-Tarrazona I, Mezquita L, López-Ovejero S, Villa H, et al. Take the money and run: psychopathic behavior in the trust game. Front Psychol. 2016;7(November):1–15.
131. Chang LJ, Smith A, Dufwenberg M, Sanfey AG. Triangulating the neural, psychological, and economic bases of guilt aversion. Neuron. 2011;70(3):560–72.
132. Yoshimura S, Okamoto Y, Onoda K, Matsunaga M, Ueda K, Suzuki S, et al. Rostral anterior cingulate cortex activity mediates the relationship between the depressive symptoms and the medial prefrontal cortex activity. J Affect Disord. 2010;122(1–2):76–85.
133. Damasio AR, Grabowski TJ, Bechara A, Damasio H, Ponto LL, Parvizi J, et al. Subcortical and cortical brain activity during the feeling of self-generated emotions. Nat Neurosci. 2000;3:1049–56.
134. Singer T, Seymour B, O'Doherty J, Kaube H, Dolan RJ, Frith CD. Empathy for pain involves the affective but not sensory components of pain. Science. 2004;303:1157–62.
135. Rilling JK, Sanfey AG. The neuroscience of social decision-making. Annu Rev Psychol. 2011;62:23–48.
136. Camerer CF. Behavioural studies of strategic thinking in games. Trends Cogn Sci. 2003;7:225–31.
137. Cherry T, Frykblom P, Shogren J. Hardnose the Dictator. Am Econ Rev. 2002;92(4):1218–22.
138. Moll J, Krueger F, Zahn R, Pardini M, de Oliveira-Souza R, Grafman J. Human fronto-mesolimbic networks guide decisions about charitable donation. Proc Natl Acad Sci U S A. 2006;103(42):15623–8.
139. Wu S-W, Delgado MR, Maloney LT. The neural correlates of subjective utility of monetary outcome and probability weight in economic and in motor decision under risk. J Neurosci. 2011;31(24):8822–31.
140. Hoffman E, McCabe K, Shachat K, Smith V. Preferences, property rights, and anonymity in bargaining games. Games Econ Behav. 1994;7:346–80.
141. Hutcherson CA, Bushong B, Rangel A. A neurocomputational model of altruistic choice and its implications. Neuron. 2015;87(2):451–62.
142. Raposo A, Vicens L, Clithero JA, Dobbins IG, Huettel SA. Contributions of frontopolar cortex to judgments about self, others and relations. Soc Cogn Affect Neurosci. 2011;6(3):260–9.
143. Kable JW, Glimcher PW. The neurobiology of decision: consensus and controversy. Neuron. 2009;63(6):733–45.
144. Ullsperger M, Fischer AG, Nigbur R, Endrass T. Neural mechanisms and temporal dynamics of performance monitoring. Trends Cogn Sci. 2014;18(5):259–67.
145. Rilling J, Gutman D, Zeh T, Pagnoni G, Berns G, Kilts C. A neural basis for social cooperation. Neuron. 2002;35:395–405.
146. McClure EB, Parrish JM, Nelson EE, Easter J, Thorne JF, Rilling JK, et al. Responses to conflict and cooperation in adolescents with anxiety and mood disorders. J Abnorm Child Psychol. 2007;35(4):567–77.

147. Sanfey AG, Rilling JK, Aronson JA, Nystrom LE, Cohen JD. The neural basis of economic decision-making in the ultimatum game. Science. 2003;300(5626):1755–8.
148. Spitzer M, Fischbacher U, Herrnberger B, Grön G, Fehr E. The neural signature of social norm compliance. Neuron. 2007;56(1):185–96.
149. Apps MAJ, Rushworth MFS, Chang SWC. The anterior cingulate gyrus and social cognition: tracking the motivation of others. Neuron. 2016;90(4):692–707.
150. Shenhav A, Straccia MA, Botvinick MM, Cohen JD. Dorsal anterior cingulate and ventromedial prefrontal cortex have inverse roles in both foraging and economic choice. Cogn Affect Behav Neurosci. 2016;16(6):1127–39.
151. Wittmann MK, Kolling N, Akaishi R, Chau BKH, Brown JW, Nelissen N, et al. Predictive decision making driven by multiple time-linked reward representations in the anterior cingulate cortex. Nat Commun. 2016;7:12327.
152. Kolling N, Wittmann MK, Behrens TEJ, Boorman ED, Mars RB, Rushworth MFS. Value, search, persistence and model updating in anterior cingulate cortex. Nat Neurosci. 2016;19(10):1280–5.
153. Wittmann MK, Kolling N, Faber NS, Scholl J, Nelissen N, MFS R. Self-other mergence in the frontal cortex during cooperation and competition. Neuron. 2016;91(2):482–93.
154. Ruff CC, Fehr E. The neurobiology of rewards and values in social decision making. Nat Rev Neurosci. 2014;15(8):549–62.
155. Apps MAJ, Lesage E, Ramnani N. Vicarious reinforcement learning signals when instructing others. J Neurosci. 2015;35(7):2904–13.
156. Abu-Akel A, Shamay-Tsoory S. Neuroanatomical and neurochemical bases of theory of mind. Neuropsychologia. 2011;49(11):2971–84.
157. Schurz M, Radua J, Aichhorn M, Richlan F, Perner J. Fractionating theory of mind: a meta-analysis of functional brain imaging studies. Neurosci Biobehav Rev. 2014;42:9–34.
158. Saxe R, Xiao D-K, Kovacs G, Perrett DI, Kanwisher N. A region of right posterior superior temporal sulcus responds to observed intentional actions. Neuropsychologia. 2004;42(11):1435–46.
159. Billeke P, Zamorano F, Chavez M, Cosmelli D, Aboitiz F. Functional network dynamics in alpha band correlate with social bargaining. PLoS One. 2014;9(10):e109829.
160. Billeke P, Armijo A, Castillo D, López T, Zamorano F, Cosmelli D, et al. Paradoxical expectation: oscillatory brain activity reveals social interaction impairment in schizophrenia. Biol Psychiatry. 2015;78(6):421–31.
161. Corbetta M, Patel G, Shulman GL. The reorienting system of the human brain: from environment to theory of mind. Neuron. 2008;58(3):306–24.
162. Margulies DS, Ghosh SS, Goulas A, Falkiewicz M, Huntenburg JM, Langs G, et al. Situating the default-mode network along a principal gradient of macroscale cortical organization. Proc Natl Acad Sci. 2016;113:12574–9.
163. Carter RM, Bowling DL, Reeck C, S a H. A distinct role of the temporal-parietal junction in predicting socially guided decisions. Science. 2012;337(6090):109–11.
164. Charness G, Rabin M. Understanding social preferences with simple tests. Q J Econ. 2002;117(3):817–69.

Bias and Control in Social Decision-Making

Paloma Díaz-Gutiérrez, Sonia Alguacil, and María Ruz

Abstract Social decisions are crucial in our life. Many of these include interactions between agents in scenarios of varying complexity, where trust and cooperation are essential and multiple sources of information influence our choices. In this chapter we review the contributions from social neuroscience to understanding the sources of bias and control mechanisms in social decisions, integrating insights from diverse methodologies and analyses. These biases include individual influences (both stable and transient) and other stimulus-driven factors, such as social stereotypes, emotion displays, or information regarding personality traits. This information modulates different stages of processing, with control-related influences playing crucial roles to override conflicts between automatic tendencies and goals.

Keywords Social neuroscience • Decision-making • Social bias • Control mechanisms • Neuroimaging

1 Introduction

Decisions of different complexity are a constant element in our life. Both simple and more thoughtful and relevant choices share the need of processing different options to choose the action that best fulfills our goals [1]. As social beings, a large part of our decisions involves other people, so that we must take into account information about others and predict their likely behavior. Accordingly, trust and cooperation are central factors in social interactions [2–4]. However, our supposedly rational decisions are fairly influenced by several factors, or biases, which generate predispositions to behave in certain ways [5–7]. The evidence to date shows that these biases are entrained not only with late decision stages related to value or response processing [8] but also with early stages of perception [9]. In addition, the

P. Díaz-Gutiérrez • S. Alguacil • M. Ruz (✉)
Mind, Brain and Behavior Research Center, University of Granada, 18071 Granada, Spain

Department of Experimental Psychology, University of Granada, 18071 Granada, Spain
e-mail: pdiaz@ugr.es; salguacil@ugr.es; mruz@ugr.es

© Springer International Publishing Switzerland 2017

A. Ibáñez et al. (eds.), *Neuroscience and Social Science*,
DOI 10.1007/978-3-319-68421-5_3

need to arbitrate among these different and complex action tendencies to make optimal decisions calls for strategic control mechanisms.

Several disciplines, such as psychology or economics, seek to understand the role of these biases on social decision-making, and the way control mechanisms are recruited to channel their influence. In this respect, social neuroscience is an innovative discipline that addresses such questions by studying the neural underpinnings of relevant phenomena, focusing on where, when, and how they take place in the brain [10, 11]. The goal of the current chapter is to provide a comprehensive overview of such contributions, integrating insights from diverse methodologies and analysis strategies [8, 12–14].

In the following sections, we first describe the methodology employed in social neuroscience to study the factors that influence social decisions. Then, we present evidence about the different sources of bias in these scenarios, which derive from individual factors and from the stimuli we perceive. Thereupon, we review how these influences are regulated by control mechanisms. Lastly, we offer some conclusions and future directions.

2 Methodological Tools

Research in social neuroscience combines various behavioral methods with modern neuroimaging techniques [15]. On the one hand, several studies rely on the use of interactive games from the field of experimental economics and classic game theory. These paradigms have been often used to derive normative descriptions of how people make economic and trust decisions while interacting with others [16]. However, the reasons for such normative behaviors can be better understood if we know their underlying sources. In this sense, the mechanisms underlying the departures from rationality that people often display in these settings can be explored at the behavioral level by paradigms developed in the field of psychology and at the neural level by modern noninvasive neuroimaging tools. Hence, this mixture of approaches promotes the explanation of human behavior at normative, mechanistic, and neural levels, which complement and nurture each other [17].

Among the tools developed in behavioral game theory, the *ultimatum game* [18] is a very popular task to study the response of people to fairness. Here one player acts as the proposer, choosing how to divide a certain amount of money. The other player, the responder, decides whether to accept or reject the offer. In the first case, both players earn their split, whereas if the responder rejects the offer, neither of them gains anything from that interaction. Reciprocation behavior has been extensively studied with the *trust game* [16]. In this case, one player (the investor) decides whether or not to share an amount of money with another partner (the trustee). If shared, this money is multiplied and transferred to the trustee, who then gets to decide whether to reciprocate or not. In the first case, both earn half of the total money, but if there is no reciprocation, the investor loses the initial sum. In this scenario, the best strategy rests with the mutual cooperation between players. The *prisoner's dilemma* [19] is similar, but here both players choose to trust the other one or not, and payoffs depend on both decisions. In addition, some studies have developed online versions of these tasks [20], whereas others have tried to implement cooperation settings in more realistic scenarios (e.g., the *apple game*; [21]).

Several paradigms developed in different fields of psychology are designed to study the mechanisms, or processes, underlying human behavior and choices. For example, the field of social psychology has developed several tasks to explore implicit biases, such as prejudice [22]. Among these, the *implicit association test* (IAT; [23]) is frequently used to explore how people associate social dimensions (e.g., gender, race) with different attributes (e.g., women are emotional vs. men are logical), which ultimately reflect automatic manifestations of prejudice. Similarly, implicit prejudices are often revealed in *sequential evaluative priming tasks*, where, for instance, participants view targets preceded by prime stimuli referring to social categories (e.g., white and black faces) and classify them as "pleasant" or "unpleasant." A variant is the *weapon identification task*, which assesses racial prejudice by asking participants to categorize guns and tools after the presentation of white and black face primes [24]. In addition, other studies use words or facial displays to assess how people form first impressions (e.g., [25]) or associate different social categories depending on their shared stereotypes (e.g., [22]). In addition, moral dilemmas [26, 27], where people have to judge the moral acceptability of behaviors in complex scenarios, are used to explore how personal dispositions or induced analytical tendencies influence moral evaluations.

These behavioral paradigms offer an integrated knowledge of the different phenomena influencing our social choices at different stages of processing. Social neuroscience adds neuroimaging methods to study the neural underpinnings of these decisions. This provides a better understanding of the sources of type of information relevant for social behavior and allows analyzing the commonalities and differences between social and nonsocial phenomena [15]. Among these neuroimaging techniques, electroencephalography (EEG) and functional magnetic resonance imaging (fMRI) are the ones most frequently used to study brain activity noninvasively.

EEG, given its temporal precision, allows tracking how different cognitive processes operate in time [28]. This technique provides information about the stages of processing (e.g., perception, decision, or motor output) at which the phenomena of interest take place. Complementarily, the good spatial resolution of magnetic resonance imaging (MRI) makes it an optimal choice to explore the neural regions underlying all these processes. Additionally, functional near-infrared spectroscopy (fNIRS) measures hemodynamic activity as functional MRI (fMRI), but facilitates more natural experimental settings as it is a portable device, at the expense of lower spatial resolution compared with fMRI [29].

These methods are combined with different analytical strategies, which integrate traditional univariate with multivariate approaches adopted from machine learning. While classic univariate methods compare activation between experimental conditions for each voxel (unit of measurement in MRI), multi-voxel pattern analyses (MVPA; [30]) allow studying how information is encoded in patterns of neural activity across several voxels. Furthermore, representational similarity analysis [31] relates the structure of neural patterns with each other and also with behavioral data, offering information about the nature of representations in different brain regions and their relation to different psychological theories (e.g., [32]).

Altogether, these new approaches open new avenues to further the understanding of how biasing social information is coded in the brain and the reason for their pervasive effects in our interpersonal behavior.

3 Bias in Social Decision-Making

Influences on social decision-making stem from different sources. On the one hand, individual factors or states impact how we process information, which can bias our decisions. On the other hand, the perception of certain features in other people may also be associated with different action tendencies, judgments, or attributes, impacting how we perceive and behave toward others.

3.1 Bias in the Observer

The individual factors that influence choices include stable personal characteristics (such as gender, age, prosociality, or permanent brain lesions) and contextual, non-permanent factors (such as induced emotional states). Below we address them in turn.

Beginning with stable factors, gender has been linked to differences in social decision-making in several studies. For example, women seem to make more ethical decisions in certain social scenarios (e.g., [33]). However, altruistic behavior for each gender seems to depend on the expensiveness of the cooperation, which generates different contexts for each of them. Thus, women are more altruistic when it is most costly, whereas the opposite happens for men [34]. Gender differences in moral decisions may also be modulated by emotional empathy [35]. In this case, gender seems to influence our empathic responses to noncooperative partners. For instance, Singer et al. [36] observed that empathic responses to the pain of others, as measured in fronto-insular regions and the anterior cingulate cortex (ACC; see Fig. 1 for a visualization of the brain areas), were reduced in males when observing unfair players receiving painful electric shocks. The brain of male participants in the same situation also showed increased activation in regions related to the reward system, such as the ventral striatum, which was interpreted by the authors as a sense of "revenge." Note that this study is one of many examples of how the introduction of measurements of brain activity adds evidence that helps to understand the mechanisms underlying biases in human social behavior.

Age is another factor that has been related to differences in social decisions. At a young age, children's cooperative behavior is already dependent on the agent they are interacting with, as they are more generous with friends than non-friends and

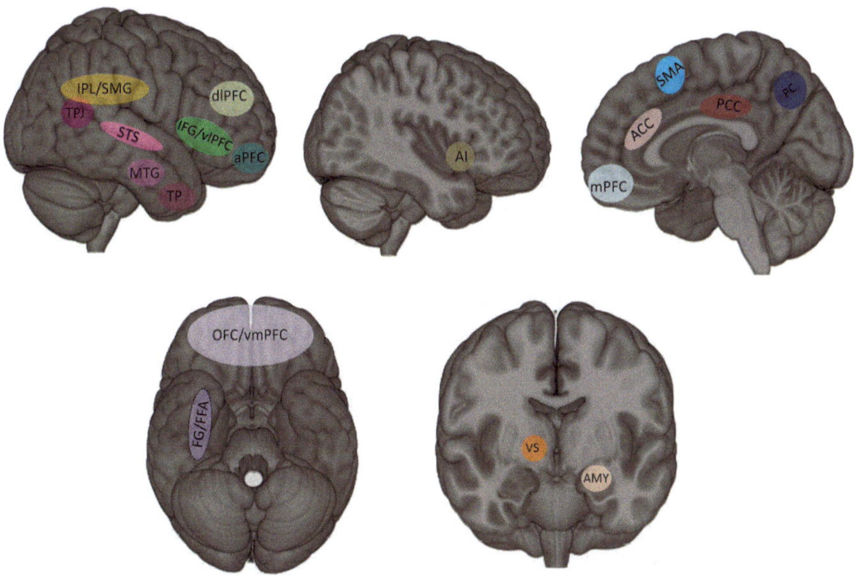

Fig. 1 Display of approximate location of the brain areas mentioned throughout the chapter. *IPL/SMG* inferior parietal lobe/supramarginal gyrus, *TPJ* temporoparietal junction, *STS* superior temporal sulcus, *MTG* middle temporal gyrus, *TP* temporal pole, *dlPFC* dorsolateral prefrontal cortex, *IFG/vlPFC* inferior frontal gyrus/ventrolateral prefrontal cortex, *aPFC* anterior prefrontal cortex, *AI* anterior insula, *mPFC* medial prefrontal cortex, *ACC* anterior cingulate cortex, *SMA* supplementary motor area, *PCC* posterior cingulate cortex, *PC* precuneus, *FG/FFA* fusiform gyrus/fusiform face area, *OFC/vmPFC* orbitofrontal cortex/ventromedial prefrontal cortex, *VS* ventral striatum, *AMY* amygdala

show cooperative tendencies toward strangers when there is no high cost involved [37]. As age increases, children attribute more positive feelings to cooperating with other children [38]. On the other hand, adults seem to be more generous than younger people in economic decisions [39]. Similarly, Rosen et al. [35] also observed that adults made more moral choices than younger participants, but this effect was again mediated by empathy, as the gender case presented above. In addition, Harlé and Sanfey [40] showed that older people appear to be more sensitive to unfairness, with higher rejection rates to unfair offers than younger participants. This unfairness effect was related to activation increases in dorsolateral prefrontal cortex (dlPFC) and decreased in the anterior insula (AI) for adults, compared to young participants. This pattern suggests higher reliance on goal maintenance and less emotional processing due to norm violation with age. However, these effects do not seem to be consistent, as manifested by Lim and Yu [41], who reviewed related literature and observed that the existing evidence proves heterogeneous and does not offer certainty about age differences in prosocial behavior.

Furthermore, individual social preferences or personal concerns for other people, such as altruism, envy, fairness, reciprocity, or inequity aversion, are another source of influence in decisions (e.g., [42]). Individual preferences have also been studied under the name of social value orientation (SVO; [43, 44]), via different tools—e.g., *decomposed games*, the *ring measure*, *social orientation choice cards*, or characteristics space theory [45]. Within the predominant SVO framework, several studies have tried to distinguish between self-oriented ("proself") and other-oriented ("prosocial") participants and how these individual differences affect cooperation tendencies. While proself subjects show increased calculating and strategizing tendencies, prosocials tend to follow social norms and have moral considerations for others, making more cooperative choices [46, 47]. Also, at a neural level, the brain of prosocials has been linked to increased activation in the precuneus, superior temporal sulcus, and medial prefrontal cortex (mPFC), showing also that this pattern correlates with increased cooperation decisions [48].

Social biases also appear in neuropsychological conditions involving damage in brain areas related to social processing. For instance, temporal lobe epilepsy patients exhibit social functioning deficits [49] in, for example, basic and complex theory of mind processes, which have an impact on social decisions. Amygdala-damaged patients display higher cooperation rates, especially when interacting with untrustworthy partners [50]. This pattern could reflect a deficit in the integration of different social signals that takes place in the amygdala, which would disable proper indications to guide successful social interactions. Moreover, utilitarian judgments in moral dilemmas increase in patients with ventromedial prefrontal cortex (vmPFC) lesions, which has been taken as evidence for the role of this region in the representation of the emotional value of stimuli [51, 52]. Frontotemporal dementia patients also show altered social decisions, with increased impulsiveness and risky behavior, which could be partly related to damage in the orbitofrontal cortex (OFC; [53]). Additionally, these patients make more utilitarian choices in moral dilemmas, which seems to be related to theory of mind deficits [54]. For example, during social bargaining, they manifest altered prosociality and punishment behavior, due to a failure to incorporate information about the perspective of others [55].

Apart from individual factors, a large part of the literature on biases employs experimental settings to induce transient mental states in the agent. A cornerstone source of influences on decisions is the framing effect, which refers to how decisions are affected by the way the scenario is presented [56]. For instance, working with moral dilemmas, De Martino et al. [57] showed that when the problem was framed in a "gain" context, participants tended to choose the safe option, whereas in a "loss" situation they chose the risky alternative to a higher extent. In similar scenarios, positive framing in moral dilemmas has been associated with risk aversion choices, accompanied by increased activation in a cluster involving the ACC and the vmPFC compared to negative framing [58]. Conversely, risk-seeking behavior under negative framing of social cues has been related to activation in the inferior frontal gyrus (IFG; [59]).

Furthermore, the induction of mood states is also a tool frequently used to explore how incidental emotions bias our choices. The affect infusion model [60] claims that incidental emotions prime mood-congruent dispositions, positing that behavioral effects in decision-making tasks depend on the participants' mood [61]. For instance, positive moods prime positive information and have been related to growing confidence, friendliness, and cooperative tendencies during interpersonal interactions. In this context, positive moods lead to a greater joint gain seeking, interpersonal trust behavior, and cooperative choices [62–64] and also generate an increased preference for avoiding losses (e.g., [65]). Similarly, social reward can serve as another bias in cooperative behavior, as people tend to act more generously when they know they are being watched. When feeling observed, people want to be socially acknowledged about their behavior, which itself constitutes a larger social reinforcement associated with greater activation in the ventral striatum [4].

Conversely, negative moods can have different effects. In economic games, whereas sad affection has been associated with generous behavior [66], it has also been related to a decrement of acceptance of unfair offers, which could be the reflection of a mood-congruent framing for negative outcomes [61]. At a neural level, this bias has been related to increased activity in the bilateral AI, which was thought to mediate between mood and choices. It was also accompanied by higher activity in the dorsal ACC (dACC) for unfair offers, indicating a possible affective conflict. According to the affect infusion model, negative moods would induce a sensitive disposition to detect social violations. This negative mood appears to be coupled with lower activity in the reward system (e.g., ventral striatum) to fair offers. Additionally, "harm to save" dilemmas tend to induce negative emotions such as fear or disgust, each of these biasing participants' response toward different responses. When participants experience fear, they show deontological bias (do nothing), while disgust seems to enhance utilitarian responses (e.g., kill one in order to save five; [67]). Moreover, the application of emotion regulation strategies can also modulate behavioral and neural responses during social decisions. This regulation has been associated with the involvement of the IFG, temporoparietal junction (TPJ), and the AI [68]. The implementation of downregulation (a more positive interpretation) entails higher acceptance rates for unfair offers, while upregulation (a more negative interpretation) elicits more rejection decisions [69].

As we have just described, a variety of individual factors bias people's choices in social decisions. Nonetheless, external factors, mostly originated from the agents we interact with, also exert varying degrees of influence on our choices, as shown in the next section.

3.2 *Bias in the Stimuli*

Biases in social decision-making also stem from different features of the stimuli we perceive. These choices frequently involve perception and social categorization, as well as the generation of expectations. Faces often provide rich information in these contexts, such as the gender, social group, emotion, and trustworthiness of the people we interact with. This information is highly valuable to generate expectations about others to guide successful decisions. Below, we will first focus on the mechanisms underlying social judgments about other people and then examine how emotion displays and personal information bias choices.

3.2.1 Social Categorization

When we first interact with others, we tend to form impressions about how they are, what they like, or how we expect them to behave, which is a case of social categorization. We form an initial idea of others very quickly, based on the information we can gather in a few milliseconds [70]. These rapid impressions have been related to activity in the posterior cingulate cortex (PCC) and the amygdala [25], both involved in social cognition. The amygdala has also been studied in connection with other regions in terms of the context-relevant representation of social stimuli, especially faces. Its ventrolateral region belongs to a network specialized in social perception [71], in connection with sensory regions of the temporal lobe – the superior temporal sulcus, the temporal pole, the fusiform gyrus, and the OFC.

Categorization judgments are closely related to stereotypes and expectations [72, 73]. Some of such stereotypes refer to biases related to gender, as people tend to assign attributes and internal dispositions differently to women and men. Regions related to evaluative processes and representation of knowledge [74], such as the vmPFC and the amygdala, together with the supramarginal gyrus and the middle temporal gyrus, seem to be at the basis of these judgments. Additionally, contextual influences on face categorization appear mediated by retrosplenial and prefrontal cortices [75].

Furthermore, some biases relate to racial stereotypes. Traditionally, the amygdala has been set as a racial prejudice marker [76], showing higher activation in participants facing a member of a racial outgroup. This involvement has been explained appealing to different roles: activity in this structure could act as a marker for a threat of an outgroup, as an indicator of fear of being considered prejudicial, or as a motivational response [24]. It has also been suggested that the amygdala may be in charge of the representation of relevant social information, while the striatum, which participates in the computation of valence, would represent these stereotypes to guide decisions toward positive interactions and trust behavior with the racial outgroup [77]. Moreover, the AI has been related to negative reactions to a disliked racial outgroup when it has been rewarded. However, this region has also been linked to empathy toward the ingroup [24]. On the other hand, neural representations in the OFC seem to underlie affect-based judgments depending on race, while neural patterns in the anterior portion of the PFC (aPFC) differentiate stereotype-based judgments [78].

At a perceptual level, race influences visual face processing and attention at early stages [79–81]. In this regard, Tortosa et al. [81] observed larger amplitude in the N170 during the processing of black versus white faces, a negative potential related to face encoding [82], which seems related to implicit racial bias [83]. The variations in this potential seem to be originated in early visual processing in the fusiform gyrus [84]. In addition, different studies have also reported varying neural patterns in the fusiform face area depending on the race of faces (e.g., [85]) and how these differences may rely on implicit racial bias [86].

Race bias additionally acts at the decision point. For example, some reports show higher punishment to members of one's own racial group, because they, unlike outsiders, are expected to cooperate [87]. Moreover, others (e.g., [88]) have observed that participants offer more money and show increased trust toward white versus black partners. However, Tortosa et al. [81] observed that Caucasian participants cooperated more with black than white partners while presenting implicit race bias, which may be explained by participants' desire to counteract their implicit biases.

Interestingly, Stolier and Freeman [22] have recently shown how different social categories are entangled with each other, in the sense that one category activates stereotypes shared with another. Even more, employing novel representational similarity analyses, the authors suggested that the stereotypes related to different categories represented in the OFC modulate activity in earlier visual processing areas of the fusiform gyrus. This results in a greater perceptual similarity between representations of faces sharing the same stereotypes, even if they are of different gender or race. According to the dynamic interactive model [9, 89], social perception is highly dynamic, based on an interactive system in which bottom-up perceptual information activates categorization, which in turn activates stereotypes. Additionally, top-down factors, such as expectations or goals, can modulate lower processing stages in a dynamic fashion.

3.2.2 Emotional Expressions and Trustworthiness

A large part of judgments about others is related to the emotions we perceive in them. Emotional expressions are rapidly processed, even in the absence of awareness (e.g., [90]). In this way, emotional displays have a significant effect on trustworthiness judgments (e.g., [91]), friendliness, or dominance [92], given that they provide information that can be used to decode the intentions, beliefs, and desires of others in social scenarios [93].

Positive expressions tend to induce trust and cooperation [94, 95], whereas negative emotions are associated with uncooperative behavior [16]. However, these emotional expressions may not have the same interpretation in all contexts [93, 96, 97]. For instance, de Melo et al. [98] found that, after mutual cooperation, happiness increased cooperation expectations, whereas in noncooperative scenarios, smiles decreased such expectations. Alternatively, when partners defected, their positive expressions could be considered redundant to their behavior, thus not affecting cooperation expectations. Conversely, when people consider their partner's emotions, anger expressions can induce cooperative decisions (e.g., [99]).

These biases are not only reflected in the type of decision participants make but also in the time they need to make up their minds. Some studies have found "emotional conflict" effects, where participants take longer to choose an option contrary to the automatic response elicited by ignored and non-predictive emotions. For instance, responses in a trust game were slower when emotion and identity information did not lead to the same responses, even when participants were told to ignore these emotions [100]. Moreover, responses are also slowed down when emotions predict consequences opposite to their "natural consequences" [101, 102]. In this scenario, when emotional expressions are predictive of their natural consequences, activity increases in the precuneus [101], a region associated with the representation of personal information [103] and trust in cooperative scenarios [104].

Furthermore, facial expressions seem to be associated with trustworthiness judgments along a continuum, where untrustworthy faces are linked to anger expressions, whereas trustworthiness is related to happiness [92]. These trust judgments correlate with amygdala activity, as this region presents a higher response to untrustworthy agents [105]. Interestingly, such behavioral and neural sensitivity to trustworthiness may occur even with no perceptual awareness [106, 107]. In this regard, several studies have shown that trustworthiness can indeed impact our decisions in different ways. During trust and economic games, people manifest higher cooperation rates and acceptance of offers from trustworthy agents [108, 109]. People invest more money with partners who have been rated as trustworthy even when there is no objective relationship between ratings and actual behavior [110]. Moreover, rejection of offers based on facial trustworthiness correlates with activity in the OFC, and its functional connectivity with the AI correlates with individual rejection decisions from untrustworthy partners [108].

3.2.3 Personal Information

In certain cases, interactions among strangers take place at distance, without physical information about others. Nonetheless, even in these cases we can obtain information about them that may bias our decisions, even if this knowledge is unrelated to their actual behavior. In this regard, initial research showed that positive and negative moral information about others influence decisions and reduce reliance on feedback for learning [111].

First, we can assume several characteristics when interacting with people who are familiar to us. Thus, closeness with partners is associated with higher trustworthiness judgments and cooperation decisions, accompanied by higher response in the striatum and mPFC when friends reciprocate [112]. Also, striatal activity seems related to reputation learning of agents with different closeness [112]. Yet, there are situations where we need to make decisions involving unknown people, which is a frequent scenario in experimental settings. In this regard, Hackel et al. [113] showed that the striatum supports feedback-based instrumental learning, integrating different sources of social information, while vmPFC activation correlates with behavioral decisions according to trait-learned information about generosity during social exchanges.

Moreover, our choices can also be influenced by knowledge about our partners' personal characteristics. For instance, participants reject more offers from partners associated with negative descriptions compared to those described by positive information [114]. These influences are stronger when offers are unfair, as well as in uncertain contexts [115]. Negative descriptions of partners compared to positive ones increase the amplitude of the medial frontal negativity (MFN), a potential associated with the emotional evaluation of negative outcomes [116]. However, this negative polarity is reversed when unfair offers come from a friend, a scenario that is also associated with fewer rejection rates [117]. This may reflect that personal information about the partners, as well as social distance, bias the evaluation of *objective* offers differently, making them look more adverse when the partners are associated with negative personal information. In addition, previous information can influence competence expectations, related to choices whether to continue or not a social interaction with a specific partner [21].

In conclusion, several individual factors carry a heavy impact on social decision-making. In nonnatural controlled scenarios, these sources of bias can also be evaluated through the manipulation of motivational and emotional elements in the experimental setting. In these contexts, biases relate to stereotypes built on the characteristics of others, which are represented at several stages that take place during the analysis of perceptual and social representations about others. To avoid such information when it conflicts with internal goals, control mechanisms become essential.

4 Control Mechanisms During Social Decision-Making

Adaptive social interactions need control mechanisms to regulate actions in scenarios where biases conflict with short- or long-term goals. Here we review part of the evidence on the functioning of these mechanisms. Our focus is on regulation mechanisms involved in economic and moral decisions as well as in contexts where automatic responses must be controlled or our expectations clash with other agents' behavior.

A large part of the biases reviewed so far are studied in relation to the control mechanisms that steer the organism toward context-appropriate actions. For instance, in a classic study, Sanfey et al. [8] employed the ultimatum game to explore reactions to unfair offers. Here, they observed increased activity to unfair offers in the ACC, a region related to conflict of different types, which suggests the existence of interference between emotional reactions and the monetary goals of the task. In addition, they also observed a trade-off between the activity of the AI and the dlPFC to unfair offers. Specifically, the activation in the insula was larger than in the dlPFC when unfair offers were rejected, which may reflect the negative reactions associated with unfairness. On the other hand, activity in the dlPFC was higher when unfair offers were accepted, supporting the function of this region in the control of social behavior.

Similarly, Knoch et al. [118] showed that the disruption of the dlPFC reduces rejection rates to unfair offers. In this vein, Baumgartner et al. [119], observed that dlPFC and vmPFC activity, as well as their effective communication, was needed to make costly normative choices, that is, to reject offers and, therefore, lose earnings. However, the role of these regions in social decisions is not clear, as other studies have shown that people with damage to the vmPFC seem more likely to reject unfair offers [51]. These divergent results might be explained by different dynamics in the PFC in healthy participants and neurological patients, and they also suggest the importance of the communication between these prefrontal regions to regulate social behaviors.

In moral dilemmas, reasoning processes also influence our choices [120]. For example, performing the *cognitive reflection task* induces a decrement of confidence in intuition, related to an increment in utilitarian judgments [120]. These utilitarian decisions have been associated with activity in the dlPFC, inferior parietal lobe, and PCC [121]. In addition, the disruption of activity of the dlPFC after transcranial magnetic stimulation increases utilitarian choices [122]. Taken together, these data add support to the role of the PFC in overcoming emotional reactions in moral scenarios [123].

Ochsner and Gross [124] proposed the mediation of two routes in this control. The dorsal PFC, which has been related to orientation to task context and goals, would be in charge of changing stimuli-emotional response associations. On the other hand, the ventral PFC would maintain the representation of the emotional value of stimuli according to the context and jointly with the OFC and would impact emotional reactions through its reciprocal connections with the nucleus accumbens and the amygdala. These regions would, in turn, modulate the representation of relevant information in higher control areas (e.g., PFC, OFC). In addition, rational behavior in framing tasks, in which decisions are not influenced by framing effects, is correlated with enhanced activation of the OFC and vmPFC [57].

Moreover, biases that derive from stereotypes of prejudice toward others can also be modulated by control. Top-down processes can attenuate this influence [76] through the detection of conflict between goals and biases by the dACC and implementation by the PFC of domain-general control. In addition, the mPFC and the rostral ACC seem to be in charge of more specific representations of social cues to orient regulatory processes to suppress behavior opposite to social norms [24]. Another type of control-demanding situations takes place when facing emotional conflict during decision-making, where need to route the emotional information displayed by faces and attend to the relevant information to fulfill our goals. The resolution of emotional interference has been associated with the activation of the IFG [125] and top-down modulation of the amygdala by the ACC [126].

In social decisions, control mechanisms are recruited when we hold expectations about other people that are not matched by their actual behavior. In this regard, when emotional expressions do not predict their "natural consequences" (happiness = cooperation, anger = no cooperation), there is an increment in the N1 potential, related to attentional processes [102]. Moreover, when emotional displays interfere with identity expectations, Alguacil et al. [127] observed an early conflict effect during face processing, associated with higher amplitude in N170 potential, associated with structural encoding of faces. Later stages linked to response selection were also affected, as reflected by increments in the amplitude of the P3 potential.

The violation of expectations, when we need to overcome the automatic response associated with the expectations induced by emotional expressions, has been shown to engage activity in the PFC, ACC, and AI [101]. This study also observed different coupling of the ACC depending on the level of conflict. While in low-conflict contexts the ACC showed greater interaction with the precuneus and the vmPFC, high conflict was associated with greater coupling with control-related regions, such as the supplementary motor area and the middle region of the cingulate cortex. This agrees with data indicating that emotional conflict engages the increment of task-relevant information processing, including high-level areas involved in non-emotional tasks [128].

Furthermore, the ability to respond accordingly to previous expectations, even in the presence of behavior which conflicts with that information, seems to be in charge of the ventrolateral prefrontal cortex (vlPFC). In this regard, Fouragnan et al. [129] observed that deactivations in the ventral striatum when trust was violated were functionally correlated with vlPFC activation. Therefore, the vlPFC modulated striatal activity to orient decisions to match expectations when these conflicted with the observed behavior. In addition, [111] observed increased activation in the ACC when participants offered responses that contradicted previous information they held about their partners.

In addition, research in the field of cognitive control suggests the existence of two different networks linked to control. The frontoparietal and cingulo-opercular networks act at different timescales to orient our behavior according to our goals [130]. The increase of activation in these networks has been traditionally associated with a deactivation in the default mode network (DMN), which has been interpreted as an indicator of this network's absence of functionality during difficult tasks [131]. Interestingly, although these mechanisms have been more extensively studied in nonsocial contexts, Cáceda et al. [132] have observed similar neural patterns related to prosocial behavior. These authors reported that enhanced intrinsic connectivity between the salience and the central executive networks (insula/ACC and dlPFC/posterior parietal cortex, respectively) predicted increased cooperation decisions. Moreover, multivariate approaches have been employed to explore control mechanisms encoding the response in social scenarios. For example, Hollmann et al. [133] employed real-time MVPA to explore control mechanisms during economic decisions. They observed that participants' decisions to reject the offers could be decoded in the AI and lateral portion of the PFC (lPFC). Taken together, results add further evidence to the need of control mechanisms to successful social functioning.

In this section, we have reviewed how control mechanisms are recruited to overcome interference. Such conflictive situations tend to arise from contradictions related to personal information or from the incompatibility between personal goals and non-appropriate or automatic responses, which may appear very early in time. Through coordinated activations, frontal regions participate in the evaluation of stimuli and expectations, and they also contribute to maintain neural representations of relevant goals to flexibly adjust behavior.

5 Final Remarks and Conclusions

We have reviewed some of the contributions of social neuroscience to the understanding of the sources of bias in social decisions. We introduced the methodologies that allow the study of the behavior and neural underpinnings of these phenomena, reviewed internal and external sources of biases, and considered the control mechanisms engaged during conflictive situations.

Altogether, the evidence underscores the relevance of the amygdala and the vmPFC in the integration of emotional and social signals relevant to guide our behavior in social scenarios. The amygdala may enhance processing of social relevant stimuli, while the vmPFC has been related to the representation of other's intentions. Furthermore, positive mood seems to foster cooperation through the reward system (e.g., striatum), which is also in charge of reputational learning according to observed behavior. Conversely, negative states engage areas associated with conflict and the emotional value of negative outcomes, such as the AI or the ACC. Moreover, the OFC appears crucial to represent expectations, especially based on stereotypical information. Interestingly, these expectations also dictate representations in lower-level regions, such as the fusiform cortex, which suggests the importance of top-down modulations in the representation of social information.

The evidence suggests the presence of common pathways of biases on perception and on decisions. For instance, Amodio [24] proposed a neural circuitry for stereotyping, which included mainly the vmPFC, amygdala, AI, and OFC. As we have seen above, these regions also are involved in other biasing contexts. The mPFC has been associated with prosocial dispositions as well as with the representation of a partner's personal traits. Its ventral part also seems relevant for the integration of emotional stimuli in moral dilemmas, including framing effects, as well as in categorization processes [74, 108]. Moreover, the amygdala is necessary to regulate interpersonal trust and facial categorizations [25, 50]. The AI appears to be involved in the emotional evaluation of negative outcomes, which can be guided by negative mood, prejudice, or trustworthiness [61, 101]. Furthermore, the OFC seems to be in charge of representing expectations of others based on stereotypes and emotions [22, 108] and to guide adaptive behavior in social contexts [53]. In addition to these areas, the ventral striatum has a central role in reward-related processing, learning in social scenarios about the valence of the interactions, and fostering interactions associated with positive outcomes [112, 113].

As regards control mechanisms, the evidence points to the relevance of regions such as the ACC, the AI, and several regions of the PFC to maintain goals and suppress deviant responses and to modulate regions involved in social processes to increase attention to the task. Furthermore, data suggest that these regions work at a network level, where frontoparietal and cingulo-insular networks seem to foster prosocial behavior. This highlights the importance of control mechanisms in cooperative scenarios, not only to overcome automatic or undesirable responses but also to behave adaptively in our social environment. Crucially, the data shows that there is important similarity between control mechanisms involved during social decisions and those that have been extensively studied in nonsocial domains (e.g., [134]).

Likewise, in this review we have presented some evidence noting the relevance for cooperative behavior of some regions associated with the DMN, which comprises areas such as the mPFC, precuneus, PCC, angular gyrus, and some temporal areas. This network has been considered until very recently as functionally inactive during effortful tasks, being involved in mind wondering and self-referential processes [131]. However, recent data seem to indicate that the DMN encodes task-relevant information, even in complex settings and nonsocial tasks [135, 136]. In addition, it has also been related to socials tasks [137, 138] and emotional engaging in social interactions [129]. Unraveling the processes underlying this network is a field of intensive current research (e.g., [139]).

Taking all this into consideration, the use of different methodologies turns crucial to understand how social information is represented in the brain and how different mechanisms coordinate with each other to regulate human social activity and orient our behavior toward goals. Given the complexity of social scenarios, more realistic paradigms are being developed to be implemented in laboratories, in more natural settings [140]. In this regard, the use of methodologies, such as fNIRS in social scenarios (e.g., [141]), may be an interesting approach to study the influences on social decisions in real life.

Social neuroscience is an interdisciplinary and vibrant field. It incorporates methodologies from complementary fields to generate a description of the variety of factors that can influence our interactions and how the different biases operate from early to late stages of processing. In this context, social decisions are key to understand interpersonal exchanges, which are crucial in our life. These processes are important to analyze group dynamics, social perception, or how rational decisions such as economic ones are modulated by different factors. Furthermore, this field may aid to develop interventions for patients with some sort of neural damage that affects their social functioning. Finally, social contexts can extend our knowledge about how our brain works in a large diversity of scenarios filled with rich social stimuli, where decisions take place. Hence, current research efforts provide a comprehensive view of the mechanisms underlying core processes in our daily social life.

References

1. Sanfey AG. Social decision-making: insights from game theory and neuroscience. Science. 2007;318(5850):598–602. https://doi.org/10.1126/science.1142996.
2. Adolphs R, Anderson D. Social and emotional neuroscience. Curr Opin Neurobiol. 2013;23(3):291–3. https://doi.org/10.1016/j.conb.2013.04.011.
3. Fehr E, Gachter S. Cooperation and punishment in public goods experiments. Am Econ Rev. 2000;90(4):980–94. https://doi.org/10.1257/aer.90.4.980.
4. Ruff CC, Fehr E. The neurobiology of rewards and values in social decision making. Nat Rev Neurosci. 2014;15(8):549–62. https://doi.org/10.1038/nrn3776.
5. Frith CD, Frith U. Implicit and explicit processes in social cognition. Neuron. 2008;60(3):503–10. https://doi.org/10.1016/j.neuron.2008.10.032.
6. van Kleef GA. How emotions regulate social life. Curr Dir Psychol. 2009;18(3):184–8. https://doi.org/10.1111/j.1467-8721.2009.01633.x.
7. Dunne S, O'Doherty JP. Insights from the application of computational neuroimaging to social neuroscience. Curr Opin Neurobiol. 2013;23(3):387–92. https://doi.org/10.1016/j.conb.2013.02.007.
8. Sanfey AG, Rilling JK, Aronson JA, Nystrom LE, Cohen JD. The neural basis of economic decision-making in the ultimatum game. Science. 2003;300(5626):1755–8. https://doi.org/10.1126/science.1082976.

9. Stolier RM, Freeman JB. Functional and temporal considerations for top-down influences in social perception. Psychol Inq. 2016;27(4):352–7. https://doi.org/10.1080/1047840X.2016.1216034.
10. Cacioppo JT, Berntson GG. Social psychological contributions to the decade of the brain. Doctrine of multilevel analysis. Am Psychol. 1992;47(8):1019–28.
11. Ochsner KN, Lieberman M. The emergence of social cognitive neuroscience. Am Psychol. 2001;56(9):717–34.
12. Fehr E, Fischbacher U, Kosfeld M. Neuroeconomic foundations of trust and social preferences neuroeconomic foundations of trust and social preferences. Am Econ Rev. 2005;95(2):346–51. https://doi.org/10.1257/000282805774669736.
13. Sanfey AG, Chang LJ. Multiple systems in decision making. Ann N Y Acad Sci. 2008;1128:53–62. https://doi.org/10.1196/annals.1399.007.
14. Leotti LA, Delgado MR. The value of exercising control over monetary gains and losses. Psychol Sci. 2014;25(2):596–604. https://doi.org/10.1177/0956797613514589.
15. Berkman ET, Cunningham WA, Lieberman MD. Research methods in social and affective neuroscience. In: Reis HT, Judd CM, editors. Handbook of research methods in personality and social psychology. 2nd ed. New York: Cambridge University Press; 2014. p. 123–58.
16. Camerer CF. Behavioral game theory: experiments in strategic interactions. Princeton: Princeton University Press; 2003.
17. Ruz M, Acero JJ, Tudela P. What does the brain tell us about the mind? Psicológica. 2006;29:149–57.
18. Güth W, Schmittberger R, Schwarze B. An experimental analysis of ultimatum bargaining. J Econ Behav Organ. 1982;3(4):367–88.
19. Sally D. Conversation and cooperation in social dilemmas: a meta-analysis of experiments from 1958 to 1992. Ration Soc. 1995;7(1):58–92.
20. Gilam G, Lin T, Raz G, Azrielant S, Fruchter E, Ariely D, et al. Neural substrates underlying the tendency to accept anger-infused ultimatum offers during dynamic social interactions. NeuroImage. 2015;120:400–11. https://doi.org/10.1016/j.neuroimage.2015.07.003.
21. Heijne A, Sanfey AG. How social and nonsocial context affects stay/leave decision-making: the influence of actual and expected rewards. PLoS One. 2015;10(8):e0135226. https://doi.org/10.1371/journal.pone.0135226.
22. Stolier RM, Freeman JB. Neural pattern similarity reveals the inherent intersection of social categories. Nat Neurosci. 2016;19(6):795–7. https://doi.org/10.1038/nn.4296.
23. Greenwald AG, Mcghee DE, Schwartz JLK. Measuring individual differences in implicit cognition. J Pers Soc Psychol. 1998;74(6):1464–80.
24. Amodio DM. The neuroscience of prejudice and stereotyping. Nat Rev. 2014;15(10):670–82. https://doi.org/10.1038/nrn3800.
25. Schiller D, Freeman JB, Mitchell JP, Uleman JS, Phelps EA. A neural mechanism of first impressions. Nat Neurosci. 2009;12(4):508–14. https://doi.org/10.1038/nn.2278.
26. Greene J, Haidt J. How (and where) does moral judgement work? Trends Cogn Sci. 2002;6(12):517–23.
27. Greene JD, Morelli SA, Lowenberg K, Nystrom LE, Cohen JD. Cognitive load selectively interferes with utilitarian moral judgment. Cognition. 2008;107(3):1144–54. https://doi.org/10.1016/j.cognition.2007.11.004.
28. Luck SJ. An introduction to the event-related potential technique. Cambridge, MA: The MIT Press; 2005.
29. Cutini S, Basso Moro S, Bisconti S. Functional near infrared optical imaging in cognitive neuroscience: an introductory review. J Near Infrared Spectrosc. 2012;20:75–92. https://doi.org/10.1255/jnirs.969.
30. Haxby JV, Gobbini MI, Furey ML, Ishai A, Schouten JL, Pietrini P. Distributed and overlapping representations of faces and objects in ventral temporal cortex. Science. 2001;293(5539):2425–30. https://doi.org/10.1126/science.1063736.

31. Kriegeskorte N, Mur M, Bandettini P. Representational similarity analysis – connecting the branches of systems neuroscience. Front Syst Neurosci. 2008;2:4. https://doi.org/10.3389/neuro.06.004.2008.
32. Tamir DI, Thornton MA, Contreras JM, Mitchell JP. Neural evidence that three dimensions organize mental state representation: rationality, social impact, and alence. Proc Natl Acad Sci U S A. 2016;113(1):194–9. https://doi.org/10.1073/pnas.1511905112.
33. Glover SH, Bumpus MA, Sharp GF, Munchus GA. Gender differences in ethical decision making. Women Manag Rev. 2002;17(5):217–27. https://doi.org/10.1108/09649420210433175.
34. Andreoni J, Vesterlund L. Which is the fair sex? Gender differences in altruism. Q J Econ. 2001;116(1):293–312. https://doi.org/10.1162/003355301556419.
35. Rosen JB, Brand M, Kalbe E. Empathy mediates the effects of age and sex on altruistic moral decision making. Front Behav Neurosci. 2016;10:67. https://doi.org/10.3389/fnbeh.2016.00067.
36. Singer T, Seymour B, Doherty JPO, Stephan KE, Dolan RJ, Frith CD. Empathic neural responses are modulated by the perceived fairness of others. Nature. 2006;439(7075):466–9. https://doi.org/10.1038/nature04271.
37. Moore C. Fairness in children's resource allocation depends on the recipient. Psychol Sci. 2009;20(8):944–8. https://doi.org/10.1111/j.1467-9280.2009.02378.x.
38. Weller D, Hansen Lagattuta K. Helping the in-group feels better: children's judgments and emotion attributions in response to prosocial dilemmas. Child Dev. 2013;84(1):253–68. https://doi.org/10.1111/j.1467-8624.2012.01837.x.
39. Bailey PE, Ruffman T, Rendell PG. Age-related differences in social economic decision making: the ultimatum game. J Gerontol Ser B Psychol Sci Soc Sci. 2012;68(3):356–63. https://doi.org/10.1093/geronb/gbs073.
40. Harlé KM, Sanfey AG. Social economic decision-making across the lifespan: an fMRI investigation. Neuropsychologia. 2012;50(7):1416–24. https://doi.org/10.1016/j.neuropsychologia.2012.02.026.
41. Lim KTK, Yu R. Aging and wisdom: age-related changes in economic and social decision making. Front Aging Neurosci. 2015;7:120. https://doi.org/10.3389/fnagi.2015.00120.
42. Fehr E, Schmidt KM. The economics of fairness, reciprocity and altruism – experimental evidence and new theories. In: Kolm SC, Ythier JM, editors. Handbook of the economics of giving, altruism and reciprocity. Amsterdam: Elsevier; 2006. p. 615–91.
43. Murphy R, Ackermann K. A review of measurement methods for social preferences. ETH Zurich Chair of decision theory and behavioral game theory, working paper. http://vlab.ethz.ch/svo/SVO_rev_paper.pdf.
44. Murphy RO, Ackermann KA, Handgraaf MJJ. Measuring social value orientation. Judgm Decis Mak. 2011;6(8):771–81. https://doi.org/10.2139/ssrn.1804189.
45. Au WT, Kwong JY. Measurements and effects of social-value orientation in social dilemmas. In: Suleiman R, Budescu DV, Fischer I, Messick DM, editors. Contemporary research on social dilemmas. New York: Cambridge University Press; 2004. p. 71–98.
46. Bogaert S, Boone C, Declerck C, Bogaert Boone C, Declerck CHS. Social value orientation and cooperation in social dilemmas: a review and conceptual model. Br J Soc Psychol. 2008;47(3):453–80. https://doi.org/10.1348/014466607X244970.
47. Emonds G, Declerck CH, Boone C, Vandervliet EJM, Parizel PM. Comparing the neural basis of decision making in social dilemmas of people with different social value orientations, a fMRI study. J Neurosci Psychol Econ. 2011;4(1):11–24. https://doi.org/10.1037/a0020151.
48. Emonds G, Declerck CH, Boone C, Seurinck R, Achten R. Establishing cooperation in a mixed-motive social dilemma. An fMRI study investigating the role of social value orientation and dispositional trust. Soc Neurosci. 2014;9(1):10–22. https://doi.org/10.1080/17470919.2013.858080.

49. Wang WH, Shih YH, Yu HY, Yen DJ, Lin YY, Kwan SY, et al. Theory of mind and social functioning in patients with temporal lobe epilepsy. Epilepsia. 2015;56(7):1117–23. https://doi.org/10.1111/epi.13023.
50. Koscik TR, Tranel D. The human amygdala is necessary for developing and expressing normal interpersonal trust. Neuropsychologia. 2011;49(4):602–11. https://doi.org/10.1016/j.neuropsychologia.2010.09.023.
51. Koenigs M, Young L, Adolphs R, Tranel D, Cushman F, Hauser M, et al. Damage to the prefrontal cortex increases utilitarian moral judgements. Nature. 2007;446(7138):908–11. https://doi.org/10.1038/nature05631.
52. Moll J, de Oliveira-Souza R. Moral judgments, emotions and the utilitarian brain. Trends Cogn Sci. 2007;11(8):319–21. https://doi.org/10.1016/j.tics.2007.06.001.
53. Gleichgerrcht E, Ibanez A, Roca M, Torralva T, Manes F. Decision-making cognition in neurodegenerative diseases. Nat Rev Neurol. 2010;6(11):611–23. https://doi.org/10.1038/nrneurol.2010.148.
54. Gleichgerrcht E, Torralva T, Roca M, Pose M, Manes F. The role of social cognition in moral judgment in frontotemporal dementia. Soc Neurosci. 2011;6(2):113–22. https://doi.org/10.1080/17470919.2010.506751.
55. O'Callaghan C, Bertoux M, Irish M, Shine JM, Wong S, Spiliopoulos L, et al. Fair play: social norm compliance failures in behavioral variant frontotemporal dementia. Brain. 2016;139(1):204–16. https://doi.org/10.1093/brain/awv315.
56. Tversky A, Kahneman D. The framing of decisions and the psychology of choice. Science. 1981;211(4481):453–8.
57. De Martino B, Kumaran D, Seymour B, Dolan RJ. Frames, biases, and rational decision-making in the human brain. Science. 2006;313(5787):684–7. https://doi.org/10.1126/science.1128356.
58. Wang XT, Rao L, Zheng H. Neural substrates of framing effects in social contexts: a meta-analytical approach. Soc Neurosci. 2016;28:1–12. https://doi.org/10.1080/17470919.2016.1165285.
59. Zheng H, Wang XT, Zhu L. Framing effects: behavioral dynamics and neural basis. Neuropsychologia. 2010;48(11):3198–204. https://doi.org/10.1016/j.neuropsychologia.2010.06.031.
60. Forgas JP. Mood and judgment: the affect infusion model (AIM). Psychol Bull. 1995;117(1):39–66.
61. Harlé KM, Chang LJ, van't Wout M, Sanfey AG. The neural mechanisms of affect infusion in social economic decision-making: a mediating role of the anterior insula. NeuroImage. 2012;61(1):32–40. https://doi.org/10.1016/j.neuroimage.2012.02.027.
62. Forgas J. On feeling good and getting your way: mood effects on negotiator cognition and bargaining strategies. J Pers Soc Psychol. 1998;74(3):565–77.
63. Mislin A, Williams LV, Shaughnessy BA. Motivating trust: can mood and incentives increase interpersonal trust? J Behav Exp Econ. 2015;58:11–9. https://doi.org/10.1016/j.socec.2015.06.001.
64. Rand DG, Kraft-Todd G, Gruber J. The collective benefits of feeling good and letting go: positive emotion and (dis) inhibition interact to predict cooperative behavior. PLoS One. 2015;10(1):1–12. https://doi.org/10.1371/journal.pone.0117426.
65. Nygren TE, Isen AM, Taylor PJ, Dulin J. The influence of positive affect on the decision rule in risk situations: focus on outcome (and especially avoidance of loss) rather than probability. Organ Behav Hum Decis Process. 1996;66(1):59–72.
66. Tan HB, Forgas JP. When happiness makes us selfish, but sadness makes us fair: affective influences on interpersonal strategies in the dictator game. J Exp Soc Psychol. 2010;46(3):571–6. https://doi.org/10.1016/j.jesp.2010.01.007.
67. Szekely RD, Miu AC. Incidental emotions in moral dilemmas: the influence of emotion regulation. Cognit Emot. 2015;29(1):64–75. https://doi.org/10.1080/02699931.2014.895300.

68. Grecucci A, Giorgetta C, Bonini N, Sanfey AG. Reappraising social emotions: the role of inferior frontal gyrus, temporo-parietal junction and insula in interpersonal emotion regulation. Front Hum Neurosci. 2013;7:523. https://doi.org/10.3389/fnhum.2013.00523.
69. Grecucci A, Giorgetta C, Van't Wout M, Bonini N, Sanfey AG. Reappraising the ultimatum: an fMRI study of emotion regulation and decision making. Cereb Cortex. 2013;23(2):399–410. https://doi.org/10.1093/cercor/bhs028.
70. Bar M, Neta M, Linz H. Very first impressions. Emotion. 2008;6(2):269–78. https://doi.org/10.1037/1528-3542.6.2.269.
71. Bickart KC, Dickerson BC, Feldman Barrett L. The amygdala as a hub in brain networks that support social life. Neuropsychologia. 2014;63:235–48. https://doi.org/10.1016/j.neuropsychologia.2014.08.013.
72. Tajfel H, Billig MG, Bundy RP, Flament C. Social categorization and intergroup behavior. Eur J Soc Psychol. 1971;1(2):149–78. https://doi.org/10.1002/ejsp.2420010202.
73. Fiske ST. Stereotyping, prejudice, and discrimination. In: Gilbert DT, Fiske ST, Lindzey G, editors. The handbook of social psychology. New York: McGraw-Hill; 1998. p. 357–411.
74. Quadflieg S, Turk DJ, Waiter GD, Mitchell JP, Jenkins AC, Macrae CN. Exploring the neural correlates of social stereotyping. J Cogn Neurosci. 2009;21(8):1560–70. https://doi.org/10.1162/jocn.2009.21091.
75. Freeman JB, Ma Y, Barth M, Young SG, Han S, Ambady N. The neural basis of contextual influences on face categorization. Cereb Cortex. 2015;25(2):415–22. https://doi.org/10.1093/cercor/bht238.
76. Frith CD, Frith U. How we predict what other people are going to do. Brain Res. 2006;1079(1):36–46. https://doi.org/10.1016/j.brainres.2005.12.126.
77. Stanley DA, Sokol-Hessner P, Fareri DS, Perino MT, Delgado MR, Banaji MR, et al. Race and reputation: perceived racial group trustworthiness influences the neural correlates of trust decisions. Philos Trans R Soc B Biol Sci. 2012;367(1589):744–53. https://doi.org/10.1098/rstb.2011.0300.
78. Gilbert SJ, Swencionis JK, Amodio DM. Evaluative vs. trait representation in intergroup social judgments: distinct roles of anterior temporal lobe and prefrontal cortex. Neuropsychologia. 2012;50(14):3600–11. https://doi.org/10.1016/j.neuropsychologia.2012.09.002.
79. Ito T, Urland G. Race and gender on the brain. J Pers Soc Psychol. 2003;85(4):616–26. https://doi.org/10.1037/0022-3514.85.4.616.
80. Ofan RH, Rubin N, Amodio DM. Seeing race: N170 responses to race and their relation to automatic racial attitudes and controlled processing. J Cogn Neurosci. 2011;23(10):3153–61. https://doi.org/10.1162/jocn_a_00014.
81. Tortosa M, Lupiáñez J, Ruz M. Race, emotion and trust: an ERP study. Brain Res. 2013;1494:44–55. https://doi.org/10.1016/j.brainres.2012.11.037.
82. Bentin S, Allison T, Pruce A, Perez E, Mccarthy G. Electrophysiological studies of face perception in humans. J Cogn Neurosci. 1996;8(6):551–65.
83. Ibáñez A, Gleichgerrcht E, Hurtado E, Gonzalez R, Haye A, Manes F. Early neural markers of implicit attitudes: N170 modulated by intergroup and evaluative contexts in IAT. Front Hum Neurosci. 2010;4:188. https://doi.org/10.3389/fnhum.2010.00188.
84. Haxby JV, Hoffman EA, Gobbini MI. The distributed human neural system for face perception. Trends Cogn Sci. 2000;4(6):223–33. https://doi.org/10.1016/S1364-6613(00)01482-0.
85. Contreras JM, Banaji MR, Mitchell JP. Multivoxel patterns in fusiform face area differentiate faces by sex and race. PLoS One. 2013;8(7):e69684. https://doi.org/10.1371/journal.pone.0069684.
86. Brosch T, Bar-David E, E a P. Implicit race bias decreases the similarity of neural representations of black and white faces. Psychol Sci. 2013;24(2):160–6. https://doi.org/10.1177/0956797612451465.
87. Mendoza SA, Lane SP, Amido DM, Amodio DM. For members only: ingroup punishment of fairness norm violations in the ultimatum game. Soc Psychol Personal Sci. 2014;5(6):662–70. https://doi.org/10.1177/1948550614527115.

88. Stanley DA, Sokol-Hessner P, Banaji MR, Phelps EA. Implicit race attitudes predict trustworthiness judgments and economic trust decisions. Proc Natl Acad Sci U S A. 2011;108(19):7710–5. https://doi.org/10.1073/pnas.1014345108.
89. Freeman JB, Ambady N. A dynamic interactive theory of person construal. Psychol Rev. 2011;118(2):247–79. https://doi.org/10.1037/a0022327.
90. Adolphs R. Perception and emotion: how we recognize facial expressions. Curr Dir Psychol Sci. 2006;15(5):222–6. https://doi.org/10.1111/j.1467-8721.2006.00440.x.
91. Oosterhof NN, Todorov A. The functional basis of face evaluation. Proc Natl Acad Sci U S A. 2008;105(32):11087–92. https://doi.org/10.1073/pnas.0805664105.
92. Todorov A, Said CP, Engell AD, Oosterhof NN. Understanding evaluation of faces on social dimensions. Trends Cogn Sci. 2008;12(12):455–60. https://doi.org/10.1016/j.tics.2008.10.001.
93. G a v K. The emerging view of emotion as social information. Soc Personal Psychol Compass. 2010;4(5):331–43. https://doi.org/10.1111/j.1751-9004.2010.00262.x.
94. Scharlemann JPW, Eckel CC, Kacelnik A, Wilson RK. The value of a smile: game theory with a human face. J Econ Psychol. 2001;22(5):617–40. doi: https://ora.ox.ac.uk/objects/uuid:60f12fb9-4fea-4f5f-9837-7d1982010b76.
95. Mussel P, Göritz AS, Hewig J. The value of a smile: facial expression affects ultimatum-game responses. Judgm Decis Mak. 2013;8(3):1–5.
96. Hareli S, Hess U. What emotional reactions can tell us about the nature of others: an appraisal perspective on person perception. Cognit Emot. 2010;24(1):128–40. https://doi.org/10.1080/02699930802613828.
97. Ibañez A, Kotz SA, Barrett L, Moll J, Ruz M. Situated affective and social neuroscience. Front Hum Neurosci. 2014;8:547. https://doi.org/10.3389/fnhum.2014.00547.
98. de Melo CM, Carnevale PJ, Read SJ, Gratch J. Reading people's minds from emotion expressions in interdependent decision making. J Pers Soc Psychol. 2014;106(1):73–88. https://doi.org/10.1037/a0034251.
99. van Kleef GA, De Dreu CKW, Manstead ASR. The interpersonal effects of anger and happiness in negotiations. J Pers Soc Psychol. 2004;86(1):57–76. https://doi.org/10.1037/0022-3514.86.1.57.
100. Alguacil S, Tudela P, Ruz M. Ignoring facial emotion expressions does not eliminate their influence on cooperation decisions. Psicológica. 2015;36(2):309–35. http://www.redalyc.org/articulo.oa?id=16941182006.
101. Ruz M, Tudela P. Emotional conflict in interpersonal interactions. NeuroImage. 2011;54(2):1685–91. https://doi.org/10.1016/j.neuroimage.2010.08.039.
102. Ruz M, Madrid E, Tudela P. Interactions between perceived emotions and executive attention in an interpersonal game. Soc Cogn Affect Neurosci. 2013;8(7):838–44. https://doi.org/10.1093/scan/nss080.
103. Gobbini MI, Haxby JV. Neural systems for recognition of familiar faces. Neuropsychologia. 2007;45(1):32–41. https://doi.org/10.1016/j.neuropsychologia.2006.04.015.
104. Fett AKJ, Shergill SS, Joyce DW, Riedl A, Strobel M, Gromann PM, et al. To trust or not to trust: the dynamics of social interaction in psychosis. Brain. 2012;135(3):976–84. https://doi.org/10.1093/brain/awr359.
105. Todorov A, Baron SG, Oosterhof NN. Evaluating face trustworthiness: a model based approach. Soc Cogn Affect Neurosci. 2008;3(2):119–27. https://doi.org/10.1093/scan/nsn009.
106. Todorov A, Pakrashi M, Oosterhof NN. Evaluating faces on trustworthiness after minimal time exposure. Soc Cogn. 2009;27(6):813–33. https://doi.org/10.1521/soco.2009.27.6.813.
107. Freeman JB, Stolier RM, Ingbretsen ZA, Hehman EA. Amygdala responsivity to high-level social information from unseen faces. J Neurosci. 2014;34(32):10573–81. https://doi.org/10.1523/JNEUROSCI.5063-13.2014.

108. Kim H, Choi M-J, Jang I-J. Lateral OFC activity predicts decision bias due to first impressions during ultimatum games. J Cogn Neurosci. 2012;24(2):428–39. https://doi.org/10.1162/jocn_a_00136.
109. Rezlescu C, Duchaine B, Olivola CY, Chater N. Unfakeable facial configurations affect strategic choices in trust games with or without information about past behavior. PLoS One. 2012;7(3):e34293. https://doi.org/10.1371/journal.pone.0034293.
110. van't Wout M, Sanfey AG. Friend or foe: the effect of implicit trustworthiness judgments in social decision-making. Cognition. 2008;108(3):796–803. https://doi.org/10.1016/j.cognition.2008.07.002.
111. Delgado MR, Frank RH, Phelps EA. Perceptions of moral character modulate the neural systems of reward during the trust game. Nat Neurosci. 2005;8(11):1611–8. https://doi.org/10.1038/nn1575.
112. Fareri DS, Chang LJ, Delgado MR. Computational substrates of social value in interpersonal collaboration. J Neurosci. 2015;35(21):8170–80. https://doi.org/10.1523/JNEUROSCI.4775-14.2015.
113. Hackel LM, Doll BB, Amodio DM. Instrumental learning of traits versus rewards: dissociable neural correlates and effects on choice. Nat Neurosci. 2015;18(9):1233–5. https://doi.org/10.1038/nn.4080.
114. Gaertig C, Moser A, Alguacil S, Ruz M. Social information and economic decision-making in the ultimatum game. Front Neurosci. 2012;6(July):1–8. https://doi.org/10.3389/fnins.2012.00103.
115. Ruz M, Moser A, Webster K. Social expectations bias decision-making in uncertain inter-personal situations. PLoS One. 2011;6(2):e15762. https://doi.org/10.1371/journal.pone.0015762.
116. Moser A, Gaertig C, Ruz M. Social information and personal interests modulate neural activity during economic decision-making. Front Hum Neurosci. 2014;8(February):31. https://doi.org/10.3389/fnhum.2014.00031.
117. Campanhã C, Minati L, Fregni F, Boggio PS. Responding to unfair offers made by a friend: neuroelectrical activity changes in the anterior medial prefrontal cortex. J Neurosci. 2011;31(43):15569–74. https://doi.org/10.1523/JNEUROSCI.1253-11.2011.
118. Knoch D, Pascual-Leone A, Meyer K, Treyer V, Fehr E. Diminishing reciprocal fairness by disrupting the right prefrontal cortex. Science. 2006;314(5800):829–32. https://doi.org/10.1126/science.1129156.
119. Baumgartner T, Knoch D, Hotz P, Eisenegger C, Fehr E. Dorsolateral and ventromedial prefrontal cortex orchestrate normative choice. Nat Neurosci. 2011;14(11):1468–74. https://doi.org/10.1038/nn.2933.
120. Paxton JM, Ungar L, Greene JD. Reflection and reasoning in moral judgment. Cogn Sci. 2012;36(1):163–77. https://doi.org/10.1111/j.1551-6709.2011.01210.x.
121. Greene JD, Nystrom LE, Engell AD, Darley JM, Cohen JD. The neural bases of cognitive conflict and control in moral judgment. Neuron. 2004;44(2):389–400. https://doi.org/10.1016/j.neuron.2004.09.027.
122. Tassy S, Oullier O, Duclos Y, Coulon O, Mancini J, Deruelle C, et al. Disrupting the right prefrontal cortex alters moral judgement. Soc Cogn Affect Neurosci. 2012;7(3):282–8. https://doi.org/10.1093/scan/nsr008.
123. Frith C, Singer T. The role of social cognition in decision making. Philos Trans R Soc Lond Ser B Biol Sci. 2008;363(1511):3875–86. https://doi.org/10.1098/rstb.2008.0156.
124. Ochsner KN, Gross JJ. The cognitive control of emotion. Trends Cogn Sci. 2005;9(5):242–9. https://doi.org/10.1016/j.tics.2005.03.010.
125. Levens SM, Phelps EA. Insula and orbital frontal cortex activity underlying emotion interference resolution in working memory. J Cogn Neurosci. 2010;22(1978):2790–803. https://doi.org/10.1162/jocn.2010.21428.

126. Etkin A, Egner T, Peraza DM, Kandel ER, Hirsch J. Resolving emotional conflict: a role for the rostral anterior cingulate cortex in modulating activity in the amygdala. Neuron. 2006;51(6):871–82. https://doi.org/10.1016/j.neuron.2006.07.029.
127. Alguacil S, Madrid E, Espín AM, Ruz M. Facial identity and emotional expression as predictors during economic decisions. Cogn Affect Behav Neurosci. 2016:1–15. https://doi.org/10.3758/s13415-016-0481-9.
128. Egner T, Hirsch J. Cognitive control mechanisms resolve conflict through cortical amplification of task-relevant information. Nat Neurosci. 2005;8(12):1784–90. https://doi.org/10.1038/nn1594.
129. Fouragnan E, Chierchia G, Greiner S, Neveu R, Avesani P, Coricelli G. Reputational priors magnify striatal responses to violations of trust. J Neurosci. 2013;33(8):3602–11. https://doi.org/10.1523/JNEUROSCI.3086-12.2013.
130. Dosenbach NUF, Fair DA, Cohen AL, Schlaggar BL, Petersen SE. A dual-networks architecture of top-down control. Trends Cogn Sci. 2008;12(3):99–105. https://doi.org/10.1016/j.tics.2008.01.001.
131. Raichle M. The brain's default network. Ann N Y Acad Sci. 2015;8(38):433–47. https://doi.org/10.1146/annurev-neuro-071013-014030.
132. Cáceda R, James GA, Gutman DA, Kilts CD. Organization of intrinsic functional brain connectivity predicts decisions to reciprocate social behavior. Behav Brain Res. 2015;292:478–83. https://doi.org/10.1016/j.bbr.2015.07.008.
133. Hollmann M, Rieger JW, Baecke S, Lützkendorf R, Müller C, Adolf D, et al. Predicting decisions in human social interactions using real-time fMRI and pattern classification. PLoS One. 2011;6(10):e25304. https://doi.org/10.1371/journal.pone.0025304.
134. Duncan J. The multiple-demand (MD) system of the primate brain: mental programs for intelligent behavior. Trends Cogn Sci. 2010;14(4):172–9. https://doi.org/10.1016/j.tics.2010.01.004.
135. Crittenden BM, Mitchell DJ, Duncan J. Recruitment of the default mode network during a demanding act of executive control. elife. 2015;2015(4):e06481. https://doi.org/10.7554/eLife.06481.
136. González-García C, Arco JE, Palenciano AF, Ramírez J, Ruz M. Encoding, preparation and implementation of novel complex verbal instructions. NeuroImage. 2017;148:264–73. https://doi.org/10.1016/j.neuroimage.2017.01.037.
137. Mars RB, Neubert F-X, Noonan MP, Sallet J, Toni I, Rushworth MFS. On the relationship between the "default mode network" and the "social brain". Front Hum Neurosci. 2012;6:189. https://doi.org/10.3389/fnhum.2012.00189.
138. Li W, Mai X, Liu C. The default mode network and social understanding of others: what do brain connectivity studies tell us. Front Hum Neurosci. 2014;24(8):74. https://doi.org/10.3389/fnhum.2014.00074.
139. Margulies DS, Ghosh SS, Goulas A, Falkiewicz M, Huntenburg JM, Langs G, et al. Situating the default-mode network along a principal gradient of macroscale cortical organization. Proc Natl Acad Sci. 2016;113(44):12574–9. https://doi.org/10.1073/pnas.1608282113.
140. Gilam G, Hendler T. With love, from me to you: embedding social interactions in affective neuroscience. Neurosci Biobehav Rev. 2016;68:690–701. https://doi.org/10.1016/j.neubiorev.2016.06.027.
141. Oliver D, Tachtsidis I, Hamilton AF. The role of parietal cortex in overimitation: a study with fNIRS. Soc Neurosci. 2017:1–12. https://doi.org/10.1080/17470919.2017.1285812.

Neurobiological Approaches to Interpersonal Coordination: Achievements and Pitfalls

Carlos Cornejo, Zamara Cuadros, and Ricardo Morales

Abstract Although spontaneous interpersonal coordination was originally reported in the early 1960s, the accurate measurement of this phenomenon is very recent. Sophisticated methods used by dynamic systems theory and social neuroscientific perspectives have allowed capturing and analyzing patterns of neural and bodily coordination between interactants, favoring a deeper understanding of the factors and processes involved. In the present chapter, we review neurobiological evidence on interpersonal coordination and acknowledge that, despite the use of cutting-edge technology, extant findings have not yet resulted in an understanding of real-life interpersonal coordination. Theoretical and methodological efforts in social neuroscience aimed to explore interpersonal dynamics through joint tasks have been tacitly based on an individualistic approach to social cognition that underestimates the social nature of interactional phenomena. In turn, dynamic systems theory tends to approach human interaction in the same way as any complex system, disregarding the specific features of social life. We argue instead that interpersonal coordination should be studied under the assumption that people engage in meaningful interactions, so that its study requires the design of more ecological paradigms integrating the benefits of high-precision temporal recordings and a holistic account of the brain and bodily dynamics that occur during real human interaction.

Keywords Interpersonal coordination • Interactional synchrony • Motion capture • Hyperscanning • Social cognition • Dynamic systems theory

C. Cornejo (✉) • Z. Cuadros • R. Morales
Laboratorio de Lenguaje, Interacción y Fenomenología (LIF), Escuela de Psicología, Pontificia Universidad Católica de Chile,
Avenida Vicuña Mackenna 4860, 7820436 Santiago, Chile
e-mail: cca@uc.cl; zcuadros@uc.cl; rimorales@uc.cl

© Springer International Publishing Switzerland 2017

A. Ibáñez et al. (eds.), *Neuroscience and Social Science*,
DOI 10.1007/978-3-319-68421-5_4

1 Introduction

More than 50 years of interdisciplinary research in the cognitive sciences has revealed that interpersonal coordination is a pervasive phenomenon in face-to-face human interactions. When interacting in social settings, individuals spontaneously tend to temporally synchronize their behaviors at different levels [1, 2]. For example, during a walk in the woods, it is likely that people will synchronize not only the trajectory, rhythm, and frequency of their limb movements but also their heart rhythms, breathing rhythms, speech rhythms, and even body language, gestures, and feelings. Accordingly, most research in this field has inquired into how individuals involved in social settings coordinate with each other at linguistic, psychophysiological, neurophysiological, and behavioral levels. Findings from linguistic research in conversational contexts have shown the existence of synchronization patterns at multiple scales of linguistic structure [3]. For instance, when people chat, they align their accent [4, 5], vocal intensity [6], length and placement of pauses [7, 8], descriptive schemes and utterances [9, 10], utterance length [7], response latency [8], speaking rate [11, 12], phoneme productions [13], and syntactic constructions [14, 15]. Psychophysiological studies have further revealed that people, when interacting naturally or playing together, coordinate their breathing [16], heartbeats [17–20], and galvanic skin response [21–23]. Moreover, neurophysiological evidence has allowed characterizing how neural activity becomes coupled as people solve coordination or imitation tasks in real time with another participant [24–34], with a computer program [35], or with a prerecorded video [36, 37]. Also, at a behavioral level, studies indicate that people synchronize their body movements with those of others with whom they interact in social settings [1, 38–45].

Although these phenomena have been studied from different perspectives [46], the most prolific explanations of the factors and processes involved come mainly from two research programs on interpersonal coordination: (a) the dynamical systems perspective and (b) social neuroscience. The dynamical systems approach assumes that interpersonal coordination is governed by the universal laws of self-organization of natural systems [46–50]. Therefore, much of the research in this approach has attempted to evidence whether the dynamic principles governing the coordinated movement of fireflies, schools of fish, and human limbs can also predict and explain the temporal synchronization of bodily movements among people performing joint tasks [49]. Empirical evidence reveals that, indeed, motor coordination patterns of individuals performing highly structured joint tasks are constrained by the same mechanical [51–58] and perceptual [55–60] factors that limit other movements in other natural systems via personal [45, 61–65] and contextual constraints [42, 66–69].

Using the tools and theories of cognitive neuroscience, social neuroscience seeks to understand the cognitive processes that allow people to properly understand and store personal information about each other [70]. For this approach, a truly comprehensive theory of social phenomena must consider the biological, cognitive, and social levels of organization that constitute social phenomena as well as the different relations among them [70, 71]. Consequently, neuroscientists have inquired into the cerebral structures and brain dynamics that support human social abilities. For social neuroscience, interpersonal coordination is a particular case of social cognition comprising different cognitive mechanisms that allow a person to synchronize his/her movements with some referent, in this case, another human being. In this sense, empirical evidence highlights relevant neural networks as revealed via simultaneous cerebral recordings of two subjects as they perform similar tasks or engage in social activities [72], using the same tools and techniques typically employed to describe individual brain activity—such as functional magnetic resonance imaging (fMRI) [73], electroencephalography (EEG) [28], and near-infrared spectroscopy (NIRS) [30].

In this chapter, we review empirical evidence from both perspectives and highlight a set of theoretical and methodological pitfalls that obstruct understanding of interpersonal coordination as a social and affective phenomenon occurring in naturalistic settings. Finally, we will propose that interpersonal coordination should be studied with the assumption that people engage in mutually constructed and meaningful interactions. We will thus argue that the study of interpersonal coordination should focus on emergent properties of interaction, which do not pertain to individuals, but rather emerge as a holistic organization of changes between subjects situated in a meaningful context.

2 The Dynamical Systems Perspective

Thirty years of research on the coordination of movements among people performing tasks individually or jointly has favored the emergence of the dynamical systems perspective. This framework conceptualizes interpersonal coordination as a complex, interactive, and dynamic system governed and limited by the self-organization laws of natural systems [46–50]. This approach is often referred to as "ecological and dynamical systems perspective," because it entails the recognition of reciprocal interactive effects between multiple levels of organization of perception-action systems interacting in environments.[1] This viewpoint hinges on at least four basic assumptions about interpersonal coordination. The first one is that interpersonal coordination should be understood as a complex and multilevel system. This means that it is a phenomenon composed of several elements that reciprocally interact and organize at different levels of complexity. Coordination of bodily movements involves synchronization of multiple elements that shape intra- and interindividual perception-action systems. The interacting elements begin to couple, producing temporally stable states of synchronized activity between people. Interpersonal coordination is thus conceived as a collection of patterns that emerge

[1] Note that, for the dynamic systems perspective, the concept of "ecology" is far removed from the traditional notion that denotes the study of the way human beings conceive, value, use, and impact their environment.

in the course of connection experiences between different levels of organization. This avoids the inclination to fragment the phenomenon into discrete units contained in the body (e.g., representations, neural networks, single limb movements). However, we presume that the study of bodily movement—and its coordination—in terms of the "outcome" of reciprocal interaction processes between elements of systems does not necessarily elude a solipsistic approach to the phenomenon.

The second principle assumes that interpersonal coordination emerges from reciprocal relationships among people's bodies and environments. Since "others 'moor' us in space and time" [49] p. 323, synchronization between people can be understood as part of a spontaneous tendency by which they are physically and socially pulled or attracted into the activity field of another's movements. Supposedly, this axiom highlights the relational and ecological nature of coordination phenomena between people, but it should be noted that relations are described in terms of natural laws and that the environments are devoid of meanings and values. In addition, within the framework of this principle, the perspective of dynamic systems states that the analysis unit is not the internal processes nor the movement of a particular body, but rather the reciprocal relationships between people's brains and bodies, while they interact in their environments. Therefore, the analysis unit is social in nature, as shared movement between people reveals their feelings of connection and social bond with others [48, 50]. In this respect, note that the social aspect of this perspective is reduced to the mere copresence of an interaction partner. We are sure that such a condition is necessary but not sufficient for interpersonal coordination to occur, and we are less convinced that an interaction described in such terms can account for a true social unit.

The third principle states that synchrony patterns of movements change over time, configuring temporally stable orders of motor coactivity. The spontaneous formation of these orders can occur at different time scales (e.g., milliseconds, seconds, minutes, weeks, etc.), as "time defines the frames of reference for our past, present, and future behavior" [49] p. 323. The last principle states that changes in coordinating patterns of movement can be explained by self-organization laws of natural systems [49, 50, 74]. This means that recursive interactions among the components of the system give rise to increasingly complex motor coactivity patterns (e.g., shift from no coordination to time-delay coordination or zero-lag coordination) [49, 75]. Reorganization of these temporally stable coordinated motion patterns occurs in phase transitions, that is, abrupt and nonlinear changes in the organization of the system. Thus, stability periods are followed by a phase transition that is characterized by an imbalance of the established patterns. After this period of fluctuation, the system stabilizes, giving rise to new patterns. Sensitivity to changes in the structure of coordinated patterns of movement is precisely due to multilevel relations underlying these phenomena. However, a tendency toward stabilization prevails. The degree to which movements are synchronized during a phase transition is variable, flexible, and sensitive to disturbances. However, before and after the phase transitions, the pattern is less variable, tending to remain relatively unchanged for a period of time.

Initially, the fourth of these principles led to a vast and productive line of research on the dynamics underlying intrapersonal coordination of movements. The ensuing evidence not only revealed that coordination of the limbs in a single person performing bimanual tasks is governed and limited by the self-organization laws of natural systems but also allowed the mathematical modeling of such dynamics [76–81]. Indeed, the HKB model [77] characterizes dynamical phase transitions (e.g., switching from antiphase mode to inphase mode due to an increase in movement frequency) and dynamic constraints that increase lags and variability in coordination patterns (e.g., differences in oscillator frequencies). Later, dynamic systems research concentrated its efforts on verifying whether dynamic constraints modeled by the HKB equation for intrapersonal coordination also governed coordination of movements between people [46–48, 50]. Thereby, the first studies on interpersonal coordination focused on the dynamic constraints underlying coordination of movements among people. Subsequent studies have further considered the conditions under which coordination of movements between people entails social connection. In the following subsections, we present the conceptual approaches and empirical evidence on each of these lines of research.

2.1 Dynamic Constraints of Coordinated Movement

A dynamic approach understands that coordinated movement between people is constrained by inherent dynamics at their perception-action systems [50, 82]. This principle is supported by abundant research that found the same dynamic constraints for intrapersonal and interpersonal coordination [55, 56, 82–84]. Studies in this field have traditionally used experimental tasks in which pairs of subjects, sitting side by side, are asked to swing one of their limbs at the rhythm of a metronome while trying to synchronize with symmetric or alternate movements of their interaction partner. For example, Schmidt [82] conducted frame-by-frame analyses of two subjects' leg movements in studies that manipulated the type of movement requested and the metronome oscillation frequency. In a first study, participants were asked to move the lower part of their legs at the metronome oscillation frequency while simultaneously trying to maintain the same movement (inphase mode condition) or alternate movement (antiphase mode condition) with respect to the interaction partner; participants were also asked to try to return to the initial phase after a coordination failure. The results of this study reveal less stability in the coordination of alternating movements as the metronome frequency increases, while the stability of symmetrical movement coordination remains constant. In a second study, instructions were similar to those of study 1, except that the participants were asked to maintain the new phase mode once a coordination failure occurred if this new phase mode was easier to maintain. The results of the second study revealed that as the metronome frequency increases, a phase transition occurs from the antiphase mode to the inphase mode but not the other way around. This observed phase transition possesses the physical bifurcation properties previously reported by Haken [77] and

Schöner [85] in intrapersonal coordination: coordination of alternating movements gradually weakens, goes through a period of critical fluctuations, and finally arrives at a new state characterized by symmetrical coordination of movements. Another dynamic constraint of intrapersonal coordination was evidenced by Schmidt and Turvey [84] with pairs of participants who swung pendulums of different length under conditions of uncoupled or coordinated movements. They found greater decoupling in participants' movements as the difference in the lengths of the pendulums increased. Schmidt and O'Brien [83] corroborated previous findings during unintentional interpersonal coordination. These authors found that when pairs of subjects were asked to move the pendulums to their preferred frequency, a phase transition occurred toward a state of greater symmetry in the coordination of movements, but this coordination was never absolute. They also found greater stability in coordination when couples moved pendulums of similar length.

Findings on dynamic constraints have allowed a more complete understanding of mechanistic conditions favoring the emergence of temporally stable patterns of coordinated movement between people. However, the highly structured nature of the tasks calls into question the ecological validity of such findings [50, 86]. In real life, coordinated movements occur in situations that surpass the complexity involved in laboratory tasks requiring couples to stereotypically move at metronome rhythm while attempting to synchronize (inphase or antiphase) with specific limb movements by their interaction partner. This lack of naturalness in experimental environments is also characteristic of studies on dynamic constraints underlying unintentional interpersonal coordination. Although these studies allow subjects to move at their preferred frequency [55, 56, 87], the type of activities requested and the number of repetitions are distant from the conditions under which coordinated movements typically occur in real social interactions [88].

2.2 Socio-environmental Constraints of Coordinated Movement

The socio-environmental approach not only assumes that interpersonal coordination can be predicted by dynamic laws of individual perception-action systems; it also claims that interpersonal coordination can be predicted from constraints resulting from situated interaction between multiple perception-action systems. According to this view, interpersonal coordination is more than simple mechanical coordination of movements. It configures a "social unit" in which patterns of synchronized movements describe the linkages between people [48, 50]. As long as the phenomenon emerges from interactions with others, it cannot be studied independently of the set of exchanges in which patterns of coordinating activity emerge, organize, and reorganize. Studies in this area have explored at least three types of socio-environmental factors as predictors of interpersonal coordination: (1) perceptual access to the interaction partner, (2) personal characteristics of the interaction partner, and (3) features of the interactional situation. Mainly, the experimental settings of such studies include the use of joint action paradigms in which pairs of participants are asked to perform simple limb movements (such as finger tapping, rocking in a rocking chair, postural swaying, swinging pendulums, walking, jumping, dancing, climbing stairs), play a musical instrument, play a video game, or, to a lesser extent, engage in conversations. In general, movements of individual limbs or displacements of objects by joint task participants are recorded through motion-tracking devices such as accelerometers, potentiometers, electrogoniometers, and optical and magnetic capture systems. Such devices yield time-series measurements of movement variation in space (e.g., angles, velocity, acceleration, and distance) and allow implementing linear and nonlinear analysis methods (e.g., cross-recurrence quantification analysis, circular variance of the relative phase, cross-correlations, cross-spectral coherence, and distribution of relative phase angles).

Concerning the effects of perceptual constraints on interpersonal coordination, the movement of pairs of subjects participating in joint tasks is usually compared under conditions in which they have and do not have visual, auditory, or haptic access to their interaction partner [55–60, 83, 89, 90]. For instance, Oullier [90] studied the influence of visual coupling on spontaneous social coordination in pairs of people participating in a finger-tapping task under conditions in which they could or could not see each other's fingers. The results revealed that finger coordination between pairs occurs as soon as they exchange visual information. Richardson [55] contrasted the movement of dyads rocking in rocking chairs under conditions in which they could see the total or peripheral movement of their partner. Their results suggest a major stability of unintentional interpersonal coordination when an individual focuses visual attention directly on the partner's movement, compared to instances in which individuals have peripheral access to that information. Using the same paradigm, but this time contrasting visual and verbal constraints, Richardson [56] found greater unintended interpersonal coordination when participants had access to visual information compared to the condition where they only had access to verbal information from the partner. Demos [59] compared visual (vision, no vision) and auditory (no sound, rocking sound, music) conditions between dyads rocking in rocking chairs. Their results suggest that spontaneous coordination occurs under conditions of both seeing and hearing the other person rocking, but "coupling with the music was weaker than with the partner, and the music competed with the partner's influence, reducing coordination" [59] p. 49. The impact of access to peer visual information on interpersonal coordination and its prevalence compared with other types of perceptual information has been reported in other studies [87, 89, 91, 92]. However, in the case of people with musical training, Nowicki [60] found greater interpersonal coordination under conditions in which they had access to auditory feedback on the partner's musical performance, compared to a condition in which they had access to visual feedback. Other studies also highlight the relevance of access to haptic information in the consolidation of coordinated movement patterns between dyads swaying rhythmically [57, 58]. Taken together, these studies have made it possible to understand the impact of informational dynamic constraints on interpersonal synchronization. However, it is noteworthy that in these studies, the social and environmental aspects are reduced to the exchange of perceptual information between the interactants. Similar to studies of dynamic constraints, studies of informational constraints do not pay much attention to the truly social aspects underlying coordinated patterns of movement, that is, the values and meanings involved in synchronized motor actions.

Other studies have been conducted to ascertain the effect of personal characteristics on interpersonal coordination. For example, to study the influence of pro-social and pro-self orientation on interpersonal coordination, Lumsden [61] executed a study with individuals participating in an arm curl coordination task (to the rhythm of a metronome) with a virtual confederate (a prerecorded video). The results revealed that participants with a pro-social orientation were more coordinated with the virtual confederate than those with a pro-self orientation. In another study, Schmidt [93] found higher levels of synchronization in pendulum swinging tasks performed by dyads with heterogeneity in their social competence (high-low), compared to couples with homogeneity in their social competence (high-high and low-low). Recently, Zhao [65] reported higher levels of synchronization in individuals who believed they were performing a motor coordination task with a physically attractive virtual confederate, in contrast to individuals who believed they were interacting with a less attractive virtual confederate.

Research has also been conducted on personal characteristics that reduce the probability of consolidating patterns of coordinated movement with others. Marsh [62] reported a lower degree of motor coordination between the rocking of children diagnosed with autism spectrum disorders and an adult (both sitting on rocking chairs side by side during story reading) in comparison with typically developing children in the same experimental situation. Similar results were found by Varlet [94] in adults diagnosed with social anxiety disorder. Patients presented less motor coordination with their interaction partner in a pendulum oscillation task than the healthy control group. This line of research has allowed a broader understanding of personal factors that promote or inhibit coordination between people. However, this approach still neglects the study of the social and environmental nature of the phenomenon to the extent that the emphasis is on how individual variables impact or determine patterns of coordination between people.

Another group of studies has demonstrated that some characteristics of social contexts differentially impact coordination levels among interactants. Experimental studies via classic paradigms involving the movement of objects or individual limbs of joint task participants have shown that interpersonal coordination occurs in competitive, collaborative, and recreational contexts [66, 95, 96]. Such studies have also shown that engaging in emotionally negative contexts could decrease or extinguish coordinated behavior. For example, Miles [42] asked individuals to partake in a stepping task with a female confederate, who half of the times arrived 15 min late. The results evidence that inphase synchrony was significantly reduced when participants interacted with the confederate who arrived late. These results are consistent with evidence from more naturalistic studies that highlight higher levels of interpersonal coordination in affiliative conversational contexts than in argumentative conversational contexts [68, 70]. However, the scope of these studies' conclusions is limited by the lack of accurate and fine measurements of the movements of participants in naturalistic conversations; these studies typically use automated video analysis techniques, such as frame differencing, motion energy analysis, and correlation map analysis. Although research on interpersonal coordination in conversational contexts has opened up a promising outlook for understanding the socio-environmental nature of this phenomenon, studies that accurately measure movements in more naturalistic contexts are urgently needed.

3 Social Neuroscience

With the emergence of the so-called interactive turn in cognitive science [97], social neuroscience has begun to study the dynamics of interpersonal coordination. This pursuit has been undertaken with the tools and theories offered by studies of social cognition. Empirical evidence from a wide variety of studies on social cognition has illuminated the roles of specific brain regions in social cognition tasks. For example, different neural networks that operate during social cognition tasks have been identified. Kennedy and Adolphs [72] highlight four core neural networks that can be described in the brain when it engages in social activities: (1) the amygdala network, (2) the mentalizing network, (3) the empathy network, and (4) the mirror-simulation network.

With the goal of generating a comprehensive account of social phenomena, Cacioppo and Berntson [71] have outlined several principles that should guide the empirical and theoretical aspects of social neuroscience. The first principle is multilevel determinism, which specifies that behaviors can have multiple antecedents across various levels of organization. This principle highlights that a truly comprehensive theory of social phenomena requires consideration of multiple levels of organization underlying social cognition phenomena and that the mappings among elements across proximal levels of organization become more complex as the number of intervening levels increases [70]. The second principle is nonadditive determinism, which specifies that the properties of the whole are not always predictable by the sum of the recognized properties of the individual levels. The last principle is reciprocal determinism, which highlights the mutual influences between biological and social factors in explaining behavior [71]. A consequence of the above-outlined principles is that a comprehensive account of human social behavior cannot be achieved taking into account only the biological, cognitive, or social level. To give a fully comprehensive and non-reductive view of the social cognition, multiple levels (personal, biological, cognitive, and social) should be addressed assuming their nonadditive, mutually influencing, and multi-layered nature.

Nevertheless, in spite of the integrative approach, social neuroscience has seen interpersonal coordination as a particular case of social cognition. Social cognition approaches different social phenomena as cognitive processes that occur within the mind of an individual, who constructs models of other people's mental states and who uses these models to predict and explain others' behaviors and intentions (see [98]).

Under this assumption, interpersonal coordination is understood as the set of internal mechanisms that allows a person to synchronize his/her movements with some referent who, in the particular case of interpersonal coordination, happens to be another human being. In what follows, we will present the two main conceptual approaches that have been proposed to understand this phenomenon: representationalism and interactivism.

3.1 Representationalist Approaches to Interpersonal Coordination

A representationalist theory conceives social cognition as a cognitive process that occurs within the mind of an individual, who constructs models of other people's mental states. This approach assumes that the cognitive processes necessary for social interaction are internal and individual, such that one can understand social life by studying individual minds in isolation. A large amount of research in social neuroscience has embraced this view. Common experimental paradigms in social neuroscience typically place human participants in fMRI scanners, devices that constrain the natural movement of the subjects. Once in the scanners, participants are asked to respond to "social" stimuli by observing pictures or videos of other people. These studies have identified several brain areas that respond in social settings, such as the amygdala, the orbitofrontal cortex, the temporal cortex, and the medial prefrontal cortex [17].

Many fMRI paradigms have employed this kind of pseudo-interactive setting. In these cases, the experimental situation relies on scanning one person at a time or on telling participants that they are interacting with a real person, while they are actually interacting with a computer. In a study conducted by Earls [36], Caucasians showed higher peak activation while observing (via a recorded video) and imitating the hand movements of Caucasian actors, relative to observing and imitating the hand movements of African–American actors, in key areas of the previously defined action simulation network: the inferior frontal gyrus, the inferior parietal lobule, the superior parietal lobule, and the superior temporal sulcus. In a study conducted by Cacioppo [35], participants inside a fMRI scanner played a game called "bexting" (beat-texting), which consisted of simple back-and-forth keyboard tapping as if two people were texting each other. Participants were told that they were exchanging texts with another person in the room, whereas they were really interacting with a computer programmed to respond synchronously (in the same rhythm) or asynchronously (in a different rhythm) to the player tapping. The synchronous tapping condition was characterized by greater response in the left inferior parietal lobule, the parahippocampal gyrus extending to the amygdala, the ventromedial prefrontal cortex, and the anterior cingulate cortex.

In sum, the major achievement of individualistic approaches is that they have identified those brain areas that regularly become more active with social stimuli, such as the left inferior parietal lobule, the parahippocampal gyrus, the amygdala, the ventromedial prefrontal cortex, the inferior frontal gyrus, and the inferior parietal lobule [35, 36]. Nevertheless, the representationalist approach to social interaction and interpersonal coordination has been criticized, as the studied social situation does not consist of a true and ecologically valid interaction with another person. Such experimental paradigms severely constrain mutual information exchange and continuous adaptation among interacting participants. Social interaction seems to be substantially different in situations wherein people are engaged in a social unit, compared with situations in which people are acting alone [99, 100].

3.2 Interactivist Approaches to Interpersonal Coordination

Claims about ecological validity have led to an alternative approach to understand social interaction. This perspective considers social cognition as a process that occurs between dyads or among people interacting together, coordinating their actions in a common space and time. Real-life social cognition requires two or more subjects in live interaction [17]. This "interactivist" view has moved away from studying brains in isolation, toward the study of more than one brain in live interaction. Empirically, this perspective implies the study of people during coordinative actions, which requires measuring brain dynamics during live interaction.

Accordingly, social neuroscience has recently examined interpersonal coordination processes under constructs such as "brain coherence" [30], "brain activity coupling" [37], "interbrain coupling" [28], "interbrain synchronization" [26], and "inter-subject neural synchronization" [31]. Researchers have used the term "hyperscanning" when any fMRI, electroencephalography (EEG), or near-infrared spectrometry (NIRS) setup is used to simultaneously track two or more brains [29, 73, 101]. The goal of hyperscanning techniques is to provide simultaneous recordings of brain activity in interactional settings that involve two or more subjects [101].

The first hyperscanning of cerebral activity during interactions between subjects was reported by Montague [73]. In their work, two participants were scanned using two different fMRI devices during a simple game. One participant was assigned to the role of sender; the other, to the role of receiver. Black or white stimuli were presented on the screen of the sender, who could decide which color to transmit to the receiver through a computer screen. The receiver had to determine whether the sender was sharing the true color presented on her screen. Montague et al. [73] observed common activity in the supplementary motor areas of both the sender and the receiver.

In recent times, EEG and NIRS have also been used to study the neuronal dynamics of more than one brain, while different participants perform a given activity [27, 28, 102]. For example, Astolfi et al. [26] obtained EEG recordings from two pairs of subjects playing a card game to measure the neural dynamics of cooperation during face-to-face interaction. They found functional connectivity in the alpha, beta, and gamma bands between the cooperating pairs but not the competing pairs, showing different patterns of cortical activity in different interactional situations. Konvalinka et al. [28] conducted an EEG hyperscan to explore the neural mechanism underlying coordinative and complementary behavioral patterns during joint action. They had participants (seated with their backs to one another) tap together synchronously or to follow a computer metronome in the control condition. The degree of tapping coordination between participants was used to measure leader-follower behavior in each pair. They assessed the adaptability of one member in relation to the other; for example, if member A was leading, member B would change the speed of his/her movements to adapt to A's rhythm. When participants interacted with another person, but not with the computer metronome, the researchers found suppression of alpha and low-beta oscillations over motor and frontal areas. They also found asymmetric brain-coupling patterns or complementary patterns of individual brain mechanisms. Specifically, they found frontal alpha-suppression, especially for the leader, during the anticipation and execution of the task. Their results suggest that leader-follower behavior can emerge spontaneously in dyadic interactions and that leaders invest more resources in prospective planning and control.

In a NIRS study performed by Cui et al. [30], participants sat side by side and played a computer game in which they had to either cooperate or compete. Each trial began with a hollow gray circle at the center of the screen, visible for a random interval between 0.6 and 1.5 s. Subsequently, a green cue signaled participants to press keys simultaneously using the index or middle finger of their right hands. If the difference between their response times was smaller than a threshold, both participants were rewarded with one point; otherwise, both participants lost one point. The competition task was similar to the cooperation task, except that each participant was rewarded for responding faster than his/her partner. The authors found interbrain coherence in the frequency band between 3.2 and 12.8 or between 0.3 and 0.08 Hz in the superior frontal cortex during cooperation but not in the competition condition.

Both "isolated brain" experiments [35, 36] and "interactional experiments" [26, 28, 30] explore the mechanisms underlying interpersonal coordination. Nevertheless, they explore different aspects. The isolated brain approach inquiries into individual processes involved in processing social stimuli, exploring which brain areas or neuronal networks became active during observation of (or judgment about) others or during pseudo-interactions in which there is no real-time feedback between the interactants [37]. In turn, the interactive approach explores the mechanisms needed to interact with another person, during task of mutual coordination. The two perspectives complement each other in quantifying different properties of social interactions [17]. These approaches have allowed the scientific community to achieve a better grasp of the neuronal level of interpersonal coordination processes.

3.3 Psychophysiological Measures of Interpersonal Coordination

In the study of interpersonal coordination, brain activity corresponds to one important level of a phenomenon that involves the whole person—an important level, yet not the only one. Psychophysiological measures of interpersonal coordination have also been used since the 1980s [103], revealing the centrality of the affective dimension involved in social interactions. For example, heart rate and galvanic skin response are relatively unobtrusive methods that have been used to capture the bodily dynamics that occur among people in different kinds of interactions, on time scales as short as minutes or even seconds. Synchrony of involuntary and automatic psychophysiological responses has been found across a broad range of contexts. For instance, Levenson and Gottman [103] evidenced heart rate synchrony between spouses engaged in conversation. More recently, Chatel-Goldman [22] observed that touching each other increases skin conductance synchrony in couples. Additionally, Mønster [23] found evidence of skin conductance synchrony among team members during a cooperative task.

Heart rate and skin conductance have also been used to address interpersonal coordination in groups. Strang [21] aimed to identify the relationship between physio-behavioral coupling and team performance. Dyads played cooperatively and were assigned to the roles of rotator or locator in a variant of the Tetris video game. The researchers measured physio-behavioral coupling by means of the coupling strength between cardiac inter-beat intervals and used a self-report questionnaire that assessed group cohesion, team trust, effectiveness of team communication, and collective efficacy. They found that physio-behavioral coupling exhibited negative relationships with team performance and team attributes, such as cohesion, team trust, and effectiveness of team communication. These findings imply that team attributes generally increased with decreases in physio-behavioral coupling, reflecting a complementary process of coordination (as opposed to mirroring coordination) during task performance, potentially due to different team roles, such as rotator or locator.

3.4 Common Coding Theory

Even though there are many empirical findings about neuronal correlates of interpersonal coordination, there has been little theoretical or conceptual consideration of this phenomenon [17, 101]. One main conceptual approach that has been used in the study of interpersonal coordination holds that coordination is based on a "common coding mechanism" [104–106]. From this perspective, successful interactions between people depend on their capacity to attribute mental states to others.

Because of the centrality of the mirror neuron network in this theoretical approach, here we briefly review its central aspects and address its relevance for research on interpersonal coordination. Mirror neurons, first discovered in nonhuman primates in the premotor cortex, are said to be activated when subjects engage in instrumental actions and when one participant sees another person engage in those actions [107, 108]. The activation of this neuron assembly is related to grasping the intention of the acting individual (thus supporting a form of mind reading). Different studies note that this system discriminates among physically identical movements according to the pragmatic contexts in which these movements occur [109–111]. The evidence that links the mirror neuron system with interpersonal coordination is the finding that people rely on their own motor system when perceiving and predicting others' actions [112].

According to common coding theory [105], the links between mirror neurons and interpersonal coordination explain how interpersonal coordination occurs among people. More precisely, it explains how people predict the action of others to allow a successful pattern of coordinated behaviors. The discovery of the mirror neuron system is said to provide a neural substrate for interpersonal coordination. Coordination processes would be based on the coding and integration of the outcomes of the actions of others and one's own actions. To engage in coordinated behaviors with others, we must understand what others are doing and predict what they will do [105]. For interpersonal coordination to happen, people must predict three aspects of the behavior of others. First, predictions must indicate what kind of action the other will perform as well as the intention that drives the action. Second, predictions should provide information about the temporal unfolding of the action to allow swift, effective interpersonal coordination of actions. Finally, predictions should provide information about the spatial unfolding of the actions of others to effectively distribute a common space to avoid collisions and optimize movement.

In making these predictions, the brain is theorized to rely on the mechanisms of its own motor system. These mechanisms are supported by feed-forward models of sensory feedback in various modalities [105, 113]. Thus, the prediction models are based on the internal motor commands that the observer would use for performing the action himself [113, 114]. Therefore, the same processes underlying individual action planning are involved in predicting the actions of the other person.

4 A Critique of the Theoretical Models of Interpersonal Coordination

Even though interpersonal coordination was initially documented more than 50 years ago at behavioral level [2], the first report of interbrain synchrony appeared only in the last decade [73]. This delay is due partially to the considerable technical difficulties that needed to be overcome to enable recording and analysis of the brain activity of two (or more) interacting people. If the mathematical processing of the brain activity of one individual is complex, the task of identifying synchrony between two or more brains is doubtlessly more difficult. However, it is worth noting that cognitive neuroscience faced questions of similar mathematical difficulty years ago, such as olfactory bulb modeling [115] and intrabrain synchrony [116]. Thus, the main factor to explain such a delay should be sought at a conceptual rather than a methodological level.

Since cognitive neuroscience inherits the same philosophy of mind that originally inspired the cognitive revolution, some of its substantive assumptions continue in contemporary neuroscience. One of these is the idea that the cognizing agent operates while radically isolated from others. Knowledge is originated and stored in individual entities, which encounter the environment isolated from their fellows. Even more, the others like me are in principle another kind of things, whose specific features (e.g., having minds) must first be proven. Thus, the fact that other persons are mind-endowed entities is not a starting point but rather the result of a calculation occurring over the first years of life, from which the cognizing entity infers that the complexity of the other's behavior cannot be explained unless proper desires, intentions, and beliefs are ascribed. Considering this inherited view of mind, it is not difficult to understand why the study of socio-interactional phenomena, such as interpersonal coordination, took time to enter the focus of cognitive neuroscience.

The solipsist bias is still recognizable in several socio-neuroscientific approaches to interactional phenomena. For example, despite its focus on joint actions, common coding theory, paradoxically enough, assumes an individualistic approach to social cognition. From a philosophical perspective, the emphasis on predicting the mental states of others has been put into question [117, 118]. Common coding theory holds several assumptions about social interaction. The clearest one is the mentalizing supposition, which assumes that to understand and coordinate with others, we must infer their mental states and future actions. This assumption entails that people must be observers and adopt a third-person attitude toward other people as a condition to explain and predict their behavior.

By denying access to other minds, common coding theory assumes a priori the opacity of others. It is precisely because of the alleged absence of experiential access to other minds that we need to rely on and employ internal simulations. Hidden mental entities should be inferred to predict the actions of others [105] from the actions of publicly observable bodies. Nevertheless, there is a difference between arguing that the mental models are a way to understand the experience of others and claiming that mental models are the only way for understanding the experience of others [117]. This difference is disregarded in common coding theory, which assumes that social cognition processes occur in the isolated minds of people by generating feed-forward models.

Furthermore, there are empirical facts on interpersonal coordination that can hardly be explained if one assumes that the core of social understanding lies in predicting the future actions of others. In particular, evidence shows that people synchronize their movements simultaneously when interacting socially. Cornejo et al. [86] studied interpersonal coordination through an experimental paradigm in which people talked and moved rather spontaneously. Bodily movements were tracked by an optical motion capture system. They conducted two studies aiming to describe patterns of interpersonal coordination in situations of trust and distrust. The results of both studies show a simultaneous coordination of the participants' movements during the conversations. This strongly suggests the presence of a kind of interpersonal coordination that occurs with no time delay between the participants' movements. These findings highlight that zero-lag coordination occurs on a faster time scale than simple human reaction times, which implies that it cannot be interpreted as an imitative movement by one participant with respect to the other. The findings of Cornejo et al. [86] also reveal that speakers coordinate their movements with listeners' movements—both simultaneously and with a delay. Speakers also react to their listeners in a chain of dynamic coordination patterns affected by interactants' immediate disposition and long-term relationship. Thus, interaction dynamics implies complex processes of coupling and mutual adaptation. It is not clear how common coding theory [105], whose explanatory factor resides on predictive mechanisms, can explain zero-lag coordination, in which coordinative movements among interactants are perfectly simultaneous.

Dynamic systems approaches are possibly in a better position to overcome the solipsistic bias still present in social neuroscience. As described above, this set of theories overcomes the inherited idea that a social interaction is no more than the encounter of two encapsulated, mutually inaccessible individualities. On the contrary, they propose as a unit of analysis the complex system that emerges from the interaction among the individuals: interacting people would constantly and unintentionally configure a "coupled system" [68]. As long as the coupled system existed, the rules for dynamic complex systems would apply. Although this approach succeeds in dealing with the individualistic bias of traditional cognitive neuroscience by avoiding the burden of the concept of representation, it falls into another pitfall of a different sort. By modeling human interaction as another type of dynamic complex system, it blurs the substantive differences between human social life and any other complex system in the physical world. From the fact that the atmospheric movement of gases, the stock market, and the immune system exhibit complex behavior, it does not follow that these entities are ontologically the same. From the fact that a certain explanans (in this case, a certain mathematical model) is helpful to describe a certain explanandum (in this case, human interaction), it does not follow that both are the same thing. Human interaction is not a dynamic complex system, just because nothing is per se a dynamic system. Rather, certain phenomena can be described as such. It may well be the case that human interaction displays features described through nonlinear mathematics—as do several other, quite different phenomena of the natural world. If this is the case, dynamic system theories are necessary but not sufficient to explain human interaction. The task remains to explain what distinguishes this complex system from other (perhaps physical) complex systems.

Unfortunately, the specificity of human interaction is conspicuously absent in dynamic system approaches to social coordination. Most of the specifically human features of interpersonal coordination are omitted from such conceptualizations. We know, for example, that interpersonal coordination is particularly sensitive to social factors: interpersonal coordination will be stronger or more stable if interactants perceive themselves as similar [40], if they share the same social membership, or if they are cooperating rather than competing [31]. There are essential, substantive insights to be drawn from the empirical evidence thus far collected that are risk of being overlooked because they need a specifically human vocabulary—distant from the allegedly neutral vocabulary of dynamical systems theory.

In brief, the theoretical advances of the last few decades on interpersonal coordination give us two important lessons for the future. First, we need to overcome the inherited assumption that social interaction implies an encounter with opaque entities whose mentality the individual must decipher. Second, social interaction has human-specific traits whose understanding should be undertaken to capture a faithful description of human interaction.

5 Recovering the Meaning of Human Interaction

Extant evidence on interpersonal coordination underlines important features of human interaction that have been overlooked by individualistic and dynamical perspectives. One of these facts is that interpersonal coordination, far from being a brain phenomenon, involves the whole bodies of the interactants. Psychophysiological evidence is quite expressive in this respect. As presented above, we know that there is coordination of heart rates between spouses [103] as well as of skin conductance in dyads during cooperative tasks [23]. Moreover, mothers and infants coordinate their ECGs in moments of affective synchrony [119]. There is also evidence of higher heart rate synchrony in trust interactions [20]. Finally, evidence from motion capture devices shows that interpersonal coordination not only involves the whole bodies of participants in a social interaction but also, crucially, that they can be perfectly simultaneous [27, 86].

A second claim robustly supported by empirical evidence is that interpersonal coordination appears and becomes stronger whenever an activity is performed together with others [24]. Interpersonal coordination is stronger when interactants are hearing the same music [120, 121] and when they are performing a task directed toward a common goal [122]. It is relevant to note that everyday joint actions are not equivalent to coordinated movements: in social life, joint actions are deployed when the interactants understand what the common goal is. Human actions are always socially embedded; thus, interpersonal coordination never occurs in a social vacuum. In everyday life, people share an ample base of background knowledge, which makes social interactions always meaningful [123]: the individual understands others' movements not like the movements of objects but rather as actions, i.e., as meaningful movements. This social background provides a substratum that cuts across sensorial, motor, and cognitive processes. In our view, this is the fact that explains the constant result that interpersonal coordination becomes enhanced when interactants have visual contact [59, 60, 89, 91, 94].

A third systematic observation is the tight relation between interpersonal coordination and positive affect. We know that interpersonal coordination is strongly associated with empathy [40, 124] and with the perception of pro-social disposition in the other [61]. Interpersonal coordination is particularly enhanced whenever interactants trust each other [20, 125] or whenever interactants perceive themselves as belonging to the same reference group [36]. Finally, there is ample evidence that interpersonal coordination is higher in cooperative interactions than competitive ones [23, 27, 30, 31, 34].

From a broader viewpoint, interpersonal coordination corresponds to a basic anthropological phenomenon (behaviorally and neurophysiologically measurable) that is tightly associated with the establishment and maintenance of social bonds. It emerges with positive affect (trust, empathy, and collaboration) and tends to disappear when this affective matrix is broken. Interpersonal coordination emerges also when interactants are embedded in a "co-phenomenology" [123]—also called "we-mode" [126] or "we-relationship" [127]. It is not something that occurs in the mind of an observer but something that emerges as in an intersubjectively shared space [97, 123]. This most natural and pre-reflexive kind of interaction allows people to share a common sense within which movements are meaningful actions. It is this tacit background that makes people coordinate permanently and simultaneously and even anticipate others' movements. Its automatic, nonreflexive character is also supported by empirical evidence: interpersonal coordination tends to be higher when it is unintentional than when it is intentional [39, 45, 55, 56, 83, 128]. In addition, Konvalinka et al. [28] showed that whenever interactants are asked to lead an interaction, the symmetric brain coupling changes its dynamics, possibly due to the leader undertaking a planning process that puts her outside the natural attitude.

One aspect that should be underlined is that this interpretation of interpersonal coordination assumes that the most natural way to interact with others is not solipsistically but intersubjectively. Schütz [127] notes that in social relations our consciousness is interlocked, with each person's mental states immediately affecting the other, and in such situations, there is a form of immediate interpersonal understanding. In the most basic way to interact, we do not approach them from a third-person perspective. People are primordially not things for us. They can, under certain circumstances, become like things, when we are forced to abandon the we-relationship and theorize about their real intentions. In those circumstances, we are reflecting on the other individual's behavior, and it is likely that no interpersonal coordination will be perceptible anymore.

6 Conclusion

Given the wide availability of brain-imaging techniques and methods to measure interpersonal coordination, perhaps the most important challenge in this area is to build a coherent theoretical framework for integrating the existing results. Here, we proposed that instead of assuming that interpersonal coordination requires prediction mechanisms or that it is another physical-like dynamical system, a theoretical framework should focus on the construction of a common social and affective space.

We stated that the study of interpersonal coordination has been advanced basically by the dynamical systems perspective and by social neuroscience. However, despite the use of sophisticated methods to capture and analyze neural and bodily synchrony, the methodological efforts of both perspectives were still detached from real-life human interactions. In most studies, emphasis is placed on the accurate measurement of dyads' actions (movements or neural activity) but only during highly structured tasks, focusing on the individual brain/mind or paying little attention to the affective and social nature of face-to-face encounters. This bias is particularly strong in social neuroscience, since it inherits the axiom that social interaction can be explained as the encounter of two individual minds attempting to decipher each other's mentality: first comes the individual mind, then social life. This axiom produces several anomalies, such as simultaneous coordination, that social neuroscience is in no condition to adequately explain. On the other hand, dynamic systems theory, while avoiding the problems of solipsism, dismisses the specificities of human interaction in favor of understanding it as any other dynamic complex system—including physical ones. The consequence of the complexity approach is neglect for the meaning of social life.

We advanced a theoretical alternative that satisfies both necessities: (1) studying interactions as such (and not as individual mental puzzles) and (2) recovering the meaning in social interaction. In this framework, interpersonal coordination is the behavioral/neurophysiological correlate of the most basic form of interaction, the we-relationship, in which an authentic co-phenomenology is felt and lived. This is the reason why interpersonal coordination is unintentional, strongly affective, bodily, and highly sensitive to a sense of common belonging.

Certainly, findings on interpersonal coordination have opened a new space to study the interactional context in which human actions occur. Future research needs to focus on integrating the different levels of analysis at which this phenomenon occurs while respecting the ecology of social life. The challenge is to build paradigms that reproduce real-life situations as much as possible, integrating the benefits of high-precision temporal recordings and a whole-body account of the brain and bodily dynamics that occur during a real human interaction.

Acknowledgments This chapter was supported by the Chilean National Fund for Scientific and Technological Development, FONDECYT (Grant number 1141136), and the Scientific and Technological Equipment Program, FONDEQUIP (Grant number EQM120026).

References

1. Bernieri FJ, Reznick JS, Rosenthal R. Synchrony, pseudosynchrony, and dissynchrony: measuring the entrainment process in mother-infant interactions. J Pers Soc Psychol. 1988;54(2):243–53. https://doi.org/10.1037/0022-3514.54.2.243.
2. Condon WS, Ogston WD. Sound film analysis of normal and pathological behavior patterns. J Nerv Ment Dis. 1966;143(4):338–47. https://doi.org/10.1097/00005053-196610000-00005.

3. Pickering MJ, Garrod S. Toward a mechanistic psychology of dialogue. Behav Brain Sci. 2004;27(2):169–190.-226. https://doi.org/10.1017/S0140525X04000056.
4. Giles H. Accent mobility: a model and some data. Anthropol Linguist. 1973;15(2):87–105. https://doi.org/10.1111/j.1460-2466.2008.00398.x.
5. Bourhis RY, Giles H. The language of intergroup distinctiveness. In: Giles H, editor. Language, ethnicity and intergroup relations. London: Academic; 1977. p. 119–35.
6. Natale M. Convergence of mean vocal intensity in dyadic communication as a function of social desirability. J Pers Soc Psychol. 1975;32(5):790–804. https://doi.org/10.1037/0022-3514.32.5.790.
7. Bilous FR, Krauss RM. Dominance and accommodation in the conversational behaviours of same- and mixed-gender dyads. Lang Commun. 1988;8(3-4):183–94. https://doi.org/10.1016/0271-5309(88)90016-X.
8. Cappella JN, Planalp S. Talk and silence sequences in informal conversations III: interspeaker influence. Hum Commun Res. 1981;7(2):117–32. https://doi.org/10.1111/j.1468-2958.1981.tb00564.x.
9. Garrod S, Anderson A. Saying what you mean in dialogue: a study in conceptual and semantic co-ordination. Cognition. 1987;27(2):181–218. https://doi.org/10.1016/0010-0277(87)90018-7.
10. Garrod S, Doherty G. Conversation, co-ordination and convention: an empirical investigation of how groups establish linguistic conventions. Cognition. 1994;53(3):181–215. https://doi.org/10.1016/0010-0277(94)90048-5.
11. Street RL. Speech convergence and speech evaluation in fact-finding interviews. Hum Commun Res. 1984;11(2):139–69. https://doi.org/10.1111/j.1468-2958.1984.tb00043.x.
12. Giles H, Coupland J, Coupland N. Contexts of accommodation: developments in applied sociolinguistics. Cambridge: Cambridge University Press; 1991. https://doi.org/10.1017/CBO9780511663673.
13. Pardo JS. On phonetic convergence during conversational interaction. J Acoust Soc Am. 2006;119(4):2382–93. https://doi.org/10.1121/1.2178720.
14. Bock JK. Syntactic persistence in language production. Cogn Psychol. 1986;18(3):355–87. https://doi.org/10.1016/0010-0285(86)90004-6.
15. Branigan HP, Pickering MJ, Cleland AA. Syntactic co-ordination in dialogue. Cognition. 2000;75(B):13–25. https://doi.org/10.1016/S0010-0277(99)00081-5.
16. Mcfarland DH. Respiratory markers of conversational interaction. J Speech Lang Hear Res. 2001;44(44):128–43. https://doi.org/10.1044/1092-4388(2001/012).
17. Konvalinka I, Roepstorff A. The two-brain approach: how can mutually interacting brains teach us something about social interaction? Front Hum Neurosci. 2012;6:215. https://doi.org/10.3389/fnhum.2012.00215.
18. Gottman JM, Levenson RW. A valid procedure for obtaining self-report of affect in marital interaction. J Consult Clin Psychol. 1985;53(2):151–60. https://doi.org/10.1037/0022-006X.53.2.151.
19. Thomas KA, Burr RL, Spieker S, Lee J, Chen J. Mother-infant circadian rhythm: development of individual patterns and dyadic synchrony. Early Hum Dev. 2014;90(12):885–90. https://doi.org/10.1016/j.earlhumdev.2014.09.005.
20. Mitkidis P, McGraw JJ, Roepstorff A, Wallot S. Building trust: heart rate synchrony and arousal during joint action increased by public goods game. Physiol Behav. 2015;149:101–6. https://doi.org/10.1016/j.physbeh.2015.05.033.
21. Strang AJ, Funke GJ, Russell SM, Dukes AW, Middendorf MS. Physio-behavioral coupling in a cooperative team task: contributors and relations. J Exp Psychol Hum Percept Perform. 2014;40(1):145–58. https://doi.org/10.1037/a0033125.
22. Chatel-Goldman J, Congedo M, Jutten C, Schwartz J-L. Touch increases autonomic coupling between romantic partners. Front Behav Neurosci. 2014;8:95. https://doi.org/10.3389/fnbeh.2014.00095.

23. Mønster D, Håkonsson DD, Eskildsen JK, Wallot S. Physiological evidence of interpersonal dynamics in a cooperative production task. Physiol Behav. 2016;156:24–34. https://doi.org/10.1016/j.physbeh.2016.01.004.
24. Saito DN, Tanabe HC, Izuma K, et al. "Stay tuned": inter-individual neural synchronization during mutual gaze and joint attention. Front Integr Neurosci. 2010;4:1–12. https://doi.org/10.3389/fnint.2010.00127.
25. Lindenberger U, Li S-C, Gruber W, Müller V. Brains swinging in concert: cortical phase synchronization while playing guitar. BMC Neurosci. 2009;10(1):22. https://doi.org/10.1186/1471-2202-10-22.
26. Astolfi L, Toppi J, De Vico Fallani F, et al. Neuroelectrical hyperscanning measures simultaneous brain activity in humans. Brain Topogr. 2010;23(3):243–56. https://doi.org/10.1007/s10548-010-0147-9.
27. Yun K, Watanabe K, Shimojo S. Interpersonal body and neural synchronization as a marker of implicit social interaction. Sci Rep. 2012;2:959. https://doi.org/10.1038/srep00959.
28. Konvalinka I, Bauer M, Stahlhut C, Hansen LK, Roepstorff A, Frith CD. Frontal alpha oscillations distinguish leaders from followers: Multivariate decoding of mutually interacting brains. NeuroImage. 2014;94:79–88. https://doi.org/10.1016/j.neuroimage.2014.03.003.
29. Khalsa SS, Schiffman JE, Bystritsky A. Treatment-resistant OCD: options beyond first-line medications. Curr Psychiatr Ther. 2011;10(11):44–52. https://doi.org/10.1371/journal.pone.0012166.
30. Cui X, Bryant DM, Reiss AL. NIRS-based hyperscanning reveals increased interpersonal coherence in superior frontal cortex during cooperation. NeuroImage. 2012;59(3):2430–7. https://doi.org/10.1016/j.neuroimage.2011.09.003.
31. Liu T, Saito H, Oi M. Obstruction increases activation in the right inferior frontal gyrus. Soc Neurosci. 2015;919:1–9. https://doi.org/10.1080/17470919.2015.1088469.
32. Osaka N, Minamoto T, Yaoi K, Azuma M, Shimada YM, Osaka M. How two brains make one synchronized mind in the inferior frontal cortex: FNIRS-based hyperscanning during cooperative singing. Front Psychol. 2015;6:1811. https://doi.org/10.3389/fpsyg.2015.01811.
33. Cheng X, Li X, Hu Y. Synchronous brain activity during cooperative exchange depends on gender of partner: a fNIRS-based hyperscanning study. Hum Brain Mapp. 2015;36(6):2039–48. https://doi.org/10.1002/hbm.22754.
34. Nozawa T, Sasaki Y, Sakaki K, Yokoyama R, Kawashima R. Interpersonal frontopolar neural synchronization in group communication: An exploration toward fNIRS hyperscanning of natural interactions. NeuroImage. 2016;133:484–97. https://doi.org/10.1016/j.neuroimage.2016.03.059.
35. Cacioppo S, Zhou H, Monteleone G, et al. You are in sync with me: neural correlates of interpersonal synchrony with a partner. Neuroscience. 2014;277:842–58. https://doi.org/10.1016/j.neuroscience.2014.07.051.
36. Earls HA, Englander ZA, Morris JP. Perception of race-related features modulates neural activity associated with action observation and imitation. Neuroreport. 2013;24(8):410–3. https://doi.org/10.1097/WNR.0b013e328360a168.
37. Stephens GJ, Silbert LJ, Hasson U. Speaker-listener neural coupling underlies successful communication. Proc Natl Acad Sci U S A. 2010;107(32):14425–30. https://doi.org/10.1073/pnas.1008662107.
38. Bernieri FJ, Rosenthal R. Interpersonal coordination: behavior matching and interactional synchrony. In: Feldman RS, Rimé B, editors. Fundamentals of nonverbal behavior: studies in emotion & social interaction. New York: Cambridge University Press; 1991. p. 401–32. https://doi.org/10.1017/CBO9781107415324.004.
39. Del-Monte J, Capdevielle D, Varlet M, et al. Social motor coordination in unaffected relatives of schizophrenia patients: a potential intermediate phenotype. Front Behav Neurosci. 2013;7:137. https://doi.org/10.3389/fnbeh.2013.00137.

40. Llobera J, Charbonnier C, Chagué S, et al. The subjective sensation of synchrony: an experimental study. PLoS One. 2016;11(2):e0147008. https://doi.org/10.1371/journal.pone.0147008.
41. Marmelat V, Delignières D. Strong anticipation: complexity matching in interpersonal coordination. Exp Brain Res. 2012;222(1–2):137–48. https://doi.org/10.1007/s00221-012-3202-9.
42. Miles LK, Griffiths JL, Richardson MJ, Macrae CN. Too late to coordinate: contextual influences on behavioral synchrony. Eur J Soc Psychol. 2010;40(1):52–60. https://doi.org/10.1002/ejsp.721.
43. Ouwehand PEW, Peper CLE. Does interpersonal movement synchronization differ from synchronization with a moving object? Neurosci Lett. 2015;606:177–81. https://doi.org/10.1016/j.neulet.2015.08.052.
44. Preissmann D, Charbonnier C, Chagué S, et al. A motion capture study to measure the feeling of synchrony in romantic couples and in professional musicians. Front Psychol. 2016;7:1673. https://doi.org/10.3389/fpsyg.2016.01673.
45. Varlet M, Stoffregen TA, Chen F-C, Alcantara C, Marin L, Bardy BG. Just the sight of you: postural effects of interpersonal visual contact at sea. J Exp Psychol Hum Percept Perform. 2014;40(6):2310–8. https://doi.org/10.1037/a0038197.
46. Schmidt RC, Fitzpatrick P. The origin of the ideas of interpersonal synchrony and synergies. In: Passos P, Davids K, Chow JY, editors. Interpersonal coordination and performance in social systems. New York: Routledge; 2016. p. 17–31.
47. Nordham C, Kelso JAS. The nature of interpersonal coordination. In: Passos P, Davids K, Chow JY, editors. Interpersonal coordination and performance in social systems. New York: Routledge; 2016. p. 32–52.
48. Marsh KL, Richardson MJ, Schmidt RC. Social connection through joint action and interpersonal coordination. Top Cogn Sci. 2009;1(2):320–39. https://doi.org/10.1111/j.1756-8765.2009.01022.x.
49. Rio KW, Warren WH. Interpersonal coordination in biological systems: the emergence of collective locomotion. In: Passos P, Davids K, Chow JY, editors. Interpersonal coordination and performance in social systems. New York: Routledge; 2016. p. 3–16.
50. Schmidt RC, Richardson MJ. Dynamics of interpersonal coordination. In: Fuchs A, Jirsa VK, editors. Understanding complex systems, vol. 2008. Berlin: Springer; 2008. p. 281–308. https://doi.org/10.1007/978-3-540-74479-5_14.
51. Amazeen PG, Schmidt RC, Turvey MT. Frequency detuning of the phase entrainment dynamics of visually coupled rhythmic movements. Biol Cybern. 1995;72(6):511–8. https://doi.org/10.1007/BF00199893.
52. Coey C, Varlet M, Schmidt RC, Richardson MJ. Effects of movement stability and congruency on the emergence of spontaneous interpersonal coordination. Exp Brain Res. 2011;211(3-4):483–93. https://doi.org/10.1007/s00221-011-2689-9.
53. Fuchs A, Jirsa VK, Haken H, Kelso JAS. Extending the HKB model of coordinated movement to oscillators with different eigen frequencies. Biol Cybern. 1996;74(1):21–30. https://doi.org/10.1007/BF00199134.
54. Jeka JJ, Kelso JA. Manipulating symmetry in the coordination dynamics of human movement. J Exp Psychol Hum Percept Perform. 1995;21(2):360–74. https://doi.org/10.1037/0096-1523.21.2.360.
55. Richardson MJ, Marsh KL, Isenhower RW, Goodman JRL, Schmidt RC. Rocking together: dynamics of intentional and unintentional interpersonal coordination. Hum Mov Sci. 2007;26(6):867–91. https://doi.org/10.1016/j.humov.2007.07.002.
56. Richardson MJ, Marsh KL, Schmidt RC. Effects of visual and verbal interaction on unintentional interpersonal coordination. J Exp Psychol Hum Percept Perform. 2005;31(1):62–79. https://doi.org/10.1037/0096-1523.31.1.62.
57. Sofianidis G, Hatzitaki V, Grouios G, Johannsen L, Wing A. Somatosensory driven interpersonal synchrony during rhythmic sway. Hum Mov Sci. 2012;31(3):553–66. https://doi.org/10.1016/j.humov.2011.07.007.

58. Sofianidis G, Elliott MT, Wing AM, Hatzitaki V. Interaction between interpersonal and postural coordination during frequency scaled rhythmic sway: the role of dance expertise. Gait Posture. 2015;41(1):209–16. https://doi.org/10.1016/j.gaitpost.2014.10.007.
59. Demos AP, Chaffin R, Begosh KT, Daniels JR, Marsh KL. Rocking to the beat: effects of music and partner's movements on spontaneous interpersonal coordination. J Exp Psychol Gen. 2012;141(1):49–53. https://doi.org/10.1037/a0023843.
60. Nowicki L, Prinz W, Grosjean M, Repp BH, Keller PE. Mutual adaptive timing in interpersonal action coordination. Psychomusicol Music Mind Brain. 2013;23(1):6–20. https://doi.org/10.1037/a0032039.
61. Lumsden J, Miles LK, Richardson MJ, Smith CA, Macrae CN. Who syncs? Social motives and interpersonal coordination. J Exp Soc Psychol. 2012;48(3):1–23. https://doi.org/10.1016/j.jesp.2011.12.007.
62. Marsh KL, Isenhower RW, Richardson MJ, et al. Autism and social disconnection in interpersonal rocking. Front Integr Neurosci. 2013;7:4. https://doi.org/10.3389/fnint.2013.00004.
63. Schmidt RC, Christianson N, Carello C, Baron R. Effects of social and physical variables on between-person visual coordination. Ecol Psychol. 1994;6(3):159–83. https://doi.org/10.1207/s15326969eco0603_1.
64. Davis E, Greenberger E, Charles S, Chen C, Zhao L, Dong Q. Emotion experience and regulation in China and the United States: how do culture and gender shape emotion responding? Int J Psychol. 2012;47(3):230–9. https://doi.org/10.1080/00207594.2011.626043.
65. Zhao Z, Salesse RN, Gueugnon M, Schmidt RC, Marin L, Bardy BG. Attractive moving virtual agent elicits more stable interpersonal coordination. Hum Mov Sci. 2010;12(1976):90073. https://doi.org/10.1002/Ejsp.721.
66. Hammal Z, Cohn JF, George DT. Interpersonal coordination of headmotion in distressed couples. IEEE Trans Affect Comput. 2014;5(2):155–67. https://doi.org/10.1109/TAFFC.2014.2326408.
67. Paxton A, Dale R. Argument disrupts interpersonal synchrony. Q J Exp Psychol. 2013;66:2092–102. https://doi.org/10.1080/17470218.2013.853089.
68. Tollefsen DP, Dale R, Paxton A. Alignment, transactive memory, and collective cognitive systems. Rev Philos Psychol. 2013;4(1):49–64. https://doi.org/10.1007/s13164-012-0126-z.
69. Paxton A, Dale R. Frame-differencing methods for measuring bodily synchrony in conversation. Behav Res Methods. 2012;45(2):329–43. https://doi.org/10.3758/s13428-012-0249-2.
70. Cacioppo JT, Ortigue S. Social Neuroscience: how a multidisciplinary field is uncovering the biology of human interactions. Cerebrum. 2011;2011:17.
71. Cacioppo JT, Berntson GG. Social psychological contributions to the decade of the brain. Doctrine of multilevel analysis. Am Psychol. 1992;47(8):1019–28. https://doi.org/10.1037/0003-066X.47.8.1019.
72. Kennedy DP, Adolphs R. The social brain in psychiatric and neurological disorders Daniel. Trends Cogn Sci. 2012;16(11):559–72. https://doi.org/10.1016/j.tics.2012.09.006.The.
73. Montague P. Hyperscanning: simultaneous fMRI during linked social interactions. NeuroImage. 2002;16(4):1159–64. https://doi.org/10.1006/nimg.2002.1150.
74. Newtson D, Hairfield J, Bloomingdale J, Cutino S. The structure of action and interaction. Soc Cogn. 1987;5(3):191–237. https://doi.org/10.1521/soco.1987.5.3.191.
75. Camazine S, Deneubourg J-L, Franks NR, Sneyd J, Theraulaz G, Bonabeau E. Self-organization in biological systems. Princeton: Princeton University Press; 2003.
76. Fitzpatrick P, Schmidt R, Lockman J. Dynamical patterns in the development of clapping. Child Dev. 1996;67:2691–708. https://doi.org/10.2307/1131747.
77. Haken H, Kelso JAS, Bunz H. A theoretical model of phase transitions in human hand movements. Biol Cybern. 1985;51(5):347–56. https://doi.org/10.1007/BF00336922.
78. Kelso JAS, Jeka JJ. Symmetry breaking dynamics of human multilimb coordination. J Exp Psychol Hum Percept Perform. 1992;18(3):645–68. https://doi.org/10.1037/0096-1523.18.3.645.

79. Kelso JA, Holt KG, Rubin P, Kugler PN. Patterns of human interlimb coordination emerge from the properties of non-linear, limit cycle oscillatory processes: theory and data. J Mot Behav. 1981;13(4):226–61. https://doi.org/10.1080/00222895.1981.10735251.
80. Rosenblum LD, Turvey MT. Maintenance tendency in co-ordinated rhythmic movements: relative fluctuations and phase. Neuroscience. 1988;27(1):289–300. https://doi.org/10.1016/0306-4522(88)90238-2.
81. Schmidt RC, Shaw BK, Turvey MT. Coupling dynamics in interlimb coordination. J Exp Psychol Hum Percept Perform. 1993;19(2):397–415. https://doi.org/10.1037/0096-1523.19.2.397.
82. Schmidt RC, Carello C, Turvey MT. Phase transitions and critical fluctuations in the visual coordination of rhythmic movements between people. J Exp Psychol Hum Percept Perform. 1990;16(2):227–47. https://doi.org/10.1037/0096-1523.16.2.227.
83. Schmidt RC, O'Brien B. Evaluating the dynamics of unintended interpersonal coordination. Ecol Psychol. 1997;9(3):189–206. https://doi.org/10.1207/s15326969eco0903_2.
84. Schmidt RC, Turvey MT. Phase-entrainment dynamics of visually coupled rhythmic movements. Biol Cybern. 1994;70(4):369–76. https://doi.org/10.1007/BF00200334.
85. Schöner G, Haken H, Kelso JAS. A stochastic theory of phase transitions in human hand movement. Biol Cybern. 1986;53(4):247–57. https://doi.org/10.1007/BF00336995.
86. Cornejo C, Hurtado E, Cuadros Z, et al. Dynamics of simultaneous and imitative bodily coordinations in trust, distrust and closeness. Submitted.
87. Okazaki S, Hirotani M, Koike T, et al. Unintentional interpersonal synchronization represented as a reciprocal visuo-postural feedback system: a multivariate autoregressive modeling approach. PLoS One. 2015;10(9):e0137126. https://doi.org/10.1371/journal.pone.0137126.
88. Musa R, Carré D, Cornejo C. Bodily synchronization and ecological validity: a relevant concern for nonlinear dynamical systems theory. Front Hum Neurosci. 2015;9(64):64. https://doi.org/10.3389/fnhum.2015.00064.
89. Athreya DN, Riley MA, Davis TJ. Visual influences on postural and manual interpersonal coordination during a joint precision task. Exp Brain Res. 2014;232(9):2741–51. https://doi.org/10.1007/s00221-014-3957-2.
90. Oullier O, de Guzman GC, Jantzen KJ, Lagarde J, Kelso JAS. Social coordination dynamics: measuring human bonding. Soc Neurosci. 2008;3(2):178–92. https://doi.org/10.1080/17470910701563392.
91. Varlet M, Marin L, Lagarde J, Bardy BG. Social postural coordination. J Exp Psychol Hum Percept Perform. 2011;37(2):473–83. https://doi.org/10.1037/a0020552.
92. Roerdink M, Peper CE, Beek PJ. Effects of correct and transformed visual feedback on rhythmic visuo-motor tracking: tracking performance and visual search behavior. Hum Mov Sci. 2005;24(3):379–402. https://doi.org/10.1016/j.humov.2005.06.007.
93. Schmidt RC, Bienvenu M, P a F, Amazeen PG. A comparison of intra- and interpersonal interlimb coordination: coordination breakdowns and coupling strength. J Exp Psychol Hum Percept Perform. 1998;24(3):884–900. https://doi.org/10.1037/0096-1523.24.3.884.
94. Varlet M, Marin L, Capdevielle D, et al. Difficulty leading interpersonal coordination: towards an embodied signature of social anxiety disorder. Front Behav Neurosci. 2014;8:29. https://doi.org/10.3389/fnbeh.2014.00029.
95. Rodrigues M, Passos P. Patterns of interpersonal coordination in rugby union: analysis of collective behaviours in a match situation. Sci Res. 2013;3(4):209–14. https://doi.org/10.4236/ape.2013.34034.
96. Valdesolo P, Ouyang J, DeSteno D. The rhythm of joint action: synchrony promotes cooperative ability. J Exp Soc Psychol. 2010;46(4):693–5. https://doi.org/10.1016/j.jesp.2010.03.004.
97. De Jaegher H, Di Paolo E, Gallagher S. Can social interaction constitute social cognition? Trends Cogn Sci. 2010;14(10):441–7. https://doi.org/10.1016/j.tics.2010.06.009.
98. Kumfor F, et al. Clinical studies of social neuroscience: a lesion model approach. In: Ibáñez A, Sedeño L, García AM, editors. Neuroscience and social science. New York: Springer; 2017.

99. De Jaegher H. Social understanding through direct perception? Yes, by interacting. Conscious Cogn. 2009;18(2):535–42. https://doi.org/10.1016/j.concog.2008.10.007.
100. Schilbach L. A second-person approach to other minds. Nat Rev Neurosci. 2010;11(6):449. https://doi.org/10.1038/nrn2805-c1.
101. Babiloni F, Astolfi L. Social neuroscience and hyperscanning techniques: past, present and future. Neurosci Biobehav Rev. 2014;44:76–93. https://doi.org/10.1016/j.neubiorev.2012.07.006.
102. Toppi J, Borghini G, Petti M, et al. Investigating cooperative behavior in ecological settings: an EEG hyperscanning study. PLoS One. 2016;11(4):e0154236. https://doi.org/10.1371/journal.pone.0154236.
103. Levenson RW, Gottman JM. Physiological and affective predictors of change in relationship satisfaction. J Pers Soc Psychol. 1985;49(1):85–94. https://doi.org/10.1037/0022-3514.49.1.85.
104. Sebanz N, Knoblich G, Prinz W, Wascher E. Twin peaks: an ERP study of action planning and control in co-acting individuals. J Cogn Neurosci. 2006;18(5):859–70. https://doi.org/10.1162/jocn.2006.18.5.859.
105. Sebanz N, Knoblich G. Prediction in joint action: what, when, and where. Top Cogn Sci. 2009;1(2):353–67. https://doi.org/10.1111/j.1756-8765.2009.01024.x.
106. Knoblich G, Butterfill S, Sebanz N. Psychological research on joint action: theory and data. Psychol Learn Motiv Adv Res Theory. 2011;54:59–101.
107. Rizzolatti G, Fadiga L, Gallese V, Fogassi L. Premotor cortex and the recognition of motor actions. Cogn Brain Res. 1996;3(2):131–41. https://doi.org/10.1016/0926-6410(95)00038-0.
108. Rizzolatti G, Fogassi L, Gallese V. Neurophysiological mechanisms underlying the understanding and imitation of action. Nat Rev Neurosci. 2001;2:661–70. https://doi.org/10.1038/35090060.
109. Fogassi L, Ferrari PF, Gesierich B, Rozzi S, Chersi F, Rizzolatti G. Parietal lobe: from action organization to intention understanding. Science. 2005;308:662–7. https://doi.org/10.1126/science.1106138.
110. Iacoboni M, Molnar-Szakacs I, Gallese V, Buccino G, Mazziotta JC. Grasping the intentions of others with one's own mirror neuron system. PLoS Biol. 2005;3(3):e79. https://doi.org/10.1371/journal.pbio.0030079.
111. Kaplan JT, Iacoboni M. Getting a grip on other minds: mirror neurons, intention understanding, and cognitive empathy. Soc Neurosci. 2006;1(3–4):175–83. https://doi.org/10.1080/17470910600985605.
112. Rizzolatti G, Craighero L. The mirror-neuron system. Annu Rev Neurosci. 2004;27(1):169–92. https://doi.org/10.1146/annurev.neuro.27.070203.144230.
113. Wolpert DM, Doya K, Kawato M. A unifying computational framework for motor control and social interaction. Philos Trans R Soc Lond Ser B Biol Sci. 2003;358(1431):593–602. https://doi.org/10.1098/rstb.2002.1238.
114. Wilson M, Knoblich G. The case for motor involvement in perceiving conspecifics. Psychol Bull. 2005;131(3):460–73. https://doi.org/10.1037/0033-2909.131.3.460.
115. Freeman WJ. Simulation of chaotic EEG patterns with a dynamic model of the olfactory system. Biol Cybern. 1987;56(2–3):139–50. https://doi.org/10.1007/BF00317988.
116. King-Casas B. Getting to know you: reputation and trust in a two-person economic exchange. Science. 2005;308(5718):78–83. https://doi.org/10.1126/science.1108062.
117. Zahavi D. Empathy, embodiment and interpersonal understanding: from Lipps to Schutz. Inquiry. 2010;53(3):285–306. https://doi.org/10.1080/00201741003784663.
118. Gallagher S. Direct perception in the intersubjective context. Conscious Cogn. 2008;17(2):535–43. https://doi.org/10.1016/j.concog.2008.03.003.
119. Feldman R, Magori-Cohen R, Galili G, Singer M, Louzoun Y. Mother and infant coordinate heart rhythms through episodes of interaction synchrony. Infant Behav Dev. 2011;34(4):569–77.

120. Burger B, Thompson MR, Luck G, Saarikallio SH, Toiviainen P. Hunting for the beat in the body: on period and phase locking in music-induced movement. Front Hum Neurosci. 2014;8:903. https://doi.org/10.3389/fnhum.2014.00903.
121. Toiviainen P, Alluri V, Brattico E, Wallentin M, Vuust P. Capturing the musical brain with Lasso: dynamic decoding of musical features from fMRI data. NeuroImage. 2014;88:170–80. https://doi.org/10.1016/j.neuroimage.2013.11.017.
122. Seifert L, Lardy J, Bourbousson J, et al. Interpersonal coordination and individual organization combined with shared phenomenological experience in rowing performance: two case studies. Front Psychol. 2017;8:75. https://doi.org/10.3389/fpsyg.2017.00075.
123. Cornejo C. Intersubjectivity as co-phenomenology: From the holism of meaning to the being-in-the-world-with-others. Integr Psychol Behav Sci. 2008;42(2):171–8. https://doi.org/10.1007/s12124-007-9043-6.
124. Chartrand TL, Bargh JA. The chameleon effect: the perception-behavior link and social interaction. J Pers Soc Psychol. 1999;76(6):893–910. https://doi.org/10.1037/0022-3514.76.6.893.
125. Launay J, Dean RT, Bailes F. Synchronization can influence trust following virtual interaction. Exp Psychol. 2013;60(1):53–63. https://doi.org/10.1027/1618-3169/a000173.
126. Gallotti M, Frith CD. Social cognition in the we-mode. Trends Cogn Sci. 2013;17(4):160. https://doi.org/10.1016/j.tics.2013.02.002.
127. Schutz A. The phenomenology of the social world. Evanston: Northwestern University Press; 1967. 255 p
128. Davis T. The ties that bind: unintentional spontaneous synchrony in social interactions. In: Passos P, Davids K, Chow JY, editors. Interpersonal coordination and performance in social systems. New York: Routledge; 2016. p. 53–64.

The Social Neuroscience of Attachment

Pascal Vrtička

Abstract Attachment theory, developed by the British psychoanalyst John Bowlby and his American colleague Mary Ainsworth (Bowlby, Attachment and loss, 1969; Ainsworth et al., Patterns of attachment, 1978), aims at explaining why early interactions with caregivers have such a pervasive and lasting effect on personality development beyond childhood. Combining aspects of Darwinian evolutionary biology with social and personality psychology, attachment theory is built upon an inherent cross talk between disciplines. Attachment is conceptualized to rely upon both a behavioral system with a biological function and a cognitive substrate in terms of mental representations of person-environment interactions. Because of its comprehensive nature, attachment theory has become one of the most heavily researched conceptual frameworks in modern psychology (Mikulincer & Shaver, Attachment in adulthood: structure, dynamics, and change, 2007) and has recently inspired growing interest in the field of social neuroscience (Vrtička & Vuilleumier, Front Hum Neurosci 6, 212, 2012). Within the context of this book concerned with the missing link between neuroscience and social science, attachment theory offers a good practical example of a fruitful dialogue between disciplines helping to better understand human development. In the present chapter, I will first describe the fundamental assumptions of attachment theory and discuss their implications from an evolutionary as well as sociocultural perspective. I will then illustrate how attachment theory has inspired applied research in the field of social neuroscience and how the insights gained so far can inform possible prevention and intervention strategies in the context of mental and physical health and policy making across disciplines. Finally, I will comment on the remaining issues and future avenues of this still very young and exciting field of research termed "the social neuroscience of attachment."

Keywords Attachment theory • Avoidant attachment style • Anxious attachment style • Social neuroscience • Emotion • Cognition • Mentalization • fMRI

P. Vrtička (✉)
Max Planck Institute for Human Cognitive and Brain Sciences, Leipzig, Germany
e-mail: vrticka@cbs.mpg.de

© Springer International Publishing Switzerland 2017

A. Ibáñez et al. (eds.), *Neuroscience and Social Science*,
DOI 10.1007/978-3-319-68421-5_5

1 Attachment Theory, Evolution, and Culture

1.1 Attachment Theory

Attachment constitutes a fundamental element of human existence. Attachment theory postulates that every child is born with an innate attachment behavioral system, "a biologically evolved neural program that organizes behavior in ways that increase the chances of an individual's survival and reproduction, despite inevitable environmental dangers and demands" ([1], p. 10). In times of danger and need, such organized behavior is aimed at establishing and maintaining proximity to significant others, with proximity seeking as the primary attachment strategy. Once distress is reduced, both physically and psychologically, the attachment system is deactivated and resources are allocated (back) to other activities. Along these lines, the attachment system is also an emotion regulation system [2].

Although almost all children are born with a normally functioning attachment system, and nearly all children become attached to significant others, the quality of attachment can vary considerably. Individual differences in attachment quality depend heavily (albeit not exclusively) on the responsiveness of particular relationship partners, also referred to as attachment figures—during childhood, these are mostly primary caregivers. If children interact with available, sensitive, and responsive caregivers, they are likely to experience felt security, a psychological sense of protection and care. Experiencing felt security helps children to develop positive views of themselves and their own capacities (to successfully elicit care and to relieve stress/regulate emotions through social proximity) as well as positive views of attachment figures (as being available and sensitive to their needs). Such positive representations of the self and others are associated with the emergence of a secure attachment style.

Conversely, if caregivers are consistently physically and emotionally unavailable or disapproving, or if caregivers' behavior is unpredictable and inconsistent in times of need, a feeling of security is not attained. This lack of felt security leads to the establishment of an insecure, either avoidant (AV) or anxious (AX) attachment style, characterized by the presence of so-called secondary attachment strategies. In the case of AV, the secondary attachment strategy represents an *escape* reaction and is associated with a deactivation of the attachment system to prevent frustration and additional distress caused by attachment figure unavailability. In the case of AX, the secondary attachment strategy consists of a *resistance* response in terms of attachment system hyperactivation aiming at intensifying proximity-seeking attempts to demand or force the attachment figure's attention, love, and support [1, 2]. Thinking of the attachment system as an emotion regulation device, deactivating secondary attachment strategies are associated with emotion downregulation or suppression, whereas hyperactivating strategies are linked to emotion upregulation or intensification. Furthermore, while AV is related to a tendency to rely on intrinsic emotion regulation, AX is associated with a tendency to rely on extrinsic emotion regulation—ideal emotion regulation functioning being understood as a balanced mixture between these two strategies in the case of attachment security [1, 2]. Besides secure attachment, AV, and AX, a fourth attachment style has been described, namely, fearful- or anxious-avoidant attachment. Furthermore, attachment theory distinguishes organized/resolved (i.e., secure, AV, and AX) from disorganized/unresolved attachment [1]. Hereafter, this chapter will mainly focus on the three organized/resolved attachment styles.

Beyond particular patterns of early infant-caregiver interactions influencing attachment system functioning during childhood in the short term, attachment theory proposes that these patterns become gradually encoded as increasingly stable cognitive schemas or mental representations of the self and others, also referred to as attachment working models (AWMs). These AWMs allow an individual to predict future interactions with others and to adjust proximity-seeking attempts without always having to rethink all previous interactions. If social interactions with caregivers during childhood are fairly consistent, "the most representative or prototypical working models of these interactions become part of a person's implicit procedural knowledge, tend to operate automatically and unconsciously, and are resistant to change." These AWMs, therefore, "become core personality characteristics, are applied in new social situations and relationships, and shape attachment-system functioning in adulthood" ([1], p. 25). AWMs thus have the power to influence behavior, cognition, and emotions during various social interactions with close and distant others, and to show long-lasting effects extending beyond childhood, and to even be transmitted across generations [3].

1.2 Evolution

As described above, attachment theory proposes that all humans are born with an innate attachment behavioral system, a biologically evolved neural program that ensures survival by means of specific attachment behaviors. From an evolutionary perspective, one may thus ask why such neural program has evolved at the first place and why attachment behaviors appear to be of high importance particularly for humans. There are, of course, no simple answers to these two questions. However, one can refer to the available literature on the topic to at least partially illustrate some of the relevant aspects.

To answer the first question of why attachment behaviors may have evolved in humans, the so-called cooperative breeding hypothesis could offer some insights. Although this hypothesis is discussed controversially in terms of the socio-cognitive consequences of cooperative breeding [4], it suggests that attachment behaviors have evolved in humans during the Pleistocene due to the emergence of allomaternal care—the principal feature of cooperative breeding [5]. More specifically, Hrdy [5] proposes that "reliance on allomaternal assistance would make maternal commitment more dependent on the mother's perception of probable support from others," and that "one artifact of such conditional maternal investment would be newborns who needed to monitor and engage mothers, as well as older infants and juveniles who needed to elicit care from a range of caretakers" (p. 9).

Interestingly, Hrdy [5] also suggests that in cooperative breeders, there was a contributing, new quality of the living arrangement that added to the incidence of attachment behaviors, namely, the *prolonged period of dependence* of the young (p. 9). Compared to other hominids and mammals, human children have a very long developmental period and are highly dependent on adult caregiving [6]. This unusual human characteristic is thought to be the consequence of an increase in the complexity of daily living, the latter in turn being linked to the fact that survival and reproduction increasingly depended on the development and maintenance of social networks. One manifestation of such increase in social complexity is thought to be its relation to brain (relative to body) size, particularly the relative size of the neocortex that supports elaborated social competencies like theory of mind and language—a relation emphasized by the *social brain hypothesis* [7]. Both a complex social life and a large brain are associated with a long juvenile period in primates [8], and the function of latter period is understood to be providing the grounds for learning about the complexities of social life in order to gain access and some level of control over resources [6]. Such social learning is most likely maintained through an interaction between genes and the environment in terms of an epigenetic process and functions best in a stable, secure social setting [9].

Within this evolutionary framework, two additional hypotheses appear relevant for human attachment. On the one hand, the *developmental immaturity* or *neoteny hypothesis* assumes that attachment among humans may be a by-product of humans' prolonged neotenous state [10]—see above. Due to the extended juvenile period, it appears plausible that:

> [...] attachment, like other infantile traits, is prolonged into early adolescence and adulthood because of the relative retardation of maturational processes. If so, then the attachment system will not become dormant as children become sexually mature, as appears to occur in many other mammalian species. Instead, the system may continue to be sensitive to certain cues and signals and readily activated in contexts that resemble the infant-parent relationship (e.g., caring, safe, or physically intimate interactions) or elicit similar feelings or behaviors. ([10], p. 733)

On the other hand, the *paternal care hypothesis* posits that attachment (equated to pair-bonding) may enhance inclusive fitness by providing an additional means of protection and care for offspring [10]. This hypothesis not only states that offspring are more likely to survive to a reproductive age if they are reared in families in which the mother and father are pair-bonded but also that paternal investment is beneficial for offspring survival.

Taken together, the neoteny and paternal care hypotheses could help shedding some light on the functions attachment plays in adolescent and adult relationships. These functions could in turn help explaining why attachment appears to be so important for humans, and why it may have such pervasive and lasting effects on personality development, with attachment behaviors persistently manifesting themselves in adolescence and adulthood. In combination with the cooperative breeding hypothesis suggesting that attachment emerged at the first place due to the appearance of allomaternal care, one central aspect that is put forward concerns the prolonged juvenile period requiring more intense and longer parental investment, or care and protection provided by a stable social surrounding more generally.

1.3 Culture

Attachment theory provides a common framework describing which strategies people employ in social relationships and what the evolutionary origins and functions of such behaviors could be. One should, however, always keep in mind that this attachment framework is based upon certain assumptions which reflect particular sociocultural attitudes.

It has been generally argued that "many well-established theoretical positions in psychology cannot be as widely generalized as their authors assume" ([11], p. 2), because these positions were developed in, and mainly tested on, individuals from *Western*, *educated*, *industrialized*, *rich*, and *democratic*—in short, *WEIRD*—cultures [12]. In the context of attachment theory, particularly the terms "bonding," "attachment," "critical period," and so forth have been referred to as ill-defined, culturally decontextualized, and "inadequate to describe and to contain the experiences of mothering and nurturing under conditions of extreme scarcity and high risk of child death" ([11], p. 32).

Furthermore, the universality—and thus, cross-cultural validity—of three "core hypotheses of attachment" has been called into question [13]. These core hypotheses state that (1) caregiver sensitivity leads to secure attachment (*sensitivity hypothesis*), (2) secure attachment leads to later social competence (*competence hypothesis*), and (3) children who are securely attached use the primary caregiver as a secure base for exploring the external world (*secure base hypothesis*). By comparing security in the United States and Japan, Rothbaum et al. [13] describe fundamental cultural differences in the way sensitivity, competence, and secure base use is defined, and outline the consequences of such discrepancy on intercultural understanding. For example, they note that caregiver sensitivity is expressed differently in Western versus Eastern cultures (i.e., distal versus direct forms of contact) and that the objectives of sensitivity vary as well (fostering of exploration and autonomy versus dependency and emotional closeness). Similarly, in the United States, social competence is associated with exploration and autonomy, while the Japanese culture particularly values the conservation of social harmony through dependence and emotional restraint. Finally, while use of the secure base is mainly linked to individuation and autonomous mastery of the environment in the United States, it is primarily associated with loyalty and interdependence in Japan [13]. Altogether, these differences in the way sensitivity, social competence, and the use of the secure base are culturally defined can easily lead to misconceptions in the way that attachment patterns across cultures are not simply viewed as different but directly compared and judged based on local cultural perceptions of, and beliefs about, relationships.

Given the above considerations, attachment theory and the scientific results derived from it (see below) should always be regarded with caution, particularly regarding sociocultural aspects. This especially pertains to the tendency to regard Western patterns as "the norm" and to use such norm to interpret patterns observed in other cultures, mostly in negative terms. As Rothbaum et al. [13] put it nicely, "an awareness of different conceptions of attachment would clarify that relationships in other cultures are not inferior but instead are adaptations to different circumstances" (p. 1101).

2 The Social Neuroscience of Attachment

Social neuroscience, also called social cognitive affective neuroscience, is a relatively new research field devoted to advancing the understanding of how biological systems, and in particular the human brain, implement social processes and behavior [14]. Social neuroscience is highly interdisciplinary, which is reflected in the multilevel experimental approach that is applied to investigate the neural basis of social behavior [14]. Due to such strong heterogeneity, it may be helpful to first conceptually define what the most likely human brain substrates of attachment may be to better understand the described findings—focusing mainly on functional magnetic resonance imaging (fMRI) data—in the remainder of this part.

2.1 The Human Attachment System

Attachment theory postulates the presence of an attachment behavioral system in terms of a biologically evolved neural program that organizes behavior in times of need, particularly through proximity seeking. Accordingly, a prototypical attachment interaction "is one in which one person is threatened or distressed and seeks comfort and support from the other" ([1], p. 19). The attachment system can therefore be viewed as being made up of (at least) two different motivational components, namely, a "prevention" component aiming at "inhibiting" behaviors associated with an increased probability of danger or injury in relation to threats or stressors, and a "promotion" component seen as maintaining an approach-oriented motivation to foster closeness to others and the attainment of felt security [1]. This view accords with the *phylogenetic perspective of social engagement and attachment* proposed by Porges [15], which suggests that there is a dynamic balance between social aversion tendencies maintained by more primitive survival-enhancing systems (especially sympathetic fight-or-flight circuits), and social approach tendencies that promote a sense of safety through close social interactions [16]. Such processing of information in terms of safety versus danger, which is thought to be intrinsically linked with behavioral tendencies to either approach or avoid a stimulus, most likely occurs rapidly and automatically (sometimes even unconsciously) in core

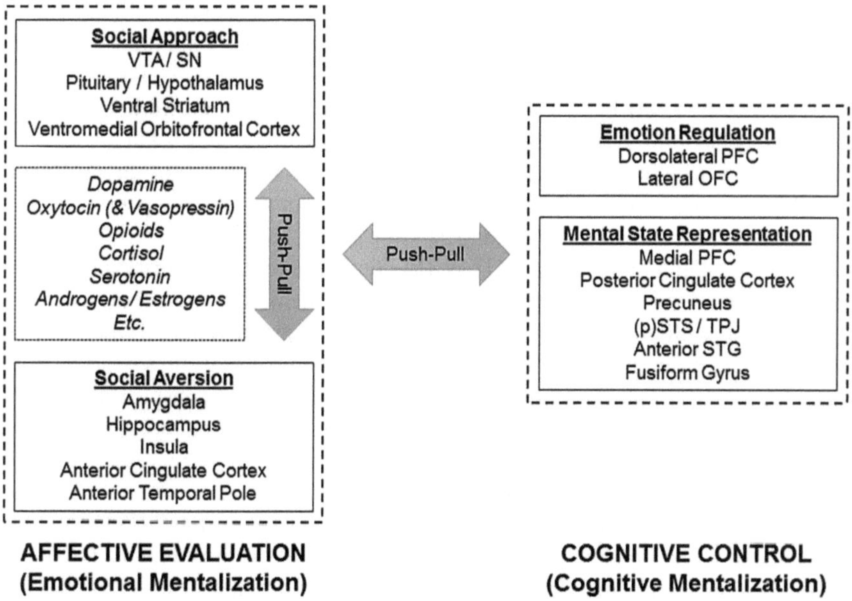

Fig. 1 Functional neuroanatomical model of brain areas and their underlying neurotransmitter/neuropeptide systems involved in human attachment. *VTA* ventral tegmental area, *SN* substantia nigra, *PFC* prefrontal cortex, *OFC* orbitofrontal cortex, *(p)STS* (posterior) superior temporal sulcus, *TPJ* temporoparietal junction, *STG* superior temporal gyrus (adapted from Vrtička and Vuilleumier [18], p. 5)

social-affective stimulus appraisal brain networks [17]—a notion also reflected in the context of AWMs as being part of a person's implicit procedural knowledge (see Sect. 1.1). Vrtička and Vuilleumier [18] have therefore proposed that the human attachment system comprises an affective evaluation network made up of a social approach and a social aversion component that are in a dynamic balance (Fig. 1).

Per the available literature and the idea of a "neuroception of safety" [15], we argue that the social approach component encodes (mutual) social interactions as innately rewarding—and thus counteracting fear tendencies—in a neural reward-related, primarily dopaminergic network including the ventral tegmental area (VTA), the substantia nigra (SN), the ventral striatum (VS), and the ventromedial orbitofrontal cortex (vmOFC) [18]. It is, however, likely that the activity within this social approach component is influenced by other neurotransmitters/neuropeptides, particularly oxytocin and vasopressin (originating from the pituitary/hypothalamus region), endogenous opioids, and serotonin, which all show strong interconnections to, and anatomical overlap with, the dopaminergic reward circuits [18]. Furthermore, the available fMRI literature suggests that this social approach component is not specifically activated during attachment interactions but by many kinds of "social interactions with beloved ones (children, parents, partners), friends, or any "significant" (e.g., contextually relevant) other person with a cooperative relationship (e.g., joint task)," which are "all associated with the experience of positive emotions and increased activity in the reward circuits" ([18], p. 6). Consequently, although it may be possible to conceptually dissociate attachment interactions from other approach-oriented social behaviors thought to constitute other behavioral systems [1, 19, 20], such dissociation may be hard to maintain on a brain area and network level.

Other behavioral systems like affiliation, the sex drive, and romantic love all conceptually have in common that they occur during the absence of threat. Hence, an affiliation interaction is defined as an interaction "in which both people are in a good mood, do not feel threatened, and have the goals of enjoying their time together or advancing common interests" ([1], p. 19). Similarly, for the sex drive and romantic love, threat is not a principal motivational component; it is the seeking of sexual gratification with the ultimate goal to ensure the propagation of the species. However, the difference between the two latter systems is that the sex drive motivates the seeking of sexual gratification nonspecifically for any conspecific, whereas romantic love promotes a focus of the mating effort on preferred conspecifics [19, 20]. Finally, caregiving/maternal love/compassion is thought to represent "a broad array of behaviors designed to reduce suffering and/or foster growth and development in a significant other such as a child or relationship partner" ([21], p. 209), particularly if the other is in need [22]. Therefore, as for attachment, there is a notion of threat for caregiving, albeit the motivation for caregiving is to alleviate another person's suffering by providing comfort and support, not by seeking comfort and support for oneself. The literature [19, 20] furthermore suggests that some of the above approach-oriented behavioral systems can be specified on the level of the underlying primary neurotransmitters/neuropeptides, with the sex drive being linked to the estrogens and androgens (i.e., testosterone), romantic love to the catecholamines and particularly dopamine, and attachment/caregiving to oxytocin and vasopressin—although already here, a strong interrelatedness of these systems is acknowledged. In addition, when testing the putative selective brain substrates of the sex drive versus romantic love versus maternal love/caregiving, there are many overlaps and common activation patterns [23], particularly within the abovementioned reward circuits, and the same areas are also found activated in fMRI studies on compassion [22] and social tasks likely tackling more basic affiliation tendencies with unknown others [24]. Overall, as pertaining to the social approach component of the attachment system, it is unlikely that the latter constitutes a specific brain network under the control of a single or a few distinct neurotransmitter(s)/neuropeptide(s). When it comes to investigating the neural basis of the promotion aspect of attachment, it therefore appears better to assess the functioning of the social approach component by means of individual differences in attachment orientations or AWM implementation, rather than attachment as a general construct.

The same notion of non-specificity also appears to hold true for the prevention aspect of attachment maintained by the social aversion component. Bowlby already acknowledged that attachment-system activation will likely occur not only through social- and attachment-related threats but through threats endangering bodily integrity or representing an immediate danger for survival broadly speaking [1]. On the brain level, this is reflected in activation of a set of regions typically associated with nonsocial negative affect, physical pain, stress, and fear but also with various aversive social responses such as psychological pain related to social exclusion/rejection, social stress, social conflict, or sadness due to a social loss [18]. Brain areas thought to mediate such negative social- and nonsocial-emotional processes include the amygdala, the hippocampus as important part of the negative feedback loop regulating the hypothalamic-pituitary-adrenal (HPA) axis (primarily mediated through cortisol), the insula, and the anterior cingulate cortex (ACC), as well as the anterior temporal pole (ATP) [18]. Hence, when assessing the social aversion component of the attachment system, dissociation between activity to more general (nonsocial) threats as opposed to socially relevant dangers in the context of attachment will be difficult, and an approach considering individual differences as pertaining to attachment may be more promising.

Besides an affective evaluation network maintaining rapid, automatic, and often unconscious appraisals of emotional information to mediate basic approach versus avoidance behaviors, the attachment system may also comprise a more controlled network maintaining conscious representations about others, as well as behavioral regulation and decision making [17], and we denoted this network as the cognitive control network [18]. In our understanding, the cognitive control network has two main functions. On the one hand, it is involved in the volitional control of emotions and social behaviors and thus fulfills a regulatory role. The corresponding "cold" cognitive computations are thought to underlie various regulatory mechanisms such as reappraisal, suppression, and distraction and to be based on activity primarily in lateral ventral, middle, and dorsal prefrontal and orbitofrontal cortex (PFC/OFC). On the other hand, the cognitive control network also maintains the representation of internally focused information about others through processes related to theory of mind (ToM) [17, 25]. As opposed to the emotional route for the understanding of others hypothesized to comprise elements such as emotion contagion/mirroring, empathy, and compassion being part of the affective evaluation network of attachment, this cognitive component representing the cognitive route for the understanding of others is thought to mainly rely on rational inferences about the mental states and intentions of others [22, 25–27]. The latter processes are suggested to be (mainly but not exclusively) encoded by an array of cortical midline areas such as the medial OFC and PFC, the posterior cingulate cortex (PCC), and the precuneus, as well as lateral temporal regions like the superior temporal sulcus (STS), the temporoparietal junction (TPJ), the anterior superior temporal gyrus (aSTG), and the fusiform gyrus (FG) [17, 18] (Fig. 1).

In our view, there is not only a dynamic balance between social approach and aversion tendencies as part of the affective evaluation network of attachment. A similar "push-pull" mechanism also appears to be present between the affective evaluation and cognitive control networks [18]. Fonagy and Luyten [25] refer to this second equilibrium as a balance between emotional and cognitive *mentalization* and state that there is a "switch point" corresponding to behavioral changes "from flexibility to automaticity, … that is from relatively slow executive functions … to faster and habitual behavior …" (p. 1367). This view accords with the proposition that stress/saliency importantly determines whether information is processed through the affective evaluation or cognitive control systems [28, 29]. In other words, the higher the stress (arousal), urgency, or novelty of a situation, the more the "switch point" between different modes of processing might be shifted toward an activation of the affective evaluation system [25]. As we have noted before [18], a shift toward emotional evaluation under threat would, in evolutionary terms, be normally adaptive because it promotes immediate and automatic self-protective (and thus socially aversive) reactions. However, "in interpersonal settings where cognitive mentalization is a necessary prerequisite and danger neither vital nor immediate [7], a too strong or exclusive reliance on affective evaluation might represent an insufficient or inappropriate strategy" ([18], p. 9). Put differently, in our modern society where there are much fewer direct and fundamental threats to survival and where the complexity of the social living arrangements is very high, the cognitive route for the understanding of others plays a much more important role, and a too strong reliance on emotional mentalizing may be associated with a higher incidence of problems in social-emotional functioning [25]. It therefore appears vital to better understand what could cause shifts in the "switch point" between the different modes of emotional versus cognitive processing.

In the context of attachment, it is interesting that individual differences in attachment-system functioning are seen as one possible determinant of such "switch point" shifts [25]. Although the corresponding theory has been developed in association with borderline personality disorder (BPD), it can be regarded as more generally predicting that a shift of the "switch point" toward emotional mentalization coincides with a low threshold of attachment-system activation. Not surprisingly, BPD patients are reliably found to be classified as anxious (or anxious avoidant) in terms of their attachment orientation [30].

Taken together, we think that there is no single attachment system in the human brain dedicated to processing information and coordinating behavior specifically related to attachment interactions. What we propose is that attachment draws upon functions of (at least two) distinct networks generally maintaining affective versus cognitive processes—the latter being further differentiable into social approach versus aversion and emotion regulation versus mental state representation components, respectively—and that individual differences in attachment orientations and/or AWMs generally influence which network and component are preferentially activated and how the networks relate to each other in terms of a dynamic balance.

2.2 *Neuroimaging Findings on Attachment-System Functioning in Humans*

Building upon Sect. 2.1, here I will mainly illustrate fMRI findings pertaining to the question of how individual differences in attachment relate to the functioning of the affective evaluation and cognitive control networks. The discussion will start with the affective evaluation network differentiated into social approach versus social aversion, followed by the cognitive control network dissociated into emotion regulation versus mental state representation.

2.2.1 Social Approach

Two fMRI studies suggest that particularly individual differences in AV may have an influence on the social approach neural system (SApNS). In a first study [24], we found that healthy adults' brain activity in the VS and VTA to positive social feedback (from unknown others) was selectively decreased as a function of the participants' AV scores. Hence, the activity within the SApNS in a context of affiliation was lower in the more avoidantly attached participants, which suggests blunted processing of social reward for this attachment orientation. One year later, a very similar pattern of reduced SApNS activation emerged in a study during which healthy mothers were shown images of their own versus unknown babies [31]. In avoidantly attached mothers, brain activity in the hypothalamus was reduced during the exposure to own versus unknown babies in general, and in the VS and medial OFC specifically to own happy babies. Furthermore, activity in the hypothalamus during the baby face task was positively correlated with peripheral oxytocin levels during an independent mother-child interaction and was generally lower during this mother-child interaction in avoidantly attached mothers. AV was therefore found to be associated with blunted reward-related activity within the SApNS also in a caregiving setting involving mothers and their own babies, and such brain activation pattern was likely (partially) mediated by oxytocin.

One additional fMRI study [32] provides further, albeit indirect support for decreased social reward-related brain activity in relation to AV, namely, in the context of interpersonal closeness as measured with the "inclusion of the other in the self" (IOS) scale, which assesses the "feeling close" and "behaving close" aspects of social interactions [33]. In this study, participants played a card guessing game for shared monetary outcomes with three partners: a computer, an unknown confederate, and a friend. They rated their excitement of winning money with each partner and provided scores on the IOS scale of their friend. Behavioral results revealed that the excitement of winning (and sharing the monetary reward) was highest for trials with the friend. The same pattern was observed in the VS and vmOFC where activity was highest for winning trials with the friend. Furthermore, there was an intriguing association between IOS scores for the friend and VS activity during winning trials as a function of the three partner types. Whereas brain activity was consistently high during winning trials for participants scoring low on IOS, a computer < confederate < friend effect was present for participants scoring high on IOS. Put differently, low interpersonal closeness seemed to have sustained or overemphasized nonsocial positive reward representation while decreasing sensitivity to social positive reward encoding in different social contexts [34].

Further neuroimaging evidence for an association between AV and reduced responsivity of the SApNS comes from a positron-emission topography (PET) study using a μ-opioid receptor (MOR) ligand [35]. The authors report a negative relation between AV and MOR availability in (among others) the dorsal striatum and OFC, which could indicate a possible role of opioids in AV related to reward. Interestingly, there are several other observations that suggest a connection between the opioid system and AV. For example, a link between the minor allele (G) of the μ-opioid receptor polymorphism OPRM1 A118G, self-reported AV, and the tendency to become engaged in affectionate relationships has been described [36]. Furthermore, the abuse of heroin (but not drugs that do not influence the opioidergic system, such as ecstasy or cannabis) has been associated with (fearful-) avoidant attachment [37]. Finally, more generally speaking, disruption of the endogenous opioid system by opiate addiction was linked to antisocial behavior [38]. Besides oxytocin, the opioids may therefore also be involved in mediating blunted reward-related responses in the SApNS in association with AV.

The above neuroimaging findings are in line with independent behavioral results showing that AV is associated with lower pleasantness and arousal ratings of positive social (but not nonsocial) images and videos [39, 40], which further corroborates the notion that AV is linked to a general deficit in the experience of positive, reward-related emotions in a social context. A corresponding pattern has also been reported by Troisi et al. [41], who found that AV was correlated with social anhedonia, i.e., a diminished capacity to experience social pleasure. We have similar unpublished data in a large adult sample linking AV particularly to diminished scores on the extraversion big five personality trait scale, with extraversion being mainly associated with a state of obtaining gratification from outside through human interactions.

Altogether, the so far available data on AV and social approach suggest that this attachment orientation may be characterized by a basic deficit in the capacity to experience social reward. The etiology of such pattern, however, remains unclear. It could be that avoidantly attached individuals have a fundamental dysfunction of the SApNS, which manifests itself in altered social behaviors that are captured by the AV dimension. At the same time, the SApNS of avoidantly attached individuals could be functional initially and only become affected by negative attachment experiences during early childhood, which would mean that the observed behavioral and neural patterns are the consequence of a learning/adaptation process [41, 42]. Future research utilizing longitudinal and gene by environment association methods is needed to resolve this question.

In the case of AX, not much data regarding social approach is available up to date. Although one study [43] reported relations between AX and brain activity during positive social information processing (i.e., masked happy faces), these relations do not anatomically fall within the SApNS.

Another study [44] described a link between AX and brain activity in the VS and vmOFC, although related to prediction-error activity in response to a social reward, and thus not social reward per se. The experiment consisted in a task during which participants' expectations for their romantic partners' positive regard of them were confirmed or violated, in either positive or negative directions. What emerged in the VS and vmOFC was a relation between AX and activity during the receipt of unexpected positive feedback. Furthermore, the authors report an inverse relation in the VS between brain activity to unexpected positive feedback and partner trust. These findings are discussed according to attachment theory in a sense that "AX represents an uncertainty about relational outcomes and the extent to which partners reciprocate romantic sentiment" ([44], p. 7). In other words, while anxiously attached participants expect or fear rejection by their partners, they at the same time hope for closeness and care, motivations which likely manifest themselves by activation of the SApNS during unexpected social confirmation. Social reward-related brain activity in anxiously attached individuals may thus not simply reflect the processing of positive (mutual) social outcomes, but rather the content of AWMs, which comprise both expectations about specific relationships with others and self-representations.

The relative absence of fMRI data on SApNS functioning associated with AX probably mirrors the fact that AX is mainly linked to hyperactivating secondary attachment strategies especially in the context of negative social information processing (see Sect. 2.2.2). However, as the goal of such hyperactivating strategies is to demand or force the attachment figure's attention, love, and support, the attainment of, or failure to attain, this goal is likely to be encoded in the SApNS—as preliminary evidence suggests (see above). More research on such aspects of AX therefore is highly encouraged.

2.2.2 Social Aversion

For AV, there is some evidence for relatively decreased social aversion neural system (SAvNS) activation in the context of social exclusion/rejection [45]. In this specific case, social exclusion was induced by a Cyberball paradigm—a virtual ball tossing game during which the participant sees two people playing with a ball either including or excluding him/her into the game. Reduced anterior insula and dorsal ACC activity in avoidantly attached participants was interpreted as reflecting the weaker social need for closeness and weaker distress elicited by social rejection in these individuals. We found a similar decrease in anterior insula, ventral ACC, and amygdala/hippocampus activation during incongruent social feedback processing (reflecting social conflict) in a population of healthy adolescents [46]. Furthermore, although not directly overlapping with the SAvNS, another study [47] found that masked sad faces induced a weaker response in the somatosensory cortex (BA 3) in avoidantly attached participants, which was attributed to their habitual unwillingness to deal with partners' distress and needs for proximity.

At the same time, AV has been observed to positively correlate with amygdala activation to negative (fearful and angry) facial expressions [48] and with lower structural integrity of the amygdala that was associated with chronic amygdala hyperactivity [49]. Also, Strathearn et al. [31] report that avoidantly attached mothers had stronger insula activation when seeing own sad babies. Furthermore, hippocampus gray matter density was found to be reduced as a function of AV scores, which was related to reduced glucocorticoid stress regulation capacity [50]. Moreover, functional resting state connectivity between the dorsal ACC and hippocampus was reported in avoidantly attached individuals after they listened to prototypical dismissive (i.e. avoidant) narratives and such pattern was linked to increased sensitivity to dismissing content [51].

The abovementioned findings suggest the presence of two opposing mechanisms in association with AV. On the one hand, deactivating secondary attachment strategies characterizing AV could entail a relative insensitivity to negative social information, for example, social rejection/exclusion, thereby preventing activation of the SAvNS. On the other hand, AV may lead to increased sensitivity to negative social information associated with decreased capacity to regulate the thereby caused distress, manifested in increased SAvNS activation as well as reduced amygdala and hippocampus structural integrity. SAvNS implication as a function of AV therefore appears to not simply reflect (de)activation to negative social information but also the ability to cope with such stressors (see Sect. 2.2.3). Moreover, the kind of negative social information processed seems to matter. Future research on SAvNS functioning in relation to AV should therefore employ paradigms where the content and degree of negativity of social negative information are directly manipulated.

What is concerning AX, this attachment orientation has been reliably and repeatedly associated with increased activation and modulation of structural integrity in the SAvNS. In our studies on social feedback in adults and adolescents [24, 46], we observed increased activity in the amygdala/hippocampus as well as anterior insula and ventral ACC for social punishment and social conflict. These findings accord with three other studies reporting that anxiously attached participants had increased amygdala activity to negative emotional faces [48, 52] and increased anterior insula and dorsal ACC activation during social exclusion induced by the Cyberball game [45]. Another fMRI investigation reported an enhanced hippocampus response in mothers who scored low on a maternal care measure when they were listening to own versus unknown baby cries [53], and the same brain area was found to have decreased gray matter volume as a function of AX scores [50]. These data generally accord with the role of the hippocampus in stress responses as part of the HPA axis [54]. High attachment-related anxiety was also found associated with increased amygdala gray matter volume [52] but decreased gray matter in the ATP in the context of affective loss [55]. The above neuroimaging data are corroborated by independent behavioral results showing increased arousal and decreased controllability/dominance ratings particularly of social negative images as a function of AX [40].

These data suggest that hyperactivating secondary attachment strategies in anxiously attached individuals are generally related to increased SAvNS activity during negative social information processing. Regarding prevention and intervention strategies, such findings may imply that in anxiously attached individuals, one promising avenue is to work on decreasing the extent of subjectively perceived negativity and/or threat of negative social clues (see also Sect. 2.2.3). What is concerning structural findings, there is evidence for an impact of AX in terms of chronic stress on the SAvNS, although both gray matter increases and decreases have been reported. More research is therefore needed to further elaborate on particularly the long-term consequences of attachment-system hyperactivation.

2.2.3 Emotion Regulation

The attachment behavioral system can also be understood as an emotion regulation device. Along these lines, both insecure attachment orientations are generally associated with impaired emotion regulation capacities [1]—although AV and AX can also be seen as specific adaptations to the particular environments they emerged in, and therefore represent viable alternatives to attachment security (see, for example, the *social defense theory* [56]). Nonetheless, successful emotion regulation is thought to depend on learning processes involving extrinsic emotion co-regulation and a subsequent internalization with the emergence of self-regulation [2]. Both processes are hampered in insecure early interactions but with different outcomes. In the case of AV, deactivating secondary attachment strategies lead to an overemphasis of intrinsic self-regulation, but the latter is only partially effective because the successful application of constructive emotion regulation strategies has not been learned through extrinsic co-regulation. Avoidantly attached individuals are therefore more likely to employ response-focused emotion regulation such as suppression, which can reduce overt emotional reactions, but is not very efficient in regulating emotion processing at the first place. Furthermore, suppression is only possible up to a certain point and/or in situations where the emotion-eliciting stimulus can be ignored or avoided, and not beneficial in the long run as emotions are just squelched but not resolved. In the case of AX, emotion co-regulation is emphasized, which precludes a proper internalization and entails a lack of self-regulation. Furthermore, anxiously attached individuals tend to upregulate their emotions through hyperactivating secondary attachment strategies, which puts them in a chronically high arousal state requiring more emotion regulation as such [1]. The question for the social neuroscience of attachment therefore not only is how such avoidant and anxious tendencies are reflected in neural activation patterns during emotion regulation but also whether the so far obtained findings can inform what kind of intervention and/or prevention may work best to overcome these emotion regulation biases.

There is evidence from several neuroimaging experiments that attachment insecurities overall are associated with less effective emotion regulation capacities (in terms of simultaneous high cognitive control and emotional evaluation area activation) within specific contexts. For example, in female participants anticipating an electric shock while holding their husbands' hand, low marital quality (likely reflecting a more insecure attachment) was associated with increased PFC as well as anterior insula and hippocampus activation pointing to emotion regulation difficulties [57]. A similar pattern of increased lateral PFC and concomitantly increased amygdala and hippocampus activation to negative attachment scenarios was observed in participants with unresolved (i.e., fearful-avoidant or disorganized) attachment [58]. Finally, in a Stroop task using attachment-related words, an insecure attachment style was linked to high dorsolateral OFC and PFC activity but poor task performance, which was associated with less efficient cognitive control capacities and thus heightened vulnerability to distraction by attachment-relevant emotional information [59].

Conversely, in the study mentioned above during which female participants were anticipating an electric shock while holding their husbands' hand, high marital quality (likely reflecting a more secure attachment) was associated with lower SAvNS activity [57]. Similar evidence is available from another study [60] during which female participants saw images of their partner while they received painful stimuli, which lead to decreased pain ratings and weaker SAvNS activity. Such data suggest that the "prevention" aspect of attachment may (at least partially) be maintained through endogenous opioids (involved in endogenous pain analgesia). Furthermore, there is preliminary evidence that using prototypical secure attachment information through priming procedures can decrease distress-related activity particularly in the SAvNS. In a fMRI experiment using social (i.e., negative emotional faces) and linguistic threat, Norman and colleagues [48] observed increased amygdala activation as such, as well as a positive correlation between amygdala activation and the degree of participants' attachment insecurity (AV and AX). However, in a control group that underwent secure attachment priming, amygdala activity to threat was strongly decreased, and there was no longer an association between amygdala activity and attachment insecurity. Although it is still not precisely known how and where attachment security primes as such are neurally processed, whether they are more effective when perceived explicitly versus implicitly, and how the processing of security primes itself is modulated by individual differences in attachment style [61], working with prototypical attachment security information appears to represent one promising avenue for enhancing emotion regulation capacities related to attachment insecurity.

When more specifically looking into the neural patterns associated with emotion regulation in avoidantly attached individuals, preliminary evidence supports the view of an association with preferential use of suppression [18]. In an fMRI experiment including positive and negative social versus nonsocial images [62], we asked participants to either attend naturally to emotional scenes (NAT) or to use one of two emotion regulation strategies. In some trials, participants were asked to suppress any visible expression of internally arising emotion elicited by the images (ESUP). In other trials, they were asked to cognitively reassess the meaning of emotional images through cognitive reappraisal (REAP). We observed heightened cognitive and emotional conflict (ACC activation) in combination with increased regulatory inhibition (lateral and medial dorsal PFC) during spontaneous viewing of social-emotional scenes. Furthermore, during REAP, amygdala activation to negative social images decreased for low but not high avoidantly attached participants. Finally, during ESUP, AV was associated with stronger responses to positive social images in the supplementary motor area (SMA) and caudate, implying stronger regulatory efforts with the successful use of suppression. These fMRI data suggest that AV is associated with a preferential use of emotion suppression in interpersonal/social contexts and that reappraisal may not work as regulatory strategy. This pattern may help understanding why avoidantly attached individuals tend to become highly emotional when their preferred regulation strategy of suppression fails or cannot be employed.

Another investigation [63] examined brain activity in participants who were told to either think or stop thinking about negative relationship scenarios. In avoidantly attached participants, the authors report stained activity in subcallosal cingulate and medial frontal gyrus (BA 9) during both experimental conditions, which they interpret as a failure of task-induced deactivation. Although it is not entirely clear how such deactivation was maintained (i.e., through suppression, reappraisal, and/or other emotion regulation strategies), these findings generally corroborate the notion that AV is associated with concomitantly increased activity in the SAvNS and emotion regulation areas, which again points to a relative ineffectiveness in emotion regulation during the exposure to social negative information—information that avoidantly attached individuals usually try to literally avoid.

Finally, a longitudinal study [64] revealed that infant attachment status at 18 months predicted neural responding during the upregulation of positive affect 20 years later. More precisely, predominantly avoidantly classified adults showed greater activation in prefrontal regions involved in cognitive control and reduced co-activation of nucleus accumbens with prefrontal cortex, consistent with relative inefficiency in the neural regulation of positive affect.

Overall, these findings pertaining to AV suggest that this attachment orientation indeed entails altered emotion regulation capacities and a bias toward the use of suppression. Interestingly, suppression appears to be employed by avoidantly attached individuals not only for social negative but also for social positive information. While the reported findings are still preliminary, they may already be informative for therapeutical purposes. For improving regulation of negative social emotions in avoidantly attached individuals, it may be beneficial to focus on weakening the bias toward the use of suppression by training other emotion regulation strategies. Yet, positive social emotions should be considered as well, as the latter also seem to be suppressed and/or poorly regulated, but vital for successful and mutually agreeable social interactions.

What is regarding AX, the so far available neuroimaging data are less conclusive. The study [63] asking participants to either think or stop thinking of negative relationship scenarios also revealed a neural pattern for AX. More specifically, the authors report that anxiously attached participants showed increased activity in the ATP, the hippocampus, and the dorsal ACC when thinking about negative emotions but less activity in the OFC when suppressing these thoughts. Moreover, activity in the ATP and the OFC was inversely correlated. These results imply that AX entails stronger activity in the SAvNS during "normal" processing of attachment-related information and altered regulatory capacities to inhibit such processing during emotion regulation. Although generally according with the employment of hyperactivating secondary attachment strategies in association with AX, these data can unfortunately not provide any specific information on emotion regulation strategy use. Similarly, in our own investigation [62] comprising two emotion regulation strategies (ESUP, REAP) and one natural emotion processing condition (NAT), we only found evidence for increased amygdala activation for negative social images during NAT in anxiously attached participants, but no specific brain activation patterns during ESUP and/or REAP. In contrast to AV, our findings pertaining to AX may nonetheless suggest that anxiously attached participants could successfully employ ESUP and REAP strategies if properly instructed how to do so. However, more research is clearly needed to further specify the neural mechanisms underlying emotion processing and regulation in association with AX.

2.2.4 Mental State Representation

Attachment theory postulates that attachment behavioral system functioning relies on AWMs which comprise prototypical mental representations of the self and others. Furthermore, it has been suggested that individual differences in attachment influence the degree to which incoming information is processed, either more cognitively relying on mechanisms like ToM by the cognitive control network or more emotionally by the affective evaluation network. So far, we have seen how individual differences in attachment may influence activity in the social approach and aversion neural systems as parts of the affective evaluation network and how they could affect emotion regulation as one part of the cognitive control network—the latter two components being interconnected through modulatory influences. The remaining question is how attachment orientations may shape mental state representation about the self and others and how this could relate back to emotional evaluation and/or cognitive regulation.

There is general evidence that emotional and cognitive mentalization may be in a dynamic balance with each other during processes relevant to attachment like romantic and maternal love [18]. However, both increases and decreases in cognitive mentalization (versus emotional mentalization) have been reported, and it remains poorly understood how such processes specifically relate to AV and AX.

In a recent study [65], we for the first time explicitly addressed this question by asking participants (adolescents) to attribute positive and negative trait adjectives to themselves or their best (same-sex) friend. Adjective attribution (i.e., mental state representation) was reliably associated with activity in an extended cognitive and emotional mentalizing network comprising cortical midline structures, lateral anterior and superior temporal cortex, as well as VS/caudate and amygdala/hippocampus. By subsequently looking for correlations between brain activity and attachment-derived self- and other-models, we observed significant effects for the self-model associated with AX. The more negative the participants' self-model was, the more activity we observed in the amygdala/hippocampus, the ATP/aSTG, the (pre)cuneus, the dorsolateral PFC, the fusiform face area (FFA), and the cerebellum during positive and negative adjective self-attribution, but the less activity was present in those areas during negative adjective attribution to their best friend. What these findings suggest is that thinking about the self and a close other entailed concomitant activation decreases and increases in both the affective evaluation (especially the SAvNS) and cognitive control (emotion regulation and mental state representation components) networks associated with AX. Interestingly, self-representations (both positive and negative) appeared to have been enhanced, while negative close other-representations were reduced. However, it remains to be seen how such findings generalize across other populations, and whether they also hold for adjective attribution to different, closer versus more distant others. Also, the above data pertain to internally driven self- and other-representations and not to more complex ToM processes employed during the exposure to external stimuli. Finally, the context within which adjective attribution was carried out in our study was relatively stress-free. In terms of the "push-pull" between cognitive and emotional mentalization, future investigations should also look at different degrees of stress that may affect the switch point, either as such or as a function of individual differences in attachment.

2.2.5 Summary

The proposition we put forward in 2012 [18] of the attachment system in humans not representing a single neural entity, but drawing upon an extended network of brain areas composed of (at least) an affective evaluation and a cognitive control network, is maintained here (Fig. 1). Accumulating new evidence bolsters and further specifies the basic patterns suggested previously (Table 1) but also raises new questions and sustains the need for more research. Some of the potential future avenues are outlined below.

3 Remaining Issues and Future Avenues

So far, most research on the neural basis of attachment has focused on the investigation of relations between (trait) attachment and brain activity in cross-sectional participant samples. While this approach revealed many interesting and valuable findings, the etiology of the observed patterns in humans remains poorly understood. Because attachment is probably best described as an environment x gene interaction process [9], and longitudinal designs are complicated and difficult to implement, future studies employing epigenetic methods appear well suited to close this gap. In the context of stress and anxiety, animal models have already shown epigenetic modification of the glucocorticoid receptor gene in offspring as a function of early caregiver interactions [66]. In humans, a possible link between methylation of the promoter region of OXT (a precursor protein that is synthetized to produce oxytocin and neurophysin I and thus presumably linked to higher oxytocin expression) associated with several overt measures of sociability and a greater incidence of secure attachment has also been reported [67]. Generally speaking, early adverse experiences (likely reflected by an insecure attachment style) are thought to be linked to alterations in the functioning of particularly the affective evaluation network, and within this network especially the VS and amygdala, probably related to altered gene expression of the dopamine, oxytocin, and glucocorticoid systems [68, 69]. More research is also needed on the endogenous opioid

Table 1 Summary of activation decreases and increases within the human attachment system as a function of individual differences in attachment avoidance and anxiety

		Attachment avoidance	Attachment anxiety
Affective evaluation	Social approach	Decreased neural representation of social reward in several contexts. Likely mediated through dopamine, oxytocin, and/or endogenous opioids	Probably increased neural representation of social reward per se and likely also reward prediction errors
	Social aversion	Activation decreases and increases reported in different contexts. Probably a combination of deactivating secondary attachment strategies and emotion regulation capacities (i.e., deactivation is applied but can only be maintained up to a certain degree)	Activation increases as well as modulation of structural integrity likely related to hyperactivating secondary attachment strategies. Probably mediated through stress/arousal (HPA axis and cortisol/glucocorticoid signaling)
Cognitive control	Emotion regulation	"Impaired" emotion regulation (concomitantly high activity in emotion regulation and affective evaluation areas). Evidence for preferential use of suppression for both positive and negative emotions. Relative efficiency but high emotionality if preferential emotion regulation strategy cannot be employed or fails	"Impaired" emotion regulation (concomitantly high activity in emotion regulation and affective evaluation areas). Mainly associated with increased activity during natural viewing when no emotion regulation strategy is employed. However, preliminary evidence that emotion regulation can be efficient if properly instructed
	Mental state representation	No clear evidence so far	Increased positive and negative self-representation as well as decreased negative other-representation (in adolescents). However, no clear evidence regarding theory of mind (ToM) during exposure to external information so far

HPA hypothalamic-pituitary-adrenal (axis), *ToM* theory of mind. "Impaired" emotion regulation refers to the fact emotion regulation related to AV and AX can be found decreased within certain contexts, but nonetheless be operational in other contexts—again emphasizing that AV and AX represent specific adaptations to the environment they emerged in and may thus constitute viable alternatives to attachment security [56].

system, as the latter also seems to be altered in insecurely attached individuals, particularly in the case of AV [35].

In addition, most research on the neural basis of attachment has employed experimental designs only measuring one person at a time, despite attachment being an interpersonal process from the very beginning. The emergence of a secure versus insecure attachment style is conditional to the caregiver's reactions to the child, and "normal" functioning of the attachment system as an emotion regulation device depends on successful co-regulation experiences that are later on internalized and used for self-regulation [2]. Individual differences in attachment should therefore not only manifest themselves in altered neural responses to social-emotional information if the latter is presented in the form of sounds, images, or videos but even more strongly during interpersonal processes. Because a secure attachment style is thought to arise through interactions with available and sensitive caregivers, and the degree of sensitivity is often associated with the degree of synchrony of the infant's and mother's behavior [70, 71], behavioral and brain-to-brain synchrony appears a valid and promising interpersonal marker of attachment [72–74]. Being less susceptible to motion artifacts and allowing for ecologically more valid experimental designs, functional near-infrared spectroscopy (fNIRS) appears to be one method of choice for such so-called hyperscanning experiments [75, 76].

Finally, as mentioned in Sect. 1.3, there is a long-standing debate on the universality versus culture specificity of many well-established theoretical positions in psychology, including attachment theory [11–13]. This debate naturally also affects the interpretation of experimental findings emerging from the social neuroscience of attachment. Consequently, there is a clear need for more cross-cultural research in the future. If data pertaining to the neural underpinnings of attachment are to inform new intervention and prevention strategies and policy making at the societal level, they should adequately reflect both generally valid as well as locally maintained perceptions and beliefs about relationships. As nicely summarized by Keller [77], such process should involve interdisciplinary research programs that systematically conceptualize and empirically analyze differing cultures of attachment, ideally by linking the Bowlby/Ainsworth tradition with newly emerging knowledge from "evolution as well as cultural conceptions of socialization, parenting, and children's development" (p. 187). A result of such interdisciplinary research may also be the emergence of new ways of measuring attachment in different contexts within and between cultures through ongoing "field work," rather than by solely relying upon already extant procedures like the strange situation paradigm. To close with Keller's [77] words, "creating different conceptions of attachment on these grounds would not only help understanding development as the cultural solution of universal developmental tasks but also pave the way for the improvement of clinical and educational programs as defined by the needs of people" (p. 187).

The social neuroscience of attachment is still a very young field of research. Many of the reported findings remain preliminary and are thus in the need of being replicated and further extended. Nonetheless, the so far obtained results are providing exciting new insights into the social brain from an attachment theory perspective and allow for refinement and elaboration of attachment theory from the point of view of social neuroscience. Because attachment theory as such is built upon an inherent cross talk between disciplines, I very much hope that this cross talk keeps continuing to also include social neuroscience methods and will ultimately contribute to making the world a better place for the next generations to come.

References

1. Mikulincer M, Shaver PR. Attachmen in adulthood: structure, dynamics, and change. New York: The Guilford Press; 2007.
2. Mikulincer M, Shaver PR, Pereg D. Attachment theory and affect regulation: the dynamics, development, and cognitive consequences of attachment-related strategies. Motiv Emot. 2003;27(2):77–102.
3. Shah PE, Fonagy P, Strathearn L. Is attachment transmitted across generations? The plot thickens. Clin Child Psychol Psychiat. 2010;15(3):329–45.
4. Thornton A, McAuliffe K, Dall SRX, Fernandez-Duque E, Garber PA, Young AJ. Fundamental problems with the cooperative breeding hypothesis. A reply to Burkart & van Schaik. J Zool. 2016;299(2):84–8.
5. Hrdy SB. Evolutionary context of human development: The cooperative breeding model. In: Carter CS, editor. Attachment and bonding: a new synthesis. Cambridge: MIT Press; 2005.
6. Geary DC, Flinn MV. Evolution of human parental behavior and the human family. Parent SciPract. 2001;1(1–2):5–61.
7. Dunbar RIM. The social brain hypothesis. Evol Anthropol. 1998;6(5):178–90.
8. Joffe TH. Social pressures have selected for an extended juvenile period in primates. J Hum Evol. 1997;32(6):593–605.
9. Fonagy P. The human genome and the representational world: The role of early mother-infant interaction in creating an interpersonal interpretive mechanism. Bull Menn Clin. 2001;65(3):427–48.
10. Fraley RC, Brumbaugh CC, Marks MJ. The evolution and function of adult attachment: A comparative and phylogenetic analysis. J Pers Soc Psychol. 2005;89(5):731–46.
11. Lancy DF. The anthropology of childhood: cherubs, chattel, changelings. 2nd ed. Cambridge: Cambridge University Press; 2015.
12. Henrich J, Heine SJ, Norenzayan A. The weirdest people in the world? Behav Brain Sci. 2010;33(2–3):61–83.
13. Rothbaum F, Weisz J, Pott M, Miyake K, Morelli G. Attachment and culture - Security in the United States and Japan. Am Psychol. 2000;55(10):1093–104.
14. Cacioppo JT, Berntson GG. Social psychological contributions to the decade of the brain - Doctrine of multilevel analysis. Am Psychol. 1992;47(8):1019–28.
15. Porges SW. Social engagement and attachment - a phylogenetic perspective. Ann N Y Acad Sci. 2003;1008:31–47.
16. MacDonald K, MacDonald TM. The peptide that binds: a systematic review of oxytocin and its prosocial effects in humans. Harvard Rev Psychiatry. 2010;18(1):1–21.
17. Lieberman MD. Social cognitive neuroscience: a review of core processes. Annu Rev Psychol. 2007;58:259–89.
18. Vrtička P, Vuilleumier P. Neuroscience of human social interactions and adult attachment style. Front Hum Neurosci. 2012;6:212.
19. Fisher HE. Lust, attraction, and attachment in mammalian reproduction. Human Nature-an Interdisciplinary Biosocial. Perspective. 1998;9(1):23–52.
20. Fisher HE, Aron A, Mashek D, Li H, Brown LL. Defining the brain systems of lust, romantic attraction, and attachment. Arch Sex Behav. 2002;31(5):413–9.
21. Canterberry M, Gillath O. Attachment and caregiving: functions, interactions, and implications. In: Noller P, Karantzas GC, editors. The Wiley-Blackwell handbook of couples and family relationships. 1st ed. Chichester: Blackwell Publishing Ltd.; 2012.
22. Singer T, Klimecki OM. Empathy and compassion. Curr Biol. 2014;24(18):R875–8.
23. Fisher HE, Aron A, Brown LL. Romantic love: a mammalian brain system for mate choice. Philos Transac R Soc B Biol Sci. 2006;361(1476):2173–86.
24. Vrtička P, Andersson F, Grandjean D, Sander D, Vuilleumier P. Individual attachment style modulates human amygdala and striatum activation during social appraisal. PLoS One. 2008;3(8):e2868.

25. Fonagy P, Luyten P. A developmental, mentalization-based approach to the understanding and treatment of borderline personality disorder. Dev Psychopathol. 2009;21(4):1355–81.
26. Shamay-Tsoory SG, Aharon-Peretz J, Perry D. Two systems for empathy: a double dissociation between emotional and cognitive empathy in inferior frontal gyrus versus ventromedial prefrontal lesions. Brain. 2009;132:617–27.
27. Baron-Cohen S. Autism: the empathizing-systemizing (E-S) theory. Ann N Y Acad Sci. 2009;1156:68–80.
28. Mayes LC. A developmental perspective on the regulation of arousal states. Semin Perinatol. 2000;24(4):267–79.
29. Mayes LC. Arousal regulation, emotional flexibility, medial amygdala function, and the impact of early experience - comments on the paper of Lewis et al. In: Lester BM, Masten AS, McEwen B, editors. resilience in children. Ann N Y Acad Sci. 2006;1094:178–92.
30. Levy KN. The implications of attachment theory and research for understanding borderline personality disorder. Dev Psychopathol. 2005;17(4):959–86.
31. Strathearn L, Fonagy P, Amico J, Montague PR. Adult attachment predicts maternal brain and oxytocin response to infant cues. Neuropsychopharmacology. 2009;34(13):2655–66.
32. Fareri DS, Niznikiewicz MA, Lee VK, Delgado MR. Social network modulation of reward-related signals. J Neurosci. 2012;32(26):9045–52.
33. Aron A, Aron EN, Smollan D. Inclusion of the other in the self scale and the structure of interpersonal closeness. J Pers Soc Psychol. 1992;63(4):596–612.
34. Vrtička P. Interpersonal closeness and social reward processing. J Neurosci. 2012;32(37):12649–50.
35. Nummenmaa L, Manninen S, Tuominen L, Hirvonen J, Kalliokoski KK, Nuutila P, et al. Adult attachment style is associated with cerebral mu-opioid receptor availability in humans. Hum Brain Mapp. 2015;36(9):3621–8.
36. Troisi A, Frazzetto G, Carola V, Di Lorenzo G, Coviello M, D'Amato FR, et al. Social hedonic capacity is associated with the A118G polymorphism of the mu-opioid receptor gene (OPRM1) in adult healthy volunteers and psychiatric patients. Soc Neurosci. 2011;6(1):88–97.
37. Schindler A, Thomasius R, Petersen K, Sack PM. Heroin as an attachment substitute? Differences in attachment representations between opioid, ecstasy and cannabis abusers. Attach Hum Dev. 2009;11(3):307–30.
38. Ross S, Peselow E. The neurobiology of addictive disorders. Clin Neuropharmacol. 2009;32(5):269–76.
39. Rognoni E, Galati D, Costa T, Crini M. Relationship between adult attachment patterns, emotional experience and EEG frontal asymmetry. Personal Individ Differ. 2008;44(4):909–20.
40. Vrtička P, Sander D, Vuilleumier P. Influence of adult attachment style on the perception of social and non-social emotional scenes. J Soc Pers Relat. 2012;29(4):530–44.
41. Troisi A, Alcini S, Coviello M, Nanni RC, Siracusano A. Adult attachment style and social anhedonia in healthy volunteers. Personal Individ Differ. 2010;48(5):640–3.
42. Carver CS, Johnson SL, Kim Y. Mu opioid receptor polymorphism, early social adversity, and social traits. Soc Neurosci. 2016;11(5):515–24.
43. Donges U, Kugel H, Stuhrmann A, Grotegerd D, Redlich R, Lichev V, et al. Adult attachment anxiety is associated with enhanced automatic neural response to positive facial expression. Neuroscience. 2012;220:149–57.
44. Poore JC, Pfeifer JH, Berkman ET, Inagaki TK, Welborn BL, Lieberman MD. Prediction-errror in the context of real social relationships modulates reward system activity. Front Hum Neurosci. 2012;6:218.
45. DeWall CN, Masten CL, Powell C, Combs D, Schurtz DR, Eisenberger NI. Do neural responses to rejection depend on attachment style? An fMRI study. Soc Cogn Affect Neurosci. 2012;7(2):184–92.
46. Vrtička P, Sander D, Anderson B, Badoud D, Eliez S, Debbane M. Social feedback processing from early to late adolescence: influence of age, sex and attachment style. Brain Behav. 2014;4(5):703–20.

47. Suslow T, Kugel H, Rauch AV, Dannlowski U, Bauer J, Konrad C, et al. Attachment avoidance modulates neural response to masked facial emotion. Hum Brain Mapp. 2009;30(11):3553–62.
48. Norman L, Lawrence N, Iles A, Benattayallah A, Karl A. Attachment-security priming attenuates amygdala activation to social and linguistic threat. Soc Cogn Affect Neurosci. 2015;10(6):832–9.
49. Rigon A, Duff MC, Voss MW. Structural and functional neural correlates of self-reported attachment in healthy adults: evidence for an amygdalar involvement. Brain Imag Behav. 2016;10(4):941–52.
50. Quirin M, Gillath O, Pruessner JC, Eggert LD. Adult attachment insecurity and hippocampal cell density. Soc Cogn Affect Neurosci. 2010;5(1):39–47.
51. Krause AL, Borchardt V, Li M, van Tol MJ, Demenescu LR, Strauss B, et al. Dismissing attachment characteristics dynamically modulate brain networks subserving social aversion. Front Hum Neurosci. 2016;10:77.
52. Redlich R, Grotegerd D, Opel N, Kaufmann C, Zwitserlood P, Kugel H, et al. Are you gonna leave me? Separation anxiety is associated with increased amygdala responsiveness and volume. Soc Cogn Affect Neurosci. 2015;10(2):278–84.
53. Kim P, Leckman JF, Mayes LC, Newman MA, Feldman R, Swain JE. Perceived quality of maternal care in childhood and structure and function of mothers' brain. Dev Sci. 2010;13(4):662–73.
54. Foley P, Kirschbaum C. Human hypothalamus-pituitary-adrenal axis responses to acute psychosocial stress in laboratory settings. Neurosci Biobehav Rev. 2010;35(1):91–6.
55. Benetti S, McCrory E, Arulanantham S, De Sanctis T, McGuire P, Mechelli A. Attachment style, affective loss and gray matter volume: a voxel-based morphometry study. Hum Brain Mapp. 2010;31(10):1482–9.
56. Ein-Dor T, Hirschberger G. Rethinking attachment theory. Curr Dir Psychol Sci. 2016;25(4):223–7
57. Coan JA, Schaefer HS, Davidson RJ. Lending a hand: social regulation of the neural response to threat. Psychol Sci. 2006;17(12):1032–9.
58. Buchheim A, Erk S, George C, Kachele H, Kircher T, Martius P, et al. Neural correlates of attachment trauma in borderline personality disorder: A functional magnetic resonance imaging study. Psychiat Res Neuroimag. 2008;163(3):223–35.
59. Warren SL, Bost KK, Roisman GI, Silton RL, Spielberg JM, Engels AS, et al. Effects of adult attachment and emotional distractors on brain mechanisms of cognitive control. Psychol Sci. 2010;21(12):1818–26.
60. Eisenberger NI, Master SL, Inagaki TK, Taylor SE, Shirinyan D, Lieberman MD, et al. Attachment figures activate a safety signal-related neural region and reduce pain experience. Proc Natl Acad Sci U S A. 2011;108(28):11721–6.
61. Canterberry M, Gillath O. Neural evidence for a multifaceted model of attachment security. Int J Psychophysiol. 2013;88(3):232–40.
62. Vrtička P, Bondolfi G, Sander D, Vuilleumier P. The neural substrates of social emotion perception and regulation are modulated by adult attachment style. Soc Neurosci. 2012;7(5):473–93.
63. Gillath O, Bunge SA, Shaver PR, Wendelken C, Mikulincer M. Attachment-style differences in the ability to suppress negative thoughts: exploring the neural correlates. NeuroImage. 2005;28(4):835–47.
64. Moutsiana C, Fearon P, Murray L, Cooper P, Goodyer I, Johnstone T, et al. Making an effort to feel positive: insecure attachment in infancy predicts the neural underpinnings of emotion regulation in adulthood. J Child Psychol Psychiatry. 2014;55(9):999–1008.
65. Debbane M, Badoud D, Sander D, Eliez S, Luyten P, Vrtička P. Brain activity underlying negative self- and other-perception in adolescents: The role of attachment-derived self-representations. Cogn Affect Behav Neurosci. 2017;17(3):554–76.
66. Gross C, Hen R. The developmental origins of anxiety. Nat Rev Neurosci. 2004;5(7):545–52.
67. Haas BW, Filkowski MM, Cochran RN, Denison L, Ishak A, Nishitani S, et al. Epigenetic modification of OXT and human sociability. Proc Natl Acad Sci U S A. 2016;113(27):E3816–23.

68. Kim S, Kwok S, Mayes LC, Potenza MN, Rutherford HJV, Strathearn L. Early adverse experience and substance addiction: dopamine, oxytocin, and glucocorticoid pathways. Ann N Y Acad Sci. 2017;1394:74–91.
69. Holz NE, Boecker-Schlier R, Buchmann AF, Blomeyer D, Jennen-Steinmetz C, Baumeister S, et al. Ventral striatum and amygdala activity as convergence sites for early adversity and conduct disorder. Soc Cogn Affect Neurosci. 2017;12(2):261–27.
70. Leclere C, Viaux S, Avril M, Achard C, Chetouani M, Missonnier S, et al. Why synchrony matters during mother-child interactions: a systematic review. PLoS One. 2014;9(12):e113571.
71. Nievar MA, Becker BJ. Sensitivity as a privileged predictor of attachment: A second perspective on De Wolff and van IJzendoorn's meta-analysis. Soc Dev. 2008;17(1):102–14.
72. Atzil S, Hendler T, Feldman R. The brain basis of social synchrony. Soc Cogn Affect Neurosci. 2014;9(8):1193–202.
73. Atzil S, Hendler T, Zagoory-Sharon O, Winetraub Y, Feldman R. Synchrony and specificity in the maternal and the paternal brain: relations to oxytocin and vasopressin. J Am Acad Child Adolesc Psychiatry. 2012;51(8):798–811.
74. Biro S, Alink LRA, Huffmeijer R, Bakermans-Kranenburg MJ, Van Ijzendoorn MH. Attachment quality is related to the synchrony of mother and infant monitoring patterns. Attach Hum Dev. 2017;19(3):243–58.
75. Baker JM, Liu N, Cui X, Vrtička P, Saggar M, Hosseini SMH, et al. Sex differences in neural and behavioral signatures of cooperation revealed by fNIRS hyperscanning. Sci Rep. 2016;6:26492.
76. Cui X, Bryant DM, Reiss AL. NIRS-based hyperscanning reveals increased interpersonal coherence in superior frontal cortex during cooperation. NeuroImage. 2012;59(3):2430–7.
77. Keller H. Attachment and Culture. J Cross-Cult Psychol. 2013;44(2):175–94.

Mind-Reading in Altruists and Psychopaths

Fatima Maria Felisberti and Robert King

Abstract Due to its importance in political, cultural, and clinical spheres, adult mind-reading needs to be investigated (and understood) in depth. This chapter introduces the various meanings of "mind-reading" in neurotypical adults. We highlight philosophical and psychological implications of this construct for a wide variety of specifically human social interactions, such as play, acting, and manipulation. As a general rule, humans see one another as centres of intentional gravity and are very good folk psychologists (i.e. predictors of others' behaviours). These predictive powers rest in no small part on our various abilities to mind-read. A centre of intentional gravity can be decomposed into concepts such as beliefs, desires, and motives and can have multiple orders of understanding (e.g. "he believes that she desires him to wish for…"). Such multilayered abilities underwrite a vast range of human cognitive and affective domains such as mimicry, altruism, empathy, psychopathy, and learning. Our ability to attribute independent mental states and processes to others, as well as to animals and inanimate objects, is an integral part of human social behaviour, but mind-reading alone has no necessary internal moral compass, as seen in the behaviour of altruists and psychopaths. Rather, mind-reading is presented here as an all-encompassing toolkit that enables us to navigate our *Umwelt* as effectively as possible.

Keywords Mind-reading • Theory of mind • Altruism • Empathy • Psychopathy • Social cognition • Mind attribution

1 Introduction

Complex social species require sophisticated communication systems to navigate through the intricacies of social interactions and to establish and maintain long-lasting relationships crucial for mutual fitness. To this aim, the human brain, an

F.M. Felisberti (✉)
Psychology Department, Kingston University London, Penrhyn Road, London KT1 2EE, UK
e-mail: f.felisberti@kingston.ac.uk

R. King
Applied Psychology, University College Cork, Cork T12 YN60, Ireland
e-mail: r.king@ucc.ie

© Springer International Publishing Switzerland 2017

A. Ibáñez et al. (eds.), *Neuroscience and Social Science*,
DOI 10.1007/978-3-319-68421-5_6

interconnected network of billions of neurones and glial cells, integrates externally acquired with internally stored information to render it meaningful in different social contexts. The ability to predict intention and response is observed in core social interactions (be they altruistic, mutual, selfish, or spiteful) generally available to all social organisms, although organisms unable to predict the intentions of conspecifics can also act in those ways. In this chapter, we discuss philosophical, psychological, neurological, and methodological aspects intrinsic to mind-reading in neurotypical adults and their links with altruism and psychopathy in social contexts.

2 Conceptual Considerations About Minds

A naïve appraisal of the importance of evolutionary insights to the understanding of brains might incline us to believe that organisms evolved to have composite perceptual systems that give truthful information about the external world. However, a moment's reflection should reveal that the possible set of truths about external reality is computationally intractable. As a result, organisms do not see the "true world" because this would overwhelm them, not least in terms of energy consumption and processing speed. Instead, organisms evolved to have brains (which in these terms are primarily prediction machines) that yield useful information to increase fitness in the broad ecology of threats and opportunities in which that organism has evolved, the so-called *Umwelt*—a set of environmental factors affecting the behaviour of living things [1]. This distinction is crucial, and it is not unique to biology or to neuroscience. In artificial intelligence, the difficulty of filtering out what is irrelevant to focus on what is computationally tractable in its broadest sense is called "the frame problem" [2, 3]. Minds are things that brains "do", which include the wider neurophysiological system embedded in an ecology. Contemporary scholars want to ask deeper questions about these various functions, even when philosophical issues continue to bedevil such enquiries.

Do minds exist separately from bodies? While almost no one today would openly describe themselves as a Cartesian dualist regarding the mind/body problem, almost everyone is a de facto dualist when it comes to attributing minds to other humans or even certain non-humans. Why do intuitions like this persist when neuroscience repeatedly tells us that there is no "ghost in the machine" [4]? Part of the answer is that this conceptual mistake is actually a trick (sometimes called a "Baldwin effect") [5] that allows us to make fairly reliable predictions about other creatures in our *Umwelt*. For a complex eusocial species like ourselves, the most salient features of our *Umwelt* are other humans, and seeing each separate human as a seat of intentional gravity is a crucial part of our survival toolkits. In other words, our incredibly successful folk psychology keeps running up against our (now) substantial scientific knowledge, putting our intuitions under pressure.

Useful distinctions have been made between "stances" in the world: physical, design, and intentional levels [6]. A goal of science is to enable meaningful transitions between these descriptive levels, but, as a bare minimum requirement, separate valid scientific descriptions must be able to co-exist logically or, to use Wilson's [7] term, be consilient. At the most basic level is the physical description of a phenomenon. For instance, a cup of hot coffee has billions of sub-atomic particles moving randomly in a liquid and generating kinetic energy (i.e. heat). Coffee contains caffeine, which did not evolve to give us humans a morning boost but rather as an insecticide for the coffee plant (design stance); the stimulating effect it has on our brains is a by-product. If we then want to explain why many of us crave a morning coffee, we need to move from design to an intentional stance by stating our goals (e.g. getting an urgent task finished on time). These goals, desires, and intentions allow meaningful predictions of our actions in a way that a physical description of our brain cannot (and likely never will) achieve.

It should be obvious from the foregoing that most attempts to either produce or rebut so-called reductionist explanations of events are misguided. Attempts at explaining our need for a hot morning coffee in kinetic or biochemical terms would leave out our intentions and desires. This intentional level is the level of mind-reading. As Minsky [8] notes, there is no particular reason to think that we will ever be able to give a complete and useful description of human actions regarding the physical level of description of brain chemistry and neurotransmitter firing alone.

Given that the human brain evolved to solve a set of problems using whatever tools were available to evolution to build functional systems, the resulting brain consists of a vast network of complex interconnected and dynamic mechanisms. It is precisely the interface between the mechanisms evolved to promote fitness and the information that makes sense as folk psychology that needs to be understood (and implemented) to enable successful social exchanges. Integral to such social behaviour is our ability to attribute independent mental states and processes to ourselves and to others, as discussed below.

3 Mind-Reading

Mind-reading is often referred to through a range of terms, such as "theory of mind", "social intelligence", "social cognition", "mentalising", "mind attribution", "cognitive and affective mentalising", and "hot and cold empathy" (see Kumfor et al, this volume). Although each of those terms can be characterised individually, they all refer to some core and overlapping features. Zaki and Ochsner [9] grouped the features common to all those terms into (1) experience sharing, (2) mentalising, and (3) prosocial concern.

While a useful shorthand, the term mind-reading conceals a great deal of complexity. Essentially, mind-reading refers to our ability to navigate social interactions, i.e. our ability to attribute mental states to others and make conjectures about their goals, beliefs, and intentions, usually with the aim to understand, modulate, or manipulate their behaviour [10–12]. Individuals might differ in their ability to understand the mental states of others, but such differences are not associated with the recall of events and facts related to them; rather it is the complexity of other people's mental lives that imposes a cognitive limit for mind-reading [13].

Until recently, mind-reading research was underpinned by two so-called theories: theory theory (TT) and simulation theory (ST). TT refers to how mental states are interconnected when monitoring human actions, whereas ST refers to the ability to simulate the mental states of others and activate one's decision-making system, which in turn results in the attribution of beliefs and desires to the person we are trying to understand [14]. If such simulations are employed as frequently and as explicitly as ST proposes, we would expect to be aware of those mental states. Since that does not seem to be the case, mixed TT-ST models have been adopted to study mind-reading [15]. One reason for thinking that the term "theory of mind" can mislead here is that what is done by humans is so "natural" that it is nothing like the formal theories of (say) Newton, and what is meant varies considerably across domains. Consider Wittgenstein's [16] oft-requoted assertion: "My attitude towards him is an attitude towards a soul. I am not of the opinion that he has a soul". Wittgenstein is denying the idea that we set out a list of properties and capacities before deciding whether to treat someone as human. In addition, we do not need to have a formal belief in life after death to have a conception of someone that includes nested assumptions about attitudes, connections, and a moral life which can have continuity and meaning. In brief, we treat other humans as being in the intentional stance [6].

At the other end of the scale of mind ascription, the term "mind attribution" expands the definition of mind-reading to include animals, inanimate objects, and imaginary entities (e.g. gadgets, gods) [17]. For example, mind attribution can be seen in pet owners who appear to engage in long conversations with their pets, which seem to benefit the pet owner's well-being [18–20]. Some animals can indeed be understood at the intentional level (e.g. "Bilu wants to go for a walk"), and there is nothing odd or unnatural about such uses. It becomes somewhat trickier to be sure what is meant if the situation is reversed (e.g. "Bilu understands every word I say"), and there is certainly a degree of overspill of recursive attribution of mind-reading at this level.

3.1 Cognitive and Affective Mind-Reading

Mind-reading is often subdivided into its cognitive and affective aspects, making allowances for the dissociation of those two dimensions at the neural level [21–24]. The awareness of thoughts, creeds, and intentions in oneself and others is known as cognitive mind-reading, which includes different levels of metarepresentation: "first order" (e.g. I think X understands the problem) or "second order" (e.g. I think X believes that Y understands the problem) (Fig. 1). There is a ceiling to how many iterations of orders (belief about a belief about a belief) the human mind can manifest. Most researchers think that five levels are the human limit [25], although some have documented up to eight such levels [13]. Affective mind-reading, on the other hand,

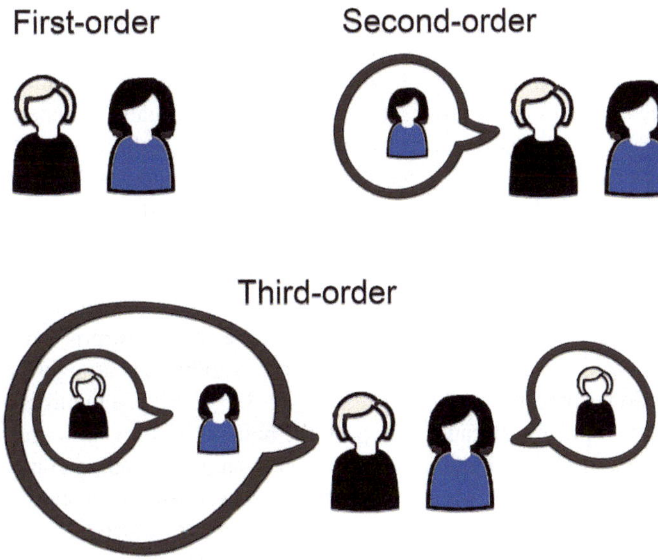

Fig. 1 Schematic illustration of first-, second-, and third-order mental states

refers to the ability to experience, to some degree, the emotional inner lives of other individuals without necessarily sharing any of their emotions or feelings [26, 27].

3.2 *Empathy and Mimicry*

Our ascription of intentional states is not merely a function of successfully predicting the behaviour of others—important though this is. It is intimately connected to our ability to learn as individuals. In many social contexts, empathy and mimicry can be intertwined with mind-reading, and it is not easy to disentangle them. Mimicry is seen as a possible precursor of or direct contributor to mind-reading, since inferences about others' mental states may have evolved from the ability to predict others' actions [28]. Empathy is a more complex construct; it refers not only to our awareness of thoughts and intentions in fellow humans but also to our ability to understand their emotional states and predict others' actions [28]. It is a multilayered ability to vicariously experience and understand mental states in oneself and others, i.e. the sharing of feelings and emotions linked to mental state attribution [29].

The role of mind-reading and empathy in social cognition (especially related to culture and politics) is underexplored given that the social environment in which one grows up is essential to the development of those abilities [30]. That role in moral judgments, actions, and deliberations has been the focus of recent and intense discussion. Mind-reading and empathic abilities have been overwhelmingly associated with prosociality and beneficial outcomes, even though that is not always the case [31–33].

The overlap between mind-reading and empathy descriptions is exemplified by studies suggesting the subdivision of empathy into two broad subtypes, namely "cold" and "hot" [34]. Cold empathy resembles "cognitive mind-reading" in that it refers to the ability to take the perspective of other individuals (to understand their feelings, problems, and sorrows) while being able to avoid sharing their emotional states. Conversely, hot empathy resembles "affective mind-reading"; individuals able to experience hot empathy share the affective mental state of others, and they seem sufficiently motivated to help others when needed.

Mimicry is critical to learning in humans and other animals. We usually expect people to be able to "read" our intentions from our actions, yet some of this ability is opaque to ourselves. This allows, for example, actors to surprise us with their superior ability to convey (or conceal) intentions. In addition to straightforward acting, professional psychics and mind-readers (in the sense of conjurors) can only entertain and surprise us because they push the boundaries of what we usually consider the limits of such intentional and informational mind-reading ability. It is not possible to perform psychic routines on other animals, however. There is no comparable version of "Was this the card you were thinking of?" which will surprise your pet [35]. The example may appear obvious—and in many ways, it is—but it underscores how naturally and regularly humans swim in a world of (circumscribed) intentionality.

3.3 *Neural Representation of Mind-Reading*

Due to the wide range of behavioural and physiological levels of processing it involves, mind-reading engages an extensive brain network of exogenous and endogenous mechanisms. Gerrans and Stone [36] point to evidence in favour of a domain-specific nervous mind-reading module with a parsimonious cognitive architecture that integrates domain-general and lower-level domain-specific mechanisms, which underlie flexible and sophisticated behaviours.

The neurobiology of intersubjectivity has revealed the existence of extended and overlapping networks during the sharing of experiences (empathy) and mind-reading. Mapping studies investigating neuronal activation during mind-reading showed that the brain areas most commonly activated were also linked to moral and social behaviours. The network involved in mind-reading is frequently reviewed and updated, usually in tandem with the advance of brain mapping technology and assorted experimental paradigms (e.g. short stories, cartoons, explicit and implicit mind-reading instructions), which did not seem to account for the variations reported in the findings.

Several studies confirmed that a wide brain network is activated during mind-reading, indicating the existence of core brain regions—including parts of the prefrontal cortex (PFC) and superior temporal sulcus (STS)—in addition to "peripheral" regions [37]. Currently, the mind-reading network includes the temporal cortex (TC), the posterior STS (pSTS), the amygdala, the dorsomedial and ventromedial prefrontal cortices (dMPFC and vMPFC), the temporal pole (TP), and the temporo-parietal junction (TPJ) [38–42]. For instance, the activity of the TPJ is linked to the understanding of our emotions in connection with specific events or individuals and such cognition-emotion link seems to modulate morally sound decision-making outcomes [43, 44]. The precuneus/posterior cingulate (PCC) was also activated during mind-reading, chiefly when one is thinking about intentions and beliefs [40, 45, 46]. Cross-cultural variability was observed in the activation of the TP and the TPJ, in line with behavioural differences across individuals from different cultures (e.g. American, French, Japanese) [47].

The mechanism underlying the ability to share experiences also relies on distributed brain networks [48]. Some of the most consistent findings related to the sharing of experiences recorded at the behavioural (self-reports) and neural levels (functional magnetic resonance imaging, fMRI) come from pain studies. The brain regions activated when we witness others suffering are the same areas activated when we are suffering ourselves: the anterior insula (AI) and the middle anterior cingulate cortex (mACC) [49, 50]. Moreover, a study using games with simulations of realistic environments (virtual reality) revealed that the right AI seems enlarged in individuals willing to risk their virtual lives attempting to rescue a person in danger [51].

Despite an extensive range of studies on the neural bases of mind-reading, on the one hand, and the related behavioural processes on the other, more studies are needed to bring together those two lines of enquiry. One example of such an extended approach can be seen in the study by Zaki et al. [52], who reported that the neural activity in areas previously associated with mind-reading matched participants' accuracy at inferring the affective state of another person. An improved understanding of mind-reading in typical individuals is essential to the understanding of the sequelae of acute brain trauma, as well social cognitive dysfunction (see Piguet, this volume), which could lead to better neurorehabilitation programmes [24].

The continuous development and improvement of methods for monitoring brain activity and social behaviour have contributed directly to the implementation of a multitude of useful experimental paradigms in mind-reading studies. Below we give some examples of the most common experimental paradigms employed in mind-reading research with typical and non-typical individuals.

4 Classical Paradigms in Mind-Reading

Below we describe some of the most common experimental paradigms used in brain imaging studies of mind-reading. They include first-, second-, and up to fifth-order mental states of mind-reading, although not all paradigms include all orders of mental state (Fig. 1).

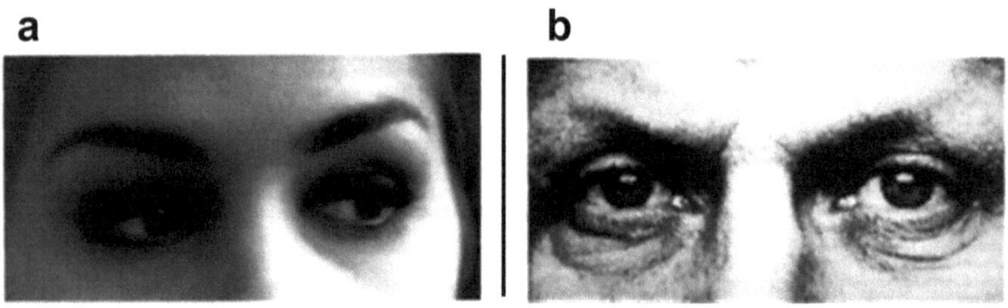

Fig. 2 Examples of eye expressions used in the RMET. Examples of the choices of descriptions for the eye expressions in the images above: (**a**) concerned vs. unconcerned and (**b**) serious message vs. playful message. The correct responses are underlined

4.1 Reading the Mind in the Eyes Test

The Reading the Mind in the Eyes Test (RMET) [53] involves the recognition of complex emotional states from photographs of faces where only the eyes area is visible. During the test, individuals see the eyes area of a face and must choose which of the two affective labels ("ashamed", "indecisive", "nervous", "suspicious", etc.) better depict the emotion displayed (Fig. 2). A more recent study showed that there are no reliable gender differences in RMET responses [54]. This task has been used to detect subtle impairments during affective processing [55, 56]. Moreover, it has been partially successful at predicting affective social deficits in children with autism spectrum disorder (ASD) and sensitive enough to be used with typical adults. Notwithstanding, it has been argued that the RMET measures the ability of participants to identify complex emotions rather than mind-reading per se [57].

4.2 False Belief Task

The false belief task (FBT) [11] has been used to empirically explore cognitive mind-reading. According to many researchers, FBT allows investigating whether individuals can distinguish between their own beliefs and those of another person (who is likely to have a different perspective) [58]. An experimental paradigm used in many FBT studies is the Sally-Anne test (Fig. 3). Like other object-transfer paradigm, it uses social vignettes to depict belief states. "Sally" (the target agent) and "Anne" are two puppets. Sally has a basket with an object in it, and Anne has an empty box. Sally leaves the room, and Anne moves the object into her previously empty box. When Sally returns to the room, the child is asked: "Where does Sally think the object is?". In other words, Sally wrongly believes that the object is in her box because she did not see that Anne has moved the object to the empty box (first-order mental state).

A typical child will realise that the action took place out of sight of Sally (second-order mental state), who should then have a mistaken belief that the object is where it was before she left the room, whereas most ASD children will conflate their own knowledge with that of Sally and maintain that she knows what they know [59]. The

Fig. 3 The Sally-Anne task

validity of the classical FBT paradigm with adults has been called into question since adults show ceiling effects (100% accuracy) when performing it [60].

It is worth noting that the FBT and the RMET might be too simple to provide a more encompassing understanding of mind-reading in typical adults. Hence, a more elaborated set of tests is still needed to investigate our ability to "mind-read" and to empathise with others in terms of moral and social behaviours [61, 62].

4.3 The Yoni Test

The Yoni test [22] is based on the "Charlie task" [59], and it incorporates visual and verbal cues. A central character, "Yoni" is represented by a happy cartoon face in the centre of an image surrounded by four images of a single category—e.g. animals, faces, and transport (Fig. 4).

Individuals must indicate by mouse-clicking the image related to the social vignettes presented (first-order level): "Yoni is close to ____", "Yoni loves____", "Yoni does not love____", "Yoni identifies with____", or "Yoni thinks about____". Second-order social vignettes refer to whose success Yoni envies, whose misfortune Yoni gloats over, and items Yoni thinks about, has, or loves that another character thinks about, has, or loves. The test is suitable for interpretations of proximity, facial expressions, and gaze direction, and it allows measures of response accuracy and latency across affective, cognitive, and physical (control) trials.

4.4 Animations: The Heider-Simmel Illusion

The Heider-Simmel illusion [63] consists of a simple animation of a large triangle, which appears to pursue two small circles around a simple virtual landscape (Fig. 5). Viewers naturally and unselfconsciously describe the scene regarding an "angry" triangle that "bullies" the smaller circles who are "frightened" of it, and so forth. Of course, at one level the viewers are perfectly well aware that triangles and circles do not have emotions and desires, but we have a natural animism (an intention ascriptor) to parallel our tendency to see faces where none exist (pareidolia). More recent versions of the Heider-Simmel illusion are seen in social animation tasks used to understand mind-reading [28].

Presumably, in the manner of the smoke detector principle, in the past it was more important to see faces and intentions (even if none existed) than to miss the ones that were present. From animism to sophisticated theologies, most humans have a deep-seated belief that things like the universe itself can have intentions and desires in relation to us. Interestingly, those on the autistic spectrum are both less likely to believe in God [64] and are less subject to pareidolic illusions [65]. Being more oriented to systematising than empathising appears to lessen the strength of this pervasive illusion [66].

4.5 Movie for the Assessment of Social Cognition

The Movie for the Assessment of Social Cognition (MASC) [67] features four realistic characters at a dinner party, who display stable traits and transient states. The relevant themes are romance and friendship, and questions about the characters' cognitive and affective mental states require the participants to interpret physical, vocal, and contextual information, as well as to understand false beliefs and metaphors (Fig. 6).

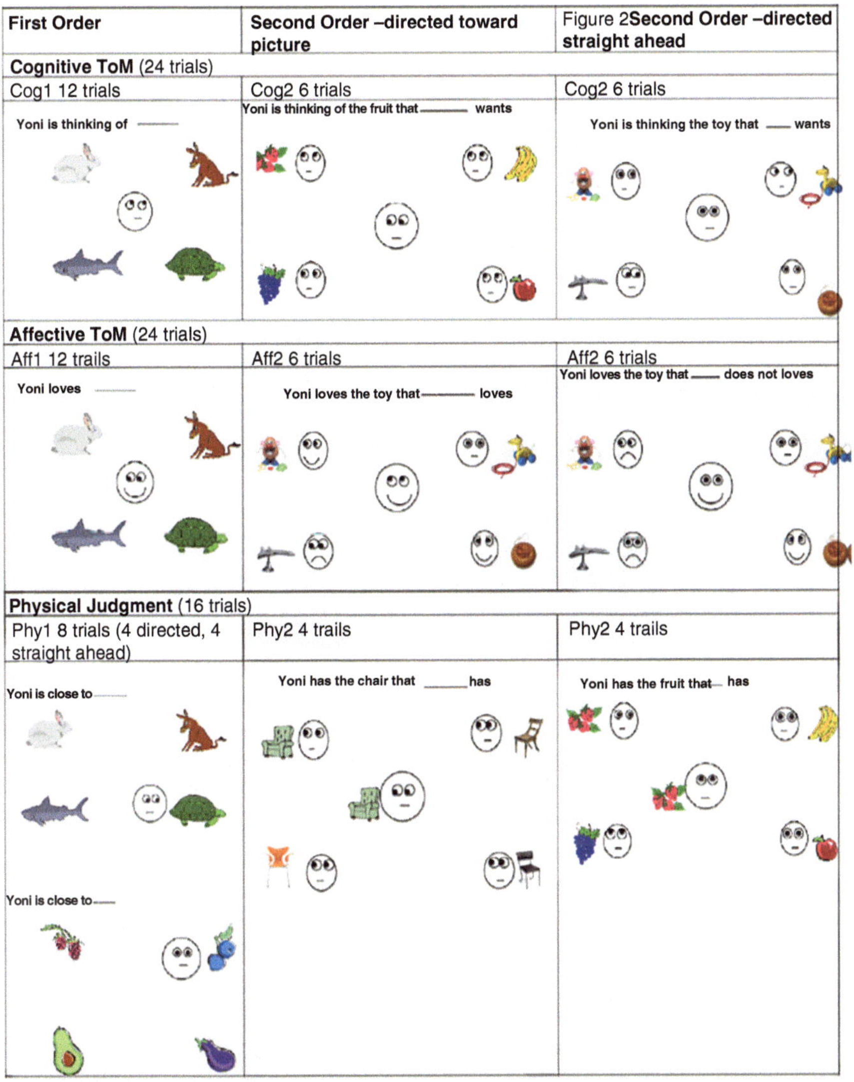

Fig. 4 The Yoni test

5 Mind-Reading in Altruism

Prosocial behaviours result from a wide variety of factors that, at first glance, seem to be polar opposites: intention or intuition, nature (instinct) or nurture (learned), value inferred from actions and their outcomes, and altruism or egoism [68]. It is important to distinguish the proximate motivations for altruism (such as empathy or concern for others) from the biological puzzle of how altruism (in its strict Hamiltonian sense) could evolve in the first place. Hamilton [69] solved the latter

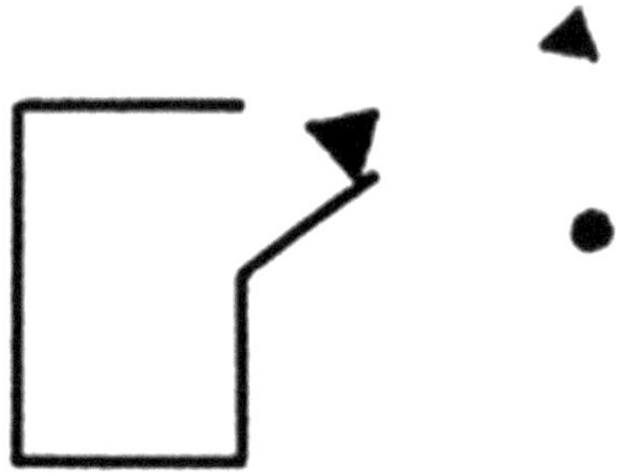

Fig. 5 A frame of the Heider-Simmel animated video clip

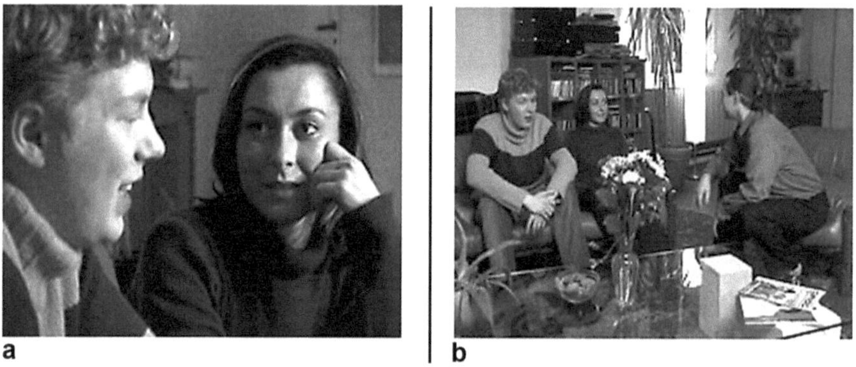

Fig. 6 Frames from the movie for the assessment of social cognition. (**a**) Cliff is the first one to arrive at Sandra's house for the dinner party. He and Sandra seem to enjoy themselves when Cliff is telling about his vacation in Sweden. (Printed with permission). (**b**) When Michael arrives, he dominates the conversation, directing his speech to Sandra alone. (Printed with permission)

puzzle regarding inclusive fitness—an axiomatic extension of Darwinian fitness to explain how sentiments and behaviours favourable to conspecifics but at the expense of the actor could evolve in the first place. However, the (ultimate) explanation is not what people typically mean when they use the term "altruism". What is meant is usually a collection of positive prosocial impulses towards others that may cost something to the actor but often result in the mutual benefit of some kind.

The altruistic motivation underlying prosocial behaviour requires explanation, since there are high costs involved in helping others, no matter whether empathically or egoistically motivated or both (i.e. feeling better about oneself for helping others, avoiding punishment, gaining rewards), though neither mind-reading nor empathy is prerequisites for prosocial behaviour [70]. The genetic (ultimate) basis for altruism is discussed in terms of kin and group selection [71], inclusive fitness [69], and reciprocal altruism [72] theories. However, empathy-altruism theory addresses the proximate relevance of such mechanisms for social cognition. The theory posits that the ability to show empathic concern underpins the altruistic motivation needed to reduce the suffering of others. In other words, the behaviour of altruists seems to be modulated by their ability to empathise with others. Furthermore, high levels of cooperation require high-level mind-reading and empathic abilities, which would have favoured altruistic behaviour in our ancestors [70].

It is widely accepted that superior mind-reading abilities facilitate group cooperation by modulating the level of understanding between team members. For example, de Vignemont and Singer [73] suggest a dual role for empathy to allow the gathering of contextual information about the future actions of others, as well as to support prosocial behaviours, cooperation, and effective social communication. The link between prosocial learning and empathy was also investigated with fMRI and revealed (not surprisingly) that individuals learned to obtain rewards for themselves faster than for others and that the variability in prosocial learning could be modelled by trait empathy: people with higher empathy learned more quickly when benefitting others than people with lower empathy [74]. Interestingly, an increased activity observed in the right pSTC during action perception, compared with action performance, was shown to be predictive of higher self-reported altruism [75].

According to Tomasello [76], socialisation via a shared culture can modulate altruism; the puzzling mixture of unselective altruism and selfish sharing behaviour observed in young children is slowly replaced during development by a more discerning and targeted type of altruism. This trend was attested in a study where children as young as 3 years old showed a more frequent sharing behaviour towards other children in their group who had been nicer to them in the past [77]. Later on, children start discerning intentionality in others by observing the direction of people's gaze and by inferring their knowledge of a given situation based on their own past actions and observations [78], which is usually referred to as "shared intentionality" [76], which in turn relies on one's mind-reading ability.

Research on the nature of the interplay between mind-reading and social behaviours such shared intentionality, altruism, and general morality in typical adults with different cultural backgrounds is still in its infancy. One should bear in mind that mind-reading per se has no internal moral compass, as reflected in the behaviour of both altruists and psychopaths. Since mind-reading can also be employed to exploit, deceive, or entertain others, behaviours such as Machiavellianism, psychopathy, narcissism, and even performances by "magicians" are considered below.

6 Mind-Reading in Psychopathy (And Other "Dark Personalities")

Psychopathy is part of the so-called dark triad of personality, which also includes narcissism and Machiavellianism. There is no clear-cut behavioural distinction between the "dark triad" personalities. To varying degrees, the elements of the triad share malevolent characteristics, which are evidenced in the propensity towards deception, self-promotion, emotional coldness, low agreeableness, and aggressiveness [79, 80].

6.1 Psychopaths

A psychopathic personality disorder (or psychopathy) is characterised by emotional detachment and assorted antisocial traits, alongside strong associations with criminal behaviour and reoffending [81, 82]. Psychopaths are often called sociopaths, and the terms have been used interchangeably. Some psychopathic behaviours are evidently criminal (i.e. murder, rape, recidivism), but there is some confusion (if not an outright contradiction) in the specification of core psychopathic behaviours; for some, psychopaths are cold-blooded and have strong self-control; for others, they are impulsive and thrill-seeking. Some are described as aggressive and very successful professionally (e.g. some CEOs are believed to have high psychopathic traits), while others are described as being emotionally superficial and reckless, but most scholars agree that psychopathic behaviour is mean, bold, and lacking in moral inhibition [83, 84]. Unlike altruists, psychopaths show poor empathic concern. Comparable to conmen and torturers, psychopaths also need to have well-developed mind-reading ability (even if only cognitive mind-reading skills such as inferring others' intentions and beliefs) to be able to fool and exploit others as effectively as they often do [85] (see Baez et al., this volume).

Mind-reading studies on individuals with high psychopathic traits revealed patterns of brain activation similar to the ones observed in altruists, even though a study employing in vivo diffusion tensor magnetic resonance imaging tractography showed abnormalities in an amygdala-orbitofrontal cortex network linked to psychopathy [86]. Nonetheless, the processes involved in mind-reading in both altruists and psychopaths were cognitively demanding and required the use of a wide range of complex and traditional executive functions, including decision-making, planning, response and conflict monitoring, working memory, and attention [50, 87, 88]. Perhaps psychopaths are good at attributing exploitative motives to others but poor at recognising that others may do the same to them? This would explain the characteristic outrage and surprise when people of this disposition are caught out [82].

6.2 Machiavellians and Narcissists

Machiavellians are seen as people with utilitarian morals who are manipulative, cynical, dominant, secretive, and suspicious [89, 90]. They think in both concrete and pragmatic terms and tend to be emotionally unstable and anxious about relationships [12]. Machiavellians routinely assume that they will be exploited by others, to whom they attribute negative intentions and unwillingness to cooperate [91]. They also show decreased motivation regarding "affective" mind-reading or "hot" empathy [92], and a negative correlation between Machiavellianism level and affective face recognition was observed [93, 94]. Although Machiavellians seem to be able to infer the thoughts and intentions of other people, they fail to grasp emotional states such as guilt, shame, or sympathy and lack the motivation to feel what others are feeling [12, 94]. A narcissistic personality disorder shares many of the features observed in psychopaths and Machiavellians, but it is dominated by a heightened sense of self-worth and superiority, a propensity to self-deception, and a link with antisocial behaviour [95].

6.3 *Con Artists and Magicians*

Con artists are different again, although they have considerable overlap with Machiavellian personality types. They rely on the fact that most people, most of the time, do not regularly lie about certain sorts of events. In other words, we normally assume that we can make reasonable predictions about intentions and the con artists can only survive using frequency-dependent selection: if enough people were untrustworthy, then working on confidence would no longer work because confidence in general would have broken down. It is said that it is impossible to con an honest man. This is untrue. However, it is impossible to con someone without their attributing malicious intentions to us.

While not Machiavellians in the true sense, magicians (especially those who claim genuine abilities) do exhibit traits that shed remarkable light on the complexity of intention reading in humans. For magicians, the term "mind-reading" has a somewhat different (albeit overlapping) meaning compared to the definition used by psychologists and neuroscientists. In the context of a magic show, mind-reading equates to a series of displays of the apparently impossible, such as plucking thoughts from a person's mind or seeing what they have written and sealed in an envelope. Modern stage magicians can be sharply divided over the ethics of performing mind-reading effects on stage (as opposed to other forms of conjuring) and how these should be presented. There is a good reason for this, namely, that lay audiences are inclined to believe that what they are witnessing is real. This belief can be (and has been) exploited by the unscrupulous to pretend that they can read the intentions (the minds) of potential lovers and lost children, for example.

By contrast with other forms of magical performance, audiences do not typically seriously entertain the hypothesis that the performer has actual powers. There are exceptions to this (e.g. the spoon bending of Uri Geller). Yet, a typical audience member does not believe that David Copperfield can fly (one hopes) but that he or she is watching a surprising and mystifying illusion. All the stage psychic is doing is pushing at the bounds of a belief that already has some very porous edges. As mentioned above, one key contrast is the mutual social construction of shared reality that occurs in psychic performances of mind-reading but not with other forms of magical performance. One way to see this contrast is to appreciate that a non-human animal can be brought to respond with surprise and attention to some forms of prestidigitation. Presumably, this is because their theory of the continued existence of unseen objects is something they share with us. However, there is no equivalent of surprising a dog with a mind-reading trick. "Revealing the card (or treat) they were thinking of" has no meaning to a dog because our shared reality does not involve this level of mind-reading. Other primates seem to show an attenuated sense of being able to fool others or showing expectations of being fooled [96].

7 Final Considerations

This chapter addressed conceptual and methodological aspects of mind-reading, which is a highly flexible human ability since vicarious responses are (more often than not) successfully adapted to a wide range of contexts. The mind-reading toolkit encompasses a complex range of behaviours such as altruism, empathy/psychopathy, and cognitive abilities from more general domains, which are rooted in cognitive and affective processes evolved to facilitate social interactions. As suggested by many researchers, more studies are needed to elucidate the underpinnings of mind-reading in neurotypical adults during assorted social exchanges (which include verbal communication). Furthermore, the understanding of the extent to which mind-reading is modulated by culture is of utmost relevance for science and society.

References

1. von Uexküll J. Theoretical biology. New York: Harcourt, Brace & Co; 1926.
2. McCarthy J, Hayes PJ. Some philosophical problems from the standpoint of artificial intelligence. In: Ginsberg ML, editor. Readings in nonmonotonic reasoning. Los Altos, CA: Morgan Kaufmann; 1987. p. 26–45.
3. King RI. Can't get no (Boolean) satisfaction: a reply to Barrett et al. (2015). Front Psychol. 2016;7:1880.
4. Ryle G. The concept of mind. London: Hutchinson; 1949.
5. Simpson GG. The Baldwin effect. Evolution. 1953;7(2):110–7.
6. Dennett DC. The intentional stance. Cambridge: MIT Press; 1989.
7. Wilson EO. Consilience: the unity of knowledge. New York: Vintage; 1999.
8. Minsky M. Society of mind. New York: Simon and Schuster; 1988.
9. Zaki J, Ochsner K. The neuroscience of empathy: progress, pitfalls and promise. Nat Neurosci. 2012;15(5):675–80.
10. Leslie AM. Pretense and representation: the origins of "Theory of Mind". Psychol Rev. 1987;95(4):412–26.
11. Wimmer H, Perner J. Beliefs about beliefs: representation and constraining function of wrong beliefs in young children's understanding of deception. Cognition. 1983;13:103–128.
12. Ináncsi T, Láng A, Bereczkei T. Machiavellianism and adult attachment in general interpersonal relationships and close relationships. Europe J Psychol. 2015;11(1):139–54.
13. Kinderman P, Dunbar RIM, Bentall RP. Theory-of-mind deficits and causal attributions. Br J Psychol. 1998;89:191–204.
14. de Bruin L, Strijbos D, Slors M. Early social cognition: alternatives to implicit mindreading. Rev Philos Psychol. 2011;2:499–517.
15. Gallagher S. Simulation trouble. Soc Neurosci. 2007;2:353–65.

16. Wittgenstein L. In: PMS H, Schulte J, editors. Philosophical investigations. 4th ed. Oxford: Wiley-Blackwell; 2009.
17. Waytz A, Gray K, Epley N, Wegner DM. Causes and consequences of mind perception. Trends Cogn Sci. 2010;14(8):383–8.
18. Eyssel FA, Pfundmair M. Predictors of psychological anthropomorphization mind perception and the fulfillment of social needs: a case study with a zoomorphic robot. In: 24th IEEE International Symposium 2015; 2015. pp. 827–32.
19. Nejati V, Zabihzadeh A, Maleki G, Tehranchi A. Mind reading and mindfulness deficits in patients with major depression disorder. Procedia Soc Behav Sci. 2012;32:431–7.
20. Tan LBG, Lo BCY, Macrae CN. Brief mindfulness meditation improves mental state attribution and empathizing. PLoS One. 2014;9(10):e110510.
21. Sebastian CL, Fontaine NMG, Bird G, Blakemore S-J, De Brito SA, McCrory EJP, et al. Neural processing associated with cognitive and affective theory of mind in adolescents and adults. Soc Cogn Affect Neurosci. 2012;7:53–63.
22. Shamay-Tsoory SG, Aharon-Peretz J. Dissociable prefrontal networks for cognitive and affective theory of mind: a lesion study. Neuropsychologia. 2007;45:3054–67.
23. Kalbe E, Schlegel M, Sack AT, Nowak DA, Dafotakis M, Bangard C, et al. Dissociating cognitive from affective theory of mind: a TMS study. Cortex. 2010;46(6):769–80.
24. Henry JD, von Hippel W, Molenberghs P, Lee T, Sachdev PS. Clinical assessment of social cognitive function in neurological disorders. Nat Rev Neurol. 2016;12(1):28–39.
25. Liddle B, Nettle D. Higher-order theory of mind and social competence in school-age children. J Cult Evol Psychol. 2006;4:231–46.
26. Duval C, Piolino P, Bejanin A, Eustache F, Desgranges B. Age effects on different components of theory of mind. Conscious Cogn. 2011;20:627–42.
27. Brothers L, Ring B. A neuroethological framework for the representation of minds. J Cogn Neurosci. 1992;4:107–18.
28. Castelli F, Happe F, Frith U, Frith C. Movement and mind: a functional imaging study of perception and interpretation of complex intentional movement patterns. NeuroImage. 2000;12:314–25.
29. Singer T. The neuronal basis and ontogeny of empathy and mind reading: review of literature and implications for future research. Neurosci Biobehav Rev. 2006;30:855–63.
30. Suddendorf T, Whiten A. Mental evolution and development: evidence for secondary representation in children, great ages, and other animals. Psychol Bull. 2001;127(5):629–50.
31. Decety J, Cowell JM. The complex relation between morality and empathy. Trends Cogn Sci. 2014;18(7):337–9.
32. Bloom P. Empathy and its discontents. Trends Cogn Sci. 2016;21(1):24–31.
33. Prinz JJ. Is empathy necessary for morality? In: Coplan A, Goldie P, editors. Empathy: philosophical and psychological perspectives. Oxford: Oxford University Press; 2011.
34. Kang M, Camerer C. fMRI evidence of a hot-cold empathy gap in hypothetical and real aversive choices. Front Neurosci. 2013;7:104.
35. Macknik S, Martinez-Conde S, Blakeslee S. Sleights of mind: what the neuroscience of magic reveals about our everyday deceptions. New York: Henry Holt and Company; 2010.
36. Gerrans P, Stone VE. Generous or parsimonious cognitive architecture? Cognitive neuroscience and theory of mind. Br J Philos Sci. 2008;59:121–41.
37. Carrington SJ, Bailey AJ. Are there theory of mind regions in the brain? A review of the neuroimaging literature. Hum Brain Mapp. 2009;30:2313–35.
38. Vogeley K, Bussfeld P, Newen A, Herrmann S, Happé F, Falkai P, et al. Mind reading: neural mechanisms of theory of mind and self-perspective. NeuroImage. 2001;14:170–81.
39. Lindquist KA, Barrett LF. A functional architecture of the human brain: emerging insights from the science of emotion. Trends Cogn Sci. 2012;16(11):533–40.
40. Lieberman MD. Social cognitive neuroscience: a review of core processes. Annu Rev Psychol. 2007;58:259–89.

41. Decety J, Jackson PL, Sommerville JA, Chaminade T, Meltzoff AN. The neural bases of cooperation and competition: an fMRI investigation. NeuroImage. 2004;23(2):744–51.
42. Amodio DM, Frith CD. Meeting of minds: the medial frontal cortex and social cognition. Nat Rev Neurosci. 2006;7:268–77.
43. Young L, Camprodon JA, Hauser M, Pascual-Leone A, Saxe R. Disruption of the right temporoparietal junction with transcranial magnetic stimulation reduces the role of beliefs in moral judgments. Proc Natl Acad Sci. 2010;107(15):6753–8.
44. FeldmanHall O, Mobbs D, Dalgleish T. Deconstructing the Brain's moral network: dissociable functionality between the temporoparietal junction and ventro-medial prefrontal cortex. Soc Cogn Affect Neurosci. 2013;9(3):297–306.
45. Saxe R, Kanwisher N. People thinking about thinking people: the role of the temporo-parietal junction in "theory of mind". NeuroImage. 2003;19(4):1835–42.
46. Bzdok D, Langner R, Hoffstaedter F, Turetsky BI, Zilles K, Eickhoff SB. The modular neuroarchitecture of social judgments on faces. Cereb Cortex. 2012;22(4):951–61.
47. Kobayashi C, Temple E. Cultural effects on the neural basis of theory of mind. Prog Brain Res. 2009;178:213–23.
48. Singer T, Kiebel SJ, Winston JS, Dolan RJ, Frith CD. Brain responses to the acquired moral status of faces. Neuron. 2004;41:653–62.
49. Kanske P, Böckler A, Trautwein F-M, Singer T. Dissecting the social brain: Introducing the EmpaToM to reveal distinct neural networks and brain–behavior relations for empathy and theory of mind. NeuroImage. 2015;122:6–19.
50. Singer T, Lamm C. The social neuroscience of empathy. Ann N Y Acad Sci. 2009;1156:81–96.
51. Pati I, Zanon M, Novembre G, Zangrando N, Chittaro L, Silani G. Neuroanatomical basis of concern-based altruism in virtual environment. Neuropsychologia. 2017. https://doi.org/10.1016/j.neuropsychologia.2017.02.015.
52. Zaki J, Weber J, Bolger N, Ochsner K. The neural bases of empathic accuracy. Proc Natl Acad Sci. 2009;106:11382–7.
53. Baron-Cohen S, Jolliffe T, Mortimore C, Robertson M. Another advanced test of theory of mind: evidence from very high functioning adults with autism or Asperger syndrome. J Child Psychol Psychiatry. 1997;38(7):813–22.
54. Baron-Cohen S, Bowen DC, Holt RJ, Allison C, Auyeung B, Lombardo MV, Smith P, Lai M-C. The "reading the mind in the eyes" test: complete absence of typical sex difference in ~400 men and women with autism. PLoS One. 2015;10(8):e0136521.
55. Adolphs R, Baron-Cohen S, Tranel D. Impaired recognition of social emotions following amygdala damage. J Cogn Neurosci. 2002;14:1264–74.
56. Gray K, Jenkins AC, Heberlein AS, Wegner DM. Distortions of mind perception in psychopathology. Proc Natl Acad Sci U S A. 2010;108:477–9.
57. Mitchell RLC, Phillips LH. The overlapping relationship between emotion perception and theory of mind. Neuropsychologia. 2015;70:1–10.
58. Dennett DC. Beliefs about beliefs. Behav Brain Sci. 1978;1:568–70.
59. Baron-Cohen S, Leslie AM, Frith U. Does the autistic child have a "theory of mind"? Cognition. 1985;21(1):37–46.
60. Bloom P, German TP. Two reasons to abandon the false belief task as a test of theory of mind. Cognition. 2000;77(1):25–31.
61. Dana S, Apperly IA, Chiavarino C, Humphreys GW. Left temporoparietal junction is necessary for representing someone else's belief. Nat Neurosci. 2004;7(5):499–500.
62. Abu-Akel A, Shamay-Tsoory S. Neuroanatomical and neurochemical bases of theory of mind. Neuropsychologia. 2011;49(11):2976.
63. Heider F, Simmel M. An experimental study of apparent behavior. Am J Psychol. 1944;57:243–59.
64. Norenzayan A, Gervais WM, Trzesniewski KH. Mentalizing deficits constrain belief in a personal God. PLoS One. 2012;7(5):e36880.
65. Ryan C, Stafford M, King RJ. Seeing the man in the moon: do children with autism perceive pareidolic faces? A pilot study. J Autism Dev Disord. 2016;46(12):3838–43.

66. Baron-Cohen S, Wheelwright S, Hill J, Raste Y, Plumb I. The "reading the mind in the eyes" test revised version: a study with normal adults, and adults with Asperger syndrome or high-functioning autism. J Child Psychol Psychiatry. 2001;42(2):241–51.
67. Dziobek I, Fleck S, Kalbe E, Rogers K, Hassenstab J, Brand M, et al. Introducing MASC: a movie for the assessment of social cognition. J Autism Dev Disord. 2006;36:623–36.
68. Gęsiarz F, Crockett MJ. Goal-directed, habitual and Pavlovian prosocial behaviour. Front Behav Neurosci. 2015;9:135.
69. Hamilton WD. The genetical evolution of social behaviour. I. J Theor Biol. 1964;7(1):1–16.
70. Batson CD, Lishner DA, Stocks EL. The empathy–altruism hypothesis. In: Schroeder DA, Graziano WG, editors. The Oxford handbook of prosocial behavior. Oxford: Oxford University Press; 2015. p. 259–81.
71. Smith JM. Group selection and kin selection. Nature. 1964;201(4924):1145–7.
72. Trivers R. The evolution of reciprocal altruism. Q Rev Biol. 1971;46:35–57.
73. de Vignemont F, Singer T. The empathic brain: how, when and why? Trends Cogn Sci. 2006;10(10):435–41.
74. Lockwood PL, Apps MA, Valton V, Viding E, Roiser JP. Neurocomputational mechanisms of prosocial learning and links to empathy. Proc Natl Acad Sci. 2016;113(35):9763–8.
75. Tankersley D, Stowe CJ, Huettel SA. Altruism is associated with an increased neural response to agency. Nat Neurosci. 2007;10(2):150–1.
76. Tomasello M. Why we cooperate. Cambridge: MIT Press; 2009.
77. Olson KR, Spelke ES. Foundations of cooperation in preschool children. Cognition. 2008;108(1):222–31.
78. Phillips A, Wellman H, Spelke E. Infants' ability to connect gaze and emotional expression as cues to intentional action. Cognition. 2002;85(1):53–78.
79. Furnham A, Richards SC, Paulhus DL. The dark triad of personality: a 10 year review. Soc Pers Psychol Compass. 2013;7(3):199–216.
80. Paulhus DL, Williams KM. The dark triad of personality: narcissism, Machiavellianism, and psychopathy. J Res Pers. 2002;36:556–63.
81. Damasio AR. A neural basis for sociopathy. Arch Gen Psychiatry. 2000;57(2):128–9.
82. Hare RD. Comparison of procedures for the assessment of psychopathy. J Consult Clin Psychol. 1985;53:7–16.
83. Gray K, Jenkins AC, Heberlein AS, Wegner DM. Distortions of mind perception in psychopathology. Proc Natl Acad Sci U S A. 2011;108(2):477–9.
84. Skeem JL, Polaschek DLL, Patrick CJ, Lilienfeld SO. Psychopathic personality: bridging the gap between scientific evidence and public policy. Psychol Sci Public Interest. 2011;12(3):95–162.
85. Frith CD, Frith U. How we predict what other people are going to do. Brain Res. 2006;1079:36–46.
86. Craig MC, Catani M, Deeley Q, Latham R, Daly E, Kanaan R, et al. Altered connections on the road to psychopathy. Mol Psychiatry. 2009;14:946–53.
87. Overwalle FV. Social cognition and the brain: a meta-analysis. Hum Brain Mapp. 2009;30:829–58.
88. Ibanez A, Huepe D, Gempp R, Gutierrez V, Rivera-Rei A, Toledo MI. Empathy, sex and fluid intelligence as predictors of theory of mind. Personal Individ Differ. 2013;54(5):616–21.
89. Byrne RW, Whiten A. Machiavellian intelligence: social expertise and the evolution of intellect in monkeys, apes, and humans. Oxford: Oxford University Press; 1988. p. 73–5.
90. Wilson DS, Near D, Miller RR. Machiavellianism: a synthesis of the evolutionary and psychological literatures. Psychol Bull. 1996;119(2):285–99.
91. Repacholi B, Slaughter V, Pritchard M, Gibbs V. Theory of mind, Machiavellianism, and social functioning in childhood. In: Individual differences in theory of mind: implications for typical and atypical development. New York: Psychology Press; 2003.
92. Andrew J, Cooke M, Muncer SJ. The relationship between empathy and Machiavellianism: an alternative to empathizing – systemizing theory. Personal Individ Differ. 2008;44:1203–11.

93. Paal T, Bereczkei T. Adult theory of mind, cooperation, Machiavellianism: the effect of mind-reading on social relations. Personal Individ Differ. 2007;43:541–51.
94. Bereczkei T, Szabo ZP, Czibor A. Abusing good intentions: Machiavellians strive for exploiting cooperators. SAGE Open. 2015;5(2):1–5.
95. Barry CT, Frick PJ, Killian AL. The relation of narcissism and self-esteem to conduct problems in children: a preliminary investigation. J Clin Child Adolesc Psychol. 2003;32(1):139–52.
96. Gomez JC. Nonhuman primate theories of (non-human primate) minds: some issues concerning the origins of mind-reading. In: Carruthers P, Smith PK, editors. Theories of theories of mind. Cambridge: Cambridge University Press; 1996. p. 330–43.

From Primary Emotions to the Spectrum of Affect: An Evolutionary Neurosociology of the Emotions

Warren D. TenHouten

Abstract Cognitive appraisal theorists, psychological constructivists, and social constructivists contend that emotions are not natural kinds and are therefore refractory to classification. Evolutionary biologists and affective neuroscientists, however, have amassed theoretical and empirical evidence suggesting basic emotions are natural kinds. We describe continuity between Maclean's fourfold ethogram, Plutchik's four life problems, and Fiske's four sociorelational models. Plutchik advanced a psychoevolutionary theory of emotions and, by splitting his four existential dimensions by valence, was able to correctly identify eight primary emotions. These are the prototypical adaptive reactions to the eight existential situations, which are also seen as natural kinds. We argue that the concept of primary emotions is an important theoretical advance and additionally enables classification of pairs and triples of primary emotions which form secondary and tertiary emotions. We present a complete classification of the 28 secondary emotions and analyze one of the 56 potential tertiary-level emotions, resentment. Affect-spectrum theory proposes causal relations between the eight valenced sociorelational models and Plutchik's inventory of eight primary emotions. We examine two hypotheses of affect-spectrum theory through a content-analytic, lexical-level study of Euro-Australians and Australian Aborigines. The analysis shows that the positive experiences of communal sharing and authority ranking are predictive of joy and anger, respectively, and joy and anger together are predictive of pride. A parallel analysis indicates that negative involvements in communally shared and authority-ranked social relations predict sadness and fear, respectively; these in turn are predictive of shame.

Keywords Emotion • Classification • Natural kinds • Primary emotions • Secondary emotions • Tertiary emotions • Relational models • Plutchik • Pride • Shame

W.D. TenHouten (✉)
Department of Sociology, University of California at Los Angeles,
405 Hilgard Ave., Los Angeles, CA 90095-1551, USA
e-mail: wtenhout@g.ucla.edu

© Springer International Publishing Switzerland 2017

A. Ibáñez et al. (eds.), *Neuroscience and Social Science*,
DOI 10.1007/978-3-319-68421-5_7

1 The Perspective of Neurosociology

The neurosciences chiefly concern the brain and central nervous system and investigate the interrelationships between mind and brain. Social neuroscience adds a third level of analysis, as it "addresses fundamental questions about the mind and its dynamic interaction with the biological systems of the brain and the social world" [1, p. 3]. These three levels of analysis—mind, brain, and the social world—also characterize the emerging interdisciplinary fields of neuropolitics [2, 3], neuroeconomics [4], neuroanthropology [5, 6], and neurosociology [7–13]. Relationships among the phenomena of mind, brain, and society, and the scientific disciplines that address them, are illustrated in Fig. 1. Social neuroscience, neurosociology, neuroanthropology, neuroeconomics, and neuropolitics share the central location in this figure, as their common topic spans mind, brain, and society.

Much of the literature on neurosociology consists of speculation about its possibilities, limitations, and promise as an interdisciplinary field of scientific inquiry. This chapter's more specific aim is to present an evolutionary neurosociology of the emotions. We propose that Fiske's [14] fourfold social relations model can be mapped onto Maclean's [15] ethogram of four communicative displays; these in turn provides experimental evidence supportive of Plutchik's [16–18] fourfold model of life problems, the basis of his psychoevolutionary model of the emotions.

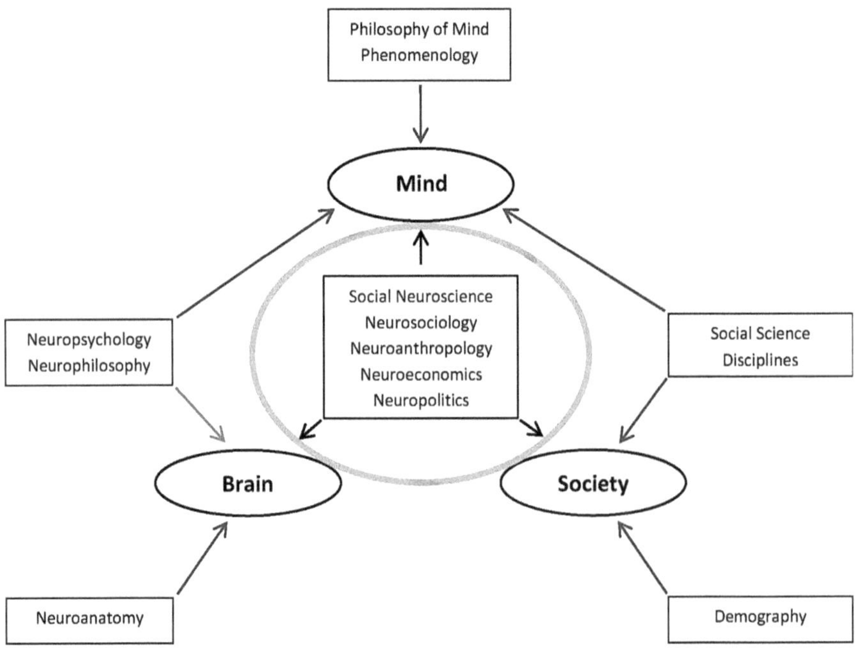

Fig. 1 Mind, brain, and society, key phenomena of concern in related fields of inquiry

The four relational models assume both positive and negative valences, resulting in eight fundamental sociorelational situations. The prototypical adaptive reactions to these sociocognitive models have led to the evolutionary development of the primary emotions. We describe these four pairs of opposite emotions as natural kinds, which can combine to form the complex emotions. We present a complete classification of the secondary emotions, together with one proposed tertiary emotion, resentment. Through an empirical analysis of two opposite secondary emotions, pride and shame, we link these secondary emotions to their opposite primary emotional components and to valenced social relations.

2 Primary Emotions

Plato was perhaps the first scholar to classify emotions. He described fear, hope, joy, and sorrow as the most basic emotions. There have been countless additional models of the putatively elementary, basic, or primary emotions [19, pp. 14–15]. Primary emotions remain a contested topic in emotions theory and research. Some see a small set of emotions as basic or primary or as natural kinds, while others, including Barrett [20, p. 28], hold that expressions of "anger, sadness, or fear" are not genuine natural kinds because they lack "specific causal mechanisms in the brain." To be considered natural kinds, a set of objects must exist as a natural "group" or "order," a "real set," that has not been placed together as an *artificial* exercise of human classification. A kind is *natural* if it corresponds to a grouping that reflects the structure of the world.

2.1 *The Case Against Primary Emotions*

Cognitive appraisal theorists, psychological constructionists, and social constructionists, among others, reject the concept of primary emotions and suggest emotions exist as psychological or social constructions. Cognitive appraisal theorists generally dismiss the claim that emotions are intrinsic to the more primitive regions of the brain [21, 22]. They instead hold that emotions emerge as neocortical regions of the brain that make cognitive sense of bodily processes and others' behaviors, in the context of social situations and events. The emotions we experience are accordingly a function of how and what we cognize to have caused the situation or event, and of how we interpret the event, including whether we interpret it as positive or negative.

Psychological constructivists (e.g., [20, 23]) have adduced neuroscientific evidence showing that discrete emotions are enabled by general brain networks that are involved in both emotion-laden and non-emotional operations, not by localized brain mechanisms dedicated to particular emotions. These theorists hold that mental states and processes classified under the vernacular category of emotions are not sufficiently similar to allow a unified theory of the emotions. Theories that explain a subset of emotions accordingly "will not adequately explain the whole range of human emotions" and even single affective states, such as anger, will require multiple theories [24]. Because emotions are not genuine "natural kinds," emotion terms should accordingly be eliminated from scientific vocabulary. In this view, emotions such as fear, anger, and joy are not natural kinds, primarily because they lack specific causal mechanisms in the brain, and have permeable boundaries impacted by culture and language use. Thus, emotions such as anger and surprise lack boundaries "carved in nature" [20, p. 28].

Social constructivists see emotions as cultural products but not as evolutionary adaptations involving brain structures [25–28]. They find little meaning in identifying an emotion as either primary or complex and tend to dismiss the view that sentiments and emotions can emerge through the combining of basic emotions. Gordon [25, p. 567] claims that sentiments are social, and emotions, psychological, so there can be a sociology of sentiments but not a sociology of emotions. He pronounces the idea of basic emotions a "fallacy" and a "reduction" of the social to the psychological. Of course, affective states can be socially constructed, and sociologists and anthropologists of the emotions have contributed greatly to understanding cultural constructions of feelings, sentiments, and emotions [29].

2.2 The Case for Primary Emotions

Emotions researchers with ecological, psychoevolutionary, and affective neuroscientific orientations [16–18, 30–36] have adduced impressive evidence indicating that a small subset of emotions are primordial, elementary, basic, or primary. Data show that emotions are innate capabilities: (1) fundamental emotions emerge in infancy while infants are still relying on subcortical behavioral mechanisms and before the onset of language [35]; (2) human babies born without cerebral hemispheres (anencephalic) cannot become intellectually developed but can grow up to be affectively vibrant if raised in nurturing and stimulating social environments [37]; (3) the first emotions of the child unfold through epigenetic programs according to precise, universal timetables [38, 39] and persist throughout the life-span [40]; and (4) Eibl-Eibesfeldt [30] shows that deaf and blind children make facial expressions similar to those of non-impaired children and inferred that several of these emotional expressions are universal, because of genetically inherited "fixed action patterns."

Abundant evidence suggests that the most basic emotions have evolved through natural selection across a wide variety of animal species [15, 41–44]. The basic emotions are neural, motivational, and expressive reactions that can occur rapidly in response to an environmental stimulus posing an opportunity or a threat [34, 44]. In humans, these primordial responses are cognitively elaborated and are crucial to the process of sharing important information with conspecifics about pressing problems of life [15, 39], [43, pp. 42–43]. They remain essential for humans' ability to meet universal survival needs, reproduce, engage the social world, and flourish (see Baez et al., Kumfor et al., and Piguet, this volume).

Primary emotions enjoy several attributes, as they (1) address fundamental problems of communication between conspecifics; (2) address the most central problems of life; (3) can be shown to have developed in a wide variety of animal species; (4) are recognized, by sight and sound, cross-culturally; (5) are not themselves mixtures or combinations of simpler emotions; (6) are able to combine with other primary emotions to form secondary emotions and with secondary emotions to form pathways to tertiary emotions; and (7) are deeply imbedded in valenced social relations models, which themselves have a deep evolutionary history and can likewise be regarded as culturally universal natural kinds.

If it can be shown that a small number of emotions have deep evolutionary roots and persist as prototypical adaptive reactions to the most fundamental kinds of life problems, then we would hardly expect them to be functions of single brain mechanisms or structures. Emotions such as joy-happiness [45] and anticipation involve widespread and highly complex cortical functions spanning ancient brain structures, limbic structures, and the neocortex; a considerable commitment of brainwork is necessary for effective functioning in social life [42, 43, 46].

In order to show that a proposed set of primary emotions are natural kinds, it would be necessary to demonstrate the following: (1) they are not themselves combinations of simpler emotions; (2) they can mix, in pairs and triples, to form second- and third-order entities that are also emotions; and (3) the resulting classification, if complete, will span the entire spectrum of the emotions. It is thus only through successful identification of the primary emotions, and subsequent classification of higher-order emotions, that the primary emotions can be demonstrated to possess boundaries that are—despite the limitations of natural language descriptions and cross-cultural variations—carved in nature.

3 Identifying Primary Emotions

There have been countless efforts to identify the basic or primary emotions, with most inventories listing from four to ten candidate emotions. These classificatory efforts suggest three possible interpretations: (1) emotions are not natural kinds, so that all efforts at classification are fruitless endeavors, and all emotions exist sui generis; (2) basic emotions *are* natural kinds, but their identities remain unknown; and (3) basic emotions are natural kinds, and one scholar has already correctly identified them (which would mean that all other inventories are incorrect). We argue below that this is just what has happened and that Robert Plutchik has gotten the primary emotions exactly right.

Numerous studies point to the existence of several cross-culturally understood emotions. Darwin [41] argued that, because of their deep evolutionary origins, facial expressions of the simplest emotions, such as joy, fear, and anger, are similar among humans, regardless of culture. Confirming Darwin's astute observations,

Ekman, Sorenson, and Friesen [47], in studies in New Guinea, Borneo, Brazil, Japan, and the United States, found that tribal-living people, with scant exposure to outsiders, were able to recognize the emotional significance of facial expressions in pictures of individuals from modern societies. Conversely, individuals in modern societies recognized emotions displayed in images of facial expressions of members of the preliterate cultures. In a comparative study of culturally isolated Namibian villagers and Westerners, Sauter et al. [48] extended cross-cultural recognition of these primary emotions (excluding surprise) to include nonverbal emotional vocalizations (including screams and laughs). Primary emotions are thus understood cross-culturally both in facial expressions and vocalizations, the two primary means of communicating social signals. On the basis of this and subsequent research, Ekman et al. [47–49] have identified six primary emotions—joy, sadness, fear, anger, disgust, and surprise.

The weight of contemporary evidence, much of it from affective neuroscience, suggests that all humans work from a common palette of affective responses [50, 51]. The *utility* of the concept of primary emotions depends to a great extent on whether it enables emotions classification, that is, the identification not only of the primary emotions but also of complex emotions whose constituent elements are primary emotions. Remarkably, little attention has been paid to this potential to use primary emotions as the basis for classifying complex emotions. If a specific set of emotions are indeed primary, then all other emotions can be conceptualized as secondary (comprised of two primaries) or tertiary (comprised of three primaries). Plutchik [17, pp. 117–118] was the first emotions researcher to attempt a classification of secondary emotions and possibly the first to suggest, albeit indirectly, the existence of tertiary emotions.

3.1 Plutchik's Psychoevolutionary Model of the Primary Emotions

While Darwin [41] saw emotions as adaptive reactions to problems of life, he did not identify these problems and made no effort to classify the many emotions he considered. Plutchik [18], however, developed a psychoevolutionary model in which he identified four such life problems—*identity*, *temporality*, *hierarchy*, and *territoriality*. Each of these life problems can be either an opportunity or a danger or threat, so that a situation is either negatively or positively valenced. Any of the eight problem-valence situations can therefore occur, each of which triggers a distinct subjective state of mind which activates an adaptive reaction. These eight prototypical adaptive reactions, Plutchik argued, constitute the primary emotions. Plutchik's wheel of primary emotions is shown in Fig. 2.

Plutchik thus identified four pairs of oppositely valenced primary emotions, with each pair addressing one of his four existential problems. Each emotion is described as a subjective state and linked to its key function: for the life problem of identity,

(A)

Life Problems	Basic functions	Emotions	Valences
identity	incorporation	acceptance	positive
.	rejection	disgust	negative
temporality	reproduction	joy–happiness	positive
.	reintegration	sadness–grief	negative
hierarchy	destruction	anger–rage	positive
.	protection	fear–terror	negative
territoriality	exploration	anticipation	positive
.	boundary defense	surprise	negative

(B)

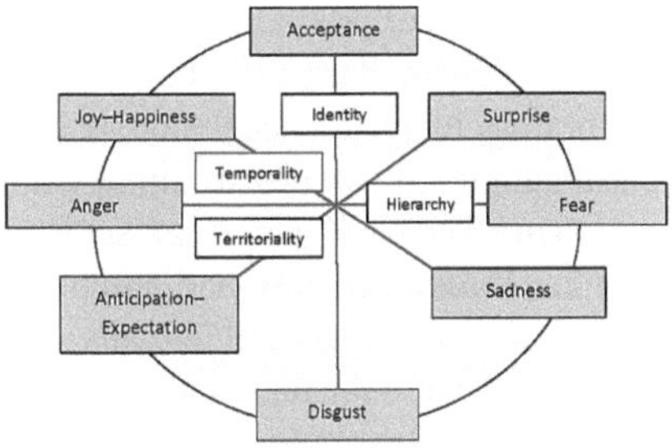

Fig. 2 (**a**) Plutchik's model of the primary emotions. (**b**) Plutchik's circumplex or "wheel" of the eight primary emotions. Source: W. D. TenHouten, Alienation and affect. New York: Routledge, p. 63, Fig. 5.1

these subjective states/functions are acceptance/incorporation and disgust/rejection; for temporality, joy/reproduction and sadness/reintegration; for hierarchy, anger/destruction and fear/protection; and for territoriality, anticipation/exploration and surprise/boundary defense. Plutchik saw that the positive experience of temporality, for example, triggers a feeling of joy-happiness and that the negative experience of a violation of one's territory or resources triggers a surprise reaction.

3.2 *MacLean's Rescue of Plutchik*

One limitation of Plutchik's primary emotions model is that he presents only limited evidence to support his identification of the most basic problems of life. He based his fourfold model of life problems not on rigorous experimental studies of animal and human brains but rather on insights gleaned from other scholars who speculatively presented similar inventories of "existential" problems. Additionally, Plutchik [52, p. 147] had little to say about the social processes through which individuals might confront these four problems of life. His conceptualization therefore suffers a sociological emptiness, which has to some extent undermined his model's appeal to practitioners of the social sciences, social psychology, social neuroscience, and neurosociology.

Plutchik's model gains validation through the evolutionary neuroethology of MacLean [15]. In his program of comparative research on lizards, rats, and humans, MacLean reached two broad conclusions: (1) the human brain has evolved a triune structure, consisting of reptilian, mammalian, and neomammalian levels of brain development, and (2) even for reptiles, four kinds of communicative displays have evolved. MacLean's triune brain model has both critics [53, 54] and defenders [32, 55], but his fourfold model of communicative displays is on solid footing. While the advent of the mammalian brain led to elaboration of the emotions as adaptive reactions, MacLean found that proto-emotional adaptive reactions are enabled by the forebrains of pre-mammalian animals, variously called the "reptilian brain," the "R-complex," and the "striatal complex." The R-complex includes the upper part of the brainstem, the diencephalon, parts of the midbrain, and the dorsal portion of the basal ganglia (the dorsal striatum, which contains as its major parts the caudate and putamen). The basal ganglia exist throughout pre-mammalian animals, including all reptiles, birds, fish, eels, and amphibians, and have been preserved and elaborated in brains of mammals and humans.

The advent of the mammalian limbic system and the human neocortex has hardly rendered the R-complex obsolete. Human basal ganglia play an important role in rational decision-making by contributing to action selection, or the process of deciding which of multiple possible actions to execute [56], and contribute to social communication, social displays, and affect-laden social relations. Following caudate damage, for example, there is degraded motivation capacity and degraded speech quality, with verbal responses becoming slow, abulic, terse, incomplete, and emotionally flat. While the exact functions of the R-complex are not fully understood, in humans this ancient brain architecture remains essential for behaviorally motivated, affect-laden social signaling and communicative displays [57].

MacLean [15] identified four kinds of communicative displays: (1) signature displays, (2) territorial displays, (3) courtship displays, and (4) challenge or dominance-submission displays. MacLean found these displays even in reptiles. There is an isomorphism between MacLean's and Plutchik's models: Maclean's signature displays underlie Plutchik's problem of identity; courtship displays, temporality, and the cycle of life and death; challenge and submission displays, hierarchy; and territorial displays, territoriality. MacLean thus provided an evolutionary neuroscientific foundation for Plutchik's fourfold model of existential problems and further showed that emotions involve social communications at the most fundamental, neurobiological level.

3.3 *Fiske's Social Relations Model Is Isomorphic to the Plutchik-MacLean Model*

The dimensionality and nature of social relations is contested in relational models theory, plural rationality theory, and related paradigms, although most contributors to this enterprise have presented four-dimensional models of human social relations. Recent fourfold conceptualizations include grid-group theory [58], plural rationality theory or cultural theory of risk [59], social-cognitive development theory [60, 61], and relational models theory [14, 62, 63]. Relational models theory asserts (1) there exist four elementary sociorelational models; (2) these four models have an evolutionary history and are, as a result, cross-culturally universal; (3) they are inseparable from their cognitive representations; and (4) they are ordered by their level of involvement in quantitative reasoning.

To enhance understanding of human emotions in their social contexts, it is helpful to generalize Plutchik's four life problems to the most elementary of social relations. Fiske [15] refers to his four sociorelational models as "communal sharing" (CS), "authority ranking" (AR), "equality matching" (EM), and "market pricing" (MP). CS relations are close and personal; they include kinship relations that enable perpetuation of the group beyond the individual, including the functions of sexual reproduction and community reintegration following the loss of a member. AR-based relations pertain to communicative displays of social power, domination, influence, and status competition in social hierarchies. EM relations involve like-mindedness, equal distribution of resources, distributive justice, and, more generally, efforts to attain conditional equality by setting aside problematic authority or dominance structures. MP-based social relations involve territory (an activity range) that provides valued resources; in humans, territoriality extends to socioeconomic behavior.

The combined MacLean-Plutchik model is isomorphic to Fiske's social relations model. Accordingly, courtship-temporality develops into CS, dominance-hierarchy into AR, signature-social-identity into EM, and territory-territoriality, that is, control of resources in the environment, into MP [9, pp. 27–42]. The resulting model is displayed in Fig. 3, which shows causal relationships between involvements in social relations models and the experience of the primary emotions. Thus, for example, an individual immersed in a positive experience of CS, CS+, who is in a close personal relationship with a significant other, can be predicted to experience joy or happiness. Similarly, an individual experiencing a negative experience of MP (MP−), upon realizing that territorial resources are threatened, will experience surprise.

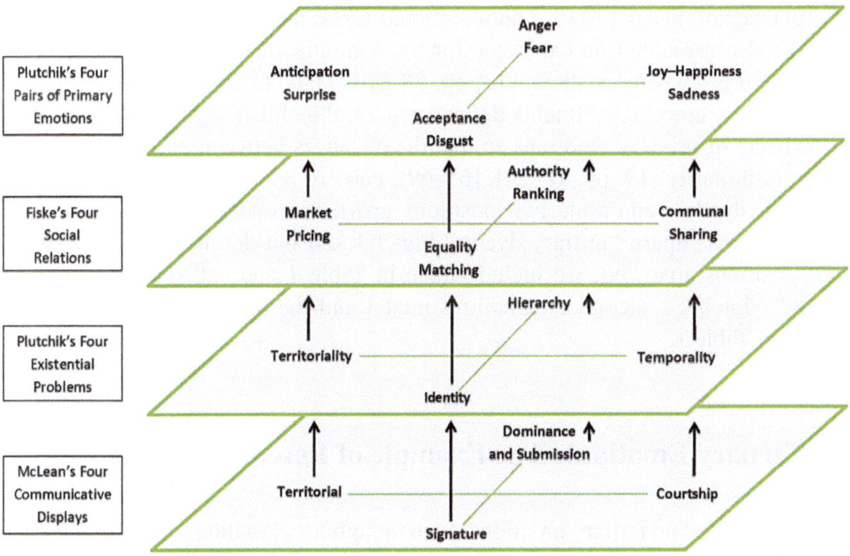

Fig. 3 Continuities in the models of MacLean, Plutchik, and Fiske. Source: W. D. TenHouten, Alienation and affect. New York: Routledge, p. 66, Fig. 5.2

4 From Primary to Secondary Emotions

> [P]assions are susceptible of an entire union; and like colours, may be blended so perfectly together, that each of them may lose itself and contribute only to vary that uniform impression which arises from the whole. Some of the most curious phenomena of the human mind are deriv'd from this property of the passions. *David Hume [1739]* [64, p. 366]

Du and colleagues [65, 66] have studied the 15 secondary emotions formed through pairings of the six emotions that Ekman and colleagues demonstrated as cross-culturally recognizable. In a study of 230 subjects, they found that the facial muscles involved in the secondary-level emotions were the same facial muscles involved in the component primary emotions. For example, the facial expression for a happy surprise (interpreted here as delight) combines muscle movements observed in both happy and surprised facial expressions. For the 21 (six primary, 15 secondary) defined categories, a computational model of face perception was used to produce facial expressions. It was found that the second-order mixed emotional expressions were visually discriminated by the subjects. These results lend important evidentiary support of the conceptual distinction between primary and second-order emotions.

Using his primary emotions model, Plutchik [17, pp. 117–18] had attempted a provisional classification of secondary emotions. Plutchik saw his effort as a development of Darwin's [41] evolutionary theory of emotions, wherein Darwin provided a "principle of antithesis," according to which opposite situations evoke opposite emotional reactions. Yet, Plutchik did not interpret the four pairs of opposite primary emotions (anger-fear, joy-sadness, acceptance-disgust, anticipation-surprise) as secondary emotions. For example, he did not consider that feeling both acceptance and disgust toward another could result in ambivalence. Plutchik [17, p. 118] also presented no candidate for the combination of surprise and disgust, interpreted elsewhere as shock [19, pp. 88–90]. Thus, of 28 possible pairings of eight primary emotions, Plutchik defined just 23. Plutchik designed his circumplex or "wheel" of primary emotions so that the distances between emotions reflected their dissimilarity [19, pp. 18–22], [67–69]. Pairs of adjacent emotions are called "primary dyads"; emotions two positions apart, "secondary dyads"; and those three positions apart, "tertiary dyads." Plutchik did not define emotions that are four positions apart, but we include them in Table 1 and call them "quaternary dyads." Plutchik's secondary emotions model and the author's revision are also shown in Table 1.

5 Tertiary Emotions: The Example of Resentment

Plutchik defined no tertiary emotions, even though his classification of primary and secondary emotions omitted complex affective states that would appear to be emotions, including jealousy, envy, discouragement, despair, regret, resentment, hatred,

Table 1 Plutchik's 1962 classification of the secondary emotions, a revision, and hypothesized associated elementary social relations

Secondary emotions	Plutchik's 1962 definitions	A revised classification	Relations
Primary dyads			
Acceptance and joy	Love	Love	EM+ and CS+
Joy and anger	Pride	Pride	CS+ and AR+
Anger and anticipation	Aggression, revenge, stubbornness	Aggressiveness	AR+ and MP+
Anticipation and disgust	Cynicism	Cynicism	MP+ and EM−
Disgust and sadness	Misery, remorse, forlornness	Loneliness	EM− and CS−
Sadness and fear	Despair, guilt	Shame	CS− and AR−
Fear and surprise	Alarm	Alarm	AR− and MP−
Surprise and acceptance	Curiosity	Curiosity	MP− and EM+
Secondary dyads			
Acceptance and anger	Dominance	Dominativenes	EM+ and AR+
Acceptance and fear	Submissiveness	Submissivencss	EM+ and AR−
Anger and disgust	Scorn, loathing, indignation, contempt, hate, resentment	Contempt	EM− and AR+
Disgust and fear	Shame, prudishness	Repugnance, abhorrence	EM− and AR−
Joy and surprise	Delight	Delight	CS+ and MP−
Sadness and surprise	Embarrassment, disappointment	Disappointment	CS− and MP−
Joy and anticipation	Optimism, courage, hopefulness	Optimism	CS+ and MP+
Sadness and anticipation	Pessimism	Pessimism	CS− and MP+
Tertiary dyads			
Anger and surprise	Outrage, resentment, hate	Outrage	AR+ and MP−
Joy and fear	Guilt	Guilt	CS+ and AR−
Acceptance and sadness	Resignation, sentimentality	Resignation	EM+ and CS−
Surprise and disgust	–	Shock	MP− and EM−
Fear and anticipation	Anxiety, caution, dread, cowardliness, distrust	Anxiety	AR− and MP+
Sadness and anger	Envy, sullenness	Sullenness	CS− and AR+
Disgust and joy	Morbidness	Derisiveness	EM− and CS+
Anticipation and acceptance	Fatalism	Fatalism, resourcefulness	
Quaternary dyads			
Acceptance and disgust	–	Ambivalence	EM+ and EM−
Joy and sadness	–	Bitter, sweetness	CS+ and CS−
Anger and fear	–	Frozenness, tonic immobility	AR+ and AR−
Anticipation and surprise	–	Confusion, discombobulation	MP+ and MP−

dread, worry, and vengefulness. Plutchik [17, p. 56] nevertheless indirectly raised the possibility of tertiary emotions by suggesting that "feelings of resentment are composed of (at least) disgust and anger." Plutchik [17, p. 118] also linked surprise to resentment, by suggesting that "anger + surprise = outrage, resentment, hate." We propose that resentment is indeed a tertiary emotion, so that "resentment_1 = anger and disgust and surprise" [69, pp. 114–117].

Resentment is an emotion of invidious social comparison. If, in comparison to other people, to groups, or even to themselves at different points in time, individuals who believe that they lack what they or their group deserve, and feel at least relative deprivation, will react with anger and resentment; anger is resentment's main ingredient and relative deprivation theorists repeatedly refer to "angry resentment" [70, pp. 217–218]. Resentment is a less ephemeral and more clearly a sociomoral emotion than is anger; the resentful person will feel anger at having been violated, mistreated, or brutalized by others and will strive to get even by seeking the misery of the violator. When the angering behavior of predatory others is seen as unjust and is rejected on moral grounds, the victim's sociomoral response includes disgust, rejection, and revulsion. An interiorized form of "pure pain" can lead to a forceful kind of resentment, which even if confused and misdirected, comprises a demand for ethical treatment and palliative intervention.

Behavior that stimulates resentment effectively penetrates one's boundaries; it possibly involves abuse of body and property and therefore contains an element of surprise (the prototypical adaptive reaction to breaches of one's territory or resources). Violations of manners, norms of interpersonal behavior, or respect for what is valued, believed in, and held to be proper or sacred can be seen as breaches of moral territory and can also stimulate resentment.

Given that resentment is a sociomoral emotion involving anger, disgust, and surprise, it can be expected to arise in complex situations in which an individual experiences EM−, AR+, and MP−, either singly or in combination. This suggests that resentment can include in its meaning the secondary emotions contempt (anger and disgust), outrage (anger and surprise), and shock (surprise and disgust). This model of resentment is shown in Fig. 4.

Resentment is likely to ensue from experiences of one or more of the three primary-secondary emotion combinations which can form pathways to resentment. Thus, given that "contempt = anger and disgust," by substitution, "resentment_2 = contempt and surprise," when resentment is triggered by another's contemptible breach of normative boundaries. Given that "shock = surprise and disgust," "resentment_3 = anger and shock"; thus, morally unacceptable or sociomorally shocking behavior with injurious results triggers angry indignation; it is provoked by something perceived as wrong, unworthy, mean, or cruel. Given "outrage = anger and surprise," we find "resentment_4 = disgust and outrage"; resentment_4 is reaction to the behavior of others adjudged to constitute disgusting and outrageous violations of social morality and ethical norms [69, pp. 105–121]. Notice that this tertiary-level emotion is deeply embedded in social comparisons and social relations.

The example of resentment shows that tertiary emotions can take three basic forms as pairings of one primary constituent and the secondary emotion comprised

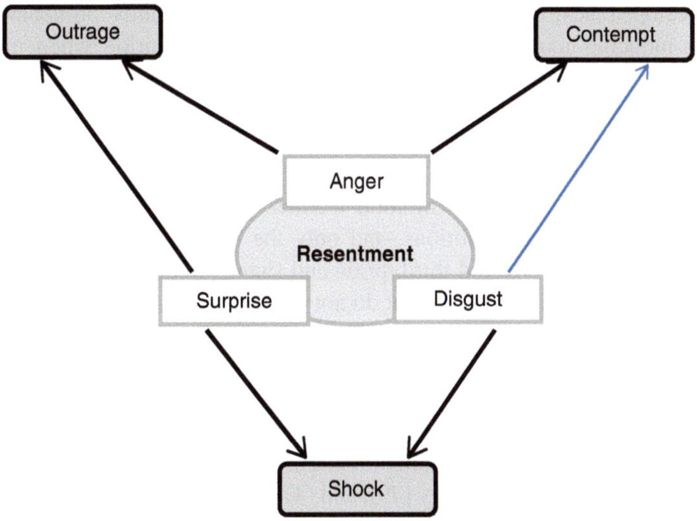

Fig. 4 Proposed primary and secondary emotions of resentment. Source: W. D. TenHouten. 2017. *Alienation and affect*. New York: Routledge, Fig. 8.1, p. 119

of the other two primary emotions. Opposite secondary emotions make natural units of analysis, which we next illustrate with a study linking the pride-shame opposition to their constituent primary emotions and to elementary social relations.

6 Pride and Shame: Two Opposite Secondary Emotions

Ethologists and primatologists have been reluctant to attribute pride and shame to nonhuman species, although nonhuman primates experience "proto-pride" [71] and chimpanzees can display "prideful" motivation to achieve dominance [72]. Many of the patterned, largely involuntary actions associated with shame in humans resemble the appeasement displays of various nonhuman primates [71, 73]. Both proto-pride and proto-shame are likely pan-primate, and these emotions' core might well be shared by all mammals and most vertebrates [74, p. 268]. This possibility contradicts longstanding conventional wisdom which holds that, while "basic" emotions such as fear and anger have evolved across animal species, more complex, cognitively elaborated, self-focused emotions such as pride and shame arose de novo in humans. There is thus growing appreciation that even the higher, self-conscious emotions that involve cognitive self-appraisal, including pride and shame [75, 76], are evolutionarily rooted in other species' simpler emotions and have emerged as complex mixtures of these emotions.

There exist vast differences between the proto-pride and proto-shame of animals and the human experiences of pride and shame. One difference is that only humans have developed abstract systems of values, ideals, standards of behavior, and ambition-inducing motivational systems that can trigger pride when upheld or shame when transgressed. Value systems can govern means as well as ends. Individuals accordingly can experience pride when, motivated by their ideals, they assertively endeavor to succeed but nonetheless fall short or "die trying." They can analogously experience shame if they succeed but become known to have done so in a disgraceful, unethical, or immoral manner. Second, for humans, pride and shame are self-conscious emotions, and only the most advanced animal species manifest self-recognition. And third, pride and shame, in humans, can be involved in competence/mastery situations that do not involve hierarchical social relations.

The observation that the emotions pride and shame have deep evolutionary roots is further supported by (1) the cross-cultural universal drive to dominance in humans [77], for humans readily learn both domination and submissive behaviors, which tend to emerge in situations of competition [30]; (2) their early developmental onset, with self-esteem, achievement-motivation, and rivalry emerging roughly around ages 3–4 [78]; (3) their spontaneity [79]; and (4) their presence in those congenitally blind [79].

6.1 *Pride*

Both pride and shame have been conceptualized as "second-order" emotions [80], but only Ribot [81] and Plutchik [17, p. 117] have attempted to identify their possible constituent "primary" or more "basic" emotions. Ribot conceptualized pride, a feeling of strength, as physiologically and psychologically based on the emotions anger and, to a lesser extent, joy and pleasure. Plutchik [17, p. 117] more explicitly proposed that "anger + joy = pride." The winner of a status competition, who has gained a position of dominance and has acquired resources and rewards, is apt to have experienced and manifested anger (behaviorally, movement toward a desired goal) and joy (satisfaction, and even celebration, of having gained resources and rewards); these combine in a straightforward manner in the inner experience and public expression of pride.

6.1.1 The Anger Component of Pride

In the authentically proud individual who has overcome obstacles and opposition, or has succeeded in a competitive endeavor, expressions of pride "show others one has achieved standards [and] show dominance/superiority" [82, p. 42]. This process involves anger, not in the sense of being irritated or annoyed with others but functionally by an action tendency Barrett calls "outward movement; inclination to show/tell others." Frijda [83, p. 88] similarly describes anger as having as its end state the "removal of obstruction," an "action tendency," and "agonistic" effort to gain or regain control. Children, ranging in ages from four to nearly 12, saw postural expressions that adults adjudged prideful as anger rather than pride [84]. This finding suggests that anger is closely related to, if not interior to, pride. Ribot [81, p. 241, 244] claimed that anger belongs to pride, which he described as a feeling of superiority, "tenacious vitality," and a "monomania of power." Ribot referred to anger as a "primary emotion" and traced pride to the "arrogant attitudes... and ostentatious displays of the peacock' and other birds in their mating behavior." Plutchik [17, p. 60, 78–84, 114] saw the basic evolutionary function of anger as "destruction" and its core stimulus event as an "obstacle." Anger is an approach-oriented, goal-directed emotion that prioritizes the attainment of favorable outcomes ([85]; see also [86]). This idea is elaborated by Panksepp [42, p. 189], who writes that "The aim of anger is to increase the probability of success in the pursuit of one's ongoing desires and competition for resources." If Plutchik was correct in contending that anger belongs to the existential problem of hierarchy and pride has evolved as an adaptive reaction to successful action in some social dominance hierarchies, then anger is interior to pride.

6.1.2 The Joy-Happiness Component of Pride

Pride is distinguishable from joy or happiness, but pride, a euphoric state of mind and body, can be seen to include happiness, a natural reaction to winning, successful effort, and the acquisition of new resources and rewards. Joy, happiness, and a sense of self-satisfaction are associated with the attainment of a high social rank, or of a conspicuous accomplishment, largely because one has gained material or social resources. Opposed to "thwarted self-assertion," Woodworth [87, p. 166] noted, is a "cheerful state of mind of one who seeks to master some person or thing and fully expects to do so, and elation, the joyful state of one who has mastered."

Michael Lewis [88, p. 168, 171] observes that "The phenomenological experience [of pride] is joy over an action, thought, or feeling well done.... We observe pride as well as happiness when children of any age succeed" in competitive endeavors. There is a natural connection between status attainment and feelings of joy. "Success, especially success in an exciting venture, triggers joy...." The affect "enjoyment–joy" is one form of "healthy pride" [89, p. 84], which, following successful goal-attainment activity, involves "effectance pleasure" [90]. Pride follows personal efficacy, a sense of uniqueness, and social distinction and is sure to include a joyful feeling of a positive evaluation of the self in comparison to others. Lazarus and Lazarus [91, p. 100] distinguish between feeling happy and feeling proud by conceptualizing pride as "enhancement of one's personal worth by taking credit for a valued object or achievement."

6.2 Shame

> The affect of shame...is much more complex and much richer in cognitive elements than affects such as rage, anxiety, or pleasure, which can be called "simple affects." In its complexity, it resembles...its own counterpart, pride, and thus belongs to the group of "compound" affects.... *Léon Wurmser* [92, p. 69]

Just as pride or proto-pride in many animals, especially primates, functions as a threat display and an assertion of dominance, so also proto-shame functions as an appeasement display, indicating acquiescence to, and acceptance of, the higher dominance ranking of a conspecific [71, 73, 93]. Shame displays, in humans, can occur following a social transgression, a failure in a competitive endeavor, or an act which is inappropriate given one's status or immoral given one's value system. Expression of shame, however indirect, can adaptively enable subdominant individuals to avoid punishment, negative appraisal, and denial of needed resources [94].

A few emotions researchers have declared shame to be a primary or fundamental emotion [33, 50, 95]. Plutchik proposed a secondary emotional combination, "fear + disgust = shame, prudishness." This formulation is questionable insofar as the combination of fear and disgust might be better defined as repugnance [9, p. 18], [19, pp. 81–82], and prudishness would appear to be a personality or character trait rather than an emotion. We propose that the complex sociorelational reality of degraded position involves the primary emotions *fear* and *sadness*, which can combine to form *shame* [96–98].

6.2.1 The Fear Component of Shame

In describing the "humble" type of individual, Ribot [81, p. 394] saw that "Their dominant note is timidity, fear, and all paralyzing modes of feeling…. They are afraid for themselves, for their families, for their small position," as they "feel…the weight of the social organism pressing against them," for "they are conscious of being weak, and without springs of action or the spirit of initiative."

Both guilt and shame involve fear, but while guilt involves fear of retribution or punishment for one's transgressions, shame involves the fear of negative evaluation, reproach, or condemnation of the self by others. Just as pride involves a sense of achievement, shame involves a fear that one is lacking or has failed. Those most fearful of failure see the latter as an unacceptable event with negative implications for their self-worth. Fear is a basic emotional reaction whose behavioral concomitant is withdrawal, avoidance, flight, and hiding. That fear is interior to shame can be seen in shame-driven behaviors, which involve looking down or away from the gaze of others, a desire to "hide" or "crawl under the rug," and engaging in various other forms of what Plutchik sees as the core behavior of fear, namely, "running or flying away" from [52, p. 289], disappearing from, or escaping the psychological pain of a shame-eliciting situation [81, pp. 196–250].

The end point of the process of shame as a complex adaptation is a reaction of "preventing dangerous exposure" [92, p. 84] and an "affective motive of defense" [99, p. 138]. Wurmser [92, p. 52] argues that the less vulnerable one feels about threats to the self, the less one will fear exposure. Plutchik [17] defined fear as an adaptive reaction to danger or threat; within shame there is always fear, which can find expression in flight—globally in running away from, or partially, in hiding or forgetting parts of one's body, life, or self. The fear within shame is expressed most radically in self-dissolution (suicide).

6.2.2 The Sadness Component of Shame

> Shame is a certain kind of sorrow which arises in one when he happens to see that his conduct is despised by others.... *Baruch Spinoza [1660]* [100, p. 76]

Fear is both toxic and debilitating and is closely linked to sadness. Izard [101, p. 197] refers to a "sadness-fear bind" that can produce a lack of physical courage and induce sadness. It results in fearful or even panicky behavior rather than the assertive action required to attain a position of high status or social dominance. The sadness-fear bind can be generalized beyond particular situations, as when dealing with life's problems becomes overwhelming and "the early and strong linkages of sadness and fear snowball" [102, p. 197].

Unacknowledged shame "can be transformed into sadness," and repeated instances of shame (which result in a more global shamefulness) advance the sadness involved in loss of self to intense sadness, even to depression. Sadness-depression is thus "an element of shame," not as "a conversion of shame but an accompanying emotion" [102, pp. 143–144]. Lewis further maintains, "When individuals experience shame..., they show behavioral characteristics of a sad person. They gaze avert, hunch their shoulders up, push their bodies inward, become inhibited, ...show problems in thinking, [and] appear to be sad." Sadness, Lewis explains, occurs around unacknowledged shame. While the self does not admit shame, its sadness component emerges in consciousness with the realization that others, through their disapproval, expose one's degraded self-image. Feeling sadness at the loss of self, the individual focuses on the social conditions of the harmful situation and the elicitors of the emotion rather than on the shame itself. Lewis suggests that sadness is more comfortable to experience than shame, so only the sadness is acknowledged; it is projected into the social encounter rather than back onto the self. In the experience of shame, there is a sense of loss. In the affective experience of shame, one senses the loss of the perceived worth of the self or, more profoundly, the loss of an intact, fully developed self.

7 Social Relations and the Primary Emotional Components of Pride and Shame: Results of an Empirical Study

To investigate causal models of pride and shame, we used a lexical-level, content-analytic analysis of life-historical interviews. The results presented here are from TenHouten [98]. After eliminating interview transcripts of less than 2000 words, the cross-cultural dataset consists of 563 interviews, with 265 Euro-Australians (45% female) and 298 Aborigines (46% female). The mean interview length was 10,534 words (standard deviation 11,659). The variables used in analysis were largely constructed using Roget's (1852) [103] *International Thesaurus*, although considerable combining and splitting of category wordlists was necessary.

7.1 Pride Analysis

The variables used for the pride analysis consist of wordlist indicators of the positive experiences of authority-ranked (AR+) and communally shared (CS+) social relations and the emotions anger, joy-happiness, and pride. A measurement model based on five manifest indicators of pride was constructed using covariance structure analysis (using SAS Calis, maximum likelihood estimation). The program converged after 15 iterations and the model fit results suggested good fit. For all five wordlists, a sum of the category's words was defined, and individual words were retained only if they had part-whole correlations of at least 0.05. Representative words for pride and the other manifest variables are shown in Table 2.

Figure 5 shows the results of the causal analysis of the five theoretical concepts. In this figure, the manifest variables are enclosed in rectangles and latent variables

Table 2 Representative words for social relations and emotions of pride

Relational models, emotions	
Indicators (number of words)	*Representative words*
Authority ranking, positive	
Dominance [20]	Dominant, dominate, dominates, predominant, supremacy
Authority [26]	Mightily, prerogative, rule, ruling, say-so, reigning, presiding
Command [46]	Commanding, dictated, directs, instructor, mandates, orders
Communal sharing, positive	
Welcome [36]	Welcome, guest, entertainment, hosts, hospitality, visited
Friends [15]	Companions, friends, friend, neighbors, neighboring, chum
Kindness [38]	Benign, considerate, gently, goodness, humanely, sympathetic
Anger	
Enmity [14]	Acrimonious, antagonize, enmity, grudge, hostility, unfriendly
Irritation [8]	Irritable, irritate, peeved, pique, provoke, rankled, umbrage
Disapproval [13]	Admonish, chided, criticized, deprecated, rebuke, reproaching
Joy-happiness	
Joyfulness [5]	Glad, gladden, joy, overjoyed
Enjoyment [7]	Enjoy, enjoyable, enjoyed, enjoys, exhilarating
Happiness [6]	Happily, happiness, heartwarming
Pride	
Self-confidence [27]	Self-assured, self-respect, self-reliant, courage, determination
Confidence [10]	Confident, confidently, reassurance, optimism, optimistic bold
Pride [15]	Pride, proud, proudly, self-satisfied, affected, dignified, dignity
Competitiveness [33]	Aims, ambitions, goals, intention, purpose, challenging, struggle
Accomplishment [12]	Initiative, wills, diligently, industrious, persevered, unrelenting

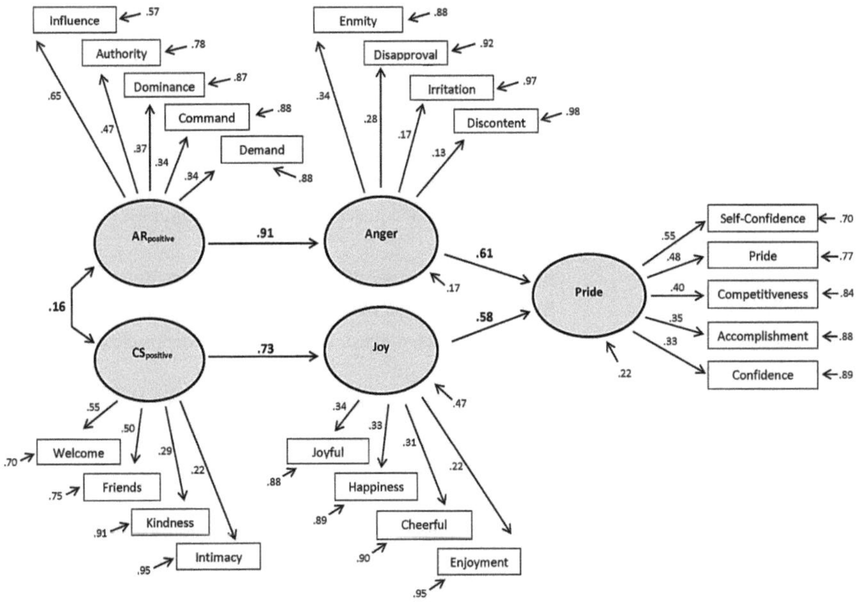

Fig. 5 Anger and joy-happiness as functions of authority-ranked (AR+) and communally shared (CS+) positively valenced social relations, pride as a function of anger and joy-happiness. Source: W. D. TenHouten. Social dominance hierarchy and the pride shame system. J Polit Power 2017;10(1):13. Fig. 3

in ellipses. The model converged after 11 iterations, and there was adequate fit between data and model: root mean square residual (RMSR) = 0.06 and adjusted goodness-of-fit index (GFI) = 0.92. The path coefficients from social relations to primary emotions were, as predicted, positive: AR+ → anger 0.91, CS+ → joy 0.73, as were the paths from primary emotions to pride, with anger → pride 0.61 and joy → pride 0.58.

7.2 *Shame Analysis*

Analysis of shame requires two sociorelational variables, AR− and CS−, the two primary emotion variables, fear and sadness, and shame. Representative words for the wordlist indicators for variables AR−, CS−, fear, sadness, and shame are shown in Table 3.

The measurement model for the latent variable showed a close fit between model and data, suggesting a unitary dimension. The standardized weights for six indicators of shame were humility 0.26, servility 0.17, modesty 0.46, disparagement 0.61, ridicule 0.26, and embarrassment 0.23. The causal model for pride is shown in Fig. 6. The data adequately fit the model (RMR 0.08, GFI 0.98). The four causal

Table 3 Representative words for social relations and emotions of shame

Relational models, emotions	
Indicators (number of words)	Representative words
Authority ranking, negative	
Lack of influence [10]	Impotent, ineffective, ineffectual, powerlessness
Invisibility [21]	Disguised, hidden, invisible, unseen, value, vaguely
Inferiority [20]	Deficiencies, inadequate, inferior, inferiority, unskilled
Communal sharing, negative	
Death [32]	Bereavement, corpse, dead, deaths, die, mortality, perish
Unkindness [6]	Brutish, beastly, wicked, vicious, unkindness, cruel
Seclusion [6]	Seclude, secrets, outcast, out-of-the-way, defenseless
Fear	
Fright, terror [22]	Afraid, fear, fright, panicked, qualms, scared, scary, terrify
Disquietude [14]	Uneasiness, worries, worried, uneasy, plaguing, disturbed
Concern [14]	Bother, concern, distressed, haunted, nerves, troubled
Sadness	
Sadness [22]	Joyless, ruefully, sad, saddest, sorrow, sullenness, unhappy
Misery [22]	Bleak, despair, despondent, forlornly, futile, beset, bother
Lamentation [12]	Bawl, crying, groaned, lament, howling, moaned, plaintive
Shame	
Humility [25]	Debase, humble, comedown, humiliation, shamed, shaming
Servility [17]	Cringe, flatter, grovel, kowtowing, lackey, obsequious, peon
Modesty [17]	Coy, meekness, modest, self-doubt, sheepish, timid, shy
Disparagement [18]	Belittle, defamation, denigrating, name-calling, slander
Ridicule [16]	Derision, insulting, put-downs, scoffed, taunting, teased
Embarrassment [8]	Chagrin, disconcerting, embarrass, embarrassed, mortification

pathways were substantial and, as predicted, positively valenced. These paths coefficients were AR− → fear 0.49, CS− → sadness 0.77, fear → shame 0.58, and sadness → shame 0.61.

8 Discussion

We have explored the evolutionary origins of the existential problems underlying the primary emotions, presented a complete classification of the secondary emotions, briefly described one hypothesized tertiary-level emotion, and presented a theoretical framework in which the causal mechanisms of emotions are found not in specific brain mechanisms but in simple or complex valenced social relations models. This theoretical approach places the entire spectrum of affect in social and cultural context.

Perhaps the two most useful models of primary emotions are those presented by Ekman and colleagues and by Plutchik. Plutchik and Ekman identify the same six

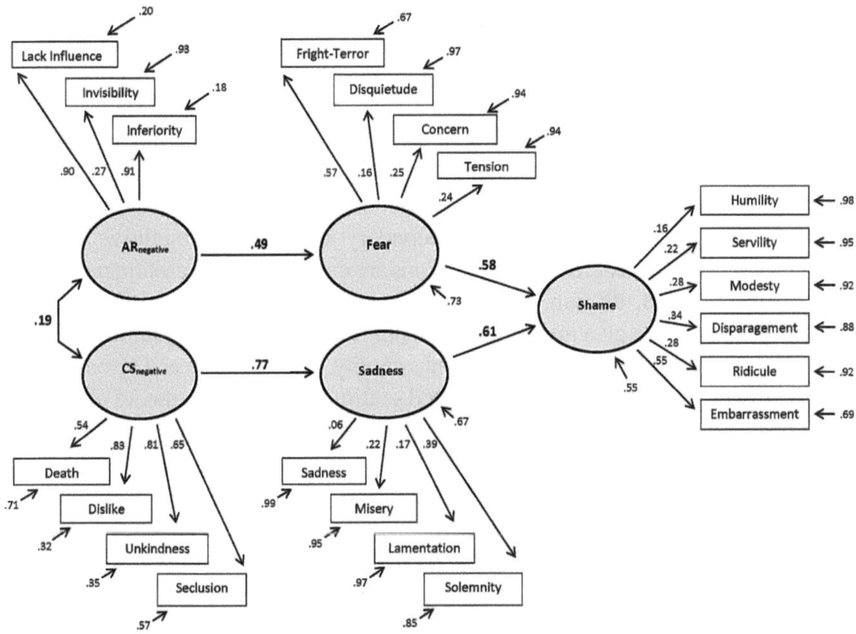

Fig. 6 Fear and sadness as functions of authority-ranked (AR) and communally shared (CS) negatively valenced social relations, shame as a function of fear and sadness. Source: W. D. TenHouten. Social dominance hierarchy and the pride shame system. J Polit Power 2017;10(1):14. Fig. 4

emotions as primary. Plutchik also identifies acceptance and anticipation as primary, even though they do not possess distinct facial expressions. Given that even secondary emotions can be recognized across cultures, we cannot isolate facial recognition as a criterion for adjudging an emotion primary or secondary.

It can be argued, of course, that secondary and tertiary emotions are social-intention states, or sentiments, which are more complex than their primary components and possess high levels of cognitive content. However, we note that (1) all emotions (with the possible exception of fear induced by a falling tree branch) and anger (upon stubbing one's toe) are social-intention states, a point that has been amply demonstrated in the sociology of emotions. And (2) it makes sense that secondary emotions would have more cognitive content than primary emotions, and it is reasonable to speculate that tertiary emotions would, in turn, have a higher level of cognitive content than secondary emotions.

While the debate about the existence, or nonexistence, of primary emotions continues unabated, less attention has focused upon the consequences of these two claims. If there are no primary emotions and all emotions therefore exist sui generis, then there can be no hierarchical classification of the emotions. However, if as claimed above primary emotions do exist, then it becomes possible to classify the complex emotions formed from pairs and triples of the primary emotions. It is additionally possible to distinguish complex emotions from affective states, feelings, and sentiments that are not emotions. Plutchik recognized this implication of his model of primary emotions, and this recognition led him to develop an innovative, if not entirely successful, classification of secondary emotions. Plutchik, however, took little subsequent interest in explaining or investigating his own classification of secondary emotions, and he only gestured toward classifying one tertiary-level emotion. There are important complex emotions that, according to the present classification, are neither primary nor secondary and therefore might well be tertiary. Among these possible tertiary emotions are envy, jealousy, resentment, hatred, bliss, despair, dread, disdain, confidence, worry, and sanguinity. Other complex affective states, while also important for social life, appear less definable as tertiary emotions; these include disillusionment, enmity, enthusiasm, and grouchiness, which might be seen as socially constructed sentiments. Without a model of primary emotions, definitional questions about higher-order affective states cannot be resolved either conceptually or empirically.

Prominent scholars have argued that emotions cannot be classified. In philosophy, Spinoza [100, p. 63] opined that, "the emotions may be compounded one with another in so many ways, and so many variations may arise therefore, as to exceed all possibility of computation." In sociology, Durkheim and Mauss, in 1903 [104, pp. 86–87], reached the remarkable conclusion that all social classifications are ultimately based on sentiments and that the "emotional value of notions…is the dominant characteristic in classification." At the same time, they lamented, "States of an emotional nature…mingle their properties in such a way that they cannot be rigorously categorized." In contrast to these pessimistic views, our premise holds that there is indeed a set of basic, or elementary, emotions with deep evolutionary roots, existing as natural kinds, and that Plutchik correctly identified them six decades ago. If this is the case, then the number of more complex emotions that can be formed from the most primordial emotions is not, as Spinoza asserted, beyond all possible computation. Instead, it can be inferred, by combinatorial logic, that if there are eight primary emotions, then there can be 28 secondary emotions, 56 tertiary emotions, and 92 in all.

If emotions do not mix or combine to form more complex emotions, then it makes little difference if primary emotions exist or not. This issue is of slight concern to affective neuroscientists, who typically focus on studies of single emotions that are not difficult to evoke in a laboratory setting. But for the social psychology, sociology, and anthropology of emotions, the possibility of primary emotions mixing and combining makes a great deal of difference, because the emotions most interior to social life are indeed complex and embedded in social relations.

By considering valence only in passing, Fiske [105, p. 9] links emotions to social relations in an ad hoc, intuitive manner. He sees "love in some CS relationships," for example, but love is rather seen here as a secondary emotion, the joyful acceptance of another, which occurs with the joint experience of CS+ and EM+, the sociorelational sources of joy and acceptance, respectively [9, pp. 50–57]. Fiske [91, pp. 11–12] sees awe/reverence in "AR (looking up)" (which implicitly means AR−), but affect-spectrum theory rather classifies awe (and alarm) as comprised of surprise and fear, from the joint occurrence of MP− and AR− [19, pp. 25–26]. Aggressiveness can be defined as a mixture of anticipation and anger reactive to the joint experience of MP+ and AR+. Awe−alarm and aggressiveness are opposites, which makes sense insofar as alarm is a defensive reaction to aggression [9, pp. 73–90,144–146]. In the above analysis, pride is linked, both theoretically and empirically, to AR+ and CS+ and shame to AR− and CS−. These examples suffice to show that only by attaching valences to the four social relations models can social relations be systematically cross-classified with the emotions.

Fiske's relational models theory has been criticized for ignoring situations of conflict and contention and complex emotions such as aggressiveness and awe [106]. More specifically, it ignores the *negatively valenced* experiences of social relations. We need to consider not only happy families, legitimate authority, social equality, and wealth but also the negative poles of the relational models: EM− can mean one is not being treated equally and is regarded as inferior; CS−, that one's place in community is disrupted; AR–, that one is powerless or subordinated and exploited; and MP−, that one is in impoverished and in economic distress. Affect-spectrum theory thus proposes that there are not four, but eight, elementary forms of sociality and that these and the emotions to which they are causally linked exist as natural kinds.

References

1. Cacioppo JT, Berntson GG. Social neuroscience. In: Cacioppo JT, Berntson GG, Adolphs R, et al., editors. Foundations in social neuroscience. Cambridge: MIT Press; 2002. p. 3–10.
2. Connolly WE. Neuropolitics: thinking, culture, space. Minneapolis: University of Minnesota Press; 2002. p. 219.
3. Vander Valk F, editor. Essays on neuroscience and political theory: thinking the body politic. New York: Routledge; 2012. p. 294.
4. Glimcher PW, Camerer CF, Fehr E, Poldrack RA, editors. Neuroeconomics: decision making and the brain. 2nd ed. Amsterdam: Elsevier; 2014. p. 562.
5. Laughlin CD, d'Aquili E. Biogenetic structuralism. New York: Columbia University Press; 1974. p. 211.
6. Lende DH, Downey G, editors. The enculturated brain: an introduction to neuroanthropology. Cambridge: The MIT Press; 2012. p. 448.
7. TenHouten WD. Neurosociology. J Soc Evolution Syst. 1997;20(1):7–37.
8. TenHouten WD. Explorations in neurosociological theory: from the spectrum of affect to time-consciousness. In: Franks DD, Smith TS, editors. Mind, brain, and society: toward a neurosociology of emotion. Stamford: JAI Press; 1999. p. 41–80.
9. TenHouten WD. Emotion and reason: mind, brain, and the social domains of work and love. London: Routledge; 2013. p. 298.
10. Franks DD. Neurosociology: the nexus between neuroscience and social psychology. New York: Springer; 2010. p. 216.
11. Franks DD, Turner JH, editors. Handbook of neurosociology. New York: Springer; 2013. p. 406.
12. Kalkhoff W, Thye SR, Lawler EJ. Biosociology and neurosociology. Bingley, UK: Emerald Grove Publishers; 2012. p. 266.

13. Verweij MT, Senior TJ, Dominguez JF, Turner R. Emotion, rationality, and decision-making: how to link affective and social neuroscience with social theory. Front Neurosci. 2015;9:745–8.
14. Fiske AP. Structures of social life: the four elementary forms of human relations. New York: The Free Press; 1991. p. 480.
15. MacLean PD. The triune brain in evolution: role in paleocerebral functions. New York: Plenum Press; 1990. p. 672.
16. Plutchik R. Outlines of a new theory of emotion. Trans N Y Acad Sci. 1958;20(5):394–403.
17. Plutchik R. The emotions: facts, theories, and a new model. 2nd ed. Lanham: University Press of America; 1991. p. 216.
18. Plutchik R. Universal problems of adaptation: hierarchy, territoriality, identity, and temporality. In: Calhoun JB, editor. Environment and population: problems of adaptation. New York: Praeger; 1983. p. 223–6.
19. TenHouten WD. A general theory of emotions and social life. London: Routledge; 2007. p. 308.
20. Barrett LF. Are emotions natural kinds? Perspect Psychol Sci. 2006;1(1):28–58.
21. Ortony A, Clore GL, Collins A. The cognitive structure of emotions. New York: Cambridge University Press; 1988. p. 207.
22. Scherer KR, Schorr A, Johnstone T. Appraisal processes in emotion: theory, methods, research. Oxford: Oxford University Press; 2001. p. 478.
23. Lindquist KA, Wager TD, Kober H, Bliss-Moreau E, Barrett LF. The brain basis of emotion: a meta-analytic review. Behav Brain Sci. 2012;35(3):121–53.
24. Griffiths PE. Emotions as normative and natural kinds. Philos Sci. 2004;21(6):759–77.
25. Gordon SL. The sociology of sentiments and emotions. In: Rosenberg M, Turner RH, editors. Social psychology: sociological perspectives. New York: Basic Books; 1981. p. 562–92.
26. Harré R, editor. The social construction of emotions. New York: Basil Blackwell; 1986. p. 316.
27. McCarthy ED. Emotions are social things: an essay in the sociology of emotions. In: Franks DD, McCarthy ED, editors. The sociology of emotions: original essays and research papers. Greenwich: JAI Press; 1989. p. 51–72.
28. Boiger M, Mesquita B. The construction of emotion in interactions, relationships, and cultures. Emot Rev. 2012;4(3):22–9.
29. Mesquita B, Boiger M, De Leersnyder J. The cultural construction of emotions. Curr Opin Psychol. 2016;8:31–6.
30. Eibl-Eibesfeldt I. Human ethology. Hawthorne: Aldine De Gruyter; 1989. p. 848.
31. Buck R. The biological affects: a typology. Psychol Rev. 1999;106(2):301–36.
32. Panksepp J. Foreword: the MacLean legacy and some modern trends in emotion research. In: Cory GA, Gardner R, editors. The evolutionary neuroethology of Paul MacLean: convergences and frontiers. Westport: Praeger; 2002. p. ix–xxx.
33. Izard CE. Human emotions. New York: Plenum Press; 1977. p. 495.
34. Izard CE. Basic emotions, natural kinds, emotion schemas, and a new paradigm. Perspect Psychol Sci. 2007;2(3):260–8.
35. Izard CE, Woodburn EM, Finlon KJ. Extending emotion science to the study of discrete emotions in infants. Emot Rev. 2010;2(2):134–6.
36. Lövheim HA. A new three-dimensional model for emotions and monoamine neurotransmitters. Med Hypotheses. 2012;78(2):341–8.
37. Shewmon DA, Holmes GL, Byrne PA. Consciousness in congenitally decorticated children: developmental vegetative state as self-fulfilling prophecy. Dev Med Child Neurol. 1999;37(5):364–74.
38. Sroufe LA. Emotional development: the organization of emotional life in the early years. New York: Cambridge University Press; 1997. p. 365.
39. LaFrenière P. Emotional development: a biosocial perspective. Wadsworth Thomson Learning: Belmont, CA; 2000. p. 331.

40. Demos EV. The dynamics of development. In: Muller JP, Brent J, editors. Self-organizing complexity in psychological systems. Lanham: Jason Aronson; 2007. p. 135–63.
41. Darwin C. The expression of the emotions in man and animals. Chicago: University of Chicago Press; 1965. p. 372.
42. Panksepp J. Affective neuroscience: the foundations of human and animal emotions. Oxford: Oxford University Press; 1998. p. 466.
43. Panksepp J, Biven L. The archeology of mind: neuroevolutionary origins of human emotions. New York: Norton; 2012. p. 562.
44. Panksepp J. The basic emotional circuits of mammalian brains: do animals have affective lives? Neurosci Biobehav Rev. 2011;35(9):1791–804.
45. Cerqueira CT, Almeida JR, Gorenstein C, Gentil V, Leite CC, Sato JR, Amaro E Jr, Busatto GF. Engagement of multifocal neural circuits during recall of autobiographical happy events. Braz J Med Biol Res. 2008;41(12):1076–85.
46. Brunia CH. Neural aspects of anticipatory behavior. Acta Psychol (Amst). 1999;101(2–3):213–32.
47. Ekman P, Sorenson ER, Friesen WV. Pan-cultural elements in facial displays of emotions. Science. 1969;164(3875):86–8.
48. Sauter DA, Eisner F, Ekman P, Scott SK. Cross-cultural recognition of basic emotions through nonverbal emotional vocalizations. Proc Natl Acad Sci U S A. 2010;107(6):2408–12.
49. Ekman P. An argument for basic emotions. Cogn Emot. 1992;6(3–4):169–200.
50. Tomkins SS. Affect imagery consciousness, vol. II, the negative effects. New York: Springer; 1963. p. 580.
51. Delgado AR. Order in Spanish colour words: evidence against linguistic relativity. Brit J Psychol. 2004;95(1):81–90.
52. Plutchik R. Emotion: a psychoevolutionary synthesis. New York: Harper & Row; 1980. p. 440.
53. Reiner A. An explanation of behavior: the triune brain in evolution: role of paleocerebral function. Paul D. MacLean. Plenum, New York, 1990 [book review]. Science. 1990;250(4978):303–5.
54. Butler AB, Hodos W. Comparative vertebrate neuroanatomy: evolution and adaptation. New York: Wiley; 1996. p. 744.
55. Cory GA. Reappraising MacLean's triune brain concept. In: Cory GA, Gardner R, editors. The evolutionary neuroethology of Paul MacLean: convergences and frontiers, vol. 2002. Westport: Praeger; 2002. p. 9–27.
56. Balasubramani PP, Chakravarthy VS, Ravindran B, Moustafa AA. A network model of basal ganglia for understanding the roles of dopamine and serotonin in reward-punishment risk based decision making. Front Comput Neurosci. 2015;9:76. https://doi.org/10.3389/fncom.2015.00076.
57. Van Lancker Sidtis D, Pachana N, Cummings JL, Sidtis JJ. Dysprosodic speech following basal ganglia insult: toward a conceptual framework for the study of the cerebral representation of prosody. Brain Lang. 2006;97(2):135–53.
58. Douglas M. Essays in the sociology of perception. London: Routledge; 1982. p. 340.
59. Thompson M, Ellis RJ, Wildavsky A. Cultural theory. Boulder: Westview Press; 1990. p. 296.
60. Damon W. Patterns of change in children's social reasoning: a two-year longitudinal study. Child Dev. 1980;51(4):1010–7.
61. Enright RD, Enright WF. Distributive justice and social class: a replication. Dev Psychol. 1981;17(6):826–32.
62. Haslam N. Categories of social relationship. Cognition. 1994;53(1):59–90.
63. Bolender J. The self-organizing social mind. Cambridge: The MIT Press; 2010. p. 190.
64. Hume D.. In: Selby-Bigge LA, editor. A treatise of human nature. Oxford: Clarendon Press; [1739] 1978. 743 p.
65. Du S, Martinez AM. The resolution of facial expressions of emotion. J Vis. 2011;11(13):1–13.

66. Du S, Tao Y, Martinez AM. Compound facial expression of emotion. Proc Natl Acad Sci U S A. 2014;111(15):e1454–62.
67. TenHouten WD. Dual symbolic classification and the primary emotions: a proposed synthesis of Durkheim's sociogenic and Plutchik's psychoevolutionary theories of emotion. Int Sociol. 1995;10(4):427–45.
68. TenHouten WD. Outline of a socioevolutionary theory of the emotions. Int J Sociol Soc Policy. 1996;1(9–10):190–208.
69. TenHouten WD. Alienation and affect. London: Routledge; 2017. p. 213.
70. Smith HJ, Pettigrew TF, Pippin GM, Bialosiewicz S. Relative deprivation: a theoretical and meta-analytic review. Pers Soc Psychol Rev. 2012;16(3):203–32.
71. Fessler DM. From appeasement to conformity: evolutionary and cultural perspectives on shame, competition, and cooperation. In: Tracy JL, Robins RW, editors. The self-conscious emotions: theory and research. New York: Guilford Press; 2007. p. 174–93.
72. Wrangham R, Peterson D. Demonic males: apes and the origins of human violence. New York: Houghton Mifflin Harcourt; 1996. p. 350.
73. Gilbert P. Depression: the evolution of powerlessness. London: Routledge; 1992. p. 559.
74. Fessler DM, Gervais M. Whence the captains of our lives: ultimate and phylogenetic perspectives on emotions in humans and other primates. In: Kappeler PM, Silk JB, editors. Mind the gap: tracing the origins of human universals. New York: Springer; 2009. p. 261–82.
75. Sznycer D, Takemura K, Delton AW, Sato K, Robertson T, Cosmides L, Tooby J. Cross-cultural differences and similarities in proneness to shame: an adaptationist and ecological approach. Evol Psychol. 2012;10(2):352–70.
76. Sznycer D, Al-Shawaf L, Bereby-Meyer Y, et al. Cross-cultural regularities in the cognitive architecture of pride. Proc Natl Acad Sci U S A. 2017;114(8):1874–9.
77. Tracy JL, Robins RW. The nonverbal expression of pride: evidence for cross-cultural recognition. J Pers Soc Psychol. 2008;94(3):516–30.
78. Stipek D. The development of pride and shame in toddlers. In: Tangney JP, Fischer KW, editors. The self-conscious emotions: the psychology of shame, guilt, embarrassment, and pride. New York: Guilford Press; 1995. p. 237–52.
79. Tracy JL, Matsumoto D. The spontaneous expression of pride and shame: evidence for biologically innate nonverbal displays. Proc Natl Acad Sci U S A. 2008;105(33):11655–60.
80. Fessler DM. Toward an understanding of the universality of second order emotions. In: Hinton AL, editor. Biocultural approaches to the emotions. Oxford: Cambridge University Press; 1999. p. 75–116.
81. Ribot T. The psychology of the emotions. 2nd ed. London: Walter Scott; 1911. p. 455.
82. Barrett KC. A functionalist approach to shame and guilt. In: Tangney JP, Fischer KW, editors. Self-conscious emotions: the psychology of shame, guilt, embarrassment, and pride. New York: Guilford; 1995. p. 25–63.
83. Frijda NH. The emotions. New York: Cambridge University Press; 1986. p. 544.
84. Nelson NL, Russell JA. Children's understanding of nonverbal expressions of pride. J Exp Child Psychol. 2012;111(3):379–85.
85. Tomarken AJ, Zald DH. Conceptual, methodological, and empirical ambiguities in the linkage between anger and approach: comment on Carver and Harmon-Jones. Psychol Bull. 2009;135(2):209–14.
86. Carver CS, Harmon-Jones E. Anger is an approach-related affect: evidence and implications. Psychol Bull. 2009;135(2):183–204.
87. Woodworth RS. Psychology: a study of mental life. New York: Henry Holt; 1924. p. 580.
88. Lewis M. The rise of consciousness and the development of emotional life. New York: Guilford Press; 2014. p. 352.
89. Nathanson DL. Shame and pride: affect, sex, and the birth of the self. New York: Norton; 1992. p. 496.
90. Broucek FJ. Efficacy in infancy: a review of some experimental studies and their possible implications for clinical theory. Int J Psychoanal. 1979;60(3):311–6.

91. Lazarus RS, Lazarus BN. Passion and reason: making sense of our emotions. Oxford: Oxford University Press; 1994. p. 321.
92. Wurmser L. The mask of shame. Baltimore: Johns Hopkins University Press; 1981. p. 345.
93. Weisfeld GE. Discrete emotions theory with specific reference to pride and shame. In: Segal NL, Weisfeld GE, Weisfeld CC, editors. Uniting psychology and biology: integrative perspectives on human development. Washington, DC: American Psychological Association; 1997. p. 419–43.
94. Shariff AF, Tracy JL, Markusoff JL. (Implicitly) judging a book by its cover: the power of pride and shame expressions in shaping judgments of social status. Per Soc Psychol Bull. 2012;38(9):1178–93.
95. Scheff T. Toward defining basic emotions. Qual Inq. 2015;21(2):111–21.
96. TenHouten WD. Normlessness, anomie, and the emotions. Sociol Forum. 2016;31(2):465–86.
97. TenHouten WD. The emotions of powerlessness. J Polit Power. 2016;9(1):83–121.
98. TenHouten WD. Social dominance hierarchy and the pride–shame system. J Polit Power. 2017;10(1):94–114.
99. Fenichel O. The psychoanalytic theory of neurosis. New York: Norton; 1945. p. 703.
100. Spinoza B. In: Elwes RH, translator, Runes DD, editor. The ethics of Spinoza. Secaucaus, NJ: Citadel Press; 1957. 215 p.
101. Izard CE. The psychology of emotions. New York: Springer; 1991. p. 451.
102. Lewis M. Shame: the exposed self. New York: The Free Press; 1992. p. 275.
103. Roget PM. Roget's international thesaurus, 4th edn, revised Chapman RL. New York: Harper & Row; 1977. 1316 p.
104. Durkheim É, Mauss M. Primitive classification. Chicago: University of Chicago Press; 1963. p. 96.
105. Fiske AP. Relational models theory 2.0. In: Haslam N, editor. Relational models theory: a contemporary overview. Mahwah: Lawrence Erlbaum Associates; 2004. p. 3–25.
106. Turner JH. The structures of social life: the four elementary forms of human relations [book review]. Contemp Sociol. 1991;21(1):126–8.

Moral Cognition and Moral Emotions

Sandra Baez, Adolfo M. García, and Hernando Santamaría-García

Abstract Moral cognition, a central aspect of human social functioning, involves complex interactions between emotion and reasoning to tell right from wrong. In this chapter, we summarize the cognitive neuroscience literature on moral cognition and moral emotions, highlighting their close relationship with other social cognition domains. We consider neuroimaging research and behavioral/neuropsychological evidence of moral impairments in patients with psychiatric and neurological conditions. We also describe cognitive neuroscience models claiming that moral cognition processes are shaped by the encompassing social context. These views

S. Baez (✉)
Grupo de Investigación Cerebro y Cognición Social, Bogotá, Colombia

Universidad de los Andes, Bogotá, Colombia

Laboratory of Experimental Psychology & Neuroscience (LPEN), Institute of Cognitive and Translational Neuroscience (INCYT), Institute of Cognitive Neurology (INECO) & CONICET, Favaloro University, Pacheco de Melo 1860, Buenos Aires, Argentina

National Scientific and Technical Research Council (CONICET), Buenos Aires, Argentina
e-mail: sj.baez@uniandes.edu.co

A.M. García
Laboratory of Experimental Psychology & Neuroscience (LPEN), Institute of Cognitive and Translational Neuroscience (INCYT), Institute of Cognitive Neurology (INECO) & CONICET, Favaloro University, Pacheco de Melo 1860, Buenos Aires, Argentina

National Scientific and Technical Research Council (CONICET), Buenos Aires, Argentina

Faculty of Education, National University of Cuyo (UNCuyo), Mendoza, Argentina
e-mail: adolfomartingarcia@gmail.com

H. Santamaría-García (✉)
Universidad Javeriana, Bogotá, Colombia

Centro de Memoria y Cognición Intellectus, Hospital Universitario San Ignacio, Bogotá, Colombia

Laboratory of Experimental Psychology & Neuroscience (LPEN), Institute of Cognitive and Translational Neuroscience (INCYT), Institute of Cognitive Neurology (INECO) & CONICET, Favaloro University, Pacheco de Melo 1860, Buenos Aires, Argentina

Grupo de Investigación Cerebro y Cognición Social, Bogotá, Colombia

National Scientific and Technical Research Council (CONICET), Buenos Aires, Argentina
e-mail: nanosanta@gmail.com

© Springer International Publishing Switzerland 2017

A. Ibáñez et al. (eds.), *Neuroscience and Social Science*,
DOI 10.1007/978-3-319-68421-5_8

emphasize how cultural and context-dependent knowledge, as well as motivational states, can be integrated to explain complex aspects of human moral cognition. Finally, we address real-life social scenarios on which available studies could make a direct impact. More generally, we analyze the extent to which moral cognition research can help to understand human social behavior and complex social-moral circumstances.

Keywords Moral cognition • Moral emotions • Moral reasoning • Moral judgment • Moral sensitivity • Neural networks • Neuroimaging • Neuropsychiatry • Neurodegenerative disease

1 Introduction

Most human acts have moral repercussions, with judgments of right and wrong depending on complex interactions between emotion and reasoning. For millennia, the role of morality in human beings has been the center of multiple discussions within philosophy, theology, and law. However, experimental research on the topic has only recently emerged. In particular, the tools of cognitive neuroscience have offered exciting opportunities to study the neural organization of processes underlying moral behavior, including moral cognition and moral emotions. Relevant evidence comes from neuroimaging research on clinical and neurotypical samples and from behavioral studies on the moral profile of patients with neurological or psychiatric disorders [1]. The aim of this chapter is to summarize available findings from both research lines. First, we present a historical perspective of the study of moral cognition. Next, we introduce three key subdomains of moral cognition (i.e., moral sensitivity, moral reasoning, and moral judgment) and review neuroimaging evidence on their complex underlying neural network. Then, we address the relationship between moral cognition and other social cognition domains—in particular, theory of mind (ToM) and empathy. Next, we highlight the relationship between moral cognition and moral emotions, focusing on the neural bases of the latter. We also consider clinical evidence on moral impairments in patients with psychiatric and neurological conditions. In addition, we describe neurocognitive models of moral cognition processes. We conclude by presenting real-life social scenarios on which available moral cognition studies could make a direct impact.

2 A Historical Perspective on the Study of Moral Cognition

For decades, moral psychology has sought to identify a rational basis of human morality [2–4]. This rationalist approach proposes conscious moral reasoning as the source of moral judgment and moral behavior. Various psychological theories of moral reasoning have been proposed. For instance, from a Piagetian perspective, moral reasoning is the coordination of all perspectives involved in a moral dilemma [5]. According to this view, moral reasoning consists in the application of a logical rule to a problem to derive a solution. Kohlberg [2] extended Piaget's ideas by formulating a model focused on how adult moral cognition progresses through developmental stages that lead to the incorporation of universal moral principles. For Kohlberg, arguably the most influential figure in the field [6], the highest moral stage requires a type of moral reasoning based on abstract and universal principles of justice.

However, the rational view has been challenged by recent approaches. Several psychological and cognitive neuroscience models [7–10] have highlighted the importance of emotions when investigating human morality. One of the earliest alternative theories to the rationalist models was the somatic marker hypothesis proposed by Damasio [10]. He postulates that emotion-based biasing signals (somatic markers) arising from the body are integrated in higher brain regions, in particular the ventromedial prefrontal cortex (vmPFC), to regulate complex decision-making situations. According to this hypothesis, we use our bodies (as represented in the brain) as sounding boards that tell us instantly, without the need for reflection, that a certain course of action is repulsive or attractive.

The role of emotions in human morality was further elaborated by Haidt [6, 8]. According to his social intuitionist model, emotion-laden hunches are the primary determinants of moral judgments. Haidt emphasized that moral decisions are driven by fast, automatic, and affect-driven intuitive processes. Thus, in this view [8], moral judgment is the consequence of quick moral intuitions, followed by slow, ex post facto moral reasoning.

A more recent explanation of moral cognition relies on dual-process theory [11–13] (see Northoff, this volume). This theory represents a synthesis of the rationalist and affective perspectives of moral cognition, emphasizing the balance of reasoning and emotions as prior to moral judgments [13]. More particularly, it focuses on two contrasting orientations to moral judgments: deontological and utilitarian. Immanuel Kant proposed the most popular deontological theory rooting morality in the logic of noncontradiction. He argued that actions were right only if they could be rationally asserted as governed by a universal rule guiding the actions of others. Thus, deontological moral judgments are based on duties, on the rightness or wrongness of actions considered independently of their consequences [6]. In contrast, utilitarians, such as Jeremy Bentham and John Stuart Mill, proposed that actions should be judged by their consequences alone, acting always in the way that they will bring about the greatest total good [6].

Most of the evidence supporting the dual-process theory comes from studies on impersonal and personal moral dilemmas. The former are epitomized by the trolley dilemma [14]. Participants have to decide whether they would flip a switch to redirect a trolley onto a man in order to save five other individuals. Such a choice is considered a utilitarian response, whereas a refrain from flipping the switch is considered deontological. On the other hand, the footbridge dilemma [15] is an example of a personal dilemma. In this case, participants also have the chance to save five people, but this time by pushing a man off a bridge in order to stop a trolley from hitting them further down the tracks. Accepting to push the man constitutes a utilitarian response, whereas failure to do so is regarded as a deontological decision. Although the two dilemmas are logically equivalent (i.e., killing one person to save five lives), numerous empirical studies [12, 13, 16–18] have demonstrated that a large majority of individuals consider it morally acceptable to sacrifice one person to save five in the impersonal dilemma, whereas they believe that it is wrong to push the man to save the five victims in the personal dilemma.

According to Greene et al. [12, 13], the reason for these contradictory responses lies in the stronger tendency of personal dilemmas, compared to the impersonal ones, to engage emotional processes, which would affect moral decisions. Following this view, moral dilemmas induce responses from two separable (and sometimes competing) neural processes, one of which is associated with fast, automatic, affective processing and the other with more conscious, deliberate, and controlled reasoning. Thus, when people face with a dilemma such as the footbridge problem, the aversive response to pushing the person overwhelms any concern about maximizing the overall good, thus generating a deontological response. In contrast, the typical utilitarian response to the trolley dilemma is explained by the impersonal nature of the scenario, which causes less emotional responses, allowing more deliberative concerns about the overall good.

Abundant research e.g., [12, 13, 16–18] has employed the trolley and the footbridge dilemmas to investigate the contributions of emotions, reasoning, and cognitive control in moral judgment. However, evidence for the dual-process theory has been empirically and conceptually challenged. For instance, Kahane et al. [19] suggested that behavioral and neural differences in responses to the classical moral dilemmas are largely due to differences in intuitiveness, not to general differences between utilitarian and deontological judgment. Specifically, they indicate that the distinction between intuitive and counterintuitive judgments is a more fundamental division in moral decision-making. According to these authors, deontological responses to the footbridge dilemma appear to be based on immediate moral intuitions. Utilitarian judgments in such dilemmas are often highly counterintuitive because they conflict with a stringent duty against harm [19]. Thus, results of this study suggest that neural differences in judgment to moral dilemmas largely depend on whether the latter were intuitive.

An alternative view has been proposed by Moll et al. [9, 20]. Their event-feature-emotion complex model [9] posits that reasoning and emotions are both primary and precede moral judgment and behavior. In contrast to the Greene's theory, this model proposes that competing representations of behavioral choices cannot be split into cognitive and emotional ones. All morally relevant experiences are considered essentially cognitive-emotional association complexes (see below). Instead of competing with each other, cognition and emotion are continuously integrated during moral decision-making [20].

As outlined in this section, regardless of their focus, current models have addressed the interplay between intuition, emotion, and reasoning for morality. However, cognitive neuroscience studies have emphasized the need to move beyond the simple dual-process dichotomy and embrace models of moral cognition that capture the rich, dynamic nature of human psychology and neuroscience [21]. Psychological models can benefit from incorporating recent findings from cognitive neuroscience research. In the next section, we present neuroimaging evidence which sheds light on a complex neural network underlying moral cognition.

2.1 Moral Cognition and Its Neural Basis

Morality can be considered as the degree of adherence to the customs and values accepted by a social group [9]. Moral cognition consists in using such codes to guide culturally adequate social behavior. Thus, moral cognition comprises all mental processes underlying our discernment of acceptable and inacceptable actions [22]. Such broad construct encompasses different but interrelated subdomains, including moral sensitivity, moral reasoning, and moral judgment.

Moral sensitivity is defined as the quick detection and evaluation of the morality of a context-bound action, including awareness of how other individuals may be affected by it [23]. This detection typically happens prior to complex moral reasoning and moral judgment [23, 24]. Thus, moral sensitivity is the first stage of a moral decision, which is associated with an instant feeling of approval or disapproval when we witness a morally laden situation [8]. In this line, functional neuroimaging research has shown that an increase in moral sensitivity when watching others being harmed positively correlates with activation of limbic regions and other brain areas implicated in processing emotions and social prompts, i.e., the orbitofrontal cortex (OFC) and the superior temporal sulcus [9, 24, 25] (Fig. 1a).

A second relevant dimension is moral reasoning, a controlled process leading to moral judgments [9]. This process includes laborious steps of deductive reasoning and cost-benefit analyses. Moral reasoning is typically associated with regions supporting interpersonal inference processes, such as ToM. These regions include the dorsomedial prefrontal cortex (dmPFC), the anterior temporal pole, and the temporoparietal junction (TPJ) [26].

Moral judgment constitutes a third key subdomain of moral cognition. It may be conceptualized as a type of evaluative judgment which is based on assessments of the adequacy of one's own and others' behaviors based on socially shaped ideas of right and wrong [9]. Moral judgment has been associated with the activity of an extended neural network [9] which includes the vmPFC, the OFC, the anterior temporal lobes (Fig. 1a), the amygdala (Fig. 1b), and the precuneus.

More specifically, knowledge about the neural basis of moral cognition has been mainly derived from research and behavioral/neuropsychological evidence of moral impairments in patients with psychiatric and neurological conditions. In particular, fMRI studies e.g., [1, 12, 27–29] have revealed that moral cognition tasks engage a distributed network whose main hubs (Fig. 1) include the vmPFC, the OFC, the ventrolateral prefrontal cortex, the amygdala, the superior temporal sulcus, the precuneus, and the TPJ.

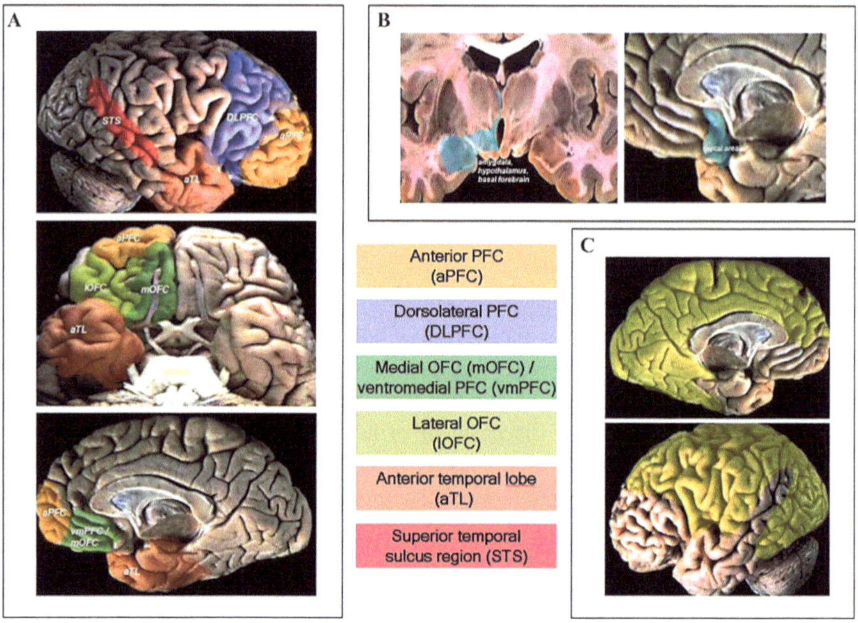

Fig. 1 Brain regions implicated in moral cognition. (**a**) Cortical regions include the anterior prefrontal cortex (PFC), the medial and lateral orbitofrontal cortices, the dorsolateral and ventromedial PFCs, the anterior temporal lobes, and the superior temporal sulcus. (**b**) Subcortical regions include the amygdala, the ventromedial hypothalamus, the septal area and nuclei, the basal forebrain, and the rostral brain-stem tegmentum. (**c**) Brain regions less consistently associated with moral cognition in patient studies include the parietal and occipital lobes, large areas of the frontal and temporal lobes, the brain stem, the basal ganglia, and additional subcortical structures. Reproduced with authorization from Moll et al. [20]

The vmPFC seems to play multiple roles in social cognitive processes. For example, it biases moral judgment by associating external stimuli with socio-emotional values and is involved in ToM and empathy processes [20, 30]. Also, the vmPFC has been proposed as a critical region for processing intention and outcome information during moral judgments [31–33]. Patients with damage to this area judge harmful intentions in the absence of harmful outcomes as more permissible than healthy subjects [32].

The OFC and the ventrolateral prefrontal cortex are implicated in the inhibition of automatic or impulsive responses and in processing social prompts [34, 35]. Specifically, the OFC is engaged by stimuli conveying rewards and punishments and in tasks requiring the integration of cognitive processes with affective values [36]. Moreover, patients with OFC lesions may develop changes in the interpersonal emotional domain and disproportionate impairments in social behavior [9].

The amygdala is involved in moral learning and responses to threats [37–39]. Indeed, perceiving an individual who intentionally hurts another person triggers an early amygdalar boost, which plays a critical role in evaluating actual or potential threats [40, 41]. Moreover, the amygdala contributes to automatic emotional evaluations of morally salient actions [42].

The precuneus subserves processing of mental states [43] and integration of self-referential stimuli in the broader emotional or moral context of the self [44]. Finally, the TPJ is involved in inferencing others' mental states [45] and integrating multi-dimensional information to establish a social context for decision-making [46]. Moreover, the right TPJ is particularly implicated in judging accidental harm. This area shows distinct spatial patterns of responses for intentional vs. accidental harm [43], and an increase in its activation is correlated with greater consideration of exculpating agents [47].

Although these particular areas have been shown to play an important role in judging moral situations, empirical [48, 49] and theoretical [9, 20, 21] works suggest that high-level social processes, such as moral cognition, should be interpreted in terms of the functioning of complex brain networks which integrate context-dependent representations. Neurocognitive models supporting this view are described below.

2.2 The Relationship between Moral Cognition and Other Social Cognition Domains

2.2.1 Moral Cognition and ToM

ToM refers to the ability to infer the beliefs, intentions, and emotions of others [50]. This skill is critical to predict the subjective consequences of our actions and to judge how people might react to them [22]. Moreover, ToM is related to both the rational and emotional facets of moral cognition. Indeed, accumulating evidence indicates that the latter domain is influenced by inferences regarding the intentional or accidental nature of an agent's action [47, 51, 52].

Moreover, adult moral judgments typically depend on the capacity to represent and integrate information about beliefs and consequences [47, 52]. In fact, individuals who inflict harm on other are usually exculpated insofar if their actions are deemed accidental [52, 53]. Such a morally loaded decision requires a robust representation of tacit intentions to override a preponderant negative response to the outcome [47]. By the same token, estimations of punishment severity depend on the assessment of the agent's implicit intentionality [51].

The relationship between moral cognition and ToM is further supported by fMRI studies highlighting their common neural basis. In particular, meta-analytical evidence [26] shows that brain activity patterns during moral cognition and ToM overlap in the vmPFC, the precuneus, and the TPJ. Thus, moral cognition processes depend on processes mediated by the TPJ [previously associated with the inferencing of mental states [45]] and, to a lesser extent, the precuneus and the medial prefrontal cortex (which are also involved in the neural network subserving ToM skills).

2.2.2 Moral Cognition and Empathy

Social interaction hinges largely on empathy, that is, the capacity to share and understand the subjective experience of others in reference to oneself [54]. This complex construct involves affective components (sharing and responding to the emotional experience of others) and cognitive components (understanding the intentions and perspectives of others) [25, 55].

The relationship between empathy and morality is well established e.g [25, 56–58]. For instance, empathy-related processes motivate prosocial behavior and caring for others, thus providing a foundation for morality [25, 56, 58]. Empathy can also interfere with morality by introducing partiality, for instance, by favoring in-group members [56]. Moreover, in moral decision-making, experiencing empathy reduces the intensity of harmful actions toward others [59]. In addition, low empathic concern levels predict utilitarian moral judgment [60].

Further support for a link between empathic concern and morality can be found in neuroimaging studies. The dmPFC seems to be a crucial convergence region subserving both moral cognition and empathy processes [26]. This direct overlap suggests that socio-emotional processes share at least some mechanisms with those supporting moral cognition and empathy.

2.2.3 Moral Cognition and Moral Emotions

As argued above, human moral behavior seems to be rooted in spontaneous and implicit emotional dispositions [6]. This complex facet of human experience relies on basic and complex emotional mechanisms that interplay with logic, reasoning, and judgment [1]. Such mechanisms are called moral emotions or social emotions, and they are crucial for implicit and explicit evaluations of interpersonal actions [1, 6, 59].

Haidt [61] defines moral emotions as those "that are linked to the interests or welfare either of society as a whole or at least of persons other than the judge or agent" (p. 276). Moral emotions seem to modulate how humans assess which behaviors are morally acceptable [6–9, 59]. They may also provide motivation to do good and bad [62, 63]. Thus, the study of moral emotions opens a window to explore a variety of complex facets of human experience, including compassion, corruption, and xenophobia, to name but a few.

Below we outline a framework to examine phenomenological features of moral emotions. In addition, we review evidence on relevant neurocognitive processes related with moral cognition and moral judgment. More particularly, we interpret the available evidence in terms of theories highlighting the role of moral emotions in the deployment and judgment of moral behaviors [1, 9].

3 Moral Emotions

Emotions pervade virtually every aspect of human life [64]. They are crucial to maintain social bonds and adequately respond to multiple interpersonal scenarios [59]. These social experiences are in part shaped by moral emotions. From an evolutionary standpoint, emotions which emerged from social interactions in our antecessors may be seen to constitute an adaptive mechanism supporting the regulation of one's own acts and the assessment of others' behaviors [6, 20, 65] (see TenHouten, this volume). Unlike basic emotions, moral emotions are linked to the interest or welfare of extended social groups, and they are evoked by circumstances that extend beyond of self-experiences and interests [61, 64].

Moral emotions, including shame, embarrassment, pride, envy, and *Schadenfreude* (a German term to refer to pleasure at others' misfortunes), hinge on the interests and welfare of others and prove decisive to encourage or inhibit behaviors depending on their social acceptability [66]. Moreover, moral emotions are prompted by the recognition and adoption of universally accepted rules and culturally defined conventions, which are crucial for group cohesiveness and social organization [66, 67].

According to Haidt [61], moral emotions are shaped under the influence of two factors: the elicitors and the action tendencies. Elicitors range from the behavior of others to events affecting them and their evaluations of our own actions. Those emotions can be triggered by joys, misfortunes, and transgressions where others are actively or passively involved. Moral emotions are also modulated by social action tendencies. Indeed, they are experienced when we are motivated to deploy other-targeted actions, modulating social order and welfare [66, 67]. Those emotions induce cognitive states increasing proclivity to engage in goal-driven social actions (e.g., revenge, affiliation, comfort).

These features, however, do not account for all the complexity of moral emotions. For instance, the same emotion may be elicited by different behaviors. A boy may experience envy when another member of his social milieu has the status he aspires to achieve or when a good action done by another is taken away. Likewise, feelings of embarrassment and shame can be associated with inappropriate behavior, though subtle differences can be found between both emotions. Embarrassment is experienced when a social code is violated, while shame is elicited by one's own attribution of reduced self-value and self-esteem upon violation of a moral norm [68, 69].

Although multiple morally loaded emotions have been described [64, 65], various aspects remain poorly understood. Yet, the evidence suffices to describe a repertoire of typical moral emotions [1, 70]. These include so-called fortune-of-other emotions (FOEs), such as envy and *Schadenfreude*, and self-conscious emotions (SCE), such as shame and guilt [1, 70]. Below we refer to these two broad categories in turn.

3.1 Fortune-of-Other Emotions

FOEs are defined as affective states emerging in response to situations affecting other people. FOEs can be evoked by others' qualities, possessions, and experiences [71]. This group of emotions is triggered by a continuous social comparison process. Festinger et al. [72, 73] described the importance of social comparison in human interaction and reported that recognizing the social role of oneself and others favors self-knowledge and self-regulation. Social comparison processes mobilize emotional and cognitive processes, modulating behavior [73].

FOEs can be divided into four categories depending on the affective reaction of the self (i.e., pleased or displeased) and the presumed value for the other individual (desirable or undesirable). Thus, two large subsets can be recognized, namely, goodwill, empathetic emotions and ill-will, counter-empathy emotions [71, 74, 75]. Goodwill emotions are experienced when a person is pleased by desirable events experienced by others ("happy for" emotion) or when a person is displeased by undesirable events in the life of others ("sorry for" emotion). Ill-will or counter-empathy emotions are experienced when a person is displeased by something desirable happening to others (resentment or envy) or when a person is pleased by others' misfortunes (*Schadenfreude*) [71, 74, 75].

3.1.1 Envy and *Schadenfreude* as Examples of FOEs

Envy is defined as discomfort associated with another's pleasant experiences, while *Schadenfreude* refers to the perceiver's pleasure at another's distressing or unfortunate situations [76, 77]. These two FOEs are relevant in maintaining stability during social interaction and in regulating social behavior [75]. Despite their differences, both emotions are grounded in social comparison processes. In particular, they are boosted when individuals make upward comparisons. Also, they seem to stabilize tensions experienced by having inferior roles in hierarchical social contexts [70, 75, 76]. Moreover, both emotions are intermingled, so that *Schadenfreude* is more likely to emerge when a misfortune happens to an envied person [77, 78]. Envy can be expressed in different ways, including dispositional or episodic envy [71]. The former is associated with a generalized reduction of self-esteem, while the latter is domain-specific. Episodic envy is largely evoked during social comparison of self-relevant traits and in the context of deservingness [79]. Thus, a person envies others when they posses relevant and desired attributes or when they experience underserved success [79]. In sum, these findings again support a crucial role of social comparison in how moral emotions are evoked and experienced [70, 74, 80].

There are other forms of envy. Sometimes, individuals admire and wish to reach the achievements of a superior person. Since this experience may prompt goal-oriented behaviors, it can be considered as a type of good envy. By contrast, a person may experience discomfort in response to the success of the superior individual and thus question their merits assuming a similar status between them. This emotional experience has been denominated by bad envy [81]. Instead, *Schadenfreude* is evoked by downward social comparisons [75, 82]. Also, *Schadenfreude* is modulated by target likeability and target deservedness [75, 82, 83]. These are present, for instance, when a person feels pleasure upon learning that a rivaling soccer team loses a crucial match, so that it is framed in a comparatively inferior place.

Finally, insights have been gained into the chemical bases of FOEs. These emotions increase following administration of oxytocin [83]. This and other peptide hormones have been implicated in the regulation of mammalian social behavior, and they regulate diverse conducts, such prosocial and empathetic dispositions [84, 85]. Oxytocin may increase the salience of social agents, hence promoting the experience of social-dependent emotions, such as envy and *Schadenfreude*.

3.1.2 Neural Correlates of FOEs

Recently, studies in cognitive neuroscience have explored the neural correlates of envy and *Schadenfreude* and the cognitive factors that modulate and impel those experiences [70, 77] (see Fig. 2). Whereas most of these studies considered neurotypical samples, some of them focused on subjects exhibiting moral behavior perturbations upon brain damage [1, 70]. Here we summarize the most important findings on the topic.

At the neuroanatomical level, envy and *Schadenfreude* are mainly supported by prefronto-striatal networks [77, 86]. In particular, experiencing envy is associated with activity in temporal regions, as well as the anterior and medial cingulate cortices [70, 74, 76, 77]. Takahashi et al. [77] reported increased anterior cingulate cortex (ACC) activations in response to envy and suggested that this may reflect the response to the painful features accompanying this FOE. This aligns with evidence of increased caudal ACC activation in response to one's own pain but not to pain in others (empathic pain), which indicates an overlap between envy and painful feelings [87]. Note that ACC activations have also been reported in response to social pain (distress of social exclusion) and in conflict monitoring [88]. Such findings warrant the speculation that envy entails a conflict between social comparisons and self-concepts.

Conversely, the experience of *Schadenfreude* has been systematically associated with activity in the ventral striatum (VS) and the medial OFC [70, 75, 89, 90]. Those brain structures have usually been involved in processing of reward information. In fact, VS activity is involved in processing altruistic punishment [91] and in observing unfair person being punished [92, 93]. Compatibly, different studies highlight the central importance of the VS in processing reciprocity of rewards in social comparisons [94, 95]. For example, Fleissbach et al. [94] investigated the neural substrates of reward processing in a context of social comparison processes. Participants performed a dot estimation task with a partner and received a reward for their answers. Authors observed that activity in the bilateral VS was sensitive to the magnitude of the partner's reward, with higher VS activity in the presence of high

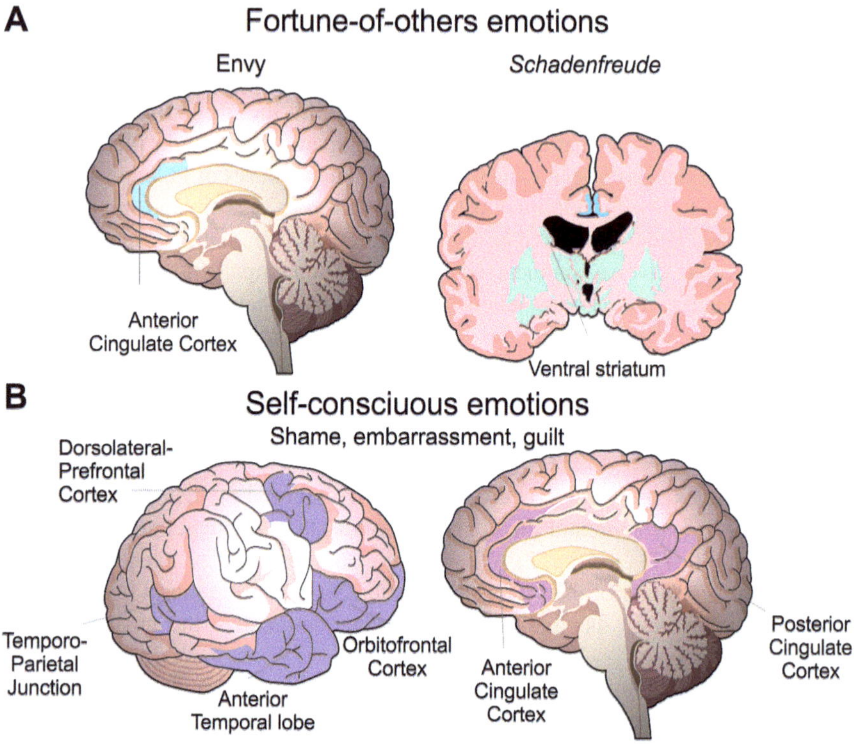

Fig. 2 Neural structures involved in the experience of moral emotions. The figure depicts the main brain regions involved in the experience of self-conscious emotions (SCEs) (**a**) and fortune-of-others emotions (FOEs) (**b**). Brain regions marked in *purple* and *pink* are associated with SCEs, including embarrassment, shame, and guilt. Brain areas marked in light blue are implicated in the experience of envy, whereas those in green are involved in experiencing *Schadenfreude*

rewards. In general, VS responses were associated with reciprocity of rewards in social contexts.

3.2 *Self-Conscious Emotions*

SCEs are evoked by the actual or expected evaluation of one individual by other persons. These emotions promote social regulation by providing information about the acceptability of one's and others' behavior [68]. According to previous studies [68, 70, 75, 89], SCEs are related to three cognitive processes: (a) self-awareness, leading to self-referential processing; (b) other awareness, underlying mental states attribution in others; and (c) social norm awareness, supporting the identification and acknowledgment of societal norms.

SCEs can be experienced since ages as early as 2 [96]. These states emerge in line with the development of some cognitive skills such as self-evaluation and self–/other distinction. Whereas some SCEs are characterized by a positive valence (e.g., pride, gratitude), others involve negative connotations (e.g., shame, embarrassment, guilt) [68, 97, 98].

3.2.1 Research on SCEs

Embarrassment, guilt, and shame are some of the most studied SCEs. Embarrassment is associated with negative self-evaluations following violations of moral or social norms [68]. It is usually evoked by non-severe social transgressions and is enhanced by public exposition [68]. This emotion is accompanied by characteristic psychophysiological responses, including flushing and changes in cardiopulmonary measures (e.g., increased heart rate or hyperpnea) [68]. Embarrassment stimulates self-regulation of one's own behavior and favors social interactions, as it helps to monitor and fit behavior avoiding non-appropriate social behaviors or nonsocial actions.

Guilt is an emotion usually evoked by drastic moral disruptions, and it is experienced in response to particular acts disrupting social values or norms [97, 99, 100]. For instance, individuals feel guilt when they offend a loved one. This emotion is strongly motivated by self-evaluation and judgment of own acts, guided by internal representations and values. Guilt encourages individuals to compensate others for unkindly acts and to avoid new inadequacies. Experiencing guilt fosters empathy and reparation [68, 97, 98].

Finally, shame is an uncomfortable SCE that occurs in response to comparisons between oneself and others [86, 96]. Thus, shame is experienced when one's own acts disrupt an idealized vision of oneself and when a person compares him–/herself with the social standard. Crucially, shame is encouraged by self-monitoring and self-evaluation, as it can be evoked when no action is performed. This emotion usually is accompanied by escape behaviors, aimed to prevent self-contempt.

3.2.2 Neural Correlates of SCEs

The experience of SCEs has been associated with activity in a network spanning temporoparietal and medial prefrontal regions [101] (see Fig. 2). Embarrassment is associated with increased brain activation in the dmPFC [86, 96], the ventrolateral prefrontal cortex [102, 103], the dorsal ACC [86, 96], the anterior insular cortex [86, 96], the anterior temporal lobes [102, 103], the posterior superior temporal sulcus [86, 96, 102, 103], and the TPJ [86, 96], among other secondary areas, such as the left hippocampus and the visual cortex [102, 103].

Activity in the ventrolateral PFC and the anterior temporal lobe is also associated with experience of guilt. Previous studies have implicated the activity of ventrolateral PFC to moral decision-making and disturbances of such skill [104, 105]. The anterior temporal lobe is believed to subserve conceptual social knowledge, the understanding of social concepts and rules [106, 107], and the recognition of situations triggering moral emotions [106, 107].

Shame is associated with activity in the ACC. In addition, this emotion is related to self-focused cognitive processes [108]. Different ACC regions have been implicated in a range of relevant functions, including the experience of negative affect [109], envy [77], social pain [59], and interoceptive awareness [110].

Guilt has been associated with the function of the ACC. The well-established role of the dorsal ACC in the experience of distress and, more particularly, social pain [111] may explain its common role in the experience of diverse moral emotions. Guilt has also been related to the activity of the dmPFC, the ventrolateral PFC, and the anterior temporal lobes. The dmPFC is involved in self-referential [112] and ToM [113] processes. The ability to evaluate others' intentions and thoughts may be related to the capacity to read emotional and social cues in others. Those processes are associated with self-blaming emotions when social/moral rules are broken.

Shame and embarrassment were both associated with hippocampal function [102]. The hippocampus has been associated with a set of brain functions such as memory, emotional integration, and stress regulation [102, 114]. The relationship between hippocampal activity and the experience of shame and embarrassment may be related to psychosocial stress [102, 114], considering that both of these emotions are associated with external threats [68, 98]. Most of the aforementioned areas have been involved in crucial social cognitive processes, including social perception, ToM, and empathy, supporting the link between emotional experience and social interaction processes [1, 110].

Recent studies in cognitive neuroscience have added empirical information about the neural processes that subsume moral emotions and how they interplay with other social, moral, and cognitive processes. Some of those advances come from studies of patients with exhibiting drastic changes in social behavior as a consequence of acquired brain lesions and in neuropsychiatric conditions.

Beyond studies of moral emotions in normal individuals, some reports have analyzed processing of moral emotions in subjects with developmental antisocial behaviors and in patients with neuropsychiatric disorders. The study of moral processing in neuropsychiatric conditions constitutes a new tool to understand the complexity of the interplay between different cognitive and moral processes.

4 Moral Cognition and Moral Emotions in Psychiatric and Neurological Conditions

Most psychiatric and neurological conditions are characterized by social cognition deficits and/or abnormal activation of "social brain areas" [115] (see Kumfor et al.; Piguet; Felisberty & Kin, this volume). Understanding the neurobiological basis of social cognitive processes is a key aim for social neuroscience. While studies in healthy individuals have undoubtedly offered important insights into moral cognition and moral emotions, research on clinical populations has contributed key information to identify relevant brain regions. Indeed, psychiatric and neurological disorders may be conceptualized as disorders of social interaction [110, 116, 117]. Next, we illustrate how moral cognition and moral emotion impairments manifest in such conditions.

Autism spectrum disorder (ASD) encompasses multiple conditions characterized by problems with reciprocal social interaction, impaired communication, repetitive behaviors, narrow interests, and impairments in aspects of social cognition necessary for proper moral reasoning. Behavioral studies have shown that adults with ASD exhibit decreased levels of emotional reaction to moral dilemmas [118] and atypical moral judgments when they need to consider both the intention to harm (accidental vs. intentional) and the outcome (neutral vs. negative) of a person's actions [119]. Moreover, individuals with ASD judge conventional and disgust transgressions as more serious than do controls while failing to distinguish between disgust and moral transgressions [120]. Adults with ASD also exhibit subtle difficulties in judging an agent's intentions in cartoons depicting aggressive actions [121]. Also, in an fMRI study [122], participants were presented with moral dilemmas followed by proposed solutions with which they could agree or disagree. Despite the absence of behavioral differences between ASD patients and healthy controls, moral reasoning in the former involved decreased activation in limbic regions, particularly the amygdala, as well as increased activation in the anterior and the posterior cingulate gyri. Thus, taken together, the evidence suggests that ASD is characterized by a failure to use relevant information about the agent's intentions, reduced emotional reactions to moral dilemmas, and abnormal brain activation during moral judgment.

Also, moral emotions seem to be differently expressed in ASD. These patients present reduced understanding of jealousy compared with neurotypicals [123]. Furthermore, previous evidence revealed a gap between a subtly reduced capacity of experiencing jealousy (a self-reflective, socially mediated emotion) and a stronger deficit in the capacity to fully reflect on the experience of such an emotion [123].

Moral cognition impairments have also been observed in adults with psychopathy. These individuals show a significantly lower distinction between moral and conventional transgressions than do healthy controls [124] and more frequently endorse utilitarian choices when facing moral dilemmas [125]. Furthermore, empirical evidence shows that although psychopaths understand the distinction between right and wrong, they do not care about such knowledge or the consequences or morally inappropriate behavior [126]. Structural neuroimaging in this population has revealed gray matter reductions in the OFC and anterior temporal cortex, the superior temporal sulcus, and the insula [127]. These structural abnormalities have been associated with reduced moral sensitivity [127].

Additionally, an fMRI study on psychopaths [128] revealed atypical activity in several regions involved in moral decision-making during the evaluation of pictures depicting moral violations. These regions include the vmPFC and the anterior temporal cortex. Moreover, results revealed a positive association between amygdalar activity and the severity of the moral violations, which was greater in non-psychopaths than psychopaths.

Subjects with antisocial traits also present pronounced emotional deficits and a notable reduction of guilt, shame, and remorse emotions, which increases their antisocial behaviors [129]. Subjects with these traits exhibit structural and functional abnormalities in brain networks subserving guilt and other prosocial emotions [130].

Moreover, individuals with addictions also show abnormal patterns of moral judgment. Relative to controls patients with alcohol dependence favor utilitarian moral judgments when faced with moral personal dilemmas [131]. Besides, an fMRI study [132] revealed that cocaine-dependent individuals show reduced activation in fronto-limbic structures during moral dilemmas. In addition, immature moral reasoning has been detected across age groups with social deprivation [58], sociopathic conditions [133], and schizophrenia [134].

Regarding moral emotions, some studies have shown that self-blaming biases, including guilt, can be found in patients with major depression disorder in melancholic forms [135], post-traumatic stress disorder [136], and obsessive-compulsive disorder (OCD) patients [137]. In particular, OCD patients show increased sensitivity to deontological, but not altruistic, guilt [137]. Also, OCD patients with hoarding disorder are prone to manifest increased guilt when they are required to discard some objects with emotional relevance [138].

OCD patients also seem sensitive to disparities of reward in social comparisons, a process highly related to envy and resentment [68]. Such sensitivity in this population is associated with feelings of discomfort and subjective experience of stress [139]. In addition this recognition affects crucial cognitive processes involved in the physiopathology of OCD, including performance monitoring and feedback evaluation [140–142]. Moreover, when playing a simple game with a simulated superior player, OCD patients exhibit significant modulations of neurophysiological components associated with performance monitoring (the so-called error-related negativity) and cognitive control (the N_2 component). This study indicates that OCD patients are sensitive to information that increases social differences with others and that their disparities probably impel FOEs.

Similar moral cognition impairments have been described in neurological disorders. For instance, moral judgment is typically impaired in behavioral variant frontotemporal dementia (bvFTD) [143]. Patients with this condition lack moral emotional reactions, favor utilitarian decisions in personal dilemmas [144], approve emotion-guided moral violations [145], and present difficulties for long-term cooperation and bargaining [146, 147]. Moreover, they tend to base their moral judgments exclusively on outcomes rather than on the integration of intentions and outcomes [48, 49]. This very atypical pattern has been only reported in extremist terrorists [148]. Structural neuroimaging [48] has revealed that this pattern of abnormal moral judgment is associated with gray matter volume of the precuneus, the amygdala, and the anterior temporal pole. In addition, a dynamic interrelationship has been described between large-scale brain networks underlying impaired moral judgments in bvFTD patients [149]. BvFTD involves reduced connectivity in the salience network [150], an anterior cingulate-frontoinsular system involved in processing emotionally significant stimuli which is inversely correlated with the default mode network (DMN) in task-free settings [151]. Moreover, when deliberating about moral dilemmas, bvFTD patients show reduced recruitment of the DMN and diminished functional connectivity from the salience network to the DMN [149].

Impairments in SCEs have also been reported in patients with frontotemporal lobar degeneration (FTLD). Although these patients preserve basic negative emotional responses in the presence of aversive stimuli, they show significantly less signs of experiencing SCEs compared to controls. Arguably, this reduction reflects alterations in frontotemporo-insular networks supporting the interaction between moral recognition, moral sensitivity, and moral reasoning [152, 153].

Abnormal patterns of moral judgment have also been observed in patients with prefrontal lesions, who rely primarily on outcome information rather than on the integration of intentions and outcomes [32, 49]. Moreover, patients with vmPFC lesions are more willing to judge personal moral violations as acceptable behaviors in personal moral dilemmas [16, 154]. Converging evidence highlights the vmPFC as a crucial area for the acquisition and maturation of moral competence [155], as well as for the processing of emotionally charged moral stimuli [27], belief valence [47], and moral violations [12].

Moral emotions are also impaired in patients with vmPFC lesions [156, 157], especially right-sided ones. Although these subjects show intact performance on a basic first-order ToM condition and relatively preserved understanding of identification, they do not recognize envy and *Schadenfreude*. Their inability to identify FOEs, therefore, seems to be independent to perspective-taking abilities and ToM processes.

Moral emotion processing has been also studied in patients with other neurological disorders. Baez et al. [89] analyzed the experience of *Schadenfreude* in patients with Huntington's disease (HD), a classical model of early structural and functional alterations of the VS, even in individuals with pre-manifest HD. While both HD patients and first-degree relatives had preserved envy experiences, they manifested lower *Schadenfreude* in response to others' misfortunes, supporting the role of the VS in the experience of this emotion.

In sum, moral cognition and moral emotions are compromised across psychiatric and neurological conditions. Indeed, social cognition deficits in both types of disorders may be partially explained by a general social-context processing impairment produced by brain network abnormalities [117, 158]. In the next section, we describe two neurocognitive models which align with this view.

5 Neurocognitive Network Models for the Study of Moral Cognition

Contextual modulations occur everywhere and social situations are not the exception. Adequate social and moral behaviors require the integration of explicit and implicit contextual cues to properly deploy politeness, humor, irony, agreement,

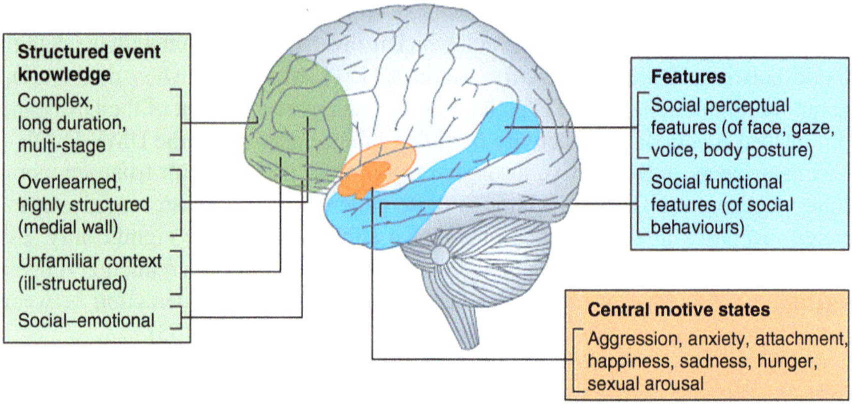

Fig. 3 The event-feature-emotion complex model. This model postulates that moral cognition arises from the binding of three main components: structured event knowledge (prefrontal regions), social perceptual and functional features (posterior and anterior regions of the temporal cortex), and central motive or basic emotional states (limbic and paralimbic regions). Reproduced with authorization from Moll et al. [9]

disagreement, or even silence [117, 158]. The role of contextual modulations has been extensively studied in basic sensory and cognitive processes, but few models have attempted to explain how specific mechanisms and brain regions contribute to contextual modulations in social cognition.

The event-feature-emotion complex model [9] (see Fig. 3) proposes that moral cognition emerges from the integration of content- and context-dependent representations in cortical-limbic networks. This model postulates three sequential moral cognition mechanisms: [1] the prefrontal cortex provides contextual event representations, [2] the temporal cortex contributes social perceptual features of the environment, and [3] limbic regions underlie emotional states. Several components of moral cognition as well as moral emotions may be explained by this model. Moral emotions would result from interactions among values, norms, and contextual elements of social situations [9].

In line with this approach, a more recent neurocognitive model has been proposed. The social context network model (SCNM) [158] describes the influence of context on social cognitive processing as dependent on a fronto-insular-temporal network. Although this model is not focused on moral cognition processes, it explains how context modulates social cognition domains in general. In terms of the SCNM, contextual associations are mediated by a cortical network (Fig. 4) engaging frontal, temporal, and insular regions [158]. The updating of ongoing contextual information and its association with episodic memory supports target-context relations driven by activity in frontal regions (e.g., OFC, lateral prefrontal cortex, superior orbital sulcus). The value of target-context associations is indexed in temporal circuits distributed throughout the amygdala, the hippocampus, and the perirhinal and parahippocampal cortices. Finally, internal and external milieus are coordinated by the insula to trigger internal motivational states.

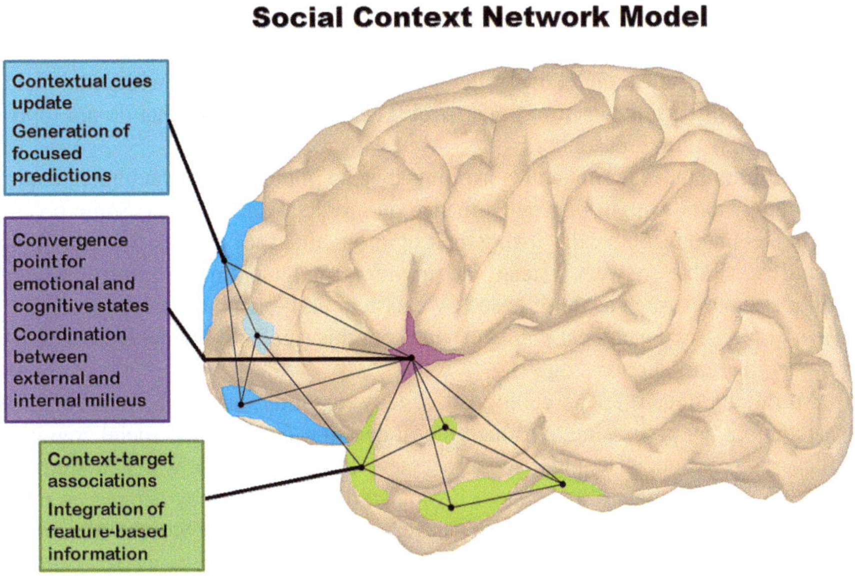

Fig. 4 The social context network model. Lateral view showing the frontotemporo-insular network postulated by the model. In this network, prefrontal areas would be involved in the generation of focused predictions by updating associations among representations in a specific context. Target-context associations subserved by temporal regions would be integrated with feature-based information processed in frontal regions. The insular cortex would support the convergence of emotional and cognitive states related to the coordination between external and internal milieus. Connected nodes represent frontotemporo-insular interactions. Reproduced with authorization from [159]

As described above, although particular brain areas have been shown to play an important role in moral cognition and moral emotions, more recent neurocognitive models suggest that such complex social cognitive domains should be interpreted in terms of the functioning of frontotemporal networks. According to this view, moral cognition impairments observed in psychiatric and neurological disorders may be explained by the disruption of such circuits. Future neuroscience research may strengthen the neurocognitive models outlined in this section by providing more refined evidence on processes and regions critically involved in contextual modulation of moral cognition.

6 Toward an Ecological Assessment of Moral Cognition and Moral Emotions

Ecological validity is especially relevant for moral cognition studies, because this process depends strongly on situational and cultural variables [9]. However, available moral cognition tasks fail to tap the ability to process contextual information.

Most experiments on moral cognition and moral emotions employ isolation paradigms [160], in which participants face pictures, words, or short histories with moral content. For example, the trolley dilemma, so often utilized within this field, is conducted in artificial settings and involves extreme situations that do not represent everyday moral reasoning.

In addition, the experimental constraints that are imposed by behavioral and neuroimaging studies might bias performance on moral cognition tasks. Some people might feel uncomfortable disclosing their opinions about sensitive issues, providing socially desirable answers instead. Passing moral judgment on extreme and unfamiliar situations, such as those posed by classic moral dilemmas, offers interesting ways to probe philosophical points of view, but this can hardly be taken as a proxy for everyday moral reasoning [9].

Moral emotions offer an ecological window to study the interactions between basic emotional processing, reasoned moral decision-making, and social cognitive processes (e.g., social comparison processes, ToM, and empathy). The study of moral emotions seems to have more ecological relevance as they depend on more realistic social human interactions. Therefore, future studies in healthy subjects as well as in psychiatric and neurological populations should consider context-dependence levels in moral cognition and moral emotions tasks, ranging from context-free to context-rich paradigms with varied manipulations of situational cues. Ecological validity could be increased through methods assessing these processes in real-time e.g., [161] and spontaneous interactions between socially engaged participants [116, 160]. In this sense, two key issues to be addressed in future research are the role of contextual information in moral cognition and moral emotions and the neural basis of mechanisms integrating information from social context frames.

6.1 Integrating Social and Cultural Perspectives

The functional role of moral emotions in regulating social behavior is highly dependent on culture. Cultural factors sculpt moral beliefs and norms, shaping the ways in which elicitors prompt interpersonal feelings according to contextual elements and social structures. Arguably, we are now more prone to experience embarrassment if someone shares our photos without our consent on Facebook than if we break some Victorian "etiquette" code at a dinner party. Also, from a psychopathological standpoint, whether emotional experiences are deemed abnormally heightened or reduced depends on what is considered "normal" in a given sociocultural niche. This highlights the need for more sophisticated, context-sensitive models of moral behavior and moral emotions to guide predictions in clinical settings.

The study of moral cognition and moral emotions reveals new perspectives in the understanding of the complex social behaviors. Societies worldwide face historical problems such as xenophobia, racism, and corruption, which require deeper translational developments rooted in transdisciplinary research. New challenges are also evident considering the new ways of socialization promoted by information technologies and globalization. Despite the ubiquity of these phenomena, we do not yet know why humans sometimes behave in nonsocial ways, given that cooperative, altruistic actions increase chances of survival and improve personal life [162, 163].

Partial contributions to our understanding of how we make moral decisions and how we experience some socially mediated emotions have been offered by cognitive neuroscience. For example, in light of evidence that moral behavior is mediated by neurocognitive mechanisms involved in empathy or ToM, a call can be made for policies and educational programs that encourage those social cognitive processes. Indeed, recent initiatives suggest that this could be the case [1, 66].

Furthermore, the study of moral emotions suggests that our brains are broadly shaped by the social world. This is evident when we analyze SCEs, which emerge when social codes are threatened. In fact, SCEs are related to frontostriatal activity, which has also been associated with the recognition and regulation of socially valued actions. This could be seen as a neural footprint of the crucial role of social world in human life. Even if we consider that nonsocial emotions, such as envy and *Schadenfreude*, also depend on social interaction, we have some certainties suggesting that the understanding of moral behavior is rooted in individual comprehension of the world around us.

Available evidence also sheds light on how children develop and deploy moral behavior. Throughout development, different within- and between-group dynamics can change how moral behaviors shape social attitudes. In fact, the preference of prosocial attitudes could vary according to group identity. Recent studies have shown that middle-aged children prefer prosocial behaviors toward the in-group and antisocial or harmful behaviors against the out-group [164–166]. This is evident when you see the dynamics of the football fans. Sometimes we are glad that the opposing team loses. However, many times our loved ones may be fans of the opposing team, but we would not like to see them enduring hardships in other circumstances. On a larger scale, we could say that our social-moral behavior could improve as we get to know and familiarize more with others whom we consider far from our social group. This is particularly relevant in some social complex scenarios, such as those marked by xenophobia.

7 Final Remarks

In this chapter, we have described convergent evidence to understand the neurocognitive mechanisms subserving moral cognition and moral emotions, highlighting the gaps between those domains and social cognitive processes. Furthermore, we have analyzed the extent to which moral cognition research can help understand human social behavior and complex social-moral circumstances by exploring real-life scenarios. Increased dialogue between social sciences and cognitive neuroscience can improve assessment, prediction, and comprehension of more morally mediated actions, potentially inspiring avenues to favor more prosocial behaviors.

Future studies will be needed to explore the neural basis of how different individuals and social groups make use of strategies and heuristics to solve moral conflicts. The implications of this new knowledge for how societies conduct business, regulate social behavior, and plan for their future remain to be explored.

Acknowledgments This work was partially supported by grants from CONICET and the INECO Foundation.

References

1. Moll J, de Oliveira-Souza R, Eslinger PJ. Morals and the human brain: a working model. Neuroreport. 2003;14(3):299–305.
2. Kohlberg L. Stage and sequence: the cognitive-developmental approach to socialization. In: Goslin DA, editor. Handbook of socialization theory and research. Chicago: Rand McNally; 1969.
3. Piaget J. The moral judgement of the child. New York: Free Press; 1965.
4. Gilligan C. In a different voice: psychological theory and women's development. Cambridge: Harvard University Press; 1982.
5. Carpendale JI. Kohlberg and Piaget on stages and moral reasoning. Dev Rev. 2000;20:181–205.
6. Haidt J. Morality. Perspect Psychol Sci. 2008;3(1):65–72.
7. Greene J, Haidt J. How (and where) does moral judgment work? Trends Cogn Sci. 2002; 6(12):517–23.
8. Haidt J. The emotional dog and its rational tail: a social intuitionist approach to moral judgment. Psychol Rev. 2001;108(4):814–34.
9. Moll J, Zahn R, de Oliveira-Souza R, Krueger F, Grafman J. Opinion: the neural basis of human moral cognition. Nat Rev Neurosci. 2005;6(10):799–809.
10. Damasio AR. Descartes' error: emotion, reason, and the human brain. New York: Avon Books; 1994.
11. Greene JD. Why are VMPFC patients more utilitarian? A dual-process theory of moral judgment explains. Trends Cogn Sci. 2007;11(8):322–3. author reply 3-4
12. Greene JD, Sommerville RB, Nystrom LE, Darley JM, Cohen JD. An fMRI investigation of emotional engagement in moral judgment. Science. 2001;293(5537):2105–8.
13. Greene JD, Nystrom LE, Engell AD, Darley JM, Cohen JD. The neural bases of cognitive conflict and control in moral judgment. Neuron. 2004;44(2):389–400.
14. Thomson J. The trolley problem. Yale Law J. 1985;94:1395–415.
15. Foot P. The problem of abortion and the doctrine of double effect. Oxford Rev. 1967;5:5–15.
16. Ciaramelli E, Muccioli M, Ladavas E, di Pellegrino G. Selective deficit in personal moral judgment following damage to ventromedial prefrontal cortex. Soc Cogn Affect Neurosci. 2007;2(2):84–92.
17. Geipel J, Hadjichristidis C, Surian L. The foreign language effect on moral judgment: the role of emotions and norms. PLoS One. 2015;10(7):e0131529.
18. Valdesolo P, DeSteno D. Manipulations of emotional context shape moral judgment. Psychol Sci. 2006;17(6):476–7.
19. Kahane G, Wiech K, Shackel N, Farias M, Savulescu J, Tracey I. The neural basis of intuitive and counterintuitive moral judgment. Soc Cogn Affect Neurosci. 2012;7(4):393–402.
20. Moll J, De Oliveira-Souza R, Zahn R. The neural basis of moral cognition: sentiments, concepts, and values. Ann N Y Acad Sci. 2008;1124:161–80.
21. Van Bavel J, FeldmanHall O, Mende-Siedlecki P. The neuroscience of moral cognition: from dual processes to dynamic systems. Curr Opin Psychol. 2015;6:167–72.

22. Casebeer WD. Moral cognition and its neural constituents. Nat Rev Neurosci. 2003;4(10):840–6.
23. Rest JR. Background: theory and research. In: Rest JR, Narvaez D, editors. Moral development in the professions: psychology and applied ethics. Hillsdale: NJ Erlbaum; 1994. p. 1–26.
24. Robertson D, Snarey J, Ousley O, Harenski K, DuBois Bowman F, Gilkey R, et al. The neural processing of moral sensitivity to issues of justice and care. Neuropsychologia. 2007;45(4):755–66.
25. Decety J, Michalska KJ, Kinzler KD. The contribution of emotion and cognition to moral sensitivity: a neurodevelopmental study. Cereb Cortex. 2012;22(1):209–20.
26. Bzdok D, Schilbach L, Vogeley K, Schneider K, Laird AR, Langner R, et al. Parsing the neural correlates of moral cognition: ALE meta-analysis on morality, theory of mind, and empathy. Brain Struct Funct. 2012;217(4):783–96.
27. Moll J, de Oliveira-Souza R, Eslinger PJ, Bramati IE, Mourao-Miranda J, Andreiuolo PA, et al. The neural correlates of moral sensitivity: a functional magnetic resonance imaging investigation of basic and moral emotions. J Neurosci. 2002;22(7):2730–6.
28. Yoder KJ, Decety J. The good, the bad, and the just: justice sensitivity predicts neural response during moral evaluation of actions performed by others. J Neurosci. 2014;34(12):4161–6.
29. Young L, Saxe R. An FMRI investigation of spontaneous mental state inference for moral judgment. J Cogn Neurosci. 2009;21(7):1396–405.
30. D'Argembeau A, Xue G, Lu ZL, Van der Linden M, Bechara A. Neural correlates of envisioning emotional events in the near and far future. NeuroImage. 2008;40(1):398–407.
31. Mendez MF. The neurobiology of moral behavior: review and neuropsychiatric implications. CNS Spectr. 2009;14(11):608–20.
32. Young L, Bechara A, Tranel D, Damasio H, Hauser M, Damasio A. Damage to ventromedial prefrontal cortex impairs judgment of harmful intent. Neuron. 2010;65(6):845–51.
33. Ciaramelli E, Braghittoni D, di Pellegrino G. It is the outcome that counts! Damage to the ventromedial prefrontal cortex disrupts the integration of outcome and belief information for moral judgment. J Int Neuropsychol Soc. 2012;18(6):962–71.
34. Baxter MG, Parker A, Lindner CC, Izquierdo AD, Murray EA. Control of response selection by reinforcer value requires interaction of amygdala and orbital prefrontal cortex. J Neurosci. 2000;20(11):4311–9.
35. Roelofs K, Minelli A, Mars RB, van Peer J, Toni I. On the neural control of social emotional behavior. Soc Cogn Affect Neurosci. 2009;4(1):50–8.
36. Amodio DM, Frith CD. Meeting of minds: the medial frontal cortex and social cognition. Nat Rev Neurosci. 2006;7(4):268–77.
37. Adolphs R, Tranel D, Damasio AR. The human amygdala in social judgment. Nature. 1998; 393(6684):470–4.
38. Berthoz S, Grezes J, Armony JL, Passingham RE, Dolan RJ. Affective response to one's own moral violations. NeuroImage. 2006;31(2):945–50.
39. Hesse E, Mikulan E, Decety J, Sigman M, Garcia Mdel C, Silva W, et al. Early detection of intentional harm in the human amygdala. Brain. 2016;139(Pt 1):54–61.
40. Decety J, Michalska KJ, Akitsuki Y. Who caused the pain? An fMRI investigation of empathy and intentionality in children. Neuropsychologia. 2008;46(11):2607–14.
41. Phelps EA. Emotion and cognition: insights from studies of the human amygdala. Annu Rev Psychol. 2006;57:27–53.
42. Shenhav A, Greene JD. Integrative moral judgment: dissociating the roles of the amygdala and ventromedial prefrontal cortex. J Neurosci. 2014;34(13):4741–9.
43. Koster-Hale J, Saxe R, Dungan J, Young LL. Decoding moral judgments from neural representations of intentions. Proc Natl Acad Sci U S A. 2013;110(14):5648–53.
44. Northoff G, Bermpohl F. Cortical midline structures and the self. Trends Cogn Sci. 2004;8(3):102–7.
45. Saxe R, Kanwisher N. People thinking about thinking people. The role of the temporo-parietal junction in theory of mind. NeuroImage. 2003;19(4):1835–42.

46. Carter RM, Huettel SA. A nexus model of the temporal-parietal junction. Trends Cogn Sci. 2013;17(7):328–36.
47. Young L, Saxe R. Innocent intentions: a correlation between forgiveness for accidental harm and neural activity. Neuropsychologia. 2009;47(10):2065–72.
48. Baez S, Kanske P, Matallana D, Montañes P, Reyes P, Slachevsky A, et al. Integration of intention and outcome for moral judgment in frontotemporal dementia: brain structural signatures. Neurodegener Dis. 2016;16(3–4):206–17.
49. Baez S, Couto B, Torralva T, Sposato LA, Huepe D, Montanes P, et al. Comparing moral judgments of patients with frontotemporal dementia and frontal stroke. JAMA Neurol. 2014;71(9):1172–6.
50. Baron-Cohen S, Leslie AM, Frith U. Does the autistic child have a "theory of mind"? Cognition. 1985;21(1):37–46.
51. Treadway MT, Buckholtz JW, Martin JW, Jan K, Asplund CL, Ginther MR, et al. Corticolimbic gating of emotion-driven punishment. Nat Neurosci. 2014;17(9):1270–5.
52. Cushman F. Crime and punishment: distinguishing the roles of causal and intentional analyses in moral judgment. Cognition. 2008;108(2):353–80.
53. Young L, Saxe R. The neural basis of belief encoding and integration in moral judgment. NeuroImage. 2008;40(4):1912–20.
54. Decety J. The neuroevolution of empathy. Ann N Y Acad Sci. 2011;1231:35–45.
55. Decety J, Jackson PL. The functional architecture of human empathy. Behav Cogn Neurosci Rev. 2004;3(2):71–100.
56. Decety J, Cowell JM. The complex relation between morality and empathy. Trends Cogn Sci. 2014;18(7):337–9.
57. Yoder KJ, Decety J. Spatiotemporal neural dynamics of moral judgment: a high-density ERP study. Neuropsychologia. 2014;60:39–45.
58. Escobar MJ, Huepe D, Decety J, Sedeno L, Messow MK, Baez S, et al. Brain signatures of moral sensitivity in adolescents with early social deprivation. Sci Rep. 2014;4:5354.
59. Eisenberg N. Emotion, regulation, and moral development. Annu Rev Psychol. 2000;51:665–97.
60. Gleichgerrcht E, Young L. Low levels of empathic concern predict utilitarian moral judgment. PLoS One. 2013;8(4):e60418.
61. Haidt J. The moral emotions. In: Handbook of affective sciences, vol. 11. Oxford: Oxford University Press; 2003. p. 852–70.
62. Kroll J, Egan E, Erickson P, Carey K, Johnson M. Moral conflict, religiosity, and neuroticism in an outpatient sample. J Nerv Ment Dis. 2004;192(10):682–8.
63. Kroll J, Egan E. Psychiatry, moral worry, and the moral emotions. J Psychiatr Pract. 2004;10(6):352–60.
64. Frijda NH. The laws of emotion. Am Psychol. 1988;43(5):349.
65. Smith A. The theory of moral sentiments. London: Penguin; 2010.
66. Moll J, de Oliveira-Souza R. "Extended attachment" and the human brain: internalized cultural values and evolutionary implications. In: The moral brain. London: Springer; 2009. p. 69–85.
67. Manstead AS, Frijda N, Fischer A. Feelings and emotions: the Amsterdam symposium. Cambridge: Cambridge University Press; 2004.
68. Tangney JP, Stuewig J, Mashek DJ. Moral emotions and moral behavior. Annu Rev Psychol. 2007;58:345–72.
69. Sinnott-Armstrong W. The neuroscience of morality: emotion, brain disorders, and development. Cambridge: MIT Press; 2008.
70. Jankowski KF, Takahashi H. Cognitive neuroscience of social emotions and implications for psychopathology: examining embarrassment, guilt, envy, and schadenfreude. Psychiatry Clin Neurosci. 2014;68(5):319–36.
71. Ortony A, Clore GL, Collins A. The cognitive structure of emotions. Cambridge: Cambridge University Press; 1990.

72. Festinger L, Hutte HA. An experimental investigation of the effect of unstable interpersonal relations in a group. J Abnorm Soc Psychol. 1954;49:512–3.
73. Lieberman MD. Social cognitive neuroscience: a review of core processes. Annu Rev Psychol. 2007;58:259–89.
74. Cikara M, Fiske ST. Bounded empathy: neural responses to outgroup targets' (mis)fortunes. J Cogn Neurosci. 2011;23:3791–803.
75. Cikara M, Fiske ST. Their pain, our pleasure: stereotype content and schadenfreude. Ann N Y Acad Sci. 2013;1299:52–9.
76. Dvash J, Gilam G, Ben-Ze'ev A, Hendler T, Shamay-Tsoory SG. The envious brain: the neural basis of social comparison. Hum Brain Mapp. 2010;31:1741–50.
77. Takahashi H, Kato M, Matsuura M, Mobbs D, Suhara T, Okubo Y. When your gain is my pain and your pain is my gain: neural correlates of envy and schadenfreude. Science. 2009;323(5916):937–9.
78. Zaki J, Ochsner KN, Ochsner KN, Plutchik R, Klöppel S, Stonnington CM, et al. Envy, politics, and age. Emotion. 2015;11:187–208.
79. Van Dijk WW, Ouwerkerk JW, Goslinga S, Nieweg M, Gallucci M. When people fall from grace: reconsidering the role of envy in Schadenfreude. Emotion. 2006;6(1):156.
80. Smith RH, Kim SH. Comprehending envy. Psychol Bull. 2007;133(1):46–64.
81. Lange J, Crusius J. Dispositional envy revisited unraveling the motivational dynamics of benign and malicious envy. Personal Soc Psychol Bull. 2014;41(2):284–94.
82. van Dijk WW, Ouwerkerk JW. Schadenfreude: understanding pleasure at the misfortune of others. Cambridge: Cambridge University Press; 2014.
83. Shamay-Tsoory SG, Fischer M, Dvash J, Harari H, Perach-Bloom N, Levkovitz Y. Intranasal administration of oxytocin increases envy and schadenfreude (gloating). Biol Psychiatry. 2009;66(9):864–70.
84. Meyer-Lindenberg A, Domes G, Kirsch P, Heinrichs M. Oxytocin and vasopressin in the human brain: social neuropeptides for translational medicine. Nat Rev Neurosci. 2011;12:524–38.
85. Israel S, Lerer E, Shalev I, Uzefovsky F, Riebold M, Laiba E, et al. The oxytocin receptor (OXTR) contributes to prosocial fund allocations in the dictator game and the social value orientations task. PLoS One. 2009;4:e5535.
86. Fontenelle LF, de Oliveira-Souza R, Moll J. The rise of moral emotions in neuropsychiatry. Dialogues Clin Neurosci. 2015;17(4):411–20.
87. Singer T, Seymour B, O'Doherty J, Kaube H, Dolan RJ, Frith CD. Empathy for pain involves the affective but not sensory components of pain. Science. 2004;303(5661):1157–62.
88. Kerns JG, Cohen JD, MacDonald AW III, Cho RY, Stenger VA, Carter CS. Anterior cingulate conflict monitoring and adjustments in control. Science. 2004;303(5660):1023–6.
89. Baez S, Santamaria-Garcia H, Orozco J, Fittipaldi S, Garcia AM, Pino M, et al. Your misery is no longer my pleasure: reduced schadenfreude in Huntington's disease families. Cortex. 2016;83:78–85.
90. Baez S, Pino M, Berrio M, Santamaria-Garcia H, Sedeno L, Garcia AM, et al. Corticostriatal signatures of schadenfreude: evidence from Huntington's disease. J Neurol Neurosurg Psychiatry. 2017. https://doi.org/10.1136/jnnp-2017-316055.
91. Fehr E, Camerer CF. Social neuroeconomics: the neural circuitry of social preferences. Trends Cogn Sci. 2007;11(10):419–27.
92. Sanfey AG, Rilling JK, Aronson JA, Nystrom LE, Cohen JD. The neural basis of economic decision-making in the ultimatum game. Science. 2003;300:1755–8.
93. Harlé KM, Sanfey AG. Social economic decision-making across the lifespan: an fMRI investigation. Neuropsychologia. 2012;50:1416–24.
94. Fliessbach K, Weber B, Trautner P, Dohmen T, Sunde U, Elger CE, et al. Social comparison affects reward-related brain activity in the human ventral striatum. Science. 2007;318(5854):1305–8.
95. Singer T, Seymour B, O'Doherty JP, Stephan KE, Dolan RJ, Frith CD. Empathic neural responses are modulated by the perceived fairness of others. Nature. 2006;439:466–9.

96. Lewis M. Self-conscious emotions: embarrassment, pride, shame, and guilt. In: Lewis WM, Haviland-Jones JM, editors. Handbook of emotions. New York: Guilford Press; 2000. p. 623–36.
97. Tangney JP. The self-conscious emotions: shame, guilt, embarrassment and pride. 1999.
98. Tangney JP, Stuewig J, Hafez L. Shame, guilt, and remorse: implications for offender populations. J Forensic Psychiat Psychol. 2011;22(5):706–23.
99. Green S, Ralph MAL, Moll J, Stamatakis EA, Grafman J, Zahn R. Selective functional integration between anterior temporal and distinct fronto-mesolimbic regions during guilt and indignation. NeuroImage. 2010;52:1720–6.
100. Moll J, de Oliveira-Souza R, Garrido GJ, Bramati IE, Caparelli-Daquer EMA, Paiva MLMF, et al. The self as a moral agent: linking the neural bases of social agency and moral sensitivity. Soc Neurosci. 2007;2:336–52.
101. FeldmanHall O, Mobbs D, Dalgleish T. Deconstructing the brain's moral network: dissociable functionality between the temporoparietal junction and ventro-medial prefrontal cortex. Soc Cogn Affect Neurosci. 2014;9(3):297–306.
102. Takahashi H, Yahata N, Koeda M, Matsuda T, Asai K, Okubo Y. Brain activation associated with evaluative processes of guilt and embarrassment: an fMRI study. NeuroImage. 2004;23(3):967–74.
103. Michl P, Meindl T, Meister F, Born C, Engel RR, Reiser M, et al. Neurobiological underpinnings of shame and guilt: a pilot fMRI study. Soc Cogn Affect Neurosci. 2014;9(2):150–7.
104. Marsh AA, Blair KS, Jones MM, Soliman N, Blair RJ. Dominance and submission: the ventrolateral prefrontal cortex and responses to status cues. J Cogn Neurosci. 2009;21:713–24.
105. Harrison BJ, Pujol J, Soriano-Mas C, Hernandez-Ribas R, Lopez-Sola M, Ortiz H, et al. Neural correlates of moral sensitivity in obsessive-compulsive disorder. Arch Gen Psychiatry. 2012;69:741–9.
106. Olson IR, McCoy D, Klobusicky E, Ross LA. Social cognition and the anterior temporal lobes: a review and theoretical framework. Soc Cogn Affect Neurosci. 2013;8(2):123–33.
107. LA R, Olson IR. Social cognition and the anterior temporal lobes. NeuroImage. 2010;49:3452–62.
108. Boehme S, Miltner WH, Straube T. Neural correlates of self-focused attention in social anxiety. Soc Cogn Affect Neurosci. 2015;10(6):856–62.
109. Morita T, Tanabe HC, Sasaki AT, Shimada K, Kakigi R, Sadato N. The anterior insular and anterior cingulate cortices in emotional processing for self-face recognition. Soc Cogn Affect Neurosci. 2013;9(5):570–9.
110. Ibanez A, Garcia AM, Esteves S, Yoris A, Munoz E, Reynaldo L, et al. Social neuroscience: undoing the schism between neurology and psychiatry. Soc Neurosci. 2016:1–39. https://doi.org/10.1080/17470919.2016.1245214.
111. Eisenberger NI. Meta-analytic evidence for the role of the anterior cingulate cortex in social pain. Soc Cogn Affect Neurosci. 2015;10(1):1–2.
112. Herold D, Spengler S, Sajonz B, Usnich T, Bermpohl F. Common and distinct networks for self-referential and social stimulus processing in the human brain. Brain Struct Funct. 2016;221(7):3475–85.
113. Molenberghs P, Johnson H, Henry JD, Mattingley JB. Understanding the minds of others: a neuroimaging meta-analysis. Neurosci Biobehav Rev. 2016;65:276–91.
114. Kennedy S. Psychosocial stress, health, and the hippocampus. J Undergrad Neurosci Educ. 2016;15(1):R12–r3.
115. Kennedy DP, Adolphs R. The social brain in psychiatric and neurological disorders. Trends Cogn Sci. 2012;16(11):559–72.
116. Schilbach L, Timmermans B, Reddy V, Costall A, Bente G, Schlicht T, et al. Toward a second-person neuroscience. Behav Brain Sci. 2013;36(4):393–414.
117. Baez S, Garcia AM, Ibanez A. The social context network model in psychiatric and neurological diseases. Curr Top Behav Neurosci. 2016;30:379–96.

118. Gleichgerrcht E, Torralva T, Rattazzi A, Marenco V, Roca M, Manes F. Selective impairment of cognitive empathy for moral judgment in adults with high functioning autism. Soc Cogn Affect Neurosci. 2013;8(7):780–8.
119. Moran JM, Young LL, Saxe R, Lee SM, O'Young D, Mavros PL, et al. Impaired theory of mind for moral judgment in high-functioning autism. Proc Natl Acad Sci U S A. 2011; 108(7):2688–92.
120. Zalla T, Barlassina L, Buon M, Leboyer M. Moral judgment in adults with autism spectrum disorders. Cognition. 2011;121(1):115–26.
121. Buon M, Dupoux E, Jacob P, Chaste P, Leboyer M, Zalla T. The role of causal and intentional judgments in moral reasoning in individuals with high functioning autism. J Autism Dev Disord. 2013;43(2):458–70.
122. Schneider K, Pauly KD, Gossen A, Mevissen L, Michel TM, Gur RC, et al. Neural correlates of moral reasoning in autism spectrum disorder. Soc Cogn Affect Neurosci. 2013;8(6):702–10.
123. Bauminger N. The expression and understanding of jealousy in children with autism. Dev Psychopathol. 2004;16(1):157–77.
124. Blair RJ. A cognitive developmental approach to mortality: investigating the psychopath. Cognition. 1995;57(1):1–29.
125. Koenigs M, Kruepke M, Zeier J, Newman JP. Utilitarian moral judgment in psychopathy. Soc Cogn Affect Neurosci. 2012;7(6):708–14.
126. Cima M, Tonnaer F, Hauser MD. Psychopaths know right from wrong but don't care. Soc Cogn Affect Neurosci. 2010;5(1):59–67.
127. de Oliveira-Souza R, Hare RD, Bramati IE, Garrido GJ, Azevedo Ignacio F, Tovar-Moll F, et al. Psychopathy as a disorder of the moral brain: fronto-temporo-limbic grey matter reductions demonstrated by voxel-based morphometry. NeuroImage. 2008;40(3):1202–13.
128. Harenski CL, Harenski KA, Shane MS, Kiehl KA. Aberrant neural processing of moral violations in criminal psychopaths. J Abnorm Psychol. 2010;119(4):863–74.
129. Seara-Cardoso A, Sebastian CL, McCrory E, Foulkes L, Buon M, Roiser JP, et al. Anticipation of guilt for everyday moral transgressions: the role of the anterior insula and the influence of interpersonal psychopathic traits. Sci Rep. 2016;6:36273.
130. Glenn AL, Raine A, Schug RA. The neural correlates of moral decision-making in psychopathy. Mol Psychiatry. 2009;14(1):5–6.
131. Khemiri L, Guterstam J, Franck J, Jayaram-Lindstrom N. Alcohol dependence associated with increased utilitarian moral judgment: a case control study. PLoS One. 2012;7(6):e39882.
132. Verdejo-Garcia A, Contreras-Rodriguez O, Fonseca F, Cuenca A, Soriano-Mas C, Rodriguez J, et al. Functional alteration in frontolimbic systems relevant to moral judgment in cocaine-dependent subjects. Addict Biol. 2014;19(2):272–81.
133. Campagna AF, Harter S. Moral judgment in sociopathic and normal children. J Pers Soc Psychol. 1975;31(2):199–205.
134. Benson AL. Morality of schizophrenic adolescents. J Abnorm Psychol. 1980;89(5):674–7.
135. Green S, Lambon Ralph MA, Moll J, Deakin JF, Zahn R. Guilt-selective functional disconnection of anterior temporal and subgenual cortices in major depressive disorder. Arch Gen Psychiatry. 2012;69(10):1014–21.
136. Pugh LR, Taylor PJ, Berry K. The role of guilt in the development of post-traumatic stress disorder: a systematic review. J Affect Disord. 2015;182:138–50.
137. Basile B, Mancini F, Macaluso E, Caltagirone C, Bozzali M. Abnormal processing of deontological guilt in obsessive-compulsive disorder. Brain Struct Funct. 2014;219(4):1321–31.
138. Weingarden H, Renshaw KD. Shame in the obsessive compulsive related disorders: a conceptual review. J Affect Disord. 2015;171:74–84.
139. Santamaría-García H, Soriano-Mas C, Burgaleta M, Ayneto A, Alonso P, Menchón JM, et al. Social context modulates cognitive markers in Obsessive-Compulsive Disorder. Social Neuroscience, 2017;1–15.
140. Carter CS, Braver TS, Barch DM, Botvinick MM, Noll D, Cohen JD. Anterior cingulate cortex, error detection, and the online monitoring of performance. Science. 1998;280:747–9.

141. Melloni M, Urbistondo C, Sedeño L, Gelormini C, Kichic R, Ibanez A. The extended fronto-striatal model of obsessive compulsive disorder: convergence from event-related potentials, neuropsychology and neuroimaging. Front Hum Neurosci. 2012;6:259.
142. Endrass T, Schuermann B, Kaufmann C, Spielberg R, Kniesche R, Kathmann N. Performance monitoring and error significance in patients with obsessive-compulsive disorder. Biol Psychol. 2010;84:257–63.
143. Mendez MF, Anderson E, Shapira JS. An investigation of moral judgement in frontotemporal dementia. Cogn Behav Neurol. 2005;18(4):193–7.
144. Gleichgerrcht E, Torralva T, Roca M, Pose M, Manes F. The role of social cognition in moral judgment in frontotemporal dementia. Soc Neurosci. 2011;6(2):113–22.
145. Mendez MF, Shapira JS. Altered emotional morality in frontotemporal dementia. Cogn Neuropsychiatry. 2009;14(3):165–79.
146. Ibanez A, Billeke P, de la Fuente L, Salamone P, Garcia AM, Melloni M. Reply: towards a neurocomputational account of social dysfunction in neurodegenerative disease. Brain. 2017;140(3):e15.
147. Melloni M, Billeke P, Baez S, Hesse E, de la Fuente L, Forno G, et al. Your perspective and my benefit: multiple lesion models of self-other integration strategies during social bargaining. Brain. 2016. https://doi.org/10.1093/brain/aww231.
148. Baez S, Herrera E, Garcia A, Manes F, Young L, Ibanez A. Outcome-oriented moral evaluation in terrorists. Nat Human Behav. 2017;1:0118.
149. Chiong W, Wilson SM, D'Esposito M, Kayser AS, Grossman SN, Poorzand P, et al. The salience network causally influences default mode network activity during moral reasoning. Brain. 2013;136(Pt 6):1929–41.
150. Seeley WW, Crawford RK, Zhou J, Miller BL, Greicius MD. Neurodegenerative diseases target large-scale human brain networks. Neuron. 2009;62(1):42–52.
151. Seeley WW, Menon V, Schatzberg AF, Keller J, Glover GH, Kenna H, et al. Dissociable intrinsic connectivity networks for salience processing and executive control. J Neurosci. 2007;27(9):2349–56.
152. Levenson RW, Sturm VE, Haase CM. Emotional and behavioral symptoms in neurodegenerative disease: a model for studying the neural bases of psychopathology. Annu Rev Clin Psychol. 2014;10:581–606.
153. Sturm VE, Rosen HJ, Allison S, Miller BL, Levenson RW. Self-conscious emotion deficits in frontotemporal lobar degeneration. Brain. 2006;129(Pt 9):2508–16.
154. Koenigs M, Young L, Adolphs R, Tranel D, Cushman F, Hauser M, et al. Damage to the prefrontal cortex increases utilitarian moral judgements. Nature. 2007;446(7138):908–11.
155. Taber-Thomas BC, Asp EW, Koenigs M, Sutterer M, Anderson SW, Tranel D. Arrested development: early prefrontal lesions impair the maturation of moral judgement. Brain J Neurol. 2014;137(Pt 4):1254–61.
156. Shamay-Tsoory SG, Ahronberg-Kirschenbaum D, Bauminger-Zviely N. There is no joy like malicious joy: schadenfreude in young children. PLoS One. 2014;9(7):e100233.
157. Shamay-Tsoory SG, Tibi-Elhanany Y, Aharon-Peretz J. The green-eyed monster and malicious joy: the neuroanatomical bases of envy and gloating (schadenfreude). Brain. 2007;130(Pt 6): 1663–78.
158. Ibanez A, Manes F. Contextual social cognition and the behavioral variant of frontotemporal dementia. Neurology. 2012;78:1354–62.
159. Baez S., & Ibanez, A. (2014). The effects of context processing on social cognition impairments in adults with Asperger's syndrome. Frontiers in neuroscience, 8, 270. doi: 10.3389/fnins.2014.00270
160. Garcia AM, Ibanez A. Two-person neuroscience and naturalistic social communication: the role of language and linguistic variables in brain-coupling research. Front Psych. 2014;5:124.
161. Redcay E, Dodell-Feder D, Mavros PL, Kleiner M, Pearrow MJ, Triantafyllou C, et al. Atypical brain activation patterns during a face-to-face joint attention game in adults with autism spectrum disorder. Hum Brain Mapp. 2013;34(10):2511–23.

162. Sapolsky RM. The influence of social hierarchy on primate health. Science. 2005;308:648–52.
163. Sapolsky RM. Social status and health in humans and other animals. Annu Rev Anthropol. 2004;33:393–418.
164. Buttelmann D, Bohm R. The ontogeny of the motivation that underlies in-group bias. Psychol Sci. 2014;25(4):921–7.
165. Jordan JJ, McAuliffe K, Warneken F. Development of in-group favoritism in children's third-party punishment of selfishness. Proc Natl Acad Sci U S A. 2014;111(35):12710–5.
166. Fehr E, Bernhard H, Rockenbach B. Egalitarianism in young children. Nature. 2008;454(7208): 1079–83.

On the Cognitive (Neuro)science of Moral Cognition: Utilitarianism, Deontology, and the "Fragmentation of Value"

Alejandro Rosas

Abstract Scientific explanations of human higher capacities, traditionally denied to other animals, attract the attention of philosophers and other workers in the humanities. They are often viewed with suspicion and skepticism. Against this background, I critically examine the dual-process theory of moral judgment proposed by Greene and collaborators and the normative consequences drawn from that theory. I believe normative consequences are warranted, in principle, but I propose an alternative dual-process model of moral cognition that leads to a different normative consequence, which I dub "the fragmentation of value" (Nagel. Mortal questions. Cambridge: Cambridge University Press; 1979). This alternative model abandons the neat overlap between the deontological/utilitarian and the intuitive/reflective divides. Instead, we have both utilitarian and deontological intuitions as equally fundamental and partially in tension. Cognitive control is sometimes engaged during a conflict between intuitions. When it is engaged, the result of control is not always utilitarian; sometimes it is deontological. I describe in some detail how this version is consistent with evidence reported by many studies and what could be done to find more evidence to support it.

Keywords Cognitive control • Dual-process theory • Evolution • Intuition • Moral cognition • Moral dilemmas • Reaction times • Value pluralism

1 Introduction

Is neuropsychological research into moral judgment [1, 2] of any relevance for the humanities and the social sciences? I merge the latter two areas of knowledge because both have, presumably, an interest in understanding human morality, religiosity, aesthetic sensitivity, shared intentionality [3], and other traits widely held to

A. Rosas (✉)
Philosophy Department, National University of Colombia,
Kra. 30, #45-03, Bogotá, Colombia
e-mail: arosasl@unal.edu.co

© Springer International Publishing Switzerland 2017
A. Ibáñez et al. (eds.), *Neuroscience and Social Science*,
DOI 10.1007/978-3-319-68421-5_9

be uniquely human. The understanding they seek is not primarily explanatory and scientific. Most often, they want to know what ideals, values, and human characteristics are worth preserving and promoting. And sometimes, this interest leads them to reject scientific explanations as altogether irrelevant to concerns about values.

Our initial question can be reformulated in this way: Can we draw normative conclusions from neuropsychological theories? Can they legitimately make recommendations about what morality to accept, what type of state and government to prefer, and which laws to vote for in parliament?

A vast majority of philosophers and humanists more or less intuitively, more or less reflectively, deny any normative relevance to neuroscience. As a philosopher, I belong in the heretical (albeit growing?) minority that is open to the possibility of its normative relevance—including in this openness other empirical sciences dealing with mind and morals. If by looking at sciences like psychology, cognitive neuroscience, and evolutionary biology, we come to understand what morality is, we might get a deeper grasp of its functions and peculiar authority.

In this chapter, I discuss how normative conclusions can follow from neurocognitive research into moral judgment and how they depend, crucially, on the theoretical interpretation of the data. First, in Sect. 2, I briefly reconstruct Greene's argument [4, 5] for his normative conclusion. I concisely describe the dual-process theory of cognition, its application to moral cognition, and the evolutionary presuppositions that support the normative conclusion. In Sect. 3, I present the new data on reaction times (RTs); and in Sect. 4, I describe data from cognitive load studies suggesting an alternative version of the model. Briefly, we have both utilitarian and deontological intuitions, which are sometimes in agreement and sometimes deeply in conflict. Section 5 introduces the concepts of variable utilitarian and deontological sensitivities and explains how conflict intensity varies among individuals, some of whom might also exhibit severe weakness in one or both sensitivities. The alternative dual-process theory is presented in Sect. 6. In Sect. 7, I draw the normative conclusion.

2 Greene's Normative Claim

Greene [4, 5] complemented Greene et al.'s dual-process theory of moral judgment [1, 2]—a theory that belongs within cognitive neuroscience—with a normative claim recommending utilitarianism over deontology. His collaborative neurocognitive research had shown that utilitarian responses to moral dilemmas are connected to executive decision-making, whereas deontological ones are intuitive, automatic, and emotional. He combined this finding with the idea that deontology comprises principles of action that evolved as adaptive intuitions among our evolutionary ancestors. These intuitions, however, may produce maladaptive behavior in rapidly changing, social environments [4]. Utilitarianism corrects for these maladaptive effects. It is slow and thus inefficient when quick decisions are called for, but it is flexible and adapts rationally to varying circumstances. Initially, Greene cautiously presented this normative claim as hypothetical, as an example of how neuroscience (complemented with cognitive science and evolutionary biology) can affect our normative views [4]. Since then, he developed the theory to back up this normative claim [5]. If any, this is a serious normative conclusion to draw from research in cognitive neuroscience.

Although I am not convinced of the soundness of this normative conclusion, I must emphasize I see nothing logically or scientifically wrong with the underlying reasoning. If the neurocognitive data were as Greene and collaborators presented them in their two early papers, the normative conclusion Greene inferred would be a serious contender for the truth. But the devil is in the details (of the data). The data reported by Greene et al. [1, 2] certainly seem to support a theoretical identification of deontology with intuitive, automatic thinking on the one hand and utilitarianism with controlled, reflective, effortful thinking on the other. With additional scientific premises (widely accepted among scientists dealing with mind and morals), these data enter into an argument with the following logical structure:

1. There is a difference between automatic (intuitive) and controlled (reflexive) cognitive processes (dual-process theory in cognitive science) [6, 7].
2. Automatisms evolve to deliver fast, reliable, and therefore efficient responses. But speed is traded-off against flexibility and accuracy (a constraint in the design of organisms shaped by natural selection).
3. Controlled processes correct for inaccuracies of automatic ones (hypothesis about the function of executive control) [8].
4. In evolutionary novel situations, like those that often arise when organisms live in a complex social world, controlled processes often override—and ought to override—the fast, automatic, and intuitive responses, to keep behavior in target.
5. Deontological judgments about cases are intuitive, automatic, emotional, and fast. In contrast, utilitarian judgments are controlled and slow and work to correct intuitive judgments (the brain imaging and reaction time data from Greene et al. [1, 2, 9] interpreted in the light of dual-process theory).
6. Conclusion: Deontology ought to be overridden by utilitarianism when they conflict.

Against the scientific background of dual-process theory and evolutionary biology, Greene interprets the neurocognitive results as inviting us to endorse utilitarianism. My doubts arise in regard to premise no. 5 in the above argument, namely, the neat allocation of deontological principles to evolutionary ancient and automatic processes on the one hand and of utilitarian responses (hereafter UR) to executive or cognitive control correcting intuitive and inaccurate judgments on the other. The data strongly suggest an alternative interpretation. They could point to a different dual-process theory, where not only utilitarian but also deontological responses to moral dilemmas can claim a noble origin in the executive functions.

3 Enigmatic Reaction Time Data

According to the neuroscientific evidence reported by Greene et al. [1, 2], deontological judgments activate emotional circuits in the brain, whereas utilitarian judgments activate preferentially the dorsolateral prefrontal cortex, associated with cognitive control. Additionally, behavioral data—specifically, the RT of participants confronted with personal dilemmas—show that these are longer for UR [1], a fact that also suggests the same interpretation in terms of dual-process theory. Therefore, only utilitarianism is connected to reasoning and executive functions; deontology, in contrast, is emotional, intuitive, fast, and automatic.

The idea that deontological judgments are intuitive, automatic, and emotional is quite a challenge to the traditional philosophical view linking deontology exclusively to reason, as in Kant [10]. But new evidence alerts us against overhasty claims on this point. The new evidence came primarily from corrected measurements of RT. In the course of this chapter, I also review data coming from new cognitive load studies that support a revision of Greene et al.'s original dual-process model. The neat allocation of deontological principles to evolutionary ancient and automatic processes, on the one hand, and of UR to executive or cognitive control, on the other hand, is not as promising as it seemed to be initially. As for the fMRI data, at the end of Sect. 6, we shall see that the alternative version of the dual-process theory recommends a new design for data collection.

The original evidence suggesting a difference between the RT of deontological and UR turned out to be an artifact of including inadequate dilemmas in the battery used for testing [11–13]. Greene conceded in his reply to McGuire et al.: "The apparent RT effect was generated by the inclusion of several "dilemmas" in which a personal harm has no compelling utilitarian rationale. These dilemmas reliably elicited fast, disapproving judgments, skewing the data" [14, p. 582]. However, Greene was already aware of the problem, thanks to a personal communication with Liane Young. He reacted conducting with his collaborators a new study [9] and run the analyses only on "high-conflict" personal dilemmas. This subgroup of dilemmas does have the required structure, pitting deontological against utilitarian considerations. Greene and collaborators measured the RT for utilitarian and deontological responses in two conditions: with and without cognitive load (the load was detecting the number 5 in a row of numbers scrolling across the screen beneath the dilemmas during the deliberation time). Their results show that RT increases in the load compared to the no-load condition, but solely for the UR. Load had no effect on the RT of deontological responses. This is plausibly interpreted as implying that utilitarian, but not deontological, responses use working memory resources that are being interfered with in the load condition.

Their 2008 experiment also threw one further interesting result. In a follow-up analysis, they allotted participants to two subgroups regarding their tendency (high or low) to deliver UR. The high tendency group exhibited a surprising pattern: in the no-load condition, their UR had significantly shorter RT than their deontological responses (5350 ms vs. 6070 ms, respectively; see Fig. 1, left). On the other hand,

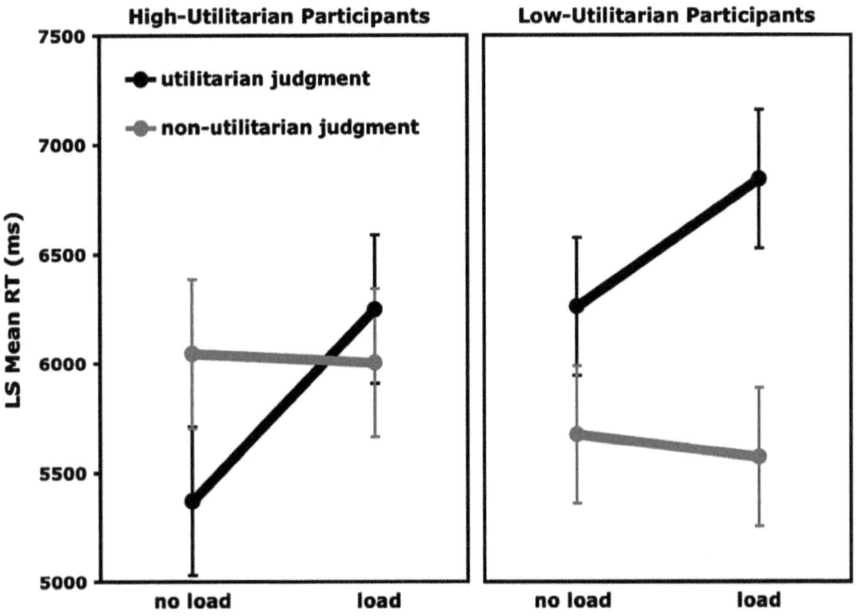

Fig. 1 Effects of load on RT for high-utilitarian (n = 41) and low-utilitarian (n = 41) groups. Original in [9, p. 1150]. Reproduced here with permission

only their utilitarian, but not their deontological, RTs were affected by load. But, precisely under load, the mean RT of their UR was not significantly higher than the RT of their deontological responses (6250 ms vs. 6000 ms, respectively; see Fig. 1, left), suggesting that some cognitive control also underlies deontological responses. So, despite the impressive result obtained comparing the load and no-load conditions, these findings about the RT are bizarre and should caution us not to endorse the dual-process theory in its original form without further investigation. Greene and collaborators grant that accounting for this result “will require a significant expansion and/or modification of our dual-process theory” [9, p. 1152].

4 Modifying the Dual-Process Theory of Moral Cognition

Greene et al. [9] ranked participants from high to low by their percentage of UR to the set of high-conflict dilemmas and divided the sample into high- and low-tendency utilitarian participants. The concept of a “tendency” to deliver UR is interesting. It could easily lead to a very different dual-process theory. A high-tendency utilitarian participant is prone to give UR easily, but deontological responses only with some difficulty. Taken to the limit, considering, e.g., only the top ranks among the high-tendency utilitarians, the “easiness” could mean that they deliver fast and intuitive utilitarian responses. Conversely, one could rank participants by percentages of deontological responses from the highest to the lowest; at the low end, we would find participants that deliver deontological responses as products of a slow, controlled, deliberative process. We would then have to admit two further types of moral judgments, impossible in the present version of the dual-process model of moral cognition but perfectly possible according to common sense: intuitive utilitarian judgments and reflective deontological judgments.

The resulting four types of judgments deliver a much messier, and less catchy, picture than the hypothesis Greene et al. proposed. This messy picture is compatible with the new evidence debunking the claim that URs have longer RT than deontological responses. Statistically, this follows from a comparison of the mean RT of both types of responses, which in this case yields no significant difference. Usually, this suggests that the RT ranges from low to high in both response types. Take, for example, a high-conflict dilemma for which the proportions of utilitarian to deontological responses are nearly equal, like *crying baby* (53.66% utilitarian response in [2]). The average RT for deontological responses ($n = 19$) is 6274 ms (range: 3199–14,445 ms). The average RT for UR ($n = 22$) is 6365 ms (range: 2453–12,456 ms) (data from [2]).[1] In principle, these data are compatible with the idea that some URs are intuitive and some reflective and the same for deontological responses. The intuitive/reflective divide would not overlap with the deontological/utilitarian divide. Reaction times alone cannot prove this, but they do suggest it. One issue raised by this possibility is this: how shall we interpret people who give intuitive deontological responses to dilemmas where the majority response is utilitarian (like impersonal dilemmas or dilemmas where killing one saves millions) or who give intuitive UR to dilemmas where the majority response is deontological (like *footbridge*)? In labeling them "intuitive," I mean delivered without conflict. What explanation could this have in terms of the moral perspective of those participants? I shall return to this question in Sect. 5, where I shall comment on the implications of individual variation disclosed in research with moral dilemmas.

Utilitarian intuitions seem to be present in participants responding to moral dilemmas. This has been suggested in a number of studies [15–18]. Some of these studies find in moral cognition signs of intuitions as placeholders for logical operations, a phenomenon observed also in reasoning tasks [19, 20]. Additionally, one paper [21] has produced experimental evidence that deontic responders faced with impersonal dilemmas (like *trolley*) do detect a conflict with utilitarian principles, despite responding deontologically. In a follow-up paper, Bialek and De Neys report that deontic responders detect conflict in an intuitive way, because the detection is not affected by load [15]. This suggests that awareness of the conflict between utilitarian and deontological principles is itself intuitive, not an effect of a controlled process. The conflict arises from the simultaneous activation of deontological and utilitarian intuitions, implying a critique of the classic default-interventionist dual-process model. In the latter, the conflict occurs between an intuitive deontological and a controlled-utilitarian process, such that only URs qualify as controlled. In the so-called hybrid dual-process model [15], the conflict occurs between two

[1] Thanks to Josh Greene for sharing the data.

intuitions. One could presume that subjects who detect a conflict give a reflective, cognitively controlled response, independently of the type of response; but at the present state of research, this can only be conjectured rather than asserted. After all, detecting a conflict is not the same as reasoning one's way out of it.

It has also been argued that dilemmas featuring extraordinary kill-save ratios, i.e., when the ratio of lives lost to lives saved is very low—e.g., kill one to save thousands—facilitate *intuitive* UR. As evidence for this claim, Trémolière and Bonnefon report that extraordinary kill-save ratios (<1:500) influence the percentage levels of UR independently of simultaneous cognitive load of the subjects solving a dilemma task [18]. Apparently, the influence of these ratios on UR occurs intuitively, not mediated by working memory. Thanks to the pioneering research of Greene et al., we also know that impersonal[2] harm drastically increases the percentage of UR. Does impersonal harm influence the response intuitively? Moore et al. [13] showed that working memory capacity does not affect increase in UR if killing is impersonal, suggesting that this feature is intuitively processed and applied to judgment with no demand on working memory.

From a commonsense perspective, we can easily conceive of intuitive UR, contradicting the default-interventionist model. Consider the cases where the utilitarian and the deontological intuitions converge on the same action, like in *preventing the spread*. Here a doctor decides to administer a deadly poison to a person who is malevolently planning to spread HIV. This dilemma (modified to make harm nonlethal) was classified in Kahane et al. [17] as "utilitarian intuitive," and indeed most participants choose the utilitarian option when judging the appropriateness of sacrificing a victim who is about to commit a criminal action. In one of Greene's classic studies, 40 from 41 participants delivered the UR to this (unmodified) dilemma in an average RT of 4646 ms (range: 2398–12,006 ms) (data from [2]). But note that we could interpret the doctor's action as third-party punishment, which is also seen as a deontological (retributive) moral attitude. Usually, malevolent people who draw pleasure from harming others are punished in order to prevent them from harming more people, among other reasons. Doing so deters future violations, generating a benefit to the group. Arguably, we face here a paradigmatic case of the partial overlap of utilitarianism and deontology. It works like this: a regard for the good of others (one's group) bans all those actions where harm to (innocent) others is used as a means to obtain selfish benefits. Disregard of this ban leads to punishment. Justice is thus born.

Another candidate for a congruent case is telling a white lie [17]. Most subjects choose to tell a lie when the truth would cause harm unnecessarily. Note, however, that *white lie* can also be read as presenting a conflict between deontological duties—"Tell the truth" vs. "Do not harm innocent people." And yet, it is plausible to claim that people prioritize the duty not to harm in this case, because it also makes utilitarian sense. It is, perhaps, a case where utilitarian and deontological intuitions are congruent.

[2] Impersonal harm is typically unintended and committed without exerting muscular force. In Sect. 6 we discuss these two aspects separately.

In dilemmas like *white lie* and *preventing the spread*, utilitarianism and deontology support the same action. It seems that the good of others (the group) is at the root of some deontological intuitions. The good of the group requires us to constrain our freedom, ultimately in attention to the welfare of the group to which we belong. These constraints are the deontological norms.

So far so good, but this is not the whole story. Congruent cases in no way deny that many moral dilemmas present a real conflict. Utilitarianism not only prescribes justice, i.e., it not only prohibits taking away from others what is theirs: their freedom, personal integrity, belongings, and reputation. Utilitarianism also requires us to give to others what is legitimately ours when others need it urgently and to give without the framework of reciprocity that usually characterizes cooperative helping. Here deontology and utilitarianism are in tension. Sacrificial dilemmas like *footbridge* bring this tension to its utmost level, because they present cases where somebody who is not doomed or guilty is forced without consent to offer his life in sacrifice for the lives of several others. This extreme form of utilitarianism is repugnant to many people. Nonetheless, when both moralities genuinely conflict, special circumstances like harm occurring unintended or extraordinary kill-save ratios [22–24] favor UR intuitively, while yet other circumstances might influence UR through controlled processes, as we shall see in Sect. 6.

5 Individual Variation in Moral Sensitivity

In the preceding section, we encountered the construct "tendency to deliver utilitarian responses." This construct was supported with a model to predict RT by Baron et al. [25] and Baron and Gürçay [26]. They modeled the probability of a UR to a given dilemma as a function of the individual ability to give UR and of the degree of difficulty of the particular dilemma. They further argued that when ability matches difficulty, the probability is 0.5 and RTs are longest. The situation is, in their opinion, analogous to the probability of giving the correct meaning of a word depending on individual word competence and word difficulty [25]. But these cases are also different in one important respect. In moral dilemmas, identifiable objective features affect the probability of an UR (e.g., death as unintended side effect, or the kill-save ratio). These objective features have to be included in the theory and in the model. In the case of word competence, there are no such features and hence the difference.

The features that affect the difficulty or easiness of a dilemma speak always to the opposition between two sensitivities in individuals: sensitivity to utilitarian considerations and sensitivity to deontological considerations. There is a complex dynamics between these two sensitivities. First, they are not always opposed to each other. In some cases, they converge on the same response. The clearest cases of convergence are the congruent cases [21, 27]. Other less obvious cases may also favor convergence: e.g., cases of punishment and white lies, as discussed above. But when these sensitivities conflict instead of synergizing, it is possible to point to objective circumstances whose presence/absence increases/decreases the probability of an UR. When we limit our scope to sacrificial dilemmas, circumstances whose presence or absence matters are:

the death caused by the maximizing action is not intended [1, 2];
or the victim would die anyway [13, 28, 29];
or extraordinary kill-save ratios [18, 22–24];
or the victim is guilty [29];
or the agent is among the saved [13, 29];
or none of the previous, but the victim is sacrificed without exerting muscular force [5, 30, 46];
or the dilemmas are presented in virtual reality rather than in text format [31, 32];
and perhaps many others yet to discover.

In all these cases, the bearing of these circumstances on UR also depends on the individual sensitivities. But given one same sensitivity level, their presence or absence weighs on the balance. Dilemmas where they are absent are easy for subjects with a strong deontological sensitivity and receive a swift deontological response. Dilemmas where one, many, or all of these circumstances are present are easy for subjects with strong utilitarian sensitivity. In some cases, they could be so easy that utilitarian responses would be intuitively issued. In the model by Baron and collaborators, circumstances of this type seem to play no role.

A paper by Krajbich and collaborators [33] explores a more suitable comparison than the comparison with semantic competence. The comparison is with public goods games (PGG). In such games, subjects are also torn between two sensitivities that oppose each other and are, when they conflict, exactly the converse of the other one: the selfish and the pro-social sensitivity. They often conflict, but not always, similar in this to the utilitarian and the deontological sensitivities. In the PGG, the difficulty refers to overcoming selfishness, which depends on objective features of the payoff structure. This is easily explained: If your contribution to the common fund generates for each group member, including yourself, a return only slightly below your contribution, it is easy to overcome the selfish inclination to contribute nothing to the public good. If on the contrary, it generates a return greatly below your contribution, it is not easy to overcome selfishness, because you risk losing virtually all your contribution if nobody else contributes [33]. People vary in the strength of their selfish and pro-social sensitivities, but this variance is always relative to those payoff structures. Krajbich et al. want to use this insight to criticize the dual-process model and favor a single process account. I believe this does not necessarily follow. Alternatively, you can argue that moral cognition depends essentially on emotional sensitivities. In particular, whether a given judgment or response to a moral dilemma or PGG is intuitive or controlled depends on the relative strength of the responder's opposed sensitivities.

A bewildering possibility is that some subjects could totally lack either the utilitarian or the deontological sensitivity. In these cases, subjects will give a response with no detection of conflict at all. Conflict-less responses can be labeled intuitive. Consider the percentage of UR to *footbridge*, which vary across studies roughly between 10% and 30%. Although consistently a minority, it is not an insignificant one. How do we interpret these participants? I see two possibilities: they feel the deontological intuition against the sacrifice and nonetheless decide that it is appropriate, or they feel no deontological intuition at all. The first case would correspond to the archetypal—though controversial—utilitarian subjects that Greene might have in mind, who out of conviction override their deontological intuitions. In the second case, however, it is hard to decide whether these participants, totally lacking a deontological sensitivity, have a moral sensitivity at all. Here several studies reporting positive correlations between UR and subclinical psychopathic tendencies become relevant. The correlations are small to moderate [34], and in all fairness, some studies have not found them [32], but in any case they might indicate that at least some subjects deliver UR score very low on empathy or high in clinical or subclinical psychopathy [24, 35–41], measured with psychometric questionnaires like the Levenson Self-Report Psychopathy Scale [42]. It is of course possible that participants lacking deontological intuitions are only a small minority within the group of up to 30% of participants that respond as utilitarians in *footbridge*. The rest are hard-core utilitarians, so to say, that override their deontological intuitions. For the sake of symmetry, one would suspect a similar situation for some deontological responses without conflict. They might reflect a cold-hearted rule following and a scant moral sensitivity [24]. Finding out if this is the case should be a goal for empirical research.

6 An Alternative Model of Moral Cognition

If we abandon the neat overlap between the deontological/utilitarian divide and the intuitive/reflexive divide, both Greene et al.'s particular dual-process model of moral cognition and Greene's normative conclusion should give way to an alternative version of the dual-process model and to a different normative conclusion. The alternative model contemplates both automatic utilitarian dispositions targeting group welfare and automatic deontological dispositions that partly conflict with them by protecting the individual against extreme group demands. When there is a conflict between utilitarian and deontological dispositions, the tension is real and cognitive control might take over (although we cannot assert with confidence that it always takes over). However, engagement of cognitive control does not necessarily lead to UR; deontological responses are also possible.

How should we picture the role of executive cognitive processes when they are engaged in tasks with moral dilemmas? In principle, cognitive control evaluates whether special circumstances speak in favor of UR or not. What kinds of circumstances are relevant? We already mentioned them above. Variables like a guilty or doomed victim, or the fact that the protagonist has stakes in the sacrifice (saves her own life), have a significant effect on the responses of participants relative to dilemmas where they are absent, like *footbridge* and *vitamins* [13, 28, 29, 43]. This increase has been confirmed with a battery that isolates the different contextual variables to different dilemmas, instead of including several in one (often the case in the items in Greene et al.'s battery and in most of its subsequent versions) and eliminating babies or children as victims [24]. The reasonable inference is that the additional circumstances (the doomed or guilty victim, or the selfish stakes in the sacrifice) are responsible for the increase, because these are the only elements that change from *footbridge* to, for example, *submarine*. In contrast, the judgment "saving five lives is better than saving one" remains constant. For this reason, if participants engage cognitive control in high-conflict dilemmas, it is probably to attend to these other variables and compute their effect on the decision. The increase in UR in the presence of these variables tells us that people pay special attention to them.

I shall now review experiments that provide evidence, sometimes indirectly, for the influence of each of these variables, beginning with doomed victims. Trémolière and Bonnefon [18] measured the UR as a function of the kill-save ratio and cognitive load. When the kill-save ratio is 1:5 cognitive, load interferes with the UR in *crying baby* and *captive soldier*. Participants under extreme load give significantly less UR than participants under light load. But when the ratio was 1:500, load did not interfere with UR in the same dilemmas. This suggests that when the kill-save ratio is not extraordinary, load interferes with processing the special circumstance of these dilemmas (doomed victim). When the kill-save ratio is extraordinary, it encourages all by itself and, intuitively [18], an increase of UR, making superfluous the controlled processing of other dilemma features. It remains to be investigated if extreme load would decrease the UR in dilemmas lacking special circumstances (like *footbridge*).

Other studies also suggest, indirectly, that participants use cognitive control to take the "doomed victim" feature into account. In an experiment designed to find evidence of the role of reflection and reasoning in moral judgment, Paxton et al. [23] tested participants with the Cognitive Reflection Test (CRT) [44] in two conditions—before and after responding to three high-conflict personal dilemmas—*footbridge*, *submarine*, and *crying baby*. Participants who responded to these dilemmas after the CRT showed a significant increase in utilitarian responses compared to participants who answered dilemmas before the CRT. Placing the CRT before the dilemmas primed participants to reflect when responding to them. But significantly, this effect was found only in *submarine* and in *crying baby*, and not in *footbridge* ([23], p. 168). They do not make much of this result, but the following explanation is plausible. When participants were primed, their reflections did not particularly target the utilitarian calculus that five is better than one (the only relevant factor present in *footbridge* and for which perhaps not much reflection is needed) but the fact that the person to be sacrificed would die anyway, a circumstance affecting *submarine* and *crying baby*, but not *footbridge*. This fact, when present, can reasonably be taken to shift the balance in favor of UR. A study by Moore et al. [13] targeted this variable directly. They investigated the effect of working memory capacity in utilitarian responses, controlling for factors like benefiting from the sacrifice or not, killing a person doomed to die or not, or killing as a means vs. killing as a side effect and without personal force. They found that participants with higher working memory scores gave significantly more UR than those with lower scores when the killing is personal and the victim is doomed to die anyway. They found no effect of working memory capacity in personal dilemmas like *footbridge*. This suggests that working memory is not engaged to compute the mere utilitarian benefit, but rather the fact that the victim is doomed to die.

Another circumstance that shifts the balance in favor of UR was disclosed in the pioneering experiments of Greene et al. They demonstrate that in impersonal dilemmas, where the loss of life results as a side effect and without exerting muscular force, most people normally condone the loss of life. Greene has argued that both features of impersonal killing are unjustified automatic settings of our moral minds. He claims, for example, that no moral difference exists between an intentional killing and one that, though not intended, is foreseen with certainty ([5], pp. 223–225). I beg to differ. I think this shows precisely how utilitarian intuitions conflicting with deontology effectively shape some of our decisions when aided by special circumstances. In this case, the special circumstance is the lack of intention to harm. To give a real-life example of a case like this one, recall Mackie's common sense explanation of why societies and states condone the loss of life statistically predicted as a side effect of motor vehicle transportation. The reason is, Mackie conjectures, that the benefits of getting faster to destination outweigh the disadvantages of lives lost, or so most of us think, consciously or not. These losses are statistically foreseen side effects, but not something that we want or intend ([45], p. 195). I think this example also brings vividly to awareness how some of our actual practices reveal a utilitarian influence that we could actually feel, after reflection, as deontologically suspect. Apparently, we humans tend to be influenced by utilitarian considerations in our moral practices and also in our judgments. Similarly, some circumstances can legitimate constraints on individual freedom—consider, for example, the measures that state and society could implement to prevent local population explosions. Those measures usually invade the (deontological) rights of the individual for the good of the group (the nation).

The other component of impersonal killing that favors UR, namely, the lack of muscular force, is certainly bizarre. Greene has insisted, correctly, that it is morally irrelevant. It could be just a hardwired and inaccurate proxy for unintended harm, functional in ancestral times, but not today. Participants in experiments do not confuse the exertion of muscular force with intention to harm, as shown by the *obstacle collide* scenario, a variant of *footbridge* where the death of the victim is caused with exertion of force but not as a means to save the five workmen ([5], pp. 218–202). But in contrast, participants seem to take the absence of muscular force for absence of intention to harm. When the victim is treated intentionally as a means to save others, but without the exertion of muscular force (Mikhail's *drop man* scenario), UR increases from 10% to 62% ([46], p. 149). The lack of muscular force increases the disposition to condone the loss of life in *drop man*, in spite of the fact that intention to harm is present in that scenario. Quite a lot of people, therefore, get things wrong and the reason seems to lie in an intuitive reaction, triggered by the automatic settings of our minds [5]. It remains to be investigated, however, whether participants scoring high in cognitive reflection, or induced to reflect before responding, are able to override its influence.

Of the variables that increase UR, one of the strangest was disclosed by two experiments that confronted subjects with virtual reality versions of personal and impersonal dilemmas. Though this mode of presentation increases emotional arousal (measured physiologically in both studies), results show, against all expectations, that it also increases UR, both in impersonal [31] and personal dilemmas [32]. In both cases, the authors explain this result with Cushman's version [47] of the dual-process model, where the processes in question concern the value of actions vs. the value of outcomes. It so happens that the virtual reality mode of presentation gives the five deaths resulting from inaction a stronger negative value than the action of killing one person. This poses an interesting challenge to interpretation, but I shall not attempt one here.

Other variables in Greene et al.'s original battery increase UR. When the victim is guilty, it is not excluded that at least some—and perhaps most—participants deliver an "intuitive" UR, as noted in the discussion of *preventing the spread* in Sect. 4 above, although I also noted that in this case it is actually difficult to distinguish it from an intuitive deontological response. It could well be a case of congruence between utilitarian and deontological intuitions, at least for some, or perhaps most participants. Another well-documented feature increasing UR is when agents benefit from the sacrifice: the fact that she is going to save her own life, not just the lives of several others—which, note, is not the case of *footbridge*—produces an increase in UR [13, 24]. Here it is plausible to postulate an automatic selfish response. Moore et al. [13] found that participants with greater working memory capacity do not give more UR in selfish dilemmas than participants with lower capacity. But Rand et al. [48] have found that pro-social responses, rather than selfish ones, are actually intuitive in the public goods game. How can we reconcile both results? Following our interpretation of Krajbich et al. [33] and the general gist of our preferred dual-process model, deontological or utilitarian responses are not per se intuitive or reflective but are one or the other depending on the particular individual sensitivities and the objective circumstances whose presence/absence speaks to those sensitivities.

We can apply this idea to all the circumstances that research has shown to increase UR. We could test each circumstance separately with the method of cognitive load, as in some papers reviewed above [15, 18, 21]. If we find that some of these circumstances increase UR independently of extreme load, this is evidence that they influence most individuals independently of working memory. If, however, the increase of UR is affected negatively by extreme load, this is evidence that most subjects need to compute them into the decision. In between, there is more individual variability, and we should not forget the possibility of cultural variability as well.

If this is how we should proceed to discern intuitive from controlled processing in moral cognition, this should also transfer to the design of experiments for collecting fMRI data. The procedure must be similar in both cases. Just as we test case by case the effect of load on the circumstances that increase UR, we should test case by case to observe how the fMRI data relate to the findings obtained from the load experiments. In this way, we can detect the instances where cognitive control attends to and ponders the circumstances that potentially justify a violation of the deontological rule. And if despite attending and pondering, the response is deontological, this should be taken as evidence that deontological responses can also arise from cognitive control.

7 The Normative Conclusion

The alternative version of the dual-process theory of moral cognition presents utilitarianism and deontology as two different moral intuitions hardwired by natural selection into our brains/minds. They are partially different and equally fundamental. This means that we are designed with a moral ambivalence. This is no surprise, for by now we know that some degree of imperfection indicates the hand of natural selection. Depending on the circumstances, some degree of interference, for the good of the group, with otherwise legitimate individual freedom will be condoned in a given society or culture. Taxes may come to mind as an example, but since taxes are so familiar to us all, no one except political philosophers would say that they violate deontological freedoms. A less familiar but not altogether distant example is the punishment that states implement to control local population explosion for the good of the group. This is a better candidate for (deontologically) illegitimate state control. Inevitably, the solutions to moral ambivalence will vary across cultural, geographical, and historical divides [49]. Thus, fundamental disagreement arises between societies and cultures, as it often arises within them.

What does our normative conclusion consist in? Greene anchored his normative conclusion in a theory over the standards of rational moral discourse. Rational moral discourse must be deliberative and argumentative in pursuit of the common good. Following singular intuitions cannot be the right track. I agree that this consideration is important and that it favors whatever moral view satisfies it. But the neurocognitive data collected in experimentation might still tell us that deontological responses satisfy it as well. I believe that we ought to recognize that deontological and utilitarian intuitions are often the boundaries within which our moral deliberations move freely and that any theory that would discount deontological principles and claims as nonrational would fail to satisfy the standards of deliberation. Counting heads is important, but several other things are important as well. The freedoms of individuals are important and so are the circumstances favoring head-count decisions in cases of conflict. But these circumstances are not written in the stars. The tension between utilitarian and deontological values is real and we have no innate guidance to resolve it. Deliberation remains a requirement for moral decisions, but deliberation trades in those two values (and possibly others). Different solutions arise in different times and places and in different heads and hearts. Normatively, there is no superiority of utilitarianism over deontology or the contrary, and no resolution of their conflict has any context-independent normative authority over any other. Thomas Nagel, not bothering to mention imperfect evolutionary design, has referred to this view as the "fragmentation of value" [50]. If my interpretation of the available neurocognitive data is correct, we are invited to embrace the "fragmentation of value," rather than full-blown utilitarian morality.

Acknowledgments I am grateful to Josh Greene for sharing the data of the 2004 *Neuron* paper, to an anonymous reviewer for helpful comments, and to Wim De Neys for sharing work in press during the writings of this paper.

References

1. Greene JD, Sommerville RB, Nystrom LE, Darley JM, Cohen JD. An fMRI investigation of emotional engagement in moral judgment. Science. 2001;293:2105–8. https://doi.org/10.1126/science.1062872.
2. Greene JD, Nystrom LE, Engell AD, Darley JM, Cohen JD. The neural bases of cognitive conflict and control in moral judgment. Neuron. 2004;44:389–400. https://doi.org/10.1016/j.neuron.2004.09.027.
3. Tomasello M, Carpenter M. Shared intentionality. Dev Sci. 2007;10(1):121–5. https://doi.org/10.1111/j.1467-7687.2007.00573.x.
4. Greene J. From neural 'is' to moral 'ought': what are the moral implications of neuroscientific moral psychology? Nat Rev Neurosci. 2003;4:847–50.
5. Greene J. Moral tribes: emotion, reason and the gap between us and them. New York: Penguin Press; 2013.
6. Kahneman D. Thinking, fast and slow. New York: Farrar, Strauss and Giroux; 2011.
7. Evans JBT. Dual-processing accounts of reasoning, judgment, and social cognition. Annu Rev Psychol. 2008;59:255–78.
8. Diamond A. Executive functions. Annu Rev Psychol. 2013;64:135–68.
9. Greene JD, Morelli SA, Lowenberg K, Nystrom LE, Cohen JD. Cognitive load selectively interferes with utilitarian moral judgment. Cognition. 2008;107:1144–54. https://doi.org/10.1016/j.cognition.2007.11.004.
10. Kant I. Critique of practical reason. Indianapolis: Hackett; 2002 [1788].
11. McGuire J, Langdon R, Coltheart M, Mackenzie C. A reanalysis of the personal/impersonal distinction in moral psychology research. J Exp Soc Psychol. 2009;45(3):581–4. https://doi.org/10.1016/j.jesp.2009.01.002.
12. Koop GJ. An assessment of the temporal dynamics of moral decisions. Judgm Decis Mak. 2013;8(5):527–39.
13. Moore AB, Clark BA, Kane MJ. Who shalt not kill? Individual differences in working memory capacity, executive control, and moral judgment. Psychol Sci. 2008;19:549–57. https://doi.org/10.1111/j.1467- 9280.2008.02122.x.
14. Greene J. Dual-process morality and the personal/impersonal distinction: a reply to McGuire, Langdon, Coltheart, and Mackenzie. J Exp Soc Psychol. 2009;45:581–4. https://doi.org/10.1016/j.jesp.2009.01.003.
15. Bialek M, De Neys W. Dual processes and moral conflict: evidence for deontological reasoners' intuitive utilitarian sensitivity. Judgm Decis Mak. 2017;12(2):148–67.
16. Dubljević V, Racine E. The ADC of moral judgment: opening the black box of moral intuitions with heuristics about agents, deeds, and consequences. AJOB Neurosci. 2014;5:3–20.
17. Kahane G, Wiech K, Shackel N, Farias M, Savulescu J, Tracey I. The neural basis of intuitive and counterintuitive moral judgment. Soc Cogn Affect Neurosci. 2012;7:393–402. https://doi.org/10.1093/scan/nsr005.
18. Trémolière B, Bonnefon JF. Efficient kill–save ratios ease up the cognitive demands on counterintuitive moral utilitarianism. Pers Soc Psychol Bull. 2014;40:923–30.
19. De Neys W. Bias and conflict: a case for logical intuitions. Perspect Psychol Sci. 2012;7:28–3.
20. Bago B, De Neys W. Fast logic? Examining the time course assumption of dual process theory. Cognition. 2017;158:90–109.
21. Białek M, De Neys W. Conflict detection during moral decision-making: evidence for deontic reasoners' utilitarian sensitivity. J Cogn Psychol. 2016;28(5):631–9. https://doi.org/10.1080/20445911.2016.1156118.

22. Nichols S, Mallon R. Moral dilemmas and moral rules. Cognition. 2006;100(3):530–42.
23. Paxton JM, Ungar L, Greene J. Reflection and reasoning in moral judgment. Cogn Sci. 2012;36:163–77.
24. Rosas A, Viciana H, Caviedes E, Arciniegas A. Hot utilitarianism and cold deontology: insights from a response-patterns approach to sacrificial and real world dilemmas. Submitted.
25. Baron J, Gürçay B, Moore AB, Starcke K. Use of a Rasch model to predict response times to utilitarian moral dilemmas. Synthese. 2012;189(S1):107–17. https://doi.org/10.1007/s11229-012-0121-z.
26. Baron J, Gürçay B. A meta-analysis of response-time tests of the sequential two-systems model of moral judgment. Mem Cogn. 2016;45:566. https://doi.org/10.3758/s13421-016-0686-8.
27. Conway P, Gawronski B. Deontological and utilitarian inclinations in moral decision making: a process dissociation approach. J Pers Soc Psychol. 2013;104(2):216–35. https://doi.org/10.1037/a0031021.
28. Huebner B, Hauser MD, Pettit P. How the source, inevitability and means of bringing about harm interact in folk-moral judgments. Mind Lang. 2011;26:210–33. https://doi.org/10.1111/j.1468-0017.2011.01416.x.
29. Rosas A, Koenigs M. Beyond 'utilitarianism': maximizing the clinical impact of moral judgment research. Soc Neurosci. 2014;9:661–7. https://doi.org/10.1080/17470919.2014.937506.
30. Greene JD, Cushman FA, Stewart LE, Lowenberg K, Nystrom LE, Cohen JD. Pushing moral buttons: the interaction between personal force and intention in moral judgment. Cognition. 2009;111:364–71. https://doi.org/10.1016/j.cognition.2009.02.001.
31. Patil I, Cogoni C, Zangrando N, Chittaro L, Silani G. Affective basis of judgment-behavior discrepancy in virtual experiences of moral dilemmas. Soc Neurosci. 2014;9(1):94–107. https://doi.org/10.1080/17470919.2013.870091.
32. Francis KB, Howard C, Howard IS, Gummerum M, Ganis G, Anderson G, Terbeck S. Virtual morality: transitioning from moral judgment to moral action? PLoS One. 2016;11(10):e0164374. https://doi.org/10.1371/journal.pone.0164374.
33. Krajbich I, Bartling B, Hare T, Fehr E. Rethinking fast and slow based on a critique of reaction-time reverse inference. Nat Commun. 2015;6:7455. https://doi.org/10.1038/ncomms8455.
34. Kahane G, Everett JAC, Earp BD, Farias M, Savulescu J. 'Utilitarian' judgments in sacrificial moral dilemmas do not reflect impartial concern for the greater good. Cognition. 2015;134:193–209. https://doi.org/10.1016/j. cognition. 2014.10.005.
35. Bartels DM, Pizarro D. The mismeasure of morals: antisocial personality traits predict utilitarian responses to moral dilemmas. Cognition. 2011;121:154–61. https://doi.org/10.1016/j.cognition.2011.05.010.
36. Djeriouat H, Trémolière B. The dark triad of personality and utilitarian moral judgment: the mediating role of honesty/humility and harm/care. Personal Individ Differ. 2014;67:11–6. https://doi.org/10.1016/j.paid.2013.12.026.
37. Duke AA, Bègue L. The drunk utilitarian: blood alcohol concentration predicts utilitarian responses in moral dilemmas. Cognition. 2015;134:121–7. https://doi.org/10.1016/j.cognition.2014.09.006.
38. Glenn AL, Koleva S, Iyer R, Graham J, Ditto PH. Moral identity in psychopathy. Judgm Decis Mak. 2010;5(7):497–505.
39. Gleichgerrcht E, Young L. Low levels of empathic concern predict utilitarian moral judgment. PLoS One. 2013;8(4):e60418. https://doi.org/10.1371/journal.pone.0060418.
40. Koenigs M, Young L, Adolphs R, Tranel D, Cushman F, Hauser M, Damasio A. Damage to the prefrontal cortex increases utilitarian moral judgments. Nature. 2007;446:908–11. https://doi.org/10.1038/nature0563.
41. Patil I. Trait psychopathy and utilitarian moral judgment: the mediating role of action aversion. J. Cogn Psychol. 2015;27(3):349–66. https://doi.org/10.1080/20445911.2015.1004334.
42. Levenson MR, Kiehl KA, Fitzpatrick CM. Assessing psychopathic attributes in a noninstitutionalized population. J Pers Soc Psychol. 1995;68(1):151–8.

43. Christensen JF, Flexas A, Calabrese M, Gut NK, Gomila A. Moral judgment reloaded: a moral dilemma validation study. Front Psychol. 2014;5:607. https://doi.org/10.3389/fpsyg.2014.00607.
44. Frederick S. Cognitive reflection and decision making. J Econ Perspect. 2005;19:25–42.
45. Mackie JL. Ethics: inventing right and wrong. London: Penguin Group; 1977.
46. Mikhail J. Universal moral grammar: theory, evidence and the future. Trends Cogn Sci. 2007;11(4):143–52.
47. Cushman F. Action, outcome, and value: a dual-system framework for morality. Personal Soc Psychol Rev. 2013;17(3):273–92. https://doi.org/10.1177/1088868313495594.
48. Rand DG, Greene JD, Nowak MA. Spontaneous giving and calculated greed. Nature. 2012;489:427–30.
49. Wong D. Natural moralities. A defense of pluralistic relativism. Oxford: Oxford University Press; 2006.
50. Nagel T. The fragmentation of value. In: Nagel T. Mortal questions. Cambridge: Cambridge University Press; 1979. p. 128–41.

The Social/Neuroscience: Bridging or Polarizing Culture and Biology?

Andrés Haye, Ricardo Morales, and Sebastián Niño

Abstract We review contemporary research on self-regulation in experimental psychology and social neuroscience in order to evaluate its conceptual foundations and discuss how theoretical assumptions about the biological dimension of this phenomenon are constructed in the crossfields of psychology and neuroscience. We argue that such a dimension is predominantly understood as a determining factor of behavior itself rooted in life structures and processes, although bearing on a restricted conception of life, characterized by dualistic, individualistic, aprioristic, adaptationistic, and anthropocentric limitations. We discuss these five features of the discursive construction of the biological dimension, building on literature reviews and critical discussions of three case examples. The focus is, first, on self-regulation theoretical models, then on emotion regulation models, and finally on attention regulation. In particular, we identify problems regarding different notions of autonomy widely at play across biological to social sciences. We argue that such a theoretical limitation compromises the link between theories of culture and biology, eventually radicalizing the very gap that the social neurosciences seek to overcome, and polarizing the relationships between biomedical discourses and psychology.

Keywords Autonomy • Attention-deficit/hyperactivity disorder • Emotion regulation • Self • Self-regulation • Theoretical biology

1 The Problematic Reception of Biological Concepts in Psychology

The program of social neuroscience, as seen by Cacioppo and Berntson [1], seeks to understand the brain processes that sustain human social abilities. To explain social behavior, social neuroscience should study the relations between the molecular, cellular, biological, cognitive, and social levels of behavior. Within this research program, basic principles for social neuroscience have been proposed. First, the

A. Haye (✉) • R. Morales • S. Niño
Pontificia Universidad Católica de Chile, Vicuña Mackenna 4860, Santiago, Chile
e-mail: ahaye@uc.cl; rimorales@uc.cl; sinino@uc.cl

© Springer International Publishing Switzerland 2017
A. Ibáñez et al. (eds.), *Neuroscience and Social Science*,
DOI 10.1007/978-3-319-68421-5_10

notion of multilevel determinism specifies that behavior can have multiple antecedents across various levels of organization. It is claimed, essentially, that multiple levels of organization should be considered in the explanation of social cognition. Second, the tenet of nonadditive determinism maintains that properties of the whole are not always predictable by the sum of the recognized properties of the individual levels. This implies a non-reductionist perspective, suggesting that the phenomenon of interest is not a mechanical but an emergent process. Third, the principle of reciprocal determinism states that there are mutual influences among biological and social factors in determining behavior. Even though that the aim of this third principle seems to be proposing a focus on the connection between biology and culture, there are some problematic theoretical aspects of this research program. Although there is an interest in the social factors underlying behavior, the unit of analysis in most research within social neuroscience is defined, as we will show, at the level of the individual subject, whose terms of intelligibility are conceived as a given (biological) reality which is then shaped by practice and culture. The missing link of biology-psychology, nature-culture, and body-mind is disclosed with the hylomorphic model of a relation of matter-form, of an ontological operation of *determination* [2]. This perspective is neither unique nor homogeneous within the social neurosciences but dominant in key research fields in which the affective, cognitive, and social psychology connect with the neurosciences. The corollary is that these different levels determine the behavior of the individual, which is the subject of the reciprocal influences of social and biological factors. A social/neuro subject is proposed therefrom. Such is, in a nutshell, the problem we address in this chapter.

Our main argument is that there is an implicit ontology of the subject as an individual and a limited conception of the biological dimension of the individual, both of which are theoretically problematic. We state that within psychology, the reception of the new neurosciences in the last two decades is conditioned by a restricted conception of life, limiting the biological dimension of psychological and social processes to the domain of the individual living being—a dimension that is already *determined*, given in the individual and given as the already realized individual reality. In turn, this biological reality of human individuals takes place as a *determining* factor of these psychological and social processes, as if biology were an external *antecedent* of mind and culture. In order to specify this thesis and present supporting evidence, we will discuss the case of self-regulation theory and research, with a particular focus on emotion and attention regulation, because these concepts have gained transversal interest in several brain and psychological sciences in relation to a wide variety of themes, from theoretical biology to education and work psychology, including topics such as addiction, affect, aging, attention, cognition and moral development, coping styles, decision-making, motivation, dementia, human failure, self, sexual behavior, and threat perception, among others [3].

We review and discuss a set of theoretical accounts within the cognitive and social psychological sciences. Our literature review covers works on self-regulation approaches (Sect. 2), the social dimension of self-regulation (Sects. 2.1 and 2.2), the brain mechanisms adduced to explain the regulation of emotions (Sect. 2.3), and the reaction to the implications of this framework for attention-deficit/hyperactivity disorder (ADHD) (Sect. 3.2). Thereon, we argue that the conceptualization of the biological that is implicitly assumed within the standard frameworks in psychology is characterized by a restriction of life to individual entities, particularly to the adaptive mechanisms and dynamics of individuals (Sect. 3.1), whose nature remains as an *a priori* that needs no further explanation beyond serving as a causal ground of the unfolding of behavior and the formation of the self (Sect. 3.3). In addition, we argue that research in self-regulation seems to paradoxically foster an anthropocentric concept of life (Sect. 4). We suggest that the ideal of autonomy is the principle of an implicit ontology of the subject, much in line with the way of Kant [4]. Finally, we give examples of contributions from philosophical and theoretical biology suggesting good reasons to criticize such a limited concept of life.

The method followed is the hermeneutic analysis of scientific discourses within key fields on intense interdisciplinary exchange between the neurobiological and the psychological sciences. For the literature review on self-regulation, we focus on cognitive and social psychology because these branches have important articulations with the neurosciences, leaving apart other relevant perspectives contributing to the link of culture and biology, such as cultural psychology and medical anthropology, although we will mention them when appropriate. The literature discussions are taken as case examples, focusing on particular issues regarding different aspects of self-regulation theory, to show how some features of the biological are discursively constructed in cognitive, affective, and social psychology and neurosciences. Our strategy is to highlight dynamic principles of explanation in order to open a road through a central problem within the program of the social neurosciences, in terms of the specific limitations of how the concept of life is received by the cognitive and social psychological sciences, as well as the potential impact in psychology more generally, as in linked areas as education and other social sciences.

If our thesis is true, then the bridge between culture and biology is at issue. Furthermore, it would also become salient that a conceptual restriction of life to individual adaptation as a normal self to a social context may be associated with the contemporary creation of new forms of biomedical knowledge that can be used to justify adjustment and cultural problems in social and mental health policies and practices—for instance, in terms of emotion and attention regulation. The modern educated subject [5], a desirable object of scientific and philosophical inquiry, has proved to be a productive site to reflect on the limitations of standard models of human development. Such is the social and political scope of relevance of our theoretical inquiry.

2 The Case of Self-Regulation Theories

We performed a literature review of theoretical accounts associated with "self-regulation" between 2009 and 2015 in Scopus, PsychInfo, Psychquest, and Google Scholar databases. This 6-year period was selected to have a range of years that included recent articles about self-regulation but leaving out the articles that were produced in the last 2 years in order to concentrate on consolidated discussions. Other keywords used were "ego depletion" and "self-regulation and goals." No filters were used. The total number of papers found was 7400, from which 20 articles were selected to analyze and track sources of theoretical elaboration.

The review suggests that there are two main theoretical approaches referred to in the literature, usually combined with different emphases in contemporary research on self-regulation, both in psychology and in neurosciences. On the one hand, Carver and Scheier [6] initially developed a cybernetic theory. Through feedback information, an agent controls behavior comparing the current state with a goal that works as a reference toward which conduct can be oriented, at least until discrepancies are salient. Self-regulation involves the capacity to establish goals, monitor their accomplishment, and operate on them, all of which can take place in a relatively automatic, nonconscious fashion. On the other hand, there is an executive-economy theory, or "strength" model of self-regulation championed by Baumeister [7], according to which the willpower to actively strive toward a goal, or to overcome an impulsive or habitual tendency, is a limited resource. Self-regulation efforts, such as the inhibition of a given conduct or its displacement by alternative behavior, may result in ego depletion and the need for a refreshing or reloading of agency, as a function of capacity and training [8]. Contrary to the former approach, this model implies that control processes are actively recruited, whereas automatic processes do not require self-regulatory efforts. Most of the discussion in this literature revolves around the issue of the automatic-unconscious-effortless versus controlled-conscious-effortful nature of self-regulation. However, both theories share important ideas, including the assumption that self-regulation is not merely a feedback-based adjustment mechanism, but involves an integration of information into a cognitive representation of the self and its goals.

For instance, Carver et al. [9] follow Higgins [10] in stating that the way in which individuals represent themselves, both their current and "possible selves," has an impact on regulatory processes by setting "self-guides," normative standards, or goals, to orient behavior. High discrepancy between perceived conduct and positive (negative) self-guides results in negative (positive) affections and then in a reaction against (toward) such emotional experience consisting in seeking for the reduction of the discrepancy (the ego-consonant distantiation from the negative self-guide). According to vanDellen and Hoyle [11], these possible selves are representations of future selves, having a self-evaluation and motivational function, linking present behavior to the future. In their study, participants were asked to write about their future, and then made them think on either their desired or their feared selves, and finally measured the level of self-regulation through a questionnaire. To think in one's feared self provoked negative emotions mainly in the case of participants with higher levels of self-regulation, because they would be more conscious about their goals and thus be more sensible to discrepancies with positive or negative ideals. Individual differences in self-regulation reflect an integration of dynamic processes by which people control their thoughts, emotions, and behavior. Morf and Horvath [12] suggest that individuals interpret and adjust to situations in characteristic ways, conducting themselves strategically in relation to goals at different levels and scales, and without conscious attention or explicit representation (i.e., automatically), thus giving this goal-orientation functions a unitary and coherent direction. Burnette et al. [13] confirm in their meta-analysis that implicit representations of the self act as cognitive frames that guide the way in which people interpret and react to the consequences of their behavior, set goals, and compare the current state with the future state.

Hagger et al. [14], in a meta-analysis tested the ego-depletion effect predicted by Baumeister et al. [7], found a robust decrease of self-regulation in a second task involving inhibition or displacement of a dominant response. Moreover, the study suggests that regulatory resources are shared across several domains—attentional control, emotional control, impulse control, thought control, decision-making, and the processing of social cues—such that self-regulation in anyone would make it harder to control behavior in any of the other domains.

Gestsdottir and Lerner [15] understand self-regulation as a general term encompassing multiple forms, from physiological functions to interpersonal processes. Intentional self-regulation involves actions actively directed to harmonize with demand and resources in the context of personal goals. Organismic self-regulation, on the contrary, are not conscious efforts, and they take place automatically. The former develops mainly during adolescence, when regulation becomes more cognitive, directed, efficient, and intentional, involving more elaborated goals. Emotional, motivational, and behavioral functions are gradually controlled by the subject rather than by the situation or by organismic regulatory mechanisms, as it is among children. Motivation and selection of goals, which are extrinsic processes during infancy, become internalized and subject to voluntary change, initiation, and maintenance. This development promotes self-directed behavior and makes the subject better in dealing with demands from the environment.

Therefore, the literature suggests that a dual account of self-regulation is the more convincing perspective. We focus now on the social dimensions of self-regulation.

2.1 *Self as a Social Regulatory Agent*

Baumeister and Vohs [16] define the self as an agent that controls its behavior, whereby self-regulation is an important component to understand the ways in which the self operates on the world and the world on the self. They speculate that self-regulation was one of the key steps in human evolution as well as one of the distinctive aspects of human psyche. As the capacity to change one's own responses and inner states, self-regulation is closely linked to self-reflection: when consciousness is directed toward its own source, subjects learn about the world and about themselves, generating a body of knowledge and beliefs about the self. These reflective processes of self-regulation give form to a self-concept, without which the very self would be unconceivable. Baumeister [17] defines the self as the unity of three interdependent processes—a network of information or self-concept, a process of interpersonal adjustment or social regulation, and the abilities to initiate and control executive functions or self-regulation. This unity is theorized as a fundamental condition of both social coexistence and brain activity. Self, and by implication self-regulation, would thus be the link between culture and biology and between the social system and the physical body.

In this connection, Fitzsimons and Finkel [18] studied interpersonal influences in self-regulatory processes. They found that other people stimulate the activation of goals, making the subject initiate new goals even unconsciously, and frame self-monitoring of goals through implicit social comparisons. However, interpersonal interactions may also cause consumption or even depletion of self-control resources. The mere empathic identification with another person executing a self-regulatory process was shown to diminish one's own self-regulatory performance. Confirming the latter idea, Ent et al. [19] found that power, the capacity to change the response of others by controlling their resources, makes subjects more motivated to achieve goals but at the same time lowers the capacity to change one's responses in order to adjust to social values and long-term goals. To exert power, to hold decision-making positions, and to take leadership over subordinated others would also tend to deplete self-regulation resources.

2.2 *Self-Regulation in Intergroup Contexts*

Amodio [20] addresses interracial interaction as a self-regulatory challenge. Even if racial prejudice is a domain in which many people hold explicit intentions to respond without prejudice, among white Americans there are implicit forms of racial bias toward black people that can influence behavior without intention or awareness [21]. Amodio bears on a standard explanation for prejudice based on a dual-process framework, whereby implicit biases are learned through repeated exposure to associations between black American cues and cues of negative concepts, whereas the controlled component of prejudice enables individuals to consciously represent and held beliefs and intentions. If implicit racial biases are activated, how are they controlled and regulated by individuals who do not want to respond in a biased way? Note that this question focuses on the inhibitory aspects of self-regulation. According to classical models of self-regulation, to override influences of bias and prejudice one should engage in self-regulatory processes that involve cognitive control. This view assumes that control is initiated intentionally, by being aware of the presence of a bias and deciding to take actions against it. Nevertheless, Amodio [20] mentions that the brain is not organized according to two simple processes and that general systems for self-regulation reflect the coordinated activity of multiple underlying systems, ranging from more automatic to more controlled. Concordantly, social neuroscience approaches should help to unpack the processes involved in the regulation of racial bias and differentiating its more deliberative from its more spontaneous aspects.

In proposing a more comprehensive view of regulatory processes, Amodio [20] assumes the more radically dual model of regulatory control proposed by Botvinick et al. [22]. In this model, it is not necessary that the individual has to be aware of its prejudice in order to override them. By postulating two independent cognitive systems, one that determines when control is needed and another that implements the intended behavior, the model should bypass that homuncular supposition of a unitary and complete reflexive self underlying each and all self-regulatory processes. This model assumes that several different response tendencies are often simultaneously activated in the brain in response to both internal and external cues. When two or more activated tendencies imply different behavioral responses, there is conflict in the system. The first component of the model monitors the degree of conflict. If this degree arises, the second system, a regulatory one, is engaged to execute deliberative forms of control. Conflict monitoring has been associated with activity of the dorsal anterior cingulate cortex (dACC) and the regulatory system lined to activity in the dorsolateral prefrontal cortex (dlPFC) [22–24]. Amodio [20] presents several studies that support this model of regulatory control, applied to the regulation of prejudice. For example, a dissociation has been observed between conflict monitoring and regulatory aspects of control in the context of race bias, providing evidence that prejudice control is a multicomponent process [25]. In another study [26], they seek to understand whether internal and external impetuses for regulatory control toward prejudice may involve different underlying mechanisms. It was expected that behavioral control driven by one's internal motivations would relate to conflict monitoring and thus dACC activity and that behavioral control motivated by social pressures would also be associated with more complex social cognitive processing in the medial prefrontal cortex (mPFC). This study offered evidence that internally versus externally driven forms of prejudice control arise from independent neural mechanisms associated with the dACC and the mPFC, respectively.

In order to further discuss this explanatory strategy pointing to brain mechanisms of social self-regulation, let us review how this is done in one of the more studied fields of self-regulation, the cognitive regulation of emotions, also organized in terms of process duality.

2.3 *Emotion Regulation and Culture*

A more focused review of theoretical accounts is associated with "emotion regulation" and "cultural regulation" between 2009 and 2015 in Scopus, PsychInfo, Psychquest, and Google Scholar databases. As before, this period of 6 years was selected to have a range of years that included recent articles about emotion regulation but excluding articles that were produced in the last 2 years in order to concentrate on consolidated discussions. Other keywords used were "emotion regulation," "cultural regulation," "regulation of feelings," "emotion regulation and culture," and "cultural emotion regulation" ("meditation" and "mindfulness" were used as filters). Under the label of "emotion regulation," at that period, were found 6931 articles, and under the labels "emotion regulation and culture" were found 130 articles, from which we selected 36 articles to analyze. For the analysis, and the presentation of the information, we selected the more influential articles in the literature, by means of either theoretical relevance or number of citations.

The literature review suggests that one of the most extended models in this field is the dual-process model of emotion regulation. According to Gross [27, 28] and Gross and Barrett [29], emotion regulation is a process that starts with an evaluation of the causes of an emotion (being internal or external), which triggers a sequence of behavioral, physiological, and experiential reactions that can shape the final form of that emotion reaction. This perspective takes its theoretical background from the work of Lazarus [30] which states that when situations or stimuli that elicit an emotional response are cognitively evaluated (which can happen consciously or unconsciously), a particular emotional meaning is attributed to a situation or stimulus.

According to the dual-process model [27, 28], there are two ways in which emotions can be regulated. The first one is the antecedent-focused emotion regulation: this process occurs when the emotion has been recently elicited. For this type of emotion regulation, people would engage in one of these different methods: situation modification, attention deployment, or cognitive reappraisal. The second type of emotion regulation is the response focused. It takes place when the expression of an emotion has already started and affected behavioral, experiential, or physiological responses. This type of emotional regulation would be engaged by means of suppression mechanisms as a primary method. Likewise, Mauss et al. [31] remark that emotion regulation involves the deliberative and automatic influences that shape emotion response. On the one hand, deliberative regulation of emotion requires attentional resources and functions according to explicit goals. On the other hand, automatic processes are elicited by the perception of environmental stimuli that activate knowledge structures shaping the emotional response by the different processes involved in the regulation of emotions. One example is that the implicit priming of concepts has a demonstrated impact in the emotion regulation processes [32].

In the experimental literature of emotion regulation, the most studied mechanisms is cognitive reappraisal [27, 33, 34], that is, the reframing of the emotion stimulus or situation in a non-emotional (or a less emotional) way, and suppression, the inhibition of the behavioral reactions caused by the stimulus. Ochsner and Gross [33] argued that due to its complex nature, reappraisal requires different cognitive processes to be implemented. When people engage in reappraisal, there is an increase in activity in cerebral regions such as the dlPFC, vmPFC, and dACC and shows less activity in the amygdala and the insula. On the other hand, suppression involves late frontal activity and an increase in the activity of the amygdala and the insula, compared to cognitive reappraisal that shows early frontal activity [33, 34].

In the last years, alternative theoretical accounts of emotion regulation have been offered. Kappas [35] argues that the experimental paradigms of emotion regulation lack the appropriate conditions to capture the spontaneous nature of this phenomenon [36], or the action tendencies associated with emotions. For Kappas [35], emotion regulation is a process in which negative stimuli trigger their own self-termination, while positive emotions trigger processes directed to their self-sustenance. Another alternative perspective about emotion regulation has been put forward by Tamir [37]. For this author, the "utilitarian" nature of emotion regulation processes has been neglected; people actively manipulate their emotions to reach some desired states, which guide the regulation of emotions. Guided by certain goals, people actively choose to remain in an emotional state that would serve them in a future purpose. For example, Tamir and Ford [38] showed that if subjects believe that they would be put in a confrontational situation, they choose to see stimulus that elicited an angry reaction, but if they believe that they would be put in a cooperative situation, they choose to see stimulus that elicited happy reactions.

Another alternative account of emotion regulation is the social baseline theory of Beckes and Coan [39]. These authors argue that the ecological niche in which people have evolved consists of other people. Therefore, the presence of other people should facilitate the process of emotion regulation. They support this claim citing evidence that the PFC is less activated in emotion regulation activities when the experimental subject is in presence of more people. Because of the social nature of emotion regulation, these authors propose to take dyads of individuals, instead of individual subjects, as the unit of analysis.

Different approaches have been taken to study emotion regulation processes in diverse cultures. One perspective is that different concepts of the self in each culture would shape the different emotion regulation strategies of each individual [40]. Experimental findings about the difference of emotion regulation processes in different cultures have focused in the variations of the expression of emotion across cultural contexts. For example, Novin et al. [41] showed that Iranian children have a tendency to hide their emotions, whereas Europeans have a tendency to show them. Also, Davis [42] reported that Chinese people show less expression and emotional intensity in the presence of images of negative valence, compared to people from the United States, who tend to engage in more cognitive strategies to maintain positive emotions compared to Asian people [43].

Overall, the literature reviews in affective, cognitive, and social psychological approaches to self-regulation strongly suggest that a dualistic perspective is the dominant feature, with several key contributions distinguishing always two main parts that can be arranged in terms of two dimensions of self-regulation, one referring to biological processes of behavioral production and another to the cultural development of more cognitive control processes. None of the alternative approaches analyzed offer a way out of this scheme.

3 Drifts of Self-Regulation Theories

After the exposition of the central models and findings giving shape to the contemporary self-regulation theory that is articulating relevant psychological experimental research within the neurosciences, in this section we discuss some of its implications. Specifically, we critically discuss the global framework underlying this literature cases in terms of three themes: (1) the importance of individual adaptation in self-regulation theories, (2) the controversy around attentional regulation, and (3) the a priori or reified nature of the biological in current discourses within the social neurosciences.

3.1 Adaptiveness

Overall, the exertion of self-control is posited as an ideal that applies to individual bodies, in the search of a functioning community. When self-regulation works well, it enables people to alter their behavior so as to conform to rules, plans, promises, ideals, and social standards. The wellness of a community appears to be rooted in the regulated and restricted participation of the different beings that compose it. The ideal of this kind of beings is a well-adjusted and well-coordinated individual that conforms to, and (re)produce, the social practices of a given social order. Failure to achieve self-regulation can be interpreted as a failure in the adaptation of that organism to its social niche.

Along with our discussion of social aspects of self-regulation (Sects. 2.1 and 2.2. above), one of the relevant definitions we found refers to the capacity for altering one's own responses, which enables a person to restrain or override behavior, making a different response possible [44]. Research on self-regulation offers diverse explanations for the evolution and development of this ability. One of the prominent explanations about the emergence of the regulation of behavior lies in its capacity to allow human beings to align their conduct to social standards such as ideals, values, moral, and social expectations and to support the pursuit of goals that benefit their society [45, 46]. Self-regulation capacities have been proposed as having evolved to solve different problems regarding the environmental niche in which human ancestors lived, such as outwitting competitors and attracting mates [47]. Nevertheless, activities that require the adaptation of the individual to major social activities, such as delayed social exchange and the formation of social coalitions, have been argued to be the principal cause of the development of regulatory systems of behavior in human beings [17]. The emergence of self-regulation, as the capacity that allows several individuals to adapt to the norms and values of their social group, has been posited as one of the central steps in human evolution [3, 6, 48]. For this view, human group life is a product of the evolutionary process that rewarded individuals who were most effective at group life [17] or the ones who adapted more successfully to the norms and constraints for the behavioral expressions. This capacity appears to be one link between the complex relationship among the individual and the broader world of social collectives.

Ranging from simple behavioral responses, such as resisting the urge to eat a marshmallow [49], to more complex tasks, like music learning [50], research on self-regulation has tapped different kinds of behavior. Research about this capacity appears to be justified because those who are better able to self-regulate themselves do better in a wide scope: they demonstrate higher levels of job success or better interpersonal relationships and mental health [51]. On the other hand, the failure to self-regulate behavior is a core feature of many social and mental health problems [52]. Failures in the capacity to exert self-regulation to oneself has been linked to behavioral and impulse-control problems, such as overeating, drug abuse, violence, overspending, sexually impulsive behavior, and smoking [3, 53–55].

Because of a special interest self-regulation as a means to adjust to social environments, the vast majority of experimental research has focused on adaptive features of regulation, specifically the inhibition of overt responses. Common experimental paradigms of self-regulation involve restraining diverse impulses [48], thought-suppression [56], prejudice suppression [57], and asking children to exert inhibitory control over certain responses, while remembering and executing a given rule for correct responding [58]. In most research procedures, inhibition is the way in which individuals have to exert control over themselves in order to follow the experimenters' instructions, and therefore it is predisposed as an expected behavioral solution to adjustment.

3.2 Attention Regulation

The relation between nature and culture in psychology and other disciplines within the social sciences has a long-standing history [59]. One of the controversies revolves around ADHD, which has undergone a history of scientific criticism from some streams of the social sciences because of the questionability of the claimed biological determination of this syndrome [60]. At the core of this debate is the problem of whether neurology helps in understanding psychological phenomena and what would be the epistemological validity that the causal explanatory reference to biology or nature has on these matters overall. It is of some interest that in this particular discussion, biomedical sciences claimed the final word at the beginning of the 2000s, with the claim that there was already a general agreement on the justification of the neurological determination of this phenomenon. This situation materialized in a Consensus signed by an ample number of active members of the American Psychiatric Association, determining the neurological basis of ADHD and encouraging researchers to solely focus on finding the most appropriate pharmacological scheme for its treatment [61].

From the critical segments of the social scientists [60], this policy was regarded as a confirmation that, first, there was no evidence or proof of the neurological nature of this so-called condition and that, second, the psychiatric and medical discourse was not interested in the business of scientifically proving points, but rather, as they viewed it, this was a case of a power discourse.

Two main perspectives can be identified within the field of psychology and other approaches from the social sciences toward ADHD. On the one hand, there are those who strongly question the biological determination of the disorder, even reaching such radical claims as proposing that ADHD is a mere fiction whose only purpose is to validate and generate certain discourses (there are different views on this matter, but the main authors referenced here are Rose [62], Conrad [63], and Clarke [64]). On the other hand, mostly within psychology, there are groups that condone the neurological basis of the disorder but integrating biology with psychology as a unity.

The first position is mainly asserted from the development of the theory of Foucault [65–68] about the subject of power and discourse, the society of control and of normalization, and the performative nature of scientific knowledge. From this point of view, the biological account of ADHD is perceived only as a means of legitimating certain specific practices that seek to (re)produce a given social order. In this way, ADHD would be placed at the structural level of social discourse that creates realities not only by naming them but also, and specially, by studying them [69]. In this interpretation, ADHD is seen as a failure in the mechanism for controlling sociopolitical normality in accordance with the hegemonic discourse. Any failure to achieve self-regulation can be interpreted as a malfunction in autonomy in accordance with the frame for standard neurological development [70]. In this scenario, the biological substrate only counts as a means of legitimating its status [71].

On the other hand, advocates of an integrated psycho-neurological paradigm, which found most acceptance in cognitive and behavioral psychologists, tend to identify a certain biological development that can be attested through demonstrable typified cognitive guided behaviors [72–74]. Here, ADHD is conceptualized as a failure in the organization and regulation of behavior that can be compensated by the strengthening of specific attention and cognitive faculties that have direct effects in the child's capacity of regulating behavior and acting in an autonomous fashion. It is also recognized that this faculties correspond to specific neurological functions that are the cause of such behaviors (typically supporting such hypothesis with a gross reference to the works of Luria [75, 76]), so these lines of treatment are often open to a mixed scheme of cognitive conditioning and pharmacological assessment.

It can be stated, therefore, that a strong link is implicitly assumed between self-regulation and autonomy in both views. First, the understanding of ADHD as a failure of the psychological development with an underlying neurological determination can be taken as meaning that there is an organic disruptive deregulation that would be the efficient cause of the disorder [72, 73]. Second, this could be further interpreted as a form of brain disorganization, determining a dysfunction in subjectivity, and ultimately, in the individual essential ability to self-regulate and, as such, to be autonomous. Here, autonomy is implicitly understood as the capacity of the subject to successfully administrate one's cognitive resources in an intentional way, without the need to recur to any source of external control. In cognitive and systemic psychology, this is commonly referred to as having an internal locus of control [77]. Both points are consistent with the assumption that autonomy is something grounded in the individual's nervous system, more specifically, located in an individual brain enclosed from its environment. It is in this sense that the subject is understood as autonomous, and it is in this way that the organism is taken into account in the cognitive view, for the cognitive closure and self-determination through the regulation of its functions rest on the nervous system.

In the Foucaultian view [62, 70, 71], the fundamental role that autonomy plays in the understanding of ADHD is even more straightforward. This is mostly because the central idea is that biology takes place in ADHD as a discourse of power legitimizing the idea of self-failure, and the disorder itself is taken into account for a disrupting subjectivity and failures in the order of the technologies of the self, an incapacity of the subject to organize herself in order to become an efficient member of the machinery of society. The way in which power operates in the subject is through this demand of autonomy, so that the individual can autonomously exercise control over herself and her capacities, managing herself in a self-determined way to conform with the ways of power, reproducing them in this very action. From this perspective, autonomy and self-regulation are intrinsically tied together, because that which the hegemonic discourse of autonomy asks from the individual is precisely to become an individual through self-regulation.

Moreover, it could be stated that even in the former cognitive account, resting largely on the neurological dimension as an essential determinant, there is no consistent conceptualization of the biological, which is therefore taken as a given reality, already determined, and being a determinant of the cognitive functions. Even though it is a core item in its argument, there is a lack of theoretical discussion focused on the neurological factor itself and its supposed relation with any of the cognitive functions. This idea of the biological as a determinant concerns precisely the relation between the biological or natural reality on the one hand, and the mental or cultural world on the other, in line with the dualistic framework predominant in several domains of self-regulation theory and research (as seen in Sect. 2 above). The Foucaultian perspective does not provide with a more complex or satisfactory explanation of the relation between the cultural and natural aspects of the subject. It is taken as the raw matter over which power operates but deserving no further discussion about its content and significance. In this sense, it seems to be lacking any causal power, because its form and place in the conception of the subject is to be determined by the way in which it is accounted for in discourse.

From all sides, then, the biological dimension remains under-conceptualized and implicitly left as an abstract ideal [62]. As a result, this common ground among different perspectives on attentional regulation seems to confirm a dualist paradigm by which the biological dimension is given as determinant of the psychological and cultural dimension.

3.3 Biology as an A Priori

The basis for the underlying dualism between mind and body, or culture and nature, that can be found at the bottom of the discussion of ADHD is addressed in the analysis of the influence of mind over body in the field of medicine, conducted by Lock and Scheper-Hughes [78]. In their research they point out a series of ways in which social mechanisms of power, control, and knowledge ultimately determine the body. The latter view is coincidentally developed through the works of Foucault, particularly in reference to his work on biopolitics, while the social dimension is worked out from his microphysics of power and the ways in which hegemonic discourses reproduce themselves through social practice. Subsequently, they organize these at different levels: individual (self-embodied), social (social and cultural representations), and politics (regulation and control). It is worth noting that the individual level of analysis is mainly approached through the idea that any western understanding of the body rests largely on the Cartesian body-mind dichotomy, where each dimension corresponds to a different realm. The authors argue that this distinction evolves to the point of assuming that while the body is a natural product, the mind is a social construction, and therefore each belong to different fields of knowledge.

This form of dualism can be identified to be even prior to the discussion about whether a disorder (mental or physical) is socially determined and if so to what extent—one of the article's main issues [78]. Theoretical discussions about either the body, the mind, or the individual, which can be understood as their unification, deal with both dimensions [79]. From the social and psychological sciences, self-regulation is seen as a cultural product based on given neurological conditions. Even though the neurological has a central role in self-regulation theory, the neurological dimension is taken externally, as a given or *a priori* determinant already pre-programmed. As a consequence, the biological is not theoretically discussed as such and in its causal relations with any of the cognitive functions beyond a mere pairing of terms.

This can verified elsewhere, for instance, in Heatherton [3], who understands the brain as a social entity that has evolved in order to direct individual impulses in compliance to social demands, developing complex mechanisms for knowing ourselves, the others, detecting social threats, and as an outcome regulating behavior to avoid social exclusion. In this sense, the brain structure is assumed as the material and causal underpinning of social interaction. Moreover, social behavior is explained in terms of the coordinated activity of several parts of the brain, which is mostly noted in the experimental observation of individuals giving them instructions to compel specific social behaviors, focusing on the underlying neurological structure to explain such behavior. This can be seen in the works of Blair [80], Petersen and Posner [81], and Kelley [82], who, in the same manner, associate the activity in some parts of the brain to certain psychological mechanisms put into play and explain the hormonal mechanics behind this activation as if it explained the relation between this activation, the specific behavior, and the supposed underlying psychological mechanism. In this sense, the straightforward pairing between psychological functions and their neurochemical basis is taken as a solution of the duality of mind-body, or culture-nature, but really it perpetrates it, rendering the matter even more obscure.

The complement of this problem can be observed in the work of Rose [62], especially when he refers to the way in which psychiatry conceptualizes mental illness as a neurochemical deviation that finds expression through different types of behavioral and adaptive dysfunctions. In these cases, a particular manner of manifestation is linked to a corresponding deficiency in brain chemistry and is subsequently treated in line with pathologies supposedly rooted in the cycles or systems of neurotransmission. The explanation provided by Rose himself to this conception is exposed in terms of the understanding of the person as a chemical self, as a result of the hegemonic domain of psychiatry over psychology. This amounts to employ the notion of the biological once again as an a priori, because it is not analyzed in order to clarify its role in cultural or psychological phenomena, not even to examine whether it has any relevant role in social construction. The biological, from the point of view of social construction, is merely criticized as a discursive way of validating certain forms of knowledge and social practices.

This very much seems to replicate the dualism that has been argued as an underlying theme. For it can be identified that the discussion between psychology and neurology is crucially rooted in the original Kantian separation between nature and culture. It can be seen that, in psychology, the biological dimension is understood as a condition of possibility for the social dimension. It is simultaneously taken as a given, as an irreducible a priori which cannot be conceptualized, and as such it remains undetermined, mysterious, and in the end unknowable. In this way, it can be seen that cognitive and social psychological sciences have persistently kept the biological in the obscure, replicating the nature-culture dualism, with nature's double form as the condition of possibility for the social but at the same time muted and deprived of its cognoscibility from the social standpoint.

4 Conclusions

In this final section we critically reflect on the reception of biological concepts in cognitive and social psychological sciences. We follow a line of discursive continuity from self-regulation in general, through emotion and attention regulation in particular, to the concept of autonomy, and suggest that in such a literature this latter concept acquires an anthropocentric and individualistic accent that can be contrasted to the wider notion of autonomy of the living in theoretical biology.

Our review and discussion suggest that self-regulation is conceived as a key system of processes through which human individuals realize their autonomy. The notion of autonomy, as applied to human individuals to account for their capacity to preserve power over their bodies so as to conduct behavior in an organized form and thus to keep their unity and self-determination, refers in biological theory to a fundamental feature of living beings. The biological concept of autonomy has been intensively worked out after World War II with the contributions of cybernetics and system theories, but the idea that living beings have an inner organization that makes them different from the environment and enables them to coordinate multiple parts as a whole is as old as the philosophy of life itself. In the self-regulation literature, we have observed a rather anthropocentric restriction of the notion of autonomy. Since Kant, humanistic philosophy, psychology, and political theory developed a strong connection between moral, political, and cognitive autonomy, on the one hand, and freedom and rational understanding, on the other. An autonomous subject, in this view, is one that thinks for herself and acts according to herself. Likewise, an autonomous society is one that rules itself, instead of being subjected to nature or other political unities. Although for these authors autonomy was not a biological condition given, but the hallmark of specifically human adults, it must be developed from the infancy of man, from the natural endowment of individuals. Overall, this notion of autonomy is internally troubled, and fueled, by the question about the continuities and discontinuities between the social and the biological. Some authors claim that autonomy is an invariant feature of living beings, while others refer to it as explaining the gap between mere biological existence and a fully human life.

The notion of autonomy, when taken from biological theory, is linked to the problem of the unity of living beings, their individuality, and coherence. The organizational framework in biological theory during the twentieth century [83, 84] offered a multilevel approach and contributed to recognizing living beings as complex systems that coordinate action through reciprocal influences among levels of analysis such as biochemical operations within the cells, physiological operations among tissues, and neural operations across the sensorimotor circuits. For a living being, to keep its unity from the multiplicity of parts, levels, and processes implies keeping organization. Organization entails dynamic relationships between different structures hierarchically arranged at different levels. Living beings are not only organized sets of organs but also organizing beings, striving to reproduce the structures of coordination and action. This means, on the one hand, to maintain a negentropic effort that must be continuous as long as the organism is alive. The concept of organization is meant to account for the fact that living beings actively deal with entropy, and, as a result, they stay alive instead of dissolving themselves in the environment. Organization, then, is what preserves the individuality of an organism, and at the same time means self-governance over multiple local operations in order to make them work together for one global end, thus emerging as an operational unity [85].

Theoretical biology elaborates on the relationship between the notions of autonomy and life in a way that is, at some point, at odds with current frameworks derived from late developments in system theory and similar perspectives, such as constructivist epistemology, strongly influential in psychology and in the broader field of the social sciences. For instance, Varela [86] defines living beings as units that produce themselves, materially making its components and borders in order to continue their existence as units. They established the concept of autopoiesis based on the model of the cell and applied to all meta-cellular organisms. A cell is a living being as long as it generates its own components, instead of importing them from the environment. Therefore, autopoiesis implies operational closure. The elements of a system can only be produced by the system, and the system produces only its elements. Open systems exchange energy with the environment, but living beings are paradoxically closed to their environment regarding their components and organization, in the sense that whatever operation takes place within it is produced by itself and for itself. Therefore, system closure is equivalent to system autonomy. Indeed, according to Varela, autopoiesis is the specific way in which living beings achieve autonomy. In other words, autopoiesis is one particular type of autonomy, whose more general concept applies not only to cells and meta-cellular organisms but also to nonliving systems as well as to human organisms and social systems. A machine endowed with cybernetic capabilities is autonomous in a particular way, limited to the generation of its operations according to its own structure. A living system is autonomous in a more radical way, changing and expanding its structures with conservation of the global organization of the system as a unity that produces itself. However, for social scientists within late system theory, autopoiesis is not a particular type of autonomy that applies only to living beings but a general definition of operationally closed systems such as cells, brains, and societies. According to Luhmann [87], also psychic and social systems produce their components. It is possible to define the "individuality of individuals as autopoiesis" [87], because the process of production of the material components of the system that keep it as a unified unit yields an individuation process supported by a selective differentiation from the environment. Although contrary to Luhmann's explicit intentions, the reception of biological and system-theory contributions to the notion of autonomy within the social sciences involves a reduction of biological self-production at the level of the individuality of the living body, of the meta-cellular individual, and the development of "higher" forms of autonomy at the levels of the psychological self and of the social group. As a matter of fact, self-regulation is currently understood as a higher-order way of realizing living autonomy, distinct of human subjects, and with a specific locus at the individuality of the human individual in her relation to a social environment [85].

According to our interpretation of the literature on self-regulation, then, there is an individualistic bias in the conception of life as a property of individual beings. However, current ideas in biological theory enable us to trouble this reduction of life to individuality, and the subsequent implication that healthy individuals are those who do well in realizing their autonomy. This strong connection among the concepts of autonomy, self-regulation, and individuality in psychology, and more generally in the social sciences, is problematic from the point of view of some biological theories within the system framework, which suggest that the individuality of living being is less adequate to account for life phenomena than the principle of organized multiplicity and collectivity [88]. From a different angle, the Human Microbiome Project Consortium [89] suggests that human individuals are not really single living beings but communities of living beings including a huge diversity of species and forming a great "part" of what at the macro-level we use to distinguish as a living individual.

In the same vein, there seems to be an anthropocentric bias in social neuroscience and psychological theories of self-regulation. The distinctive and radically social form of life of humans is taken as a teleological justification to conceive self-regulation as having fundamentally a social function [3]. To be clear, self-regulation means, firstly, the regulation of thinking, emotion, and behavior in order to adjust oneself to adaptive goals, differing personal gratification as a means to achieve mediate, or long-term goals associated with greater gratification or with social standards. In a broader sense, regulation of behavior is a feature of many different forms of living beings, whose conduct is operationally guided by processing contingent feedback from the body-environment interaction. However, some literature suggests that the notion truly applies to human beings regulation, because only human beings display authentic social goals, agency, self-concept, free will, and a "self" that involves all the former [77].

Because of the strong relations between self-regulation and individuality, human autonomy is posited as an ideal that applies to individual bodies. Moreover, the fact that autonomy is not given or granted but produced and developed helps researchers from different fields to think of autonomy as a quite general explanatory principle and even an ideal of life and development. Human beings tend to be or ought to be individual, unitary, coherent, integrated, self-referential, self-regulated, well-adjusted, and well-coordinated beings. The idea and the ideal of the biological in cognitive and social psychological sciences, conceived of as the natural foundation of individual closure and self-management, may be legitimizing contemporary forms of subjectification by loosely relying on the presumed biological roots of autonomy. We have discussed this when addressing the case of ADHD.

Finally, another consequence of the transformations undergone by the notion of autonomy within the cognitive and social psychological sciences is the epistemological implication that knowledge is always a selective simplification of the environment as part of the process of producing knowledge for the system and within itself. Radical constructivism [90] has been an influent epistemological theory in the social sciences, from which other constructivist theories of cognition and scientific knowledge derived or were renewed by taking this reference to the biological grounds of autopoiesis as a kind of scientific foundation. Beyond their differences, these theories apply the principle of autonomy to cognition stating that our knowledge starts by the information given by the environment but organizes knowledge according to the observer's organization and self-production process. Again, we recognize in Kant the first philosophical matrix of constructivist theories of knowledge, already guided by the principle of autonomy. Although not yet acquainted with the concepts of system and autopoiesis, we can speculate that Kant's *Critiques* rendered the development of these concepts historically possible. He not only argued that knowledge is configured by the structure of knowing subjects but also equated autonomous action with intrinsically self-regulated acceptance of the very principle of autonomy of individual subjects, and in his work on judgment, he also anticipated the notion of system [91]. The constructivist axiom according to which one can only perceive what my structure enables me to perceive has become a scientific commonplace in the field of psychology as well as in education, cultural studies, and discourse theory.

Overall, the reception of neuroscience discourse in psychology, at least through the self-regulation framework currently transversal in the social neuroscience and in the psychology of education, work, and mental health, seems to involve a limited and risky notion of the biological. We have disclosed dualistic, individualistic, aprioristic, adaptationistic, and anthropocentric biases. In our interpretation, such restricted conception is a concomitant of a Kantian matrix of thought that hinders the necessary bridging between the biological and the cultural dimensions that social neuroscience pretends to pursue. A philosophical project that is important for this bridging is undoubtedly the later work of Merleau-Ponty [92], although his effort is crucially limited as far as he sticks to the traditional idea that nature is the ground of culture. In his late lectures on the concept of nature, particularly, the living body is conceived as the soil in which human experience sinks its roots but at the same time as the "other side" of man that is opposed to language and culture. Such a theory is very close to Husserl's late works as well, still within the aprioristic philosophy of consciousness that we have found to be opposing culture and biology (a contemporary exemplar of the Kantian legacy relevant to this connection is Hacking [93]). A more interesting philosophical effort, based on Merleau-Ponty's but radically alternative to this aprioristic paradigm, is the work of Simondon [94]. Based also on Bergson's contribution, he offers a theory that smoothly links the physical, biological, psychic, and social dimensions of living, multilevel, nonadditive, and reciprocal regulation of becoming—not in terms of an ontology of *determination*, as it is in Cacioppo and Berntson [1], but in terms of *individuation*. In his conceptual elaboration, the animal existence and the human body are not conceived as antecedent roots of cognition and social life but the ongoing becoming of pre-individual matter into mind and culture, much in line with old Hylozoism, as well as with current neurobiological and complexity theories of the self and the dynamics of experience [85, 95–98].

References

1. Cacioppo JT, Berntson GG. Social psychological contributions to the decade of the brain. Doctrine of multilevel analysis. Am Psychol. 1992;47(8):1019–28.
2. Eberl JT. Aquinas on the nature of human beings. Rev Metaphys. 2004;58(2):333–65.
3. Heatherton TF. Neuroscience of self and self-regulation. Annu Rev Psychol. 2011;62:363–90.
4. Kant I, Friedman M. Metaphysical foundations of natural science. Cambridge: Cambridge University Press; 2004.
5. Baker B. The hunt for disability: the new eugenics and the normalization of school children. Teach Coll Rec. 2002;104(4):663–703.
6. Carver CS, Scheier MF. Control theory: a useful conceptual framework for personality–social, clinical, and health psychology. Psychol Bull. 1982;92(1):111.
7. Baumeister RF, Muraven M, Tice DM. Ego depletion: a resource model of volition, self-regulation, and controlled processing. Soc Cogn. 2000;18(2):130–50.
8. Bauer IM, Baumeister RF. Self-regulatory strength. In: Handbook of self-regulation: research, theory, and applications. New York: Guilford Press; 2011. p. 64–82.
9. Carver CS, Lawrence JW, Scheier MF. Self-discrepancies and affect: incorporating the role of feared selves. Personal Soc Psychol Bull. 1999;25(7):783–92.
10. Higgins ET. Self-discrepancy: a theory relating self and affect. Psychol Rev. 1987;94(3):319.
11. vanDellen MR, Hoyle RH. Possible selves as behavioral standards in self-regulation. Self Identity. 2008;7:295–304.
12. Morf CC, Horvath S. Self-regulation processes and their signatures. In: Handbook of personality and self-regulation. New York: Wiley; 2010. p. 115–44.
13. Burnette JL, O'boyle EH, VanEpps EM, Pollack JM, Finkel EJ. Mind-sets matter: a meta-analytic review of implicit theories and self-regulation. Psychol Bull. 2013;139:655.
14. Hagger MS, Wood C, Stiff C, Chatzisarantis NL. Ego depletion and the strength model of self-control: a meta-analysis. Psychol Bull. 2010;136:495.

15. Gestsdottir S, Lerner RM. Positive development in adolescence: the development and role of intentional self-regulation. Hum Dev. 2008;51(3):202–24.
16. Vohs KD, Baumeister RF, Schmeichel BJ. Erratum to "Motivation, personal beliefs, and limited resources all contribute to self-control" [J. Exp. Soc. Psychol. 48 (2012) 943–947]. J Exp Soc Psychol. 2013;49(1):184–8.
17. Baumeister RF. The unity of self at the interface of the animal body and the cultural system. Psychol Stud. 2011;56(1):5–11.
18. Fitzsimons GM, Finkel EJ. Interpersonal influences on self-regulation. Curr Dir Psychol Sci. 2010;19(2):101–5.
19. Ent MR, Baumeister RF, Vonasch AJ. Power, leadership, and self-regulation. Soc Personal Psychol Compass. 2012;6(8):619–30.
20. Amodio DM. Self-regulation in intergroup relations: A social neuroscience framework. In: Social neuroscience: toward understanding the underpinnings of the social mind. New York: Oxford University Press; 2011. p. 101–22.
21. Devine PG. Stereotypes and prejudice: their automatic and controlled components. J Pers Soc Psychol. 1989;56(1):5.
22. Botvinick M, Nystrom LE, Fissell K, Carter CS, Cohen JD. Conflict monitoring versus selection-for-action in anterior cingulate cortex. Nature. 1999;402(6758):179–81.
23. Carter CS, Braver TS, Barch DM, Botvinick MM, Noll D, Cohen JD. Anterior cingulate cortex, error detection, and the online monitoring of performance. Science. 1998;280(5364):747–9.
24. Van Veen V, Carter CS. The timing of action-monitoring processes in the anterior cingulate cortex. J Cogn Neurosci. 2002;14(4):593–602.
25. Amodio DM, Harmon-Jones E, Devine PG, Curtin JJ, Hartley SL, Covert AE. Neural signals for the detection of unintentional race bias. Psychol Sci. 2004;15(2):88–93.
26. Amodio DM, Kubota JT, Harmon-Jones E, Devine PG. Alternative mechanisms for regulating racial responses according to internal vs external cues. Soc Cogn Affect Neurosci. 2006;1(1):26–36.
27. Gross JJ. Antecedent-and response-focused emotion regulation: divergent consequences for experience, expression, and physiology. J Pers Soc Psychol. 1998;74(1):224.
28. Gross JJ. Handbook of emotion regulation. New York: Guilford Press; 2013.
29. Gross JJ, Barrett LF. Emotion generation and emotion regulation: one or two depends on your point of view. Emot Rev. 2011;3(1):8–16.
30. Lazarus RS. Cognition and motivation in emotion. Am Psychol. 1991;46(4):352.
31. Mauss IB, Bunge SA, Gross JJ. Automatic emotion regulation. Soc Personal Psychol Compass. 2007;1(1):146–67.
32. Koole SL, Coenen LHM. Implicit self and affect regulation: effects of action orientation and subliminal self priming in an affective priming task. Self Identity. 2007;6(2-3):118–36.
33. Ochsner KN, Gross JJ. Cognitive emotion regulation: insights from social cognitive and affective neuroscience. Curr Dir Psychol Sci. 2008;17(2):153–8.
34. Goldin PR, McRae K, Ramel W, Gross JJ. The neural bases of emotion regulation: reappraisal and suppression of negative emotion. Biol Psychiatry. 2008;63(6):577–86.
35. Kappas A. Emotion and regulation are one! Emot Rev. 2011;3(1):17–25.
36. Volokhov RN, Demaree HA. Spontaneous emotion regulation to positive and negative stimuli. Brain Cogn. 2010;73(1):1–6.
37. Tamir M. What do people want to feel and why? Pleasure and utility in emotion regulation. Curr Dir Psychol Sci. 2009;18(2):101–5.
38. Tamir M, Ford BQ. When feeling bad is expected to be good: emotion regulation and outcome expectancies in social conflicts. Emotion. 2012;12:807–16.
39. Beckes L, Coan JA. Social baseline theory: the role of social proximity in emotion and economy of action. Soc Personal Psychol Compass. 2011;5(12):976–88.
40. Trommsdorff G. Development of "agentic" regulation in cultural context: the role of self and world views. Child Dev Perspect. 2012;6(1):19–26.
41. Novin S, Banerjee R, Dadkhah A, Rieffe C. Self-reported use of emotional display rules in the Netherlands and Iran: evidence for sociocultural influence. Soc Dev. 2009;18(2):397–411.

42. Davis E, Greenberger E, Charles S, Chen C, Zhao L, Dong Q. Emotion experience and regulation in China and the United States: how do culture and gender shape emotion responding? Int J Psychol. 2012;47(3):230–9.
43. Miyamoto Y, Ma X. Dampening or savoring positive emotions: a dialectical cultural script guides emotion regulation. Emotion. 2011;11(6):1346.
44. Baumeister RF, Schmeichel BJ, Vohs KD. Self-regulation and the executive function: the self as controlling agent. In: Kruglanski AW, Higgins ET, editors. Social psychology: handbook of basic principles. New York: Guilford Press; 2007. p. 516–39.
45. Baumeister RF, Vohs KD, Tice DM. The strength model of self-control. Curr Dir Psychol Sci. 2007;16(6):351–5.
46. Carver CS, Scheier MF. Self-regulation of action and affect. In: In: Handbook of self-regulation: research, theory, and applications. New York: Guilford Press; 2011. p. 13–39.
47. Posner MI, Rothbart MK, Sheese BE, Tang Y. The anterior cingulate gyrus and the mechanism of self-regulation. Cogn Affect Behav Neurosci. 2007;7(4):391–5.
48. Baumeister RF, Heatherton TF. Self-regulation failure: an overview. Psychol Inq. 1996;7(1):1–15.
49. Mischel W, Shoda Y, Rodriguez ML. Delay of gratification in children. Science. 1989;244(4907):933.
50. McPherson GE, Zimmerman BJ. Self-regulation of musical learning. In: The new handbook of research on music teaching and learning: a project of the Music Educators National Conference. Oxford: Oxford University Press; 2002. p. 327–47.
51. Duckworth AL, Seligman MEP. Self-discipline outdoes IQ in predicting academic performance of adolescents. Psychol Sci. 2005;16(12):939–44.
52. Heatherton TF, Wagner DD. Cognitive neuroscience of self-regulation failure. Trends Cogn Sci. 2011;15(3):132–9.
53. Baumeister RF, Heatherton TF, Tice DM. Losing control: how and why people fail at self-regulation. San Diego: Academic press; 1994.
54. Gottfredson MR, Hirschi T. A general theory of crime. Stanford: Stanford University Press; 1990.
55. Tangney JP, Baumeister RF, Boone AL. High self-control predicts good adjustment, less pathology, better grades, and interpersonal success. J Pers. 2004;72(2):271–324.
56. Wyland CL, Kelley WM, Macrae CN, Gordon HL, Heatherton TF. Neural correlates of thought suppression. Neuropsychologia. 2003;41(14):1863–7.
57. Richeson JA, Shelton JN. When prejudice does not pay effects of interracial contact on executive function. Psychol Sci. 2003;14(3):287–90.
58. Blair C. Behavioral inhibition and behavioral activation in young children: relations with self-regulation and adaptation to preschool in children attending head start. Dev Psychobiol. 2003;42(3):301–11.
59. Haye AA. Living being and speaking being: toward a dialogical approach to intentionality. Integr Psychol Behav Sci. 2008;42(2):157–63.
60. Lamperd B. ADHD: 'truth discourse with a vengeance'. Australas J Am Stud. 2009:28(1):74–92.
61. Barkley RA. International consensus statement on ADHD. J Am Acad Child Adolesc Psychiatry. 2002;41(12):1389.
62. Rose N. The politics of life itself: biomedicine, power, and subjectivity in the twenty-first century. Oxford: Princeton University Press; 2007.
63. Conrad P. The medicalization of society: on the transformation of human conditions into treatable disorders. Baltimore: JHU Press; 2008.
64. Clarke AE, Shim JK, Mamo L, Fosket JR, Fishman JR. Biomedicalization: technoscientific transformations of health, illness, and U.S. biomedicine. Am Sociol Rev. 2003;68(2):161–94.
65. Foucault M, Martin LH, Gutman H, Hutton PH. Technologies of the self: a seminar with Michel Foucault. Amherst: University of Massachusetts Press; 1988.
66. Foucault M, Lagrange J, Burchell G. Psychiatric power: lectures at the college de france, 1973–1974, vol. 1. New York: Macmillan; 2006.

67. Foucault M. The birth of biopolitics: lectures at the Collège de France, 1978-1979. New York: Picador; 2010.
68. Foucault M. The courage of the truth (the government of self and others II): lectures at the Collège de France, 1983-1984. Basingstoke: Palgrave Macmillan; 2012.
69. Bianchi E. Diagnósticos psiquiátricos infantiles, biomedicalización y DSM: ¿hacia una nueva normalidad? Rev Latinoam Ciencias Soc Niñez y Juv. 2016;14(1):417–30.
70. Leavy P. "¿ Trastorno o mala educación?" Reflexiones desde la antropología de la niñez sobre un caso de TDAH en el ámbito escolar. Rev Latinoam Ciencias Soc Niñez y Juv. 2013;11(2):675–88.
71. Bianchi E. La perspectiva teórico-metodológica de Foucault. Algunas notas para investigar al " ADHD". Rev Latinoam Ciencias Soc Niñez y Juv. 2010;8(1):43–65.
72. Solovieva Y, Mata A, Rojas LQ. Vías de corrección alternativa para el trastorno de déficit de atención en la edad preescolar. Rev CES Psicol. 2014;7(1):95–112.
73. Van Kuyk JJ. Holistic or sequential approach to curriculum: what works best for young children? Rev Latinoam Ciencias Soc Niñez y Juv. 2009;7(2):949–69.
74. Quintanar L. Los trastornos del aprendizaje: Aproximación histórico-cultural. In: Quintanar RL, Eslava-Cobos J, editors. Los trastor del aprendiz perspect neuropsicol Puebla Coop Editor Magisterio. Puebla: Inst Colomb Neurociencias, Benemérita Univ Autónoma Puebla; 2008.
75. Luria AR. Human brain and psychological processes. New York: Harper & Row; 1966.
76. Luria AR. Higher cortical functions in man. New York: Basic Books/Consultants Bureau; 1980.
77. Ryan RM, Kuhl J, Deci EL. Nature and autonomy: an organizational view of social and neurobiological aspects of self-regulation in behavior and development. Dev Psychopathol. 1997;9(4):701–28.
78. Lock M, Scheper-Hughes N. A critical-interpretive approach in medical anthropology: rituals and routines of discipline and dissent. In: Medical anthropology: contemporary theory and method, vol. 3; 1990. p. 47–73.
79. Lock M. Cultivating the body: anthropology and epistemologies of bodily practice and knowledge. Annu Rev Anthropol. 1993;22(1):133–55.
80. Blair C, Raver CC. School readiness and self-regulation: a developmental psychobiological approach. Annu Rev Psychol. 2015;66:711–31.
81. Petersen SE, Posner MI. The attention system of the human brain: 20 years after. Annu Rev Neurosci. 2012;35:73–89.
82. Kelley WM, Wagner DD, Heatherton TF. In search of a human self-regulation system. Annu Rev Neurosci. 2015;38:389–411.
83. Mayr E. The growth of biological thought: diversity, evolution, and inheritance. Cambridge: Harvard University Press; 1982.
84. Rosenberg A. The structure of biological science. Cambridge: Cambridge University Press; 1985.
85. Collier J. Self-organization, individuation and identity. Rev Int Philos. 2004;2:151–72.
86. Varela F. Principles of biological autonomy. New York: Elsevier; 1979.
87. Luhmann N. Essays on self-reference. New York: Columbia University Press; 1990.
88. Moreno A, Mossio M. Biological autonomy. History, philosophy and theory. New York: Springer; 2015.
89. Human Microbiome Project Consortium. Structure, function and diversity of the healthy human microbiome. Nature. 2012;486(7402):207–14.
90. Von Foerster H. Cybernetics of cybernetics. In: Understanding understanding. New York: Springer; 2003. p. 283–6.
91. Weber A, Varela FJ. Life after Kant: natural purposes and the autopoietic foundations of biological individuality. Phenomenol Cogn Sci. 2002;1(2):97–125.
92. Merleau-Ponty M. Themes from the lectures at the Collège de France, 1952-1960. Evanston: Northwestern University Press; 1970.
93. Hacking I. The social construction of what? Cambridge: Harvard University Press; 1999.

94. Simondon G. L'individu et sa genèse physico-biologiqe: l'individuation à la lumière des notions de forme et d'information. Paris: Presses Universitaires de France; 1964.
95. Christoff K, Cosmelli D, Legrand D, Thompson E. Specifying the self for cognitive neuroscience. Trends Cogn Sci. 2011;15(3):104–12.
96. Bourbousson J, Fortes-Bourbousson M. How do co-agents actively regulate their collective behavior states? Front Psychol. 2016;7:1732.
97. Grandpierre A, Kafatos M. Biological autonomy. Philos Stud. 2012;2(9):631.
98. Jarosek S. Pragmatism, neural plasticity and mind-body Unity. Biosemiotics. 2013;6(2):205–30.

Part II
Impact of Social Neuroscience on Social Spheres

Dementia and Social Neuroscience: Historical and Cultural Perspectives

Olivier Piguet

Abstract The late nineteenth century and most of the twentieth century have seen the scientific approach applied to human cognition. Within this conceptual framework, social/emotional behaviours have often been perceived as nuisance variables in the investigations of 'higher' cognitive functions. Thus, neurodegenerative conditions associated with ageing (such as dementia), in which cognition becomes progressively affected, were diagnosed by focusing predominantly on the main domains of cognition, including memory, language, executive function, and attention. In recent years, a shift has emerged with increasing evidence that social/emotional cognition is an integral part of human cognition and needs to be apprehended as a distinct but complementary component of human behaviour. In addition, social/emotional processing has been demonstrated to be a strong modulator of cognitive performance. In this chapter, I review how the diagnosis of dementia has changed over the past 100 years to progressively include social/emotional cognition in their heuristics. I also highlight how the inclusion of social neuroscience methods in the clinical assessment of dementia patients can enhance the accuracy and specificity of the clinical diagnosis of these neurodegenerative conditions.

Keywords Alzheimer's disease • Dementia with Lewy bodies • Frontotemporal dementia • Pick's disease • Emotion processing

1 Introduction

A child born in 2017 has a predicted life expectancy of over 85 years. This contrasts with a life expectancy of less than 40 years at birth in 1850. Undoubtedly, this dramatic change is the result of a combination of factors, including the reduction in infant mortality, the development of vaccines for endemic diseases such as

O. Piguet (✉)
The University of Sydney, School of Psychology and Brain & Mind Centre, Sydney, NSW, Australia

ARC Centre of Excellence in Cognition and its Disorder, Sydney, NSW, Australia
e-mail: olivier.piguet@sydney.edu.au

© Springer International Publishing Switzerland 2017

A. Ibáñez et al. (eds.), *Neuroscience and Social Science*,
DOI 10.1007/978-3-319-68421-5_11

tuberculosis or smallpox, the consolidations of medical breakthroughs (e.g. penicillin, pasteurization), and the long-term impact of the industrial and agricultural revolutions, all resulting in improved life quality. This increased life expectancy combined with the progressive reduction in birth rate since the 1970s allowed for a marked ageing of the world population. In other words, the proportion of individuals over the age of 65 years has never been so high. It is estimated that about 15–17% of the world population is now over that age, compared to less than 4% in 1900 [1]—a trend that is likely to continue. This ageing pattern has seen an increase in the number of individuals diagnosed with dementia, progressive degenerative brain conditions which are mainly age-related. Indeed, in Australia, like in many industrialized countries, the rate of individuals diagnosed with dementia is expected to exhibit a threefold increase by 2050 [2]. How countries and communities tackle this social issue constitutes one of the major challenges of our time. This chapter reviews how ageing and dementia have been perceived over the past century in Western and some non-Western countries and how the diagnosis of dementia has evolved over that time, from a position that was predominantly preoccupied by cognition to progressively include social/emotional cognition in their heuristics. This review will highlight that even in dementia syndromes characterized by marked behaviour changes, such as frontotemporal dementia, these were downplayed initially. The chapter will also highlight how the inclusion of social neuroscience methods in the clinical assessment of dementia patients can enhance the accuracy and specificity of clinical diagnosis and contribute to predicting the rate of disease progression and underlying neuropathological patterns.

2 Ageing and Dementia Across the Ages

Awareness that changes in cognition (in particular, those affecting memory function) could happen with ageing was already present in Ancient Egypt, Greece, and Rome (e.g. Plato, Hippocrates, Galen). In Western-style societies, however, this knowledge more or less disappeared through the Middle Ages until the late eighteenth century, when Pinel (1745–1826) and Esquirol (1772–1840) reported changes in cognition and behaviour due to cerebral disease [3]. In the second half of the nineteenth century, the discovery of novel staining techniques (e.g. silver staining, Congo red) led to a rapid expansion of knowledge of the cellular morphology and organization of the brain arising from postmortem investigations. These investigations marked a rapid shift in the understanding of brain organization. In parallel, reports of localization of functions in the brain, such as language, through the work of Broca [4] and Wernicke [5], led to the progressive understanding of the topographical organization of brain functions.

The works of Arnold Pick, a Czech neurologist, and Alois Alzheimer, a German psychiatrist and pathologist, led the charge that resulted in the identification of some of the neuropathological changes underlying progressive behavioural and cognitive deficits in older individuals. Indeed, Alzheimer was credited as the first to report the abnormal aggregation of two proteins that are now pathognomonic of the disease that bears his name: senile plaques (composed of beta amyloid) and neurofibrillary tangles (composed of the tau protein) [6]. During the same period, Pick reported on a series of patients who also presented with progressive linguistic and behavioural changes and showed marked focal atrophy postmortem [7]. Subsequent investigations [8, 9] identified intraneuronal inclusions, labelled 'Pick bodies', which differed from those found in cases of Alzheimer's disease, with the syndrome and associated pathology labelled 'Pick's disease'—now known as frontotemporal dementia or frontotemporal lobar degeneration, respectively.

These seminal discoveries on frontotemporal dementia, which for the first time highlighted the clinical and pathological diversity of dementia, were somewhat forgotten with rare exceptions—e.g. [10]—until the 1970s with the work of the Geneva group [11]. Indeed, during that time, little progress was made on the clinical and pathological nomenclature of dementias, in part because of the belief that Alzheimer's disease and Pick's disease were clinically indistinguishable. In parallel, however, the distinction between the presenile and senile forms of Alzheimer's disease progressively disappeared with the discovery that they were associated with identical pathological changes in the brain [12]. Another common view during that time was related to the vascular origin of dementia caused by arteriosclerosis, reduced brain perfusion, and ministrokes [13]. While vascular dementia now accounts for a small proportion of dementia cases, contribution of vascular risk factors and vascular disease to dementia has been increasingly recognized in recent times.

3 Awareness of Ageing and Dementia in Western and Non-Western Societies

From a social viewpoint, Western societies in Europe and the USA held negative views about 'senility', as dementia was generally labelled, denoting age-related (mental) deterioration, rather than general (let alone healthy) ageing. Moving away from a religious explanation of senility as a 'sinner's accomplishment' that one brings upon oneself, senility was for the first time approached from a medical perspective, albeit a pessimistic one. Deterioration in old age was perceived to be inevitable due to the depletion of the body's vital energy [14]. Interestingly, despite the discoveries of Alzheimer and others outlining the biological bases for such disorders, the early twentieth century witnessed a focus on psychosocial causes of dementia, possibly because Alzheimer's disease was initially described as a disease of the presenium (i.e. not associated with ageing), a distinction that is no longer relevant. This position was progressively eroded with the development of new investigative techniques and the pendulum swinging back well and truly towards identifying biological causes for these progressive and abnormal declines in cognitive functions.

In countries outside Europe and North America, understanding of ageing and dementia varies widely. How older individuals are looked after also varies widely across countries, depending on many cultural aspects and how ageing is perceived—either positively or negatively—in these communities. This section provides a snapshot of this diversity and the challenges it raises when approaching management and interventions of individuals presenting with dementia. In many countries, dementia is regarded as an inherent part of the natural ageing process, rather than an abnormal brain disease that tends to become more frequent as individuals get older. Indeed, an African study conducted in the Republic of Congo showed that while symptoms of dementia and their functional impact (such as memory loss, behavioural changes) were recognized, these were not related to a disease, with little knowledge of what 'dementia' is [15]. When asked, individuals, however, displayed an awareness of the range of possible deficits and reported that as you get older, you tend to become like a child and have problems and memory loss.

Similarly to other developing countries, public awareness of dementia is low in India, where many communities do not conceive the condition as an organic or medical disease [16]. In Arabic countries and China, it may be understood as part of the normal ageing process, but it is also associated with mental illness. Often, perceived causes are multiple or not well understood, with dementia typically being ascribed to external events (such as stress, lack of activity, hard life, accidents, or emotional stress), and cognitive changes conceived a *consequence* of these events (rather than the primary cause). In China, while symptoms are recognized (memory loss, wandering, confusion), their causes are not well understood [17].

Management of dementia patients is also variable. In some African countries, religion and traditional treatments (e.g. magic, healer practices) play an important role. Interestingly, in some rural communities, tension remains between old age being seen as a source of wisdom and experience (a positive trait) even in the presence of cognitive decline and the negative effect of this cognitive decline on other individuals in the form of spell, which may require the use of sorcery [15]. In India, management is hampered by the fact that, outside of specialist centres, awareness of dementia by medical practitioners is limited. Often, signs of cognitive changes are dismissed as part of normal ageing, resulting in limited targeted interventions and management. Some recent investigations in India have highlighted the role of bilingualism as a protective factor in the development of certain forms of dementia (e.g. [18]). In China, a considerable emphasis is placed on healthy ageing, through healthy eating, regular mental, and physical exercise, among other factors. Both Chinese and Arabic communities see a very strong role of the family in looking after their elders. As such, formal support services (e.g. community care services or residential care services) are generally not well accepted. As a corollary, reluctance to admit the disease and denial are common, with family members trying to cover up emerging cognitive problems and functional difficulties [17].

4 Evolution of the Definition of Dementia

Dementia as a disease entity with clearly defined diagnostic criteria is a relatively recent construct. It first appeared in the third version of the *Diagnostic and Statistical Manual of Mental Disorders* (DSM), where it denotes a severe loss of cognitive function. Indeed, the earlier versions of this manual included 'organic brain syndrome' as a disease category (a term still used occasionally), with 'senile and presenile dementia' categorized under 'psychoses associated with organic brain syndrome'.

Following a resurgence in research and public health interest in the 1970s and 1980s, disease-specific clinical diagnostic criteria were published for the first time for Alzheimer's disease [19, 20], dementia with Lewy bodies [21], and frontotemporal dementia [22, 23]. Over the next two decades, these criteria underwent several updates, with the most recent versions published in the past 5–7 years [24–28]. During that time, most of the refinements have focused predominantly on the pathological and genetic aspects of these diseases following continuing discoveries of the pathomechanisms underlying these disorders.

The early iterations of these clinical criteria mostly outlined the presence of *progressive* changes in cognition and neurological or neuroradiological abnormalities, as well as the impact of these deficits on functional independence, with little or no reference to social cognition. For example, in Alzheimer's disease, emphasis is placed on the presence of progressive episodic memory deficits accompanied by deficits in at least one other cognitive domain. In dementia with Lewy bodies, diagnostic criteria highlight the presence of fluctuating cognition and variation in attention and alertness within the construct of extrapyramidal features (parkinsonism), visual hallucinations, and falls. Similarly, diagnostic criteria for semantic dementia and progressive non-fluent aphasia, the two language variants of frontotemporal dementia, emphasize the presence of specific and marked language deficits with other cognitive domains remaining comparatively well preserved. Noteworthy, however, is the mention of loss of empathy as a supportive diagnostic feature towards a diagnosis of semantic dementia. The only notable exception is for the behavioural variant of frontotemporal dementia, where early disturbance in the socio-emotional sphere (e.g. interpersonal conduct, emotional blunting) represents some of the core features of this syndrome, together with other behaviours such as mental rigidity, loss of insight, dietary changes, and deficits on 'frontal lobe' tests.

5 The Rise of Social Cognition in Dementia

Over the past two decades, however, decline in the integrity of social cognition skills in these dementia syndromes has become increasingly recognized (see Kumfor et al., Baez et al., this volume). Not surprisingly, because of its central place as a core diagnostic criterion, social cognition in the behavioural variant of frontotemporal dementia has been extensively investigated. Indeed, most aspects of social cognition are now known to be affected, such as emotion recognition, emotion expression, moral reasoning, theory of mind, cognitive and affective empathy, and knowledge of social rules—these alterations being associated with widespread changes in brain regions including the orbitofrontal cortex, the anterior cingulate, and other midline structures, as well as the anterior temporal lobe regions, particularly in the right hemisphere [29].

Of particular relevance is the recognition that deficits in social cognition were not limited to this frontotemporal dementia syndrome, as they are also present in both language variants of frontotemporal dementia (semantic dementia and progressive non-fluent aphasia). While loss of empathy was recognized as a supportive feature for the diagnosis of semantic dementia, recent evidence has demonstrated the presence of pervasive deficits of emotion processing, including emotion recognition and empathy, regardless of the modality of exposure (e.g. verbal, visual, auditory) [30–32]. Again, severity of these deficits appears to be related to the involvement of right anterior lobe regions. Social cognition in progressive non-fluent aphasia has been less well studied, but recent reports [33, 34] have identified subtle emotion processing deficits. This finding has important clinical implications regarding the diagnostic accuracy of this syndrome and its distinction from a second progressive non-fluent aphasia syndrome called logopenic progressive aphasia, characterized by a number of overlapping clinical features. These two syndromes, however, have very different pathologies: whereas progressive non-fluent aphasia is associated with frontotemporal lobar degeneration, logopenic progressive aphasia shows pathological changes characteristic of Alzheimer's disease. Crucially, deficits of emotion processing have been found in progressive non-fluent aphasia, but this aspect remains preserved in logopenic progressive aphasia, at least early on. This demonstrates that the addition of emotion processing tasks to an examination protocol can help improve diagnostic accuracy in the presence of a non-fluent syndrome and can also enhance management of these patients [35], some of whom may benefit from acetylcholine esterase inhibitors which are recommended in individuals with Alzheimer's disease but have been found to show no benefit in patients with frontotemporal lobar degeneration [36].

The most recent version of the diagnostic criteria for Alzheimer's disease now highlights that a minimum of two behavioural or cognitive domains need to be affected, which may or may not include memory, depending on clinical presentation (i.e. amnestic vs. nonamnestic types). One of these domains now relates to personality, behaviour, or comportment, with possible symptoms including *loss of empathy* as well as *socially unacceptable behaviours* [24]. This inclusion evinces increasing awareness of the presence of deficits in the social domain in Alzheimer's disease, and, more broadly, it denotes how widespread social cognition deficits are in dementia syndromes. Emotion processing deficits have been reported in Alzheimer's disease patients. When present, however, these appear to be mild and may vary according to the modality of presentation—e.g. [37], but see [38]. In addition, these deficits become more frequent in the later stages of the disease.

Finally, emotion processing and social cognition have not been investigated extensively in patients diagnosed with Lewy body dementia. The social domain is not considered in the most recent diagnostic criteria for this disease [28]. A recent publication, however, has reported deficits in complex social cognition skills (theory of mind) but preserved emotion recognition in individuals diagnosed with 'prodromal' Lewy body dementia—i.e. a disease state defined as diagnosed individuals in whom cognitive deficits have not yet led to functional decline [39].

6 Emotion Is Not a Nuisance Variable When Assessing Cognition

Historically, clinical investigations of cognitive functions during neuropsychological assessments have endeavoured to focus on 'pure' cognitive processes (e.g. language, memory), as these were believed to be measureable constructs. As such, over time, the preferred approach to cognitive assessment has privileged a standardized procedure, akin to lab experiments, where the examiner is controlling as many variables thought to be irrelevant or unrelated to the process of interest. An example of this approach is that of the Wechsler scales—e.g. WAIS IV [40]—where exhaustive guidelines are provided regarding not only task instructions but also environment (office layout, lighting, noise) and interactions between the examiner and the participant. This approach is by no means limited to these scales and is present in most settings conducting cognitive assessments. This artificial but well-controlled methodology has strived to enable comparisons across testing sessions and score profiles, with social interactions or social variables (e.g. mood, understanding of social cues) being minimized as much as possible as they were considered 'nuisance' variables.

In dementia settings, an unintended effect of this approach has been that emotion processing and other aspects of social cognition were thought not to be relevant within the context of cognitive examination or part of the profile of a particular clinical syndrome (with the exception of the behavioural variant of frontotemporal dementia). Another effect of this approach can be observed in the cognitive screening instruments for dementia that have been developed over the years, such as the Mini-Mental State Examination, the Addenbrooke's Cognitive Examination or the Montreal Cognitive Assessment [41–43]. Indeed, all these instruments focus only on the main cognitive domains (attention, memory, language, visuoconstructive, executive function) with emphasis on the different domains varying across the scales, but none examines social cognition. This scenario demonstrates how little attention has been paid to accruing evidence of a close relationship between aspects of social cognition, such as emotion processing, and other cognitive processes—e.g. mood congruency between encoding and retrieval sessions improves memory performance [44]. One example is that of emotional enhancement memory effect, whereby stimuli that comprise an emotional component are remembered better than neutral stimuli—e.g. [45]. Importantly, the emotional memory enhancement effect is variable across dementia syndromes: while it is preserved in Alzheimer's disease, despite the marked episodic memory deficit [46], it is reduced or absent in the behavioural variant of frontotemporal dementia but preserved in progressive non-fluent aphasia [47].

Arguably, a major contributing factor is the complexity of the social cognition construct and its assessment. Most tests of social cognition are lengthy, generally examine a single component, and may therefore not be suitable in many clinical settings where time for examination is limited. For example, tests of facial emotion recognition not uncommonly comprise in excess of 40–60 stimuli (e.g. Ekman faces). In addition, because they use static stimuli that have little resemblance to real-life situations, such tasks fail to capture the richness of the social interactions and remain relatively crude measures of social cognition integrity. In recent years, studies have used dynamic stimuli, such as video clips or movie excerpts to overcome this limitation—e.g. [48]. Most promising is The Awareness of Social Inference Test (TASIT), a battery that uses video clips depicting social interactions between one or two protagonists of various complexity [49]. Although lengthy in its original format, several short forms have been recently published that are potentially suitable as screening instruments in clinical settings—e.g. [50, 51].

Our understanding of dementia has improved tremendously over the past 40 years. Knowledge of the clinical features, disease progressive, and pathological mechanisms of the main dementia syndromes means that clinicians are better prepared to address the main questions most patients and their families ask: those concerning diagnosis, prognosis, and treatment. Recent research has also demonstrated that careful investigation of social cognition has its place in these processes, improving diagnostic accuracy. Recent evidence has also shown that the presence of social cognition deficits is associated with increased burden in carers of dementia patients, regardless of the dementia type [52]. As our arsenal to investigate social cognition continues to improve, this information will become increasingly valuable not only when planning treatment interventions with patients but also when considering provision of social or psychological support to their caregivers.

Acknowledgements OP is supported by a National Health and Medical Research Council of Australia Senior Research Fellowship (APP1103258).

References

1. US Census Bureau. U.S. Census Bureau, decennial census of population, 1900 to 2000. 2011. Available from https://www.census.gov/population/www/censusdata/hiscendata.html.
2. Brown L, Hansnata E. Economic cost of dementia in Australia 2016-2056. Alzheimer's Australia. Canberra: Institute for Governance and Policy Analysis, University of Canberra; 2017.
3. Esquirol E. Des maladies mentales considérées sous les rapports médical, hygiénique et médico-légal. Paris: Baillière; 1838.
4. Broca P. Remarques sur le siège de la faculté du langage articulé, suivies d"une observation d"aphémie (perte de la parole). Bull Soc Anat. 1861;6:330.
5. Wernicke C. The aphasia symptom complex: A psychological study on an anatomical basis. In: Eggert G, editor. Wernicke's works on aphasia. Mouton: The Hague; 1874.

6. Alzheimer A. Über eine eigenartige erkrankung der hirnrinde. Allg Z Psychiatr. 1907;64:146–8.
7. Pick A. Uber die beziehungen der senilen hirnatrophie zur aphasie. Prag Med Wochenschr. 1892;17:165–7.
8. Alzheimer A, Forstl H, Levy R. On certain peculiar diseases of old age. Hist Psychiatry. 1991; 2(5):71–3.
9. Onari K, Spatz H. Anatomische beiträge zur lehre von der pickschen umschriebenen großhirnrinden-atrophie ("Picksche krankheit"). Arbeiten aus der Deutschen Forschungsanstalt für Psychiatrie in München (Kaiser-Wilhelm-Institut). Berlin, Heidelberg: Springer; 1926. p. 546–587.
10. Van Mansvelt J. Pick's disease. A syndrome of lobar cerebral atrophy, its clinico-anatomical and histopathological types. van de Loeff: Enschede; 1954.
11. Tissot R, Constantinidis J, Richard J. La maladie de Pick. Paris: Masson; 1975.
12. Terry RD, Gonatas NK, Weiss M. Ultrastructural studies in Alzheimer's presenile dementia. Am J Pathol. 1964;44(2):269–97.
13. Jellinger KA. The enigma of vascular cognitive disorder and vascular dementia. Acta Neuropathol. 2007;113(4):349–88.
14. Ballenger JF. Self, senility, and Alzheimer's disease in Modern America. Baltimore: John Hopkins University Press; 2006.
15. Faure-Delage A, Mouanga AM, M'belesso P, Tabo A, Bandzouzi B, Dubreuil C-M, et al. Socio-cultural perceptions and representations of dementia in Brazzaville, Republic of Congo: the EDAC survey. Dement Geriatr Cogn Dis Extra. 2012;2(1):84–96.
16. Alzheimer's Related Disorders Society of India. In: Shaji KS, Jotheeswaran AT, Girish N, Srikala B, Dias A, Pattabiraman M, et al., editors. The dementia India report 2010. New Delhi: ARDSI; 2010.
17. Alzheimer's Australia Vic. Perceptions of dementia in ethnic communities. Melbourne: Alzheimer's Australia; 2008.
18. Alladi S, Bak TH, Duggirala V, Surampudi B, Shailaja M, Shukla AK, et al. Bilingualism delays age at onset of dementia, independent of education and immigration status. Neurology. 2013;81(22):1938–44.
19. McKhann G, Drachman D, Folstein M, Katzman R, Price D, Stadlan EM. Clinical diagnosis of Alzheimer's disease: report of the NINCDS-ADRDA Work Group under the auspices of Department of Health and Human Services Task Force on Alzheimer's disease. Neurology. 1984;34(7):939–44.
20. Morris JC. The clinical dementia rating (CDR): current version and scoring rules. Neurology. 1993;43(11):2412–4.
21. McKeith IG, Galasko D, Kosaka K, Perry EK, Dickson DW, Hansen LA, et al. Consensus guidelines for the clinical and pathologic diagnosis of dementia with Lewy bodies (DLB): report of the consortium on DLB international workshop. Neurology. 1996;47(5):1113–24.
22. The Lund and Manchester Groups. Clinical and neuropathological criteria for frontotemporal dementia. The Lund and Manchester Groups. J Neurol Neurosurg Psychiatry. 1994; 57(4):416–8.
23. Neary D, Snowden JS, Gustafson L, Passant U, Stuss D, Black S, et al. Frontotemporal lobar degeneration: a consensus on clinical diagnostic criteria. Neurology. 1998;51(6):1546–54.
24. McKhann GM, Knopman DS, Chertkow H, Hyman BT, Jack CR Jr, Kawas CH, et al. The diagnosis of dementia due to Alzheimer's disease: recommendations from the National Institute on Aging-Alzheimer's Association workgroups on diagnostic guidelines for Alzheimer's disease. Alzheimers Dement. 2011;7(3):263–9.
25. Rascovsky K, Hodges JR, Knopman D, Mendez MF, Kramer JH, Neuhaus J, et al. Sensitivity of revised diagnostic criteria for the behavioural variant of frontotemporal dementia. Brain. 2011;134(9):2456–77.
26. Gorno-Tempini ML, Hillis AE, Weintraub S, Kertesz A, Mendez M, Cappa SF, et al. Classification of primary progressive aphasia and its variants. Neurology. 2011;76(11):1006–14.

27. McKeith IG, Dickson DW, Lowe J, Emre M, O'Brien JT, Feldman H, et al. Diagnosis and management of dementia with Lewy bodies: third report of the DLB Consortium. Neurology. 2005;65:1863–72.
28. McKeith IG, Boeve BF, Dickson DW, Halliday GM, Taylor J-P, Weintraub D, et al. Diagnosis and management of dementia with Lewy bodies: 4th consensus report of the DLB consortium. Neurology. 2017;89(1):88–100.
29. Henry JD, Hippel von W, Molenberghs P, Lee T, Sachdev PS. Clinical assessment of social cognitive function in neurological disorders. Nat Rev Neurol. 2015;12(1):28–39.
30. Bejanin A, Chételat G, Laisney M, Pélerin A, Landeau B, Merck C, et al. Distinct neural substrates of affective and cognitive theory of mind impairment in semantic dementia. Soc Neurosci. 2017;12(3):287–302.
31. Kumfor F, Landin-Romero R, Devenney E, Hutchings R, Grasso R, Hodges JR, et al. On the right side? A longitudinal study of left- versus right-lateralized semantic dementia. Brain. 2016;139(3):986–98.
32. Hsieh S, Hornberger M, Piguet O, Hodges JR. Neural basis of music knowledge: evidence from the dementias. Brain. 2011;134(9):2523–34.
33. Couto B, Manes F, Montañes P, Matallana D, Reyes P, Velasquez M, et al. Structural neuroimaging of social cognition in progressive non-fluent aphasia and behavioral variant of frontotemporal dementia. Front Hum Neurosci. 2013;7:467.
34. Kumfor F, Miller L, Lah S, Hsieh S, Savage S, Hodges JR, et al. Are you really angry? The effect of intensity on facial emotion recognition in frontotemporal dementia. Soc Neurosci. 2011;6(5-6):502–14.
35. Piguet O, Leyton CE, Gleeson LD, Hoon C, Hodges JR. Memory and emotion processing performance contributes to the diagnosis of non-semantic primary progressive aphasia syndromes. J Alzheimers Dis. 2015;44(2):541–7.
36. Boxer AL, Knopman DS, Kaufer DI, Grossman M, Onyike C, Graf-Radford N, et al. Memantine in patients with frontotemporal lobar degeneration: a multicentre, randomised, double-blind, placebo-controlled trial. Lancet Neurol. 2013;12(2):149–56.
37. Bucks RS, Radford SA. Emotion processing in Alzheimer's disease. Aging Ment Health. 2004; 8(3):222–32.
38. Klein-Koerkamp Y, Beaudoin M, Baciu M, Hot P. Emotional decoding abilities in Alzheimer's disease: a meta-analysis. J Alzheimers Dis. 2012;32(1):109–25.
39. Kemp J, Philippi N, Phillipps C, Demuynck C, Albasser T, Martin-Hunyadi C, et al. Cognitive profile in prodromal dementia with Lewy bodies. Alzheimers Res Ther. 2017;9(1):19.
40. Wechsler D. Wechsler adult intelligence scale (WAIS–IV). 4th ed. San Antonio: NCS Pearson; 2008.
41. Hsieh S, Schubert S, Hoon C, Mioshi E, Hodges JR. Validation of the Addenbrooke's Cognitive Examination III in frontotemporal dementia and Alzheimer's disease. Dement Geriatr Cogn Disord. 2013;36(3-4):242–50.
42. Folstein MF, Folstein SE, McHugh PR. "Mini-mental state." A practical method for grading the cognitive state of patients for the clinician. J Psychiatr Res. 1975;12(3):189–98.
43. Nasreddine ZS, Phillips NA, Bédirian V, Charbonneau S, Whitehead V, Collin I, et al. The Montreal Cognitive Assessment, MoCA: a brief screening tool for mild cognitive impairment. J Am Geriatr Soc. 2005;53(4):695–9.
44. Lewis PA, Critchley HD. Mood-dependent memory. Trends Cogn Sci. 2003;7(10):431–3.
45. Kensinger EA, Schacter DL. Remembering the specific visual details of presented objects: neuroimaging evidence for effects of emotion. Neuropsychologia. 2007;45(13):2951–62.
46. Kumfor F, Irish M, Hodges JR, Piguet O. The orbitofrontal cortex is involved in emotional enhancement of memory: evidence from the dementias. Brain. 2013;136(10):2992–3003.
47. Kumfor F, Hodges JR, Piguet O. Ecological assessment of emotional enhancement of memory in progressive nonfluent aphasia and Alzheimer's disease. J Alzheimers Dis. 2014;42(1):201–10.
48. Goodkind MS, Sturm VE, Ascher EA, Shdo SM, Miller BL, Rankin KP, et al. Emotion recognition in frontotemporal dementia and Alzheimer's disease: a new film-based assessment. Emotion. 2015;15(4):416–27.

49. McDonald S, Flanagan S, Rollins J, Kinch J. TASIT: A new clinical tool for assessing social perception after traumatic brain injury. J Head Trauma Rehabil. 2003;18(3):219–38.
50. Honan CA, McDonald S, Sufani C, Hine DW, Kumfor F. The awareness of social inference test: development of a shortened version for use in adults with acquired brain injury. Clin Neuropsychol. 2016;30(2):243–64.
51. Kumfor F, Honan CA, Hazelton J, Hodges JR, Piguet O. Assessing the "social brain" in dementia: applying TASIT-S. Cortex. 2017;93:166.
52. Hsieh S, Irish M, Daveson N, Hodges JR, Piguet O. When one loses empathy: its effect on carers of patients with dementia. J Geriatr Psychiatry Neurol. 2013;26(3):174–84.

Clinical Studies of Social Neuroscience: A Lesion Model Approach

Fiona Kumfor, Jessica L. Hazelton, François-Laurent De Winter, Laurent Cleret de Langavant, and Jan Van den Stock

Abstract Understanding the neurobiological basis of complex human behaviors is a key aim for social neuroscience. Examining clinical populations with relatively circumscribed brain damage and related behavioral deficits can provide insights into brain regions which are *necessary* for abilities such as face processing, emotion recognition, theory of mind, and empathy. In this review, we reflect on the emerging body of evidence which combines experimental behavioral studies and neuroimaging analysis techniques in clinical groups, with a focus on four progressive neurodegenerative disorders: behavioral-variant frontotemporal dementia, semantic

F. Kumfor (✉)
The University of Sydney, School of Psychology, Sydney, NSW, Australia

The University of Sydney, Brain and Mind Centre, Sydney, NSW, Australia

ARC Centre of Excellence in Cognition and Its Disorders, University of Sydney, Sydney, NSW 2052, Australia
e-mail: fiona.kumfor@sydney.edu.au

J.L. Hazelton
The University of Sydney, School of Psychology, Sydney, NSW, Australia

The University of Sydney, Brain and Mind Centre, Sydney, NSW, Australia
e-mail: jessica.hazelton@sydney.edu.au

F.-L. De Winter • J. Van den Stock
Laboratory for Translational Neuropsychiatry, Department of Neurosciences, KU Leuven, Leuven, Belgium

Department of Old Age Psychiatry, University Psychiatric Center KU Leuven, Leuven, Belgium
e-mail: francoislaurent.dewinter@kuleuven.be; jan.vandenstock@kuleuven.be

L.C. de Langavant
Faculté de Médecine, Université Paris Est, 8 Rue du Général Sarrail, 94000 Créteil, France

Centre de référence maladie de Huntington, Hôpital Henri Mondor, AP-HP, 8 Rue du Général Sarrail, 9400 Créteil, France

Laboratoire de NeuroPsychologie Interventionnelle, Institut National de la Santé et Recherche Médical (INSERM) U955, Equipe 01, 8 Rue du Général Sarrail, 94000 Créteil, France

Département d'Etudes Cognitives, Ecole Normale Supérieure – PSL* Research University, 29 Rue d'Ulm, 75005 Paris, France
e-mail: laurent.cleret@gmail.com

© Springer International Publishing Switzerland 2017

A. Ibáñez et al. (eds.), *Neuroscience and Social Science*,
DOI 10.1007/978-3-319-68421-5_12

dementia, Huntington's disease, and Alzheimer's disease. These clinical syndromes are characterized by divergent patterns of neurodegeneration and variable degrees of impairment in social cognition and social behavior. Here, we review the paradigms which have been employed and the current patterns of findings in these syndromes and discuss how this line of research informs our understanding of the "social brain." In addition, we consider how our conception of these clinical phenotypes has changed, as aspects of social cognition have been incorporated into diagnostic and prognostic frameworks. Finally, we propose potential avenues for future research in these syndromes to address outstanding social neuroscience questions.

Keywords Face processing • Emotion • Empathy • Theory of mind • Alzheimer's disease • Frontotemporal dementia • Semantic dementia • Huntington's disease • Lesion models

1 Introduction

Social animals, including humans, have developed a range of communicative abilities on which their well-being and survival within society hinges. These include the ability to monitor behavior and adapt to the social signals of others to reach either a collaborative or competitive aim [1]. A long-standing assumption in the literature is that there are specific brain resources related to processing social signals, referred to as the "social brain," involving the ventromedial prefrontal cortex, insula, bilateral temporal poles, amygdala, and temporoparietal junction (e.g., [2, 3]) (Fig. 1). This use of brain resources is thought to be evident across species, which rely on conspecifics for survival [4]. This assumption is the backbone of much neurobiological social cognition research of the last two decades, where results from animal research have inspired human studies [5–7].

In this context, understanding the neurobiological basis of complex human behaviors is a key aim for social neuroscience research, in tandem with improving theoretical models of social processes and behavior. While studies in healthy humans and animals undoubtedly offer important insights into social cognition, research in clinical populations represents a key complementary technique to understand brain regions that are *necessary* for these social processes. The lesion model approach has come from a strong neuropsychological tradition, which primarily studied individuals with discrete lesions as a result of trauma, stroke, or tumor. This approach has been extended into neurodegenerative disorders, resulting in part from rapid advances in in vivo structural and functional neuroimaging techniques. Neurodegenerative disorders offer the distinct advantage of being able to provide insights into potential brain networks underpinning social behavior. Indeed, recently it has been proposed that network disruption in degenerative disorders, such as Alzheimer's disease and frontotemporal dementia, tracks the pattern of underlying pathological changes [8]. In other words, the brain regions that succumb to

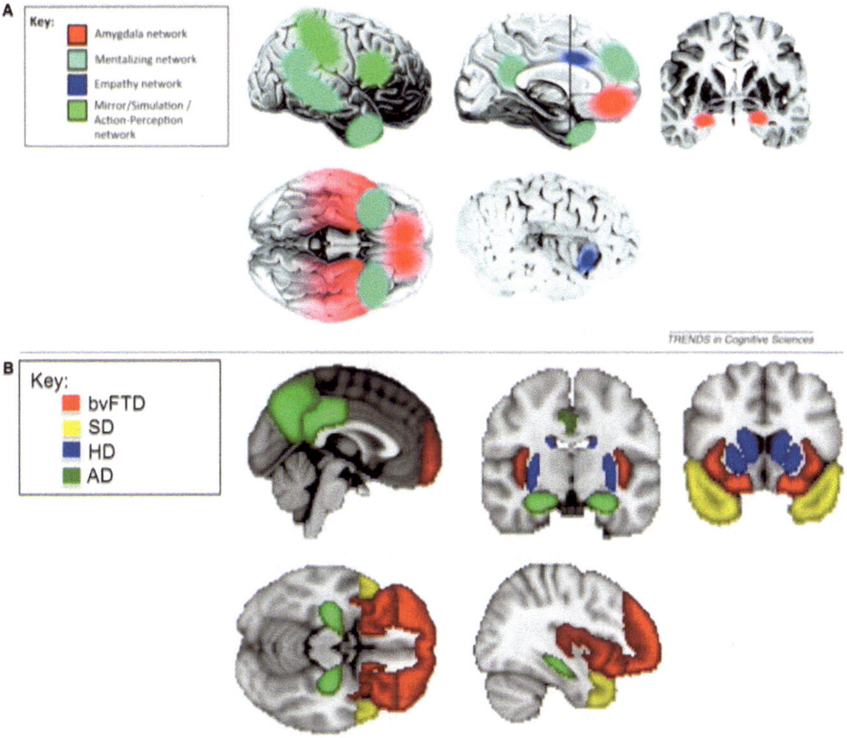

Fig. 1 The social brain and its overlap with neurodegenerative syndromes. (**a**). Brain networks subserving socio-cognitive abilities. Image reproduced with permission from Elsevier: [230]. (**b**). Typical patterns of atrophy in the dementia syndromes considered here and highlight their overlap with the social brain

pathology across dementia syndromes likely reflect regions within a shared functional network. Hence, studying different dementia syndromes with divergent patterns of atrophy offers a unique insight into how neural brain systems subserve complex human social behavior [9].

With this in mind, this chapter focuses on four of the most common neurodegenerative syndromes, which are characterized by distinct patterns of atrophy and variable decline of social cognition profile: (1) behavioral-variant frontotemporal dementia (bvFTD), (2) semantic dementia (SD), (3) Huntington's disease (HD), and (4) Alzheimer's disease (AD). First, we introduce four key subdomains of social cognition: (1) face processing, (2) emotion recognition, (3) theory of mind (ToM), and (4) empathy; then we outline the most common measures used to assess these abilities in dementia syndromes (see Piguet, this volume). Next, we describe the clinical profile of each of these syndromes—with special focus on their impairments in the four subdomains of social cognition mentioned above – and their associated pattern of neurodegeneration. In addition, we discuss how the existing evidence in these syndromes improves our understanding of the social brain and, simultaneously, how incorporating social cognition into current diagnostic criteria for dementia syndromes enriches our comprehension of these clinical conditions. Lastly, we consider areas of future research to be addressed in the field.

2 Measures of Social Cognition

2.1 Face Processing

Face processing refers to a range of abilities, from the very basic (i.e., detecting the presence of face) to more complex skills such as determining gaze, identity, age, race, and emotional expression. The ability to perceive and recognize faces represents the cornerstone of social cognition given that much of the necessary social cues, including emotional states, come from faces [10, 11]. The vast majority of research assessing face perception in neurodegenerative syndromes have employed facial identity tasks such as the Benton Facial Recognition Test [12] and/or the Warrington Recognition Memory Test for Faces [13] (Table 1), although both tests have been recently criticized. Specifically, the Benton Facial Recognition Test can be completed using a piecemeal feature matching approach, while the Warrington Recognition Memory for Faces test assesses stimulus rather than face recognition, as the test stimuli are identical to the target stimuli [14]. Moreover, a recent study found that healthy participants showed normal performance in these tests despite the removal of key facial features (e.g., eyes, mouth, and nose) [14]. In light of these findings, alternate measures of unfamiliar face processing have been employed. These include the Cambridge Face Memory Test, which assesses learning and recognition of unfamiliar faces [15], and the Cambridge Face Perception Test, which evaluates the ability to organize facial morphs according to identity [16]. These appear to be more sensitive to changes in face-processing ability and warrant further investigation in dementia syndromes.

Recognition of familiar faces is also important to consider, as this reflects the integration of face processing and semantic knowledge. Typically, this ability is assessed using famous face recognition, although in some cases, individually tailored tests have used stimuli of the participant's friends and family members. Evidently, no single famous face recognition task exists, given that "famous" individuals are highly dependent on exposure and factors such as culture, age, and personal interests (e.g., sportsmen vs. opera singers). Thus, the validity of results from these tasks may depend on the selection of relevant stimuli for the participant. Moreover, whether the task uses a recognition format (i.e., is this person famous?) vs. a naming format (i.e., who are they?) appears to tap different brain regions, with recognition typically associated with right-lateralized (or bilateral) temporal integrity, while naming is typically associated with left-lateralized temporal integrity [17].

Other relevant tasks include those that examine the holistic face perception (e.g., the composite effect), the ability to process faces under time restraints (e.g., Cambridge Face Perception Test), and the specialized nature of faces (e.g., face vs. object, considering

Table 1 Common tests used to assess social cognition in neurodegenerative syndromes, according to domain

	Ability	Test
Face processing	Face identity discrimination	Florida affect battery [232]
	Face identity recognition memory	Recognition memory test (face part) [13]; Cambridge face memory test [15]
	Face identity matching	Benton facial recognition test [233]
	Face (specific) identity matching and associated inversion effect	Facial ExpressiveAction Stimulus Test (FEAST) [18]
Emotion recognition	Facial emotion verbal categorization	Ekman 60 test; emotion hexagon; Ekman caricatures task [24]; Florida affect battery [232]; Mini-Social cognition and Emotional Assessment (Mini-SEA) [42]
	Face (specific) emotion matching	FEAST [18]; Florida affect battery [232]
	Face emotion discrimination and selection	Florida affect battery [232]
	Emotional prosody	Florida affect battery [232]
	Emotion (specific) face identity matching	Face- and emotion-processing battery [77]
	Bodily emotion matching	Bodily Expressive Action Stimulus Test (BEAST) [31]
	Face + body emotion verbal categorization	The Awareness of Social Inference Test (TASIT) [234]
Theory of mind	Detection of social norm violation	Mini-SEA [42]; Happe's stories [235]; Faux Pas Test [40]
	Social emotion verbal categorization of eye expression	Reading the Mind in the Eyes Test (RMET) [39]
	Intention attribution to abstract shape movement	Triangle animations [236]
	False belief	Cartoon task [90]; TOM-15 [231]
Empathy	Affective empathy only	The Balanced Emotional Empathy Scale (BEES) [58]
	Cognitive and affective empathy	The Interpersonal Reactivity Index (IRI) [47]; the Basic Empathy Scale (BES) [57]; the Empathy Quotient (EQ) [56]; the Multifaceted Empathy Task (MET) [61]; the Empathy-for-Pain Task (EPT) [62]
	Emotion regulation	Main focus of existing tests is not on emotion regulation; however, some aspects of available measures assess emotion regulation (e.g., personal distress scale on IRI (47), MET [61] arousal question, and EPT [62] discomfort rating)

identical cognitive load between conditions). For example, the Facial Expressive Action Stimulus Test (FEAST) [18] includes a number of subtests that assess facial identity processing across different formats (e.g., part-to-whole matching; viewpoint-independent matching). In addition, the battery includes the same tests with control stimulus categories (e.g., shoes and houses) to investigate face specificity of the effects [19–21].

Surprisingly, fewer studies have examined earlier stages of face processing in patients with dementia. This is despite a wealth of tasks developed to assess face processing in healthy individuals or other clinical populations (e.g., prosopagnosia).

2.2 Emotion Recognition

Darwin [1] first claimed that expressions of emotions are homologous across species, universal, and culturally independent (see TenHouten, this volume). The subsequent seminal studies of Paul Ekman provided empirical support for the "universal" nature of six basic emotions: anger, disgust, fear, sadness, happiness, and surprise [22]. Although there is increasing criticism regarding aspects of Ekman's methodology and the universal nature of emotion recognition [23], facial expressions have dominated research methods on emotion processing since. Ekman's original stimulus set [22, 24] is certainly the most widely used, although it is now considered somewhat outdated. This criticism, together with concerns regarding the generalizability of stimuli (regarding race, age, sex, quality of images), has led to the subsequent development of a range of different facial emotion stimulus sets including Karolinska Directed Emotional Faces stimulus set [25], FACES database [26], Radboud Faces Database [27], and NimStim Face stimulus set [28], all of which are freely available for research use. With the exception of a few isolated reports [29, 30], the literature on how bodily expressions are processed has only recently emerged. A limited number of stimulus sets are available [e.g., the Bodily Expressive Action Stimulus Test (BEAST) [31], although application of these tests in neurodegenerative syndromes is relatively scant.

The bulk of behavioral studies use explicit tasks to assess emotion recognition (see Table 1). At the most basic level, emotion *detection* involves recognizing the presence of an affective cue. Emotion detection can be assessed by simultaneously presenting a participant with two pictures, one emotional and one neutral, and asking the participant to indicate which picture is displaying an emotion. Emotion *discrimination* involves distinguishing between emotions. Here, participants are shown two emotional pictures and are asked to determine whether the pictures represent the same emotion (see Fig. 2). Emotion *matching* refers to the ability to perceive emotions as identical. A commonly used task for emotion matching is the XAB format: one emotion picture is presented on top with two emotion pictures presented underneath. Participants are then instructed to indicate which picture matches the top picture. Detection, discrimination, and matching of emotions can be assessed

Task	Face-Perception Task	Face-Matching Task	Emotion-Matching Task	Emotion-Selection Task
Stimulus Example	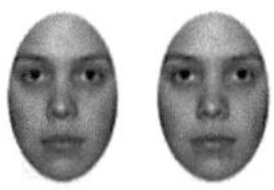	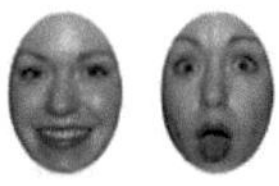	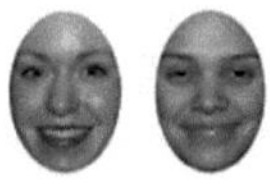	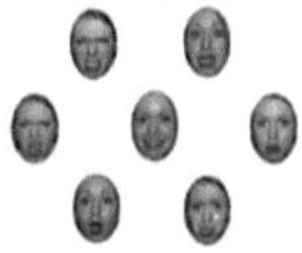
Instruction	"Is this the same picture?"	"Is this the same person?"	"Is this the same emotional expression?"	"Point to the angry face"

Fig. 2 Tasks to assess face and emotion processing. Example face- and emotion-processing stimuli. Images are from the NimStim database www.macbrain.org. Reproduced with permission from Oxford University Press: [181]

without the use of verbal labels. The following task formats typically require a more explicit use of verbal semantic processing. Emotion *selection* is evaluated by presenting a range of emotional pictures and a single verbal label (see Fig. 2). The participant indicates the picture that corresponds to the label. Emotion *categorization* refers to the classification of emotions into distinct categories and is assessed by presenting one emotional picture with a number of verbal emotion labels. In summary, there are many ways to evaluate emotion processing, and selecting one of them depends on the research question and study population.

Finally, experimental designs are moving toward the development of more valid and ecological tests combining facial emotional expressions, vocal prosody, and emotional body language. One of the most well-known is The Awareness of Social Inference Test (TASIT) that consists of short video vignettes of actors displaying different emotions [32]. Participants watch the video clips and then select which emotion was being expressed. TASIT is a very sensitive test to evaluate emotion recognition in neurodegenerative diseases. More recently, a shortened version of TASIT has been developed—TASIT-S—using Rasch analysis to minimize the number of items while maintaining the overall structure of the original task [33]. This new version is also appropriate for assessing people with dementia [34].

2.3 Theory of Mind

Theory of mind (ToM) refers to the ability to attribute beliefs and mental states to oneself and others [35]. First-order ToM is defined as the ability to understand that someone can have an inaccurate belief or "false belief" [36]. A typical scenario to evaluate false-belief ability is as follows: Person A puts an object (e.g., a candy bar) at location X (e.g., the cupboard). When Person A is not looking, Person B moves the object to location Y (e.g., the fridge). The critical question is "Where will Person A look for the object?" [37]. Participants who indicate "the fridge" are impaired in first-order false-belief processing. Second-order ToM consists of the ability to make

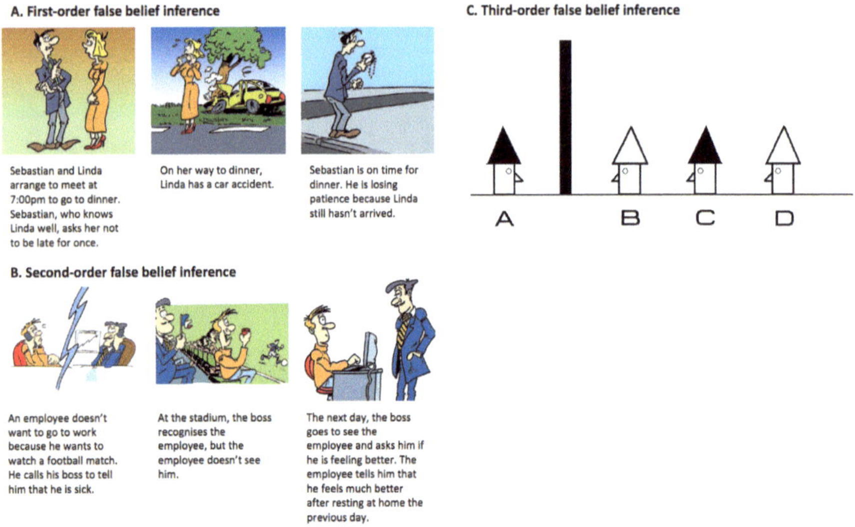

Fig. 3 Tasks to assess theory of mind (ToM). (**a**). Example of a first-order ToM task from the TOM-15. (**b**). Example of a second-order ToM task from the TOM-15: [231]. (**c**). Example of a third-order ToM Task: "Four men buried up to their necks in the ground. They cannot move, so they can only look forward. Between A and B is a brick wall which cannot be seen through. They all know that between them they are wearing four hats –two black and two white– but they do not know what color they are wearing. Each of them knows where the other three men are buried. In order to avoid being shot, one of them must call out to the executioner the color of their hat. If they get it wrong, everyone will be shot. They are not allowed to talk to each other and have 10 min to fathom it out. After 1 min, one of them calls out. Do you know which one of them? Why is he 100% certain of the color of his hat?" (https://www.mycoted.com/Four_Men_in_Hats). Reproduced with permission from Mycoted Ltd

inferences on other people's beliefs regarding the mental state of a third person (see Fig. 3a, b for examples of first- vs. second-order ToM, respectively).

Finally, third-order (and higher-order) ToM refers to the ability to infer others' mental states in complex social interactions (see Fig. 3c). Several reports suggest an association between ToM functions and executive functions, including neurological populations [38], in which executive performance should be considered when interpreting task performance. In addition to first-, second-, and third-order, ToM abilities are often further parsed into cognitive and affective components, with the former referring to the beliefs and intentions of others, and the latter to the emotional states of others. Affective ToM borders on the concept of empathy, although the emphasis in ToM is on the knowledge of the affective state of others, whereas empathy is primarily associated with the "feeling" of others' experiences.

Many ToM measures exist (Table 1), although some of the most commonly employed in dementia include the Reading the Mind in the Eyes Test (RMET) and the Faux Pas Test. The RMET assesses participants ability to deduce both basic and complex mental states after viewing a cropped image of the eye region [39]. The Faux Pas Test was originally developed in children [40] and has been modified for use in dementia research ([41], see also Mini-SEA [42]). In brief, participants read or listen to social interactions between two or three characters and identify instances where the speaker has said or done something inappropriate (faux pas). This test requires individuals to appreciate the potential difference in knowledge possessed by the speaker and the listener, as well as recognizing the emotional impact of the faux pas on the listener.

2.4 Empathy

Empathy involves understanding and responding to the emotional experience of another person [43]. Definitions of empathy vary within the literature; however, it is generally agreed that empathy can be parsed into separable components including (1) an affective component, which involves responding to the emotional experience of another person; (2) a cognitive component, which involves understanding the perspective of another person; and (3) the ability to regulate one's own emotions [43]. Cognitive empathy may be considered synonymous with affective theory of mind [44–46].

Empathy is routinely assessed via questionnaires (see Table 1). The Interpersonal Reactivity Index (IRI) is widely used in neurodegenerative disorders and consists of four subscales, which measure distinct but interrelated components of empathy: (1) perspective taking, the tendency to adopt the perspective of another person; (2) fantasy, the ability to connect with the feelings and actions of a fictitious character; (3) empathic concern, the feeling of warmth and concern for the misfortune of others; and (4) personal distress, the feeling of anxiety and unease in intense interpersonal situations [47]. Notably, while some studies employ all four subscales (e.g., [48–52]), others omit the fantasy and personal distress subscales (e.g., [53–55]). Baron-Cohen and Wheelwright [56] argue that the fantasy and personal distress subscales measure imagination and emotional control, whereas Jolliffe and Farrington [57] propose that the empathic concern subscale measures sympathy rather than empathy, per se. In light of these limitations, newer scales have been developed, such as the Empathy Quotient (EQ) [56] and the Basic Empathy Scale (BES) [57], which both measure cognitive and affective empathy. Other self-report empathy questionnaires include the Balanced Emotional Empathy Scale (BEES) [58] that focuses solely on the affective components of empathy. Although self-report measures are advantageous in that they are relatively short and easily administered, they are susceptible to subject bias and social desirability, which may influence responding. Furthermore, in neurodegenerative disorders, impaired insight needs to be considered [59, 60]. Therefore, studies often assess empathy in these patients based on carer- or informant-rated questionnaires [48, 53], as this method is considered a reliable and effective way of measuring empathy [55]. In addition, a combination of both patient- and carer-rated questionnaires may also be used to uncover possible discrepancies between points of view [59, 60].

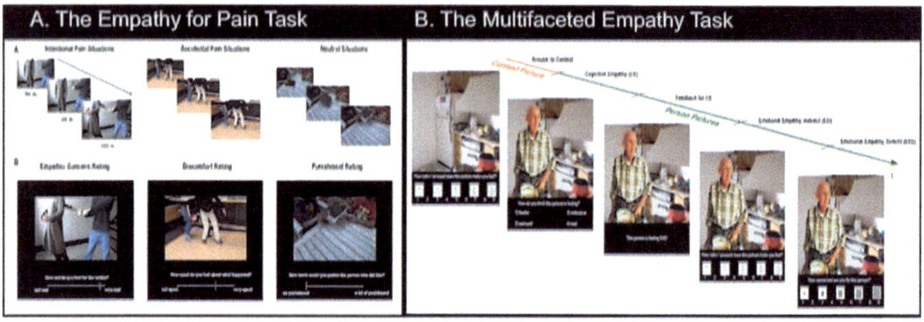

Fig. 4 Tasks to assess empathy. (**a**). Examples of the visual stimuli used in the Empathy-for-Pain-Task (EPT) and questions to probe different aspects of empathy using a computer-based visual analogue scale [65]. (**b**). Example of an item included on the Multifaceted Empathy Task (MET). Reproduced with permission from Springer: [61]

More recently, experimental tasks have been developed to assess empathy objectively, such as the Multifaceted Empathy Test (MET) [61] and the Empathy-for-Pain Task (EPT) [62] (Table 1; Fig. 4). In the MET, individuals view pairs of photographs illustrating real-life situations: first viewing a context, followed by a person embedded in the context. Individuals are then asked about the mental state of the person depicted, followed by the degree of empathic concern they feel for that person (see Fig. 4b, [61]). Another widely used method to measure empathy is assessing individuals' responses when observing pain in others (see for review [63]). These tasks provide a unique opportunity to study empathy, as there is evidence for distinct but overlapping neural mechanisms involved in the first-hand experience of pain and in empathy for pain in others (see for meta-analysis [64]). In the EPT referred here [62], empathy is assessed via individuals viewing and responding to a series of intentional and accidental harm situations (Fig. 4a, [65]). Future research combining objective and subjective measures of empathy may offer a more accurate picture of this deficit in individuals with dementia.

3 Clinical Syndromes

3.1 Behavioral-Variant Frontotemporal Dementia

bvFTD is the most common subtype of frontotemporal dementia [66] and is characterized by early and prominent deterioration of personality and social behavior. Diagnosis of this condition typically falls within the ages of 60–64 [67], with age of diagnosis ranging from the early 40s up to 80 years of age [68]. Prevalence estimates are mostly based on UK data and suggest rates of ~10 to 15/100,000 [67]. Patients with bvFTD typically present with behavioral disinhibition, apathy or inertia, and loss of sympathy or empathy. Perseverative or compulsive behavior, as well

Table 2 Clinical characteristics and cognitive profile

	Behavioral-variant frontotemporal dementia	Semantic dementia	Huntington's disease	Alzheimer's disease
Pattern of atrophy	Ventromedial prefrontal cortex, insula, and anterior temporal lobes	Bilateral, asymmetric temporal lobe atrophy (atrophy usually left > right)	Striatum (caudate and putamen)	Bilateral medial temporal lobe (including hippocampus); posterior cingulate, precuneus
Orientation	Intact	Intact	Intact	Impaired
Attention	Mild impairment	Intact	Impaired	Mild impairment
Language	Intact	Mod-severe anomia; single-word comprehension deficits	Slightly impaired	Mild-mod anomia
Visuospatial function	Intact	Intact	Intact	Moderate navigational difficulty
Episodic memory	Variable	Intact for nonverbal material	Moderate impairment	Profound impairment
Semantic memory	May be affected with disease progression	Severely impaired	Mild impairment	Mild impairment
Executive function	Impaired	Relatively intact	Impaired	Variable

as hyperorality or dietary changes, can also be a part of the clinical picture [69]. Cognitive impairment tends to be in the domains of attention and executive functioning (Table 2). Although memory was previously thought to be relatively spared, more recent evidence suggests that episodic memory deficits are present in a proportion of patients [70]. Brain atrophy is most pronounced in the medial prefrontal and orbitofrontal cortex, insula, anterior temporal regions, and striatum [71]. Distinct anatomical subtypes of bvFTD based on the relative involvement of temporal or frontal atrophy have also been reported [72].

3.1.1 Face Processing

Few studies have systematically studied face processing in bvFTD, although several have assessed facial identity matching or discrimination tasks when investigating emotion recognition (e.g., Benton Facial Recognition Task). While most studies do not show significant impairment compared to controls (e.g., [73–75]), others do (e.g., [76, 77]), with some evidence that perceptual deficits measured using a face-matching task contribute to facial emotion-matching performance in bvFTD [77]. It is possible that on some tasks, a piecemeal face-matching approach is sufficient for successful performance, and thus impairments are not detected.

Using a more systematic approach, impaired facial identity discrimination together with reduced ability to learn novel faces on the Cambridge Face Memory Test has been uncovered in bvFTD [78]. Neuroimaging analyses demonstrated that identity discrimination was associated with integrity of the left orbitofrontal region extending into the left temporal pole, the anterior cingulate, and the anterior portion of the left fusiform gyrus [78], a region associated with the core face-processing network [79]. Recognition of unfamiliar faces was related to gray matter integrity in the left temporal pole extending into the orbitofrontal cortex and the left insula [78]. Some evidence also suggests that bvFTD patients have difficulty recognizing familiar faces, with a specific deficit in familiarity and face-name matching for famous faces [19]. In contrast, the same study showed that face shape discrimination and facial identity matching were within normal limits. Both familiarity and face-name matching correlated with gray matter volume of the bilateral anterior temporal cortices [19]. Together, these findings suggest that aspects of face processing are affected in bvFTD (for review see [11]).

3.1.2 Emotion Recognition

A generalized deficit in recognition of basic emotions has been repeatedly demonstrated in bvFTD (for review see [80]). So-called negative emotions are more consistently impaired, although findings across specific emotions are inconsistent. Recognition of happiness has sometimes been found to be intact (e.g., [81]), while recognition of surprise is equivocal [74, 82]. The majority of studies have used photographs of facial expressions to assess emotion matching, emotion selection, or emotion labeling with varying levels of difficulty (e.g., [73, 75–77, 83, 84]). Importantly, the emotion recognition deficit is not modality-specific, with tasks using music, nonverbal vocalizations, and body postures yielding similar results (e.g., [21, 85, 86]). Moreover, emotion recognition performance appears to be correlated across modalities [21, 74]. Thus, although cognitive processes such as attention, executive functioning, semantic processes, or perceptual face processing probably contribute to emotion recognition performance to some extent, evidence suggests that bvFTD patients exhibit a primary emotion-processing deficit.

Emotion recognition deficits have been associated with atrophy in regions involved in face processing, such as the inferior temporal cortex [73], as well as regions involved in primary emotion processing, including the amygdala, orbitofrontal cortex, and insula [84, 87]. The latter two regions have not only been implicated in recognition of facial emotional expressions but also emotions conveyed through music [86]. The inferior frontal gyrus, a region bordering the anterior insula and associated with experience of emotion, has also been associated with face and body emotion-matching tasks [21]. In addition, a recent study using dynamic emotional body expressions as stimuli found that emotion detection and emotion categorization correlated with gray matter volume in the anterior temporal lobe and inferior frontal gyrus, respectively [88].

With regard to the functional neural correlates of emotion-processing deficits in bvFTD, implicit processing of facial emotional expressions during task-based functional magnetic resonance imaging (fMRI) was associated with decreased activation in distinct frontal and limbic regions, as well as in the ventral visual stream, specifically the fusiform cortex, compared to controls. Increased activity in posterior parietal regions was also observed and proposed to reflect an increase in attentional processes in bvFTD [89]. More recently, contrary to controls, implicit processing of emotional facial stimuli was not associated with enhanced activation in face-responsive regions in comparison with the activation for neutral stimuli in bvFTD. Notably, however, the increase of activation by emotional stimuli in the fusiform cortex was positively correlated with amygdala gray matter volume in bvFTD, showing a direct association between anterior atrophy and alterations in emotion processing in distant regions [20]. In summary, bvFTD shows widespread multimodal primary emotion recognition impairment, which is associated with structural and functional changes to key regions within the "social brain."

3.1.3 Theory of Mind

Patients with bvFTD experience difficulties on measurements for both cognitive and affective components of ToM compared to controls, including false-belief tasks, ToM cartoons and stories, faux pas comprehension, RMET, and sarcasm detection (e.g., [41, 90–92]). The magnitude of ToM impairment is large and has been demonstrated across different domains, modalities, and task types (for meta-analyses see [93, 94]). This profile of performance is evidence of a robust and generalized deficit, although impairment appears to be even greater on more advanced tasks, such as faux pas comprehension and sarcasm detection, which may in part reflect inadequate use of social norms.

Impaired ToM cannot be easily attributed to global cognitive decline, as performance on tasks matched for cognitive demands, which do not require mentalizing, are not typically impaired in bvFTD. Moreover, cognitively undemanding tasks (e.g., preference judgments based on eye gaze direction), which require mental state attribution, are also impaired [95]. Nonetheless, the relationship between executive dysfunction and ToM deficits in bvFTD has often been debated. Although executive functioning is thought to contribute to the ability to perform some ToM tasks, such as understanding a story (e.g., [90]), most studies have not found that ToM performance is dependent on executive functioning. A recent hierarchical cluster analysis demonstrated that ToM and executive functioning are largely distinct components [96]. However, some overlap between subcomponents of the Faux Pas Test and aspects of executive functioning (e.g., verbal abstraction, working memory/attention) were identified [96]. Interestingly, when subcomponents of ToM including (1) inferring someone else's belief and (2) inhibiting one's own mental perspective are assessed, bvFTD patients are selectively impaired in the latter component which appears to be strongly correlated with inhibition on a Stroop task [97].

Case studies in bvFTD have suggested that ToM performance is related to integrity of the orbitofrontal and anterior temporal cortex including the amygdala [98, 99]. These results have been largely confirmed by group studies which have demonstrated that ToM performance (RMET, cartoon task) is associated with integrity of the ventromedial and orbital regions of the prefrontal cortex as well as the anterior, lateral, and ventral temporal cortex with a right hemispheric dominance [100, 101]. In addition, Le Bouc et al. [97] showed that in a cohort including bvFTD, impaired ability to infer someone else's beliefs is correlated with hypometabolism in the left temporoparietal junction, whereas impaired self-perspective inhibition is correlated with hypometabolism in the right lateral prefrontal cortex. Thus, ToM impairment in bvFTD is widespread and related to integrity of ventromedial and anterior temporal cortices and may reflect an inability to inhibit one's own mental state/preference to make judgments about others' mental states.

3.1.4 Empathy

Empathy deficits are a well-established clinical feature of bvFTD [69]. Most studies to date have reported deficits in both cognitive and affective empathy on the IRI [53–55, 59, 90]. These findings have been interpreted as reflecting dissociable brain regions underlying cognitive versus affective empathy impairments. For instance, cognitive empathy has been associated with the integrity of a widespread neural network including the right dorsolateral prefrontal cortex [54] and bilateral fronto-insular, temporal, parietal, and occipital structures [53], whereas affective empathy has been associated with the integrity of the right superior medial prefrontal cortex and left supplementary motor cortex [54], as well as left orbitofrontal cortex, inferior frontal gyrus, insular cortex, and bilateral mid-cingulate gyrus [53].

On experimental tasks, a similar picture of deficits in cognitive and affective empathy within this syndrome exists. For instance, bvFTD patients were impaired on both cognitive and affective empathy components of the EPT [65], with lower levels of cognitive empathy associated with greater atrophy of the right amygdala and anterior paracingulate cortex and lower levels of affective empathy reflecting atrophy in the left orbitofrontal cortex [102]. Moreover, bvFTD patients showed a global cognitive empathy deficit on the MET, with impairment in affective empathy for negative stimuli [103]. Cognitive mechanisms underlying the observed empathy deficits within this syndrome are under continuing investigation. For example, some studies have proposed affective empathy is a core deficit within bvFTD, whereas deficits in cognitive empathy may reflect executive dysfunction ([65], but see [90]). More recently, disruptions in both cognitive and affective empathy have been shown to persist despite controlling for global cognitive impairment [53], suggesting an inherent empathy deficit within this group.

3.2 *Semantic Dementia*

SD also falls within the frontotemporal dementia spectrum and is characterized by the progressive deterioration of semantic knowledge [104, 105]. SD has a mean onset age of ~64 years [67, 106], although this can range from 40 to 80 years of age [106]. The prevalence of SD is difficult to estimate. Existing epidemiological studies have been primarily conducted in the UK [66, 67] and suggest that the prevalence of primary progressive aphasia is ~10/100,000, with SD accounting for approximately one-third of these cases. Clinically, SD patients present with anomia and single-word comprehension deficits and may also have impaired object knowledge, surface dyslexia, and/or dysgraphia [104] (see Table 2). SD patients may also present with behavioral dysfunction including apathy, disinhibition, irritability, depression, as well as changes in eating and repetitive and aberrant motor behaviors [106–108]. While the majority of patients present with asymmetric left-lateralized temporal lobe atrophy (referred to as left-SD), around 30% of patients present with greater right-lateralized atrophy (referred to as right-SD) [109]. With disease progression, left-SD shows cortical thinning of the temporal lobes bilaterally [110–112]. In contrast, right-SD shows widespread bilateral atrophy, affecting areas such as the right orbitofrontal cortex and anterior cingulate [112].

3.2.1 Face Processing

Overwhelming evidence indicates that recognition of famous faces is impaired in SD (e.g., [84, 113, 114]). Interestingly, prosopagnosia seems to be more common in individuals with right-sided atrophy [113, 115]. Whether this deficit reflects a loss of semantic knowledge or damage to earlier stages of visual analysis of faces (i.e., associative vs. apperceptive prosopagnosia) [116] is yet to be fully elucidated. On formal testing of famous face recognition, however, the degree of right temporal atrophy has been associated with recognition deficits, whereas the degree of left temporal atrophy appears to be associated with naming deficits only [17], suggesting that a combination of mechanisms may be involved. Early studies using the Warrington Recognition Memory Test to examine face memory for novel faces suggested that right-, but not left-SD patients were impaired [117]. More recently, however, attempts have been made to investigate learning and recognition of novel faces in SD using the Cambridge Face Memory Test and revealed a specific impairment in learning and recognizing novel faces compared to other complex objects (i.e., cars) [78]. Moreover, performance was associated with integrity of the fusiform gyrus and bilateral temporal pole. Future studies which systematically assess early stages of visual analysis of faces are warranted [11, 118].

3.2.2 Emotion Recognition

The vast majority of studies assessing emotion recognition in SD have employed facial emotion recognition tasks [73, 86, 119]. Generally, SD patients show greater impairment in recognizing negative than positive emotions; however, whether this reflects an emotion-specific impairment or task difficulty remains poorly understood (for review see [80]). Interestingly, deficits remain even when the emotional expression is intensified, which has been interpreted as evidence that abnormalities in face processing/inattention alone do not account for the observed emotion recognition deficits [120]. Indeed, despite evidence of face-processing impairments in this syndrome, emotion recognition impairments are observed across modalities. For example, patients with SD also have shown altered emotional recognition from music [85, 86], nonverbal sounds [121], words [122], and even abstract art [123] stimuli. These abnormalities appear to influence other areas of cognition, with emotional memories also being affected [124]. Furthermore, evidence suggests that emotion recognition deficits persist in SD patients, even on ecologically valid tasks, which combine facial emotional expressions, vocal prosody, and emotional body language [125].

In a mixed group of frontotemporal dementia patients, including 11 SD patients, emotion-specific neural correlates were identified, with fear recognition associated with amygdala integrity, whereas disgust recognition was associated with insula integrity ([73], see also [84, 87, 119]). Research in SD has also shed light on the special role of the right temporal pole in understanding emotion, with right-SD patients showing greater impairment in emotion recognition than left-SD patients [126]. These findings demonstrate that patients with SD experience profound and widespread deficits in the ability to recognize emotional signals across modalities. Furthermore, the presence of emotion recognition impairments is associated with degeneration of regions known to be essential for emotion processing, including the amygdala, insula, fusiform gyrus, and right temporal pole.

3.2.3 Theory of Mind

While examination of ToM in SD is scant, emerging evidence suggests that both affective and cognitive dimensions are affected in this syndrome. Duval et al. [127] conducted the first comprehensive study in a group of 15 SD patients and found widespread deficits including attribution of intentions, first- and second-order ToM on a false-belief task, lower affective ToM on an adapted version of the RMET, and impaired performance on the "Tom's taste" task. Notably, these deficits were in contrast to their performance on control tasks, suggesting that abnormal performance is not due to semantic impairments or other task demands. Moreover, SD patients demonstrated reduced insight into their ToM deficits [127]. A subsequent study examined the neural correlates of performance on a cartoon ToM task (see [90] for task details) and found that performance was related to the integrity of the right anterior temporal lobe, amygdala and left orbitofrontal cortex, and insula, when semantic deficits were accounted for [101]. More recently, affective ToM was associated with integrity of the left amygdala and hippocampal/parahippocampal regions, whereas cognitive ToM was associated with the right medial prefrontal cortex, anterior cingulate, and inferior frontal cortex [128]. In addition, analyses using resting-state fMRI found that affective ToM impairments were associated with decreased functional connectivity to the medial posterior cingulate/precuneus, whereas cognitive ToM impairment was associated with reduced functional connectivity with a distributed number of regions including left frontal, temporal, and limbic regions [128]. Together, these results demonstrate that ToM is impaired in SD, although the neural mechanisms giving rise to these impairments are only beginning to be understood.

3.2.4 Empathy

Few studies have investigated empathy in SD, despite increasing observations of behavioral changes in this syndrome. Rankin and colleagues [49] reported significantly decreased cognitive and affective empathy on the IRI in SD patients compared to an Alzheimer's disease group and healthy controls. These deficits were interpreted as reflecting atrophy of the anterior temporal lobes, amygdala, and ventromedial orbitofrontal cortex, although neuroimaging results were not reported. Other studies have shown decreases in cognitive and affective empathy in SD, although these differences did not reach statistical significance [54, 59], which may have been due to inadequate power ($n < 15$). Interestingly, the integrity of the right temporal lobe in SD has been associated with lower global empathy on the IRI [55]. Only one study to date has compared right- and left-SD patients and found lower levels of affective empathy in right-SD patients [126]. Further investigation into profiles of empathy in left-SD vs. right-SD combined with neuroimaging techniques may shed further light into the role of the right temporal pole in empathy.

3.3 Huntington's Disease

Huntington's disease (HD) is a rare neurodegenerative disorder characterized by motor symptoms including involuntary movements, cognitive impairment, and socio-behavioral symptoms. Prevalence is estimated to be between 1/10,000 and 1/20,000 in the Caucasian population [129]. HD is an autosomal dominant inherited disease caused by a trinucleotide CAG (Cytosine, Adenine and Guanine) repeat expansion (36 repeats or more) on the short arm of chromosome 4p16.3 in the Huntingtin gene. The higher the number of CAG repeats, the earlier the clinical onset of the disease [130]. Although the mean age of onset of motor symptoms is around the forties, some patients enter the symptomatic phase of the disease before the age of twenty (juvenile form), while other patients show their first symptoms in their eighties. There is currently no cure, and the evolution of the disease leads to death after 15–20 years. Although mild cognitive and behavioral symptoms can be detected before the occurrence of motor symptoms [131], the clinical onset is marked by motor disturbances including chorea, dystonia, dysarthria, or gait disturbances. Cognitive symptoms of HD are mainly characterized by executive dysfunction causing a severe loss of mental flexibility, associated with perseverative ideation and behavior. Also other cognitive domains such as language, declarative memory, and psychomotor speed can be altered [132] (see Table 2). Behavioral symptoms of HD are often severe and include depression and suicidal behavior, apathy, irritability and aggression, anxiety, obsession and compulsions, and, in rare cases, delusions and hallucinations. In order to find suitable biomarkers related to the onset of the disease, many studies compare performance in presymptomatic HD carriers (preHD) with both symptomatic patients (HD) and controls.

Anatomically, HD is considered a model of subcortical striatal degeneration involving the caudate nucleus and putamen [133]. However, brain imaging studies have shown that the neuropathology of HD also involves extrastriatal regions including the amygdala, thalamus, insula, and occipital regions from early and premanifest stages of the disease [134, 135]. In symptomatic HD patients, cerebral atrophy spreads to cortical regions including the inferior frontal, premotor, sensorimotor, mid-cingulate, frontoparietal, and temporoparietal cortices [136].

3.3.1 Face Processing

Clinically, HD patients rarely report difficulties in recognizing people in their daily life or recognizing famous faces. PreHD carriers do not show deficits on the Benton Face Recognition test [137–139] or the Warrington Recognition Memory Test [137, 139]. However, symptomatic HD patients consistently present with deficits on the Benton test [75, 138, 140–142] and Warrington Recognition Memory Test [138] in comparison to controls. In another study, a mixed group of preHD carriers and HD patients showed no deficits on the Benton test; however, lower performance was observed as the likelihood of disease onset increased [143]. In addition, HD patients exhibited impairment in recognition of neutral faces on the Karolinska institute test compared to controls [144]. It is unclear whether this deficit in face recognition is the consequence of a specific alteration in face processing or whether it reflects a more general impairment in perception [144] or difficulty in selecting answers according to the context [145]. Interestingly, preHD carriers close to disease onset show early occipital involvement, as well as widespread white matter abnormalities [146]. Further investigations are needed to better understand the nature of face-processing deficits in HD.

3.3.2 Emotion Recognition

Numerous studies have explored emotion processing in HD; however, the typical profile has evolved in recent years. Initially, several reports suggested a specific impairment in the recognition of disgust in both preHD carriers and HD patients [137, 140]. Recent studies, however, reported evidence against this view [136, 138, 139, 147, 148]. A recent meta-analysis confirmed that HD is associated with impaired processing of all emotions [149], specially the negative ones. Impaired recognition of happy faces is rather mild and may reflect task difficulty. Importantly, there is conflicting evidence from recent studies with HD patients showing an absence of emotion recognition impairments when tested with more ecological tasks providing contextual cues, such as TASIT [150, 151] (for additional discussion of contextual influences on emotion recognition in HD, see section: "Real world assessment of social cognition" below). PreHD carriers are also impaired in recognition of negative emotions, especially anger, disgust, and fear [147, 149]. Notably, longitudinal studies of preHD carriers have revealed that emotion recognition is the only cognitive measure that changes with disease progression over a 36-month follow-up [146]. Interestingly, the deficit in emotion recognition extends to voices [149, 152] and angry bodies [153]. Voxel-based morphometry studies have shown that better emotion recognition correlates with integrity of the striatum in HD patients, with some differences according to emotion [138]. No anatomical correlates of emotion processing have been found in preHD carriers using this technique [154]. FMRI studies, by contrast, suggest that distributed extrastriatal networks are involved in emotion recognition in both preHD carriers [154] and HD patients [136].

3.3.3 Theory of Mind

HD patients consistently show deficits in both affective and cognitive ToM, while preHD carriers do not (for a meta-analysis, see [149]). Deficits in ToM and executive functions are often correlated in HD (e.g., [155–157]). Furthermore, ToM impairment is associated with older age, verbal fluency deficits, and more severe motor symptoms [149]. Notably, Eddy and Rickards [158] found that preHD carriers were impaired on the Faux Pas Test and the RMET when compared to controls, despite relatively preserved executive function. Although additional studies are needed to confirm the possible impact of HD on ToM reasoning in preHD carriers, Bora et al. [149] reported a trend for a ToM deficit in this group, with more impaired performance in participants with probable disease onset within 5 years. Finally, little is known about the neuroanatomical correlates of ToM in HD. To date, only one neuroimaging study has been conducted and reported lower performance in affective ToM, which was associated with abnormal connectivity between the left amygdala and the right fusiform face area [159].

3.3.4 Empathy

Despite the social interaction deficits observed in HD, few studies have explored empathy to date. In preHD carriers, global empathy scores on the IRI [51], the BES [51], or the EQ [160] were not impaired. Yet, when analyzing IRI subscales, Eddy and Rickards [158] found reduced everyday perspective taking, lower empathic concern, and higher personal distress in preHD carriers compared to controls. HD patients did not show empathy deficits on the global score of the EQ [160], the BEES, or the IRI [148]. However, compared to matched controls, HD patients showed deficits in cognitive empathy and social skills subscores on the EQ [160]. Finally, in another study, HD patients showed impaired performance on the EPT compared to controls, but this impairment was not seen in healthy relatives of HD patients [151]. Furthermore, the results suggested that lower performance was likely due to a difficulty in intentionality detection rather than in empathic concern [151]. Interestingly, this study of HD patients did not find any correlation between emotion recognition performance and empathy-for-pain performance. Further studies are clearly needed to confirm empathy profiles in HD and explore the potential relationship with ToM, as well as the associated neural correlates.

3.4 Alzheimer's Disease

AD is the most common form of dementia and accounts for approximately 50–60% of all dementia cases. The greatest risk factor for AD is age [161]. Onset of AD typically occurs after 65 years of age, although approximately 50% of people diagnosed with dementia prior to age 65 have AD [66]. The prevalence of AD thus varies according to age. Recent statistics indicate that 1 in 9 people over age 65 have AD, which increases to one-third of people over age 85 [161, 162]. Clinically, patients with AD present with a profound inability to lay down new memories, which manifests as episodic memory impairment (see Table 2). In addition, patients may show a degree of anomia, decline in navigation, and spatial orientation as well as executive dysfunction [163–165]. From a behavioral perspective, early and prevalent behavioral changes are exclusion criteria for AD [165], although agitation, apathy, and anxiety are relatively common [166]. Usually, bilateral medial temporal lobe atrophy, including the hippocampus, as well as the posterior cingulate/precuneus, is observed, with some evidence of similar patterns prior to a confirmed clinical diagnosis [167, 168]. With disease progression, cortical thinning is observed in the bilateral parietal and occipital regions, extending more anteriorly into the medial and lateral temporal cortices [169].

3.4.1 Face Processing

Investigations of face processing in AD have predominantly focused on memory and recognition of faces. Clinically, AD patients show increased difficulty to recognize friends and family [170]. This deficit has been confirmed on formal testing, with some (but not all) AD patients showing impaired performance on famous face recognition tasks, as well as impaired ability to match unfamiliar faces (e.g., [171, 172]), which has been interpreted as reflecting episodic (or semantic) memory impairment [173, 174]. Recent evidence, however, suggests that earlier face-processing capacity may also be affected. A new study found that while healthy adults showed a greater face than object (cars) inversion effect, the magnitude was similar in AD [175], although the latter also presented slower and more errorful performance across all conditions. Indeed, electroencephalography studies have reported smaller N170 peak amplitudes in AD compared to controls ([176, 177], but see [178]) despite intact early visual processing of familiar faces in very mild AD using event-related potential [176]. Thus, evidence suggests that AD patients have difficulty learning and recognizing novel and familiar faces; however, whether this is due to impairment in earlier stages of face processing is less well established.

3.4.2 Emotion Recognition

The vast majority of research investigating emotion recognition in AD has employed facial emotion recognition tasks. On these, performance in AD patients has been mixed, with many studies reporting a decline compared to controls [77, 81, 83, 179–181]. Importantly, however, most studies have attributed the lowered performance as secondary to cognitive impairment, rather than reflecting a primary emotion-processing impairment per se (e.g., [182]), with a recent meta-analysis largely confirming this account [183]. Moreover, a recent longitudinal study has demonstrated that despite initially performing lower than controls on a facial emotion recognition task, performance remains relatively stable across the disease course [184]. While fewer studies have used non-face stimuli, existing evidence suggests that recognition of emotional prosody or emotional sounds (e.g., crying, laughing) is relatively well preserved, relative to their general cognitive decline [180, 185]. Finally, on more ecologically valid tasks of emotion perception, which use dynamic expressions of emotion (i.e., facial expressions, vocal prosody, and contextual cues), performance is intact or only mildly impaired [184, 186] and, again, does not appear to decline with disease progression [184]. Together, these findings demonstrate that, consistent with their relatively preserved social graces in day-to-day situations, patients with AD are able to recognize emotional cues and that, whenever such a skill is compromised, it is likely secondary to cognitive impairment.

3.4.3 Theory of Mind

ToM appears to be only modestly affected in AD. Early studies assessing false belief revealed impairments on second-, but not first-order tasks (e.g., [187]). Because second-order tasks are thought to place higher demands on executive functioning and working memory, this profile has been interpreted as reflecting general cognitive deficits rather than a specific ToM impairment (e.g., [187], but see [188]). More recently, Le Bouc et al. [97] used a novel ToM task which assesses: (1) representation of reality (i.e., identifying one's own belief, which would constitute the prepotent response), (2) belief inference (i.e., inference of another person's belief), and (3) self-perspective inhibition (i.e., inhibiting the prepotent response) [189]. The profile of performance suggested that AD patients have difficulty inferring someone else's beliefs, associated with lower left temporal parietal junction metabolism [97]. A recent meta-analysis, including 20 studies (402 AD patients), confirmed impaired ToM capacity in AD, although the level of impairment was less than seen in bvFTD [94]. Importantly, the meta-analysis also demonstrated that longer disease duration and greater general cognitive impairment were associated with worse ToM performance [94].

Whether patients with AD show different profiles regarding mentalizing about cognitive vs. affective states is an important consideration. Dodich et al. [190] employed the Story-based Empathy Task and found that while AD patients performed worse on both the intention attribution and emotion attribution conditions, impaired performance was at least in part accounted for by general cognitive deficits. Thus, emerging evidence suggests that in AD, ToM impairment often reflects general cognitive impairment, and therefore inclusion of control conditions is essential to correctly interpret profiles of performance.

3.4.4 Empathy

On carer-rated questionnaires, empathic concern in AD patients is rated similarly to controls [53, 55, 59, 60]. Ratings of cognitive empathy are more variable [55]; however, recent studies have revealed that when general cognitive ability is accounted for, perspective taking is within normal limits [53]. In addition, AD patients appear to have relatively good insight into their capacity for empathy, with a similar profile seen on self-report measures as carer-report measures [60, 191]. Interestingly, some evidence suggests that not only is empathic concern intact in AD, but some patients may show enhanced sensitivity to social and emotional cues [192, 193]. Paradoxically in these patients, the degree of increased "emotional contagion" (personal distress subscale of the IRI) has been associated with smaller volume of the right inferior, middle, and superior temporal gyri [192]. Such findings are complemented by objective measures of social interactions of empathy, with evidence of preserved mutual gaze during interactions between AD patients and their partner, during a conversation designed to elicit relationship conflict [194]. Thus, mounting evidence demonstrates that empathy is intact in AD. Whether these patients experience a

Table 3 Profiles of social cognition impairment in dementia syndromes

	Face processing	Emotion recognition	Theory of mind	Empathy
Behavioral-variant frontotemporal dementia	↓	↓↓	↓↓	↓↓
Semantic dementia	↓↓	↓↓	↓	↓
Huntington's Disease	↓	↓↓	↓	~
Alzheimer's Disease	↓	~[a]	↓/~[a]	~[a]/↑

Note. ↓ = impaired; ↓↓ = severely impaired; ~ = intact; ↑ enhanced
[a]After taking into account cognitive impairment

degree of enhancement in their capacity to empathize with others warrants future investigation. In summary, the findings in AD demonstrate the importance of considering the interaction between cognitive impairment and performance on tasks of social cognition when interpreting profiles of performance.

3.5 Summary of Findings

Social cognition deficits are common in individuals with neurodegenerative disorders (see Baez et al., this volume). Notably, however, subdomains of social cognition are not equally affected across disorders (Table 3). In bvFTD, impairment in social cognition is widespread and profound, reflecting the clinical phenotype, which is characterized by changes in (social) behavior and personality. In contrast, in SD, impairments in face processing and emotion recognition are common, with mild changes in theory of mind and empathy. Furthermore, in SD, the extent of social cognition impairment appears to be related to the lateralization of atrophy, with individuals with right-lateralized atrophy showing early and marked changes in social cognition, whereas these deficits are less apparent in individuals with circumscribed atrophy to the left temporal pole, although with disease progression all SD patients develop social cognition impairment [112]. In HD, the most significant deficits are observed in emotion recognition, while surprisingly, evidence for a decline in empathy is limited. Some theorists suggest that emotion recognition is necessary for empathy, which is contrary to existing evidence in HD and warrants further exploration. However, the few studies of empathy in HD to date have largely employed self-report questionnaires. Thus, it is possible that more sensitive, objective measures, which are not confounded by insight, may provide clearer understanding of the profile of social cognition deficits in this syndrome. Finally, in AD, the overwhelming evidence indicates that social cognition is relatively well preserved in the mild-moderate disease stages (with the possible exception of face processing). Hence, social cognition impairment is not inevitable in neurodegenerative syndromes. Rather, the quality and severity of social cognition impairment in these disorders directly reflects the pattern and spread of pathology over time.

4 The "Social Brain": Insights from Social Cognition Studies on Dementia

While neuroanatomical models of social cognition are typically inspired by functional imaging studies in healthy participants, structural neuroanatomical findings from the dementias provide critical information as to which areas are *necessary for* (as opposed to *involved in*) complex socio-cognitive functions (Fig. 1). Overall, the evidence from structural imaging studies in the dementias is in line with dominant models of isolated aspects of social cognition. Moreover, evidence from the dementias also adds to our theoretical and conceptual knowledge of brain behavior relationships of social cognition (see Baez, this volume).

As detailed earlier in this chapter, deficits in emotion recognition are common in dementia [119, 120, 140, 195], although it appears to be disproportionately affected in bvFTD, with the majority of studies focusing on face stimuli. The structural neuroanatomy of these emotion recognition deficits includes amygdala, insula, and inferior frontal gyrus [80], brain regions which overlap with current neuroanatomical models of (facial) emotion recognition in healthy participants [10]. Interestingly, recent evidence, particularly in bvFTD and SD, suggests that emotion recognition and face processing closely interact. While the face-processing model proposed by Haxby and colleagues [10] delineates between a "core system" for early face perception and an "extended system" which is involved in knowledge of emotional concepts (among other things), how these systems interact has been less considered. Emerging data from bvFTD and SD patients, however, suggests that disintegration of the extended system can influence functioning of the core system [11]. In this way, findings in the dementias are directly influencing social neuroscience models of healthy brain function.

Few studies to date have addressed ToM abilities in the dementias. The results indicate a necessary role for the posterior cingulate in intention attribution and a distributed set of regions (including amygdala, insula, IFG, medial prefrontal cortex, temporal pole, and thalamus) associated with recognition of ToM-related humorous cartoons [101, 196]. These brain regions show remarkable overlap with the areas that have been associated with emotion recognition. In this context, understanding of the interdependence between these subdomains of social cognition and their relative nodes within the social brain network warrants further investigation.

Finally, there appears to be some evidence that some aspects of social cognition may be enhanced in AD. These patients show increased "emotional contagion" and more appropriate mutual gaze during disagreements with their romantic partners. Recent evidence has emerged to suggest that in dementia, an enhancement of cognitive capacities can be observed, despite progressive decline in other domains (e.g., enhanced artistic skills in SD, in the context of a decline in language function) [193]. It is possible that in AD, decline in memory leads to a concurrent enhancement or reliance on intact social functioning, which helps to compensate for poor memory in social settings. At this stage, such an interpretation is speculative, but this intriguing hypothesis warrants future exploration.

5 Social Cognition in Clinical Syndromes: How Does Incorporating Social Cognition Improve Understanding of Clinical Phenotypes?

From a clinical perspective, much of the understanding of neurodegenerative syndromes has focused on general cognitive capacity and/or motor dysfunction. Indeed, consensus criteria rarely consider the relevance of formal assessment of social cognition, despite several of these syndromes presenting with changes in social behavior and emotional lability, as reviewed above. Over the last decade, with the emergence of social neuroscience, however, this has begun to change. For example, assessment of social cognition is now recognized as central in differentiating bvFTD from AD, a differential diagnosis that can be very challenging in the clinic. In bvFTD and AD, both memory impairment and executive dysfunction can be affected (e.g., [197, 198]), whereas social cognition is typically affected in bvFTD, but not AD. Patients with bvFTD show worse emotion recognition [81], sarcasm detection [199], and ToM [94] than AD. Moreover, on social cognitive screening tests, such as the mini-SEA, performance differentiates bvFTD from AD, even in patients with a similar degree of amnesia [200]. Thus, clinical assessment of aspects of social cognition can directly inform clinical diagnosis in these syndromes.

Given the progressive nature of these syndromes, more recent work has begun to understand how social cognition capacity changes with disease progression [112, 184]. This is important both from a clinical perspective, to gain a better understanding of the emergence of clinical symptoms, and also from a biological perspective, to inform how spreading of pathology through neuronal networks results in a decline in social cognition capacity. A longitudinal study investigating social cognition and social behavior in SD and AD revealed that emergence of emotion recognition deficits in SD is related to the degree of right temporal and right fusiform cortical thinning [112]. Notably, in AD, despite widespread atrophy and a decline in general cognition with disease progression, face processing, emotion recognition, and social behavior (motivation and stereotypical behaviors as measured by the Cambridge Behavioral Inventory) remained relatively intact over the disease course [112]. Similarly, when comparing bvFTD and AD groups, AD patients show relatively stable emotion recognition, although AD patients do show a decline in sarcasm detection with disease progression [184], which may reflect the cognitively demanding nature of this task. Together, these longitudinal studies demonstrate that in these syndromes, the emergence of social cognition impairment reflects spreading of pathology into the "social brain."

Importantly, social cognition assessment may help inform prognosis in some patients. In bvFTD, significant variability in the disease course has been recognized [201, 202]. Specifically, while some patients show a fairly predictable decline over 5–7 years, other patients show minimal decline over many years, despite presenting similarly at baseline. This second presentation has been referred to as the "phenocopy syndrome" and appears to have different etiology and pathology [203]. Of relevance here, this syndrome tends to be associated with minimal brain atrophy at presentation, and when bvFTD patients are divided according to the degree of brain atrophy at baseline and longitudinally assessed, bvFTD patients with limited atrophy show stable social cognition over time, whereas bvFTD patients with marked atrophy show a steady decline [184]. Thus, emerging longitudinal studies are beginning to provide important evidence that assessment of social cognition can inform the prognosis of patients with dementia.

A final important note from a clinical perspective is that changes in social cognition lead to difficulty in forming and maintaining social relationships and meaningfully participating in social interactions. Recent work has demonstrated that this loss of capacity not only affects the individual with dementia but also negatively impacts on carer burden and psychological well-being (e.g., [59, 204, 205]) (see Kemp, this volume). Moreover, impact on carer burden may be even greater in syndromes where behavioral and social changes are not emphasized, such as in SD, suggesting that psychoeducation may represent a mediating factor [59]. Interventions to improve social cognition in dementia syndromes are currently lacking, and behavioral management strategies to directly address social cognition impairment are also urgently needed. Given the widespread impact of social cognition impairment on carers, family, friends, and the wider community, research focusing on ways to manage or improve these profound and challenging symptoms is urgently needed.

6 Avenues for Future Research

6.1 The Potential Influence of Cognitive Impairment

While research to date has revealed important information about social cognitive functioning in neurodegenerative disorders, it has been postulated that these tasks have a significant executive function component and that deficits in social cognition might therefore at least partly be due to task demands, as seen in other syndromes (e.g., schizophrenia, [206]). Covariance between emotion recognition and cognitive performance can be accounted for with statistical analyses, although this approach may increase Type II error. Alternatively, inclusion of a nonsocial task that has similar cognitive demands can shed light on this issue. A deficit on one task and not the other indicates that both tasks are independent to some degree. Another alternative is the use of implicit tasks, such as oddball detection or gender recognition, with (task-irrelevant) social conditions. Although implicit tasks offer the advantage of similar executive demands across conditions, there is limited control over the (implicit) response strategy. Employment of a range of tasks and experimental designs will help to conclusively determine whether cognitive impairment, and specifically executive dysfunction, influences social cognition capacity across disorders.

6.2 *Real-World Assessment of Social Cognition: The Importance of Context*

One of the major limitations of existing tests of social cognition is that they lack ecological validity – with one noteworthy exception: TASIT [32]. In day-to-day situations, one is not shown a floating face and asked to match the emotion expressed to a verbal label. The potential impact of this lack of ecological validity has traditionally been considered of little consequence, as it was assumed that understanding of social cues (e.g., emotional expressions) was minimally influenced by contextual information [207, 208]. However, recent studies have demonstrated that contrary to these assumptions, interpretation of social cues is influenced by context [209, 210]. The potential influence of context has been recognized as a possible explanation for a long-standing contradiction in bvFTD. These patients are characterized by profound changes in behavior and social cognition. However, often in a formal clinical setting (e.g., the neurologist's office), such behaviors can be difficult to elicit. This conundrum has been proposed to reflect the high level of external control under formal testing conditions [211, 212]. While experimental studies to directly address these issues are only just emerging, evidence suggests that tests which include contextual information and have improved ecological validity can provide important insights into the profile of social cognition impairment in neurodegenerative disorders. For example in HD, despite showing lower emotion recognition of facial emotional expressions, patients are equally sensitive to contextual cues (i.e., body language) as controls [143]. This pattern of performance has been interpreted as evidence that HD patients show relatively intact processing of faces when they are embedded in context, reflecting intact low-level face processing in this syndrome. Conversely in bvFTD, patients show reduced ability to discriminate between when an individual is inflicting pain accidently or intentionally [65]. This may reflect a difficulty in interpreting ambiguous situations which depend on appropriate assessment of contextual information [213, 214]. Thus, future studies with valid ecological tasks that manipulate the degree of contextual information provided will be essential in broadening our understanding of social cognition impairment in neurodegenerative disorders.

The need for more "truly social" paradigms to assess social cognition, both in clinical syndromes and in healthy adults, is increasingly recognized (for an excellent review of second-person neuroscience approaches, see [215]). Many of the existing paradigms discussed here, and commonly used in clinical settings, take the view that participants simply *observe* other people in order to make social cognition judgments about others (see Cornejo et al., this volume). However, in real-world situations, these judgments are more likely to be based on *interactions* with another person, which enables continual updating of social judgments and the opportunity to test and update judgments dynamically. Thus, novel techniques that employ naturalistic, interactive paradigms, such as hyperscanning, virtual reality, mutual gaze paradigms, and interactions with avatars, are being developed, which are likely to offer important new insights into how people with dementia behave, in truly social situations.

6.3 Functional and Nuclear Brain Imaging of Social Cognition Across the Dementias

We are only beginning to understand how structural pathology affects functional properties of distant connected areas (e.g., [20]). The clinical symptoms in these neurodegenerative disorders undermine the implementation of task-based functional brain imaging. Yet, functional brain imaging studies are indispensable to elucidate the mechanisms underlying deterioration of social cognition (e.g., to investigate functional plasticity against the background of regional atrophy). Furthermore, considering the partial overlap between socio-cognitive phenotype and atrophic topography, studies across nosological categories may hold particular promise to reveal transdiagnostic and disease-specific characteristics of social cognition deficits. In addition, barring a handful of FDG-PET studies assessing resting-state metabolism, nuclear imaging studies investigating social cognition in dementia are lacking. Hence, knowledge about the neurotransmitter systems associated with socio-cognitive deficits in neurodegenerative disorders is nearly nonexistent. Future studies addressing this gap in the literature may also identify targets to inform the development of pharmacological interventions.

7 The Borderlands Between Social Neuroscience and Social Sciences in the Dementias

As discussed in this chapter, neurodegenerative disorders often result in disruptions to human social behavior that social cognition studies are able to explore. Importantly, investigating the mechanisms of social behavior in these syndromes may in turn enrich explorations in various domains of social sciences. Conversely, concepts and methods from social sciences may also provide new insight into the social breakdowns in dementia. In fact, merging borders between medicine and social sciences is increasingly commonplace. Here, we consider a few examples of such cross-disciplinary overlap.

7.1 Moral Judgment and Law

Moral judgment, a topic addressed in anthropology, philosophy, and social neuroscience, has been explored in patients with dementia. For example, altered moral judgment in bvFTD patients appears to be related to impaired affective ToM [216], as well as impaired integration of intentions and outcomes, which critically depends on areas beyond the ventromedial prefrontal cortex [217]. Moreover, bvFTD patients show impaired regret processing [218, 219]. Importantly, these abnormalities in moral judgment can lead to unlawful behavior in some patients, especially in bvFTD, SD [220], and HD [221]. Notwithstanding, many current legal systems may not consider these patients as lacking in responsibility for their acts because overall cognitive performance can seem preserved on formal testing. This dichotomy has important implications for medicolegal decisions relating to capacity and culpability in patients with dementia exhibiting moral judgment deficits [222]. The domains of social cognition, moral judgment, and law should therefore be examined together. In light of the aging population, issues around dealing with people with dementia in judicial and criminal situations are likely to increase. Indeed, some authors have suggested that the new onset of criminal behavior in an adult may represent frontal and/or anterior temporal brain disease, so that these individuals should be assessed for neurodegenerative conditions [220]. Further research in the social sciences is likely to improve conceptual understanding of these complex medicolegal issues (see Salles & Evers, this volume).

7.2 Stereotypes

While stereotypes have been long recognized as simplistic ideas or beliefs about others, their potential influence and relevance in dementia may have been previously underestimated. For example, negative age stereotypes, such as the culturally shared beliefs that aging is associated with cognitive decline and disease, may predict adverse outcomes among older individuals and may even influence brain health. A longitudinal study reported that participants holding more negative age stereotypes earlier in life had greater hippocampal volume loss and greater AD pathology (i.e., neurofibrillary tangles and amyloid plaques) later in life [223]. Stereotypes in the field of dementia are also being explored from a social science perspective. A recent study suggested that stigmata surrounding diagnosis of preclinical AD depends highly on the expected prognosis and highlights the need for models of Alzheimer's-directed stigmata to incorporate attributions about the condition's mutability [224]. Another study in people at risk for HD found that these individuals reported experiencing genetic discrimination and stigmatization in both institutional settings, such as when seeking employment and insurance, as well as in interpersonal relationships [225]. Studies such as these highlight how people with dementia are both at risk of being stereotyped and stigmatized by others and potentially being influenced by their own stereotypes of what to expect following a diagnosis of dementia. Collaborations with researchers investigating stereotypes in other aspects of society may help to inform the development of psychoeducation and interventions which help to foster the inclusion and positive engagement of these individuals in our societies.

7.3 Competition, Cooperation, and Communication

Competitive and cooperative behaviors are fundamental forces that shape the organization of both human and nonhuman primate societies (see Díaz-Gutiérrez et al., Billeke et al., this volume). Such behaviors have been investigated through diverse neuro-economics paradigms in dementia. In the ultimatum game paradigm, a first player receives a certain amount of money and can choose to either divide this money between him/herself and a second player; the second player can either accept or refuse this division of the money, a refusal yielding the loss of the money for both participants. On such tasks, bvFTD patients have shown deficits in the integration of social contextual information to guide normative behavior, such as prosociality and punishment [226]. Conversely, AD patients, during social games, engaged with other player, developed a friendly competition, and demonstrated social cooperation by helping and sharing knowledge with other players [227]. These findings provide further evidence of preserved social graces in this dementia syndrome. Finally, there is a growing interest in the communication abilities of patients with dementia, especially in the field of conversation analysis. Such explorations not only relate to information exchange between patients and other persons but offer a glimpse into various aspects of social interaction [228, 229].

As a whole, social science and social cognition studies tend to demonstrate that persons with dementia of various origins should be considered as minority groups with specific vulnerabilities that might preclude their proper inclusion in our normative societies. A transdisciplinary effort from both social neuroscience and social science perspectives will help to better understand, not only the challenges that face people with dementia but also the broader impact on society.

8 Final Thoughts

Social cognition impairments are common in neurodegenerative disorders. While the focus was once on traditional domains of cognition, such as memory and language, research on components of social cognition has rapidly emerged in these disorders, particularly over the last decade. In real-life situations, difficulties in participating successfully and meaningfully in social interactions directly impact on the functional capacity of patients and negatively impact on carer burden and psychological well-being. Hence, interventions and management strategies which target these skills are urgently needed (see Kemp, this volume). In addition, inclusion of clinically appropriate and valid social cognition tests is essential to better characterize these syndromes and inform diagnosis, prognosis, and management.

From a theoretical perspective, this increase in research assessing social cognition has coincided with the development of advanced neuroimaging techniques to elucidate brain behavior relationships. As this chapter demonstrates, investigation of neurodegenerative disorders is directly informing models of complex human behavior, including the perception of faces, the biological basis of emotion, and the subcomponents of theory of mind and empathy. As functional imaging tools become more widespread, our knowledge of social neuroscience will undoubtedly continue to be informed by comprehensive and systematic investigation of these syndromes, characterized by a progressive and relentless decline in these uniquely human capacities.

Acknowledgments This work was supported by funding to Forefront, a collaborative research group dedicated to the study of frontotemporal dementia and motor neurone disease, from the National Health and Medical Research Council (NHMRC) of Australia program grant (#APP1037746) and the Australian Research Council (ARC) Centre of Excellence in Cognition and its Disorders Memory Node (#CE110001021). FK is supported by an NHMRC-ARC Dementia Research Development Fellowship (#APP1097026). LCdL is supported by ANR-10-LABX-0087 IEC and ANR-10-IDEX-0001-02 PSL*. JVdS is supported by Fonds Wetenschappelijk Onderzoek-Vlaanderen (FWO), Alzheimer Research Foundation (P#14013), and KU Leuven Starting Grant.

References

1. Darwin CR. The expression of the emotions in man and animals. 1st ed. London: John Murray; 1872.
2. Adolphs R. The social brain: neural basis of social knowledge. Annu Rev Psychol. 2009;60:693–716.
3. Dunbar R. The social brain hypothesis. Brain. 1998;9(10):178–90.
4. Preston SD, de Waal FB. Empathy: its ultimate and proximate bases. Behav Brain Sci. 2002;25(1):1–20.
5. Dolan RJ. Emotion, cognition, and behavior. Science. 2002;298(5596):1191–4.
6. Phelps EA, LeDoux JE. Contributions of the amygdala to emotion processing: from animal models to human behavior. Neuron. 2005;48(2):175–87.
7. Adolphs R. Neural systems for recognizing emotion. Curr Opin Neurobiol. 2002;12(2):169–77.
8. Pievani M, de Haan W, Wu T, Seeley WW, Frisoni GB. Functional network disruption in the degenerative dementias. Lancet Neurol. 2011;10(9):829–43.
9. Kumfor F, Dermody N, Irish M. Considering the impact of large-scale network interactions on cognitive control. J Neurosci. 2015;35(1):1–3.
10. Haxby JV, Gobbini MI. Distributed neural systems for face perception. In: Calder AJ, Rhodes G, Johnson M, editors. The Oxford handbook of face perception. New York: Oxford University Press; 2011. p. 93–110.
11. Hutchings R, Palermo R, Piguet O, Kumfor F. Disrupted face processing in frontotemporal dementia: a review of the clinical and neuroanatomical evidence. Neuropsychol Rev. 2017;27(1):18–30.
12. Benton AL. Contributions to neuropsychological assessment: a clinical manual. Oxford: Oxford University Press; 1994.
13. Warrington EK. Recognition memory test. 1984.
14. Duchaine BC, Weidenfeld A. An evaluation of two commonly used tests of unfamiliar face recognition. Neuropsychologia. 2003;41(6):713–20.
15. Duchaine B, Nakayama K. The Cambridge face memory test: results for neurologically intact individuals and an investigation of its validity using inverted face stimuli and prosopagnosic participants. Neuropsychologia. 2006;44(4):576–85.
16. Bowles DC, McKone E, Dawel A, Duchaine B, Palermo R, Schmalzl L, et al. Diagnosing prosopagnosia: effects of ageing, sex, and participant–stimulus ethnic match on the Cambridge face memory test and Cambridge face perception test. Cogn Neuropsychol. 2009;26(5):423–55.

17. Gefen T, Wieneke C, Martersteck A, Whitney K, Weintraub S, Mesulam M-M, et al. Naming vs knowing faces in primary progressive aphasia a tale of 2 hemispheres. Neurology. 2013;81(7):658–64.
18. de Gelder B, Huis in 't Veld EM, Van den Stock J. The Facial Expressive Action Stimulus Test. A test battery for the assessment of face memory, face and object perception, configuration processing, and facial expression recognition. Front Psychol. 2015;6(1609):1–14.
19. De Winter FL, Timmers D, de Gelder B, Van Orshoven M, Vieren M, Bouckaert M, et al. Face shape and face identity processing in behavioral variant fronto-temporal dementia: a specific deficit for familiarity and name recognition of famous faces. Neuroimage Clin. 2016;11:368–77.
20. De Winter FL, Van den Stock J, de Gelder B, Peeters R, Jastorff J, Sunaert S, et al. Amygdala atrophy affects emotion-related activity in face-responsive regions in frontotemporal degeneration. Cortex. 2016;82:179–91.
21. Van den Stock J, De Winter FL, de Gelder B, Rangarajan JR, Cypers G, Maes F, et al. Impaired recognition of body expressions in the behavioral variant of frontotemporal dementia. Neuropsychologia. 2015;75:496–504.
22. Ekman P, Friesen WV. Pictures of facial affect. Palo Alto: Consulting Psychologists Press; 1976.
23. Gendron M, Roberson D, van der Vyver JM, Barrett LF. Perceptions of emotion from facial expressions are not culturally universal: evidence from a remote culture. Emotion. 2014;14(2):251–62.
24. Young A, Perrett D, Calder A, Sprengelmeyer R, Ekman P. Facial Expressions of Emotion – Stimuli and Tests (FEEST). Bury St Edmunds, England: Thames Valley Test Company; 2002.
25. Lundqvist D, Flykt A, Öhman A. The Karolinska Directed Emotional Faces (KDEF). Stockholm: Karolinska Institutet; 1998.
26. Ebner NC, Riediger M, Lindenberger U. FACES—A database of facial expressions in young, middle-aged, and older women and men: development and validation. Behav Res Methods. 2010;42(1):351–62.
27. Langner O, Dotsch R, Bijlstra G, Wigboldus DHJ, Hawk ST, van Knippenberg A. Presentation and validation of the Radboud faces database. Cognit Emot. 2010;24(8):1377–88.
28. Tottenham N, Tanaka JW, Leon AC, McCarry T, Nurse M, Hare TA, et al. The NimStim set of facial expressions: judgments from untrained research participants. Psychiatry Res. 2009;168(3):242–9.
29. Argyle M. Bodily communication. London: Methuen; 1988. 363 p
30. Sprengelmeyer R, Young AW, Schroeder U, Grossenbacher PG, Federlein J, Buttner T, et al. Knowing no fear. Proc Biol Sci. 1999;266(1437):2451–6.
31. de Gelder B, Van den Stock J. The Bodily Expressive Action Stimulus Test (BEAST). Construction and validation of a stimulus basis for measuring perception of whole body expression of emotions. Front Psychol. 2011;2(181):1–6.
32. McDonald S, Flanagan S, Rollins J, Kinch J. TASIT: a new clinical tool for assessing social perception after traumatic brain injury. J Head Trauma Rehabil. 2003;18(3):219.
33. Honan CA, McDonald S, Sufani C, Hine DW, Kumfor F. The Awareness of Social Inference Test: development of a shortened version for use in adults with acquired brain injury. Clin Neuropsychol. 2016;30(2):243–64.
34. Kumfor F, Honan CA, McDonald S, Hazelton JL, Hodges JR, Piguet O. Assessing the "social brain" in dementia: applying TASIT-S. Cortex. 2017;93:166.
35. Premack DG, Woodruff G. Does the chimpanzee have a theory of mind? Behav Brain Sci. 1978;1(4):515–26.
36. Wimmer H, Perner J. Beliefs about beliefs: representation and constraining function of wrong beliefs in young children's understanding of deception. Cognition. 1983;13(1):103–28.
37. Frith C, Frith U. Theory of mind. Curr Biol. 2005;15(17):R644–5.

38. Aboulafia-Brakha T, Christe B, Martory MD, Annoni JM. Theory of mind tasks and executive functions: a systematic review of group studies in neurology. J Neuropsychol. 2011;5(1):39–55.
39. Baron-Cohen S, Wheelwright S, Hill J, Raste Y, Plumb I. The "Reading the Mind in the Eyes" Test revised version: a study with normal adults, and adults with Asperger syndrome or high-functioning autism. J Child Psychol Psychiatry. 2001;42(2):241–51.
40. Baron-Cohen S, O'Riordan M, Stone V, Jones R, Plaisted K. Recognition of faux pas by normally developing children and children with Asperger syndrome or high-functioning autism. J Autism Dev Disord. 1999;29(5):407–18.
41. Torralva T, Roca M, Gleichgerrcht E, Bekinschtein T, Manes F. A neuropsychological battery to detect specific executive and social cognitive impairments in early frontotemporal dementia. Brain. 2009;132(5):1299–309.
42. Bertoux M, Delavest M, De Souza LC, Funkiewiez A, Lépine J-P, Fossati P, et al. Social cognition and emotional assessment differentiates frontotemporal dementia from depression. J Neurol Neurosurg Psychiatry. 2012;83(4):411–6.
43. Decety J, Jackson PL. A social-neuroscience perspective on empathy. Curr Dir Psychol Sci. 2006;15(2):54–8.
44. Eslinger PJ. Neurological and neuropsychological bases of empathy. Eur Neurol. 1998;39(4):193–9.
45. Davis MH. A multidimensional approach to individual differences in empathy. JSAS Cat Sel Doc Psychol. 1980;10:85–104.
46. Shamay-Tsoory S, Tomer R, Goldsher D, Berger B, Aharon-Peretz J. Impairment in cognitive and affective empathy in patients with brain lesions: anatomical and cognitive correlates. J Clin Exp Neuropsychol. 2004;26(8):1113–27.
47. Davis MH. Measuring individual differences in empathy: evidence for a multidimensional approach. J Pers Soc Psychol. 1983;44(1):113–26.
48. Hazelton JL, Irish M, Hodges JR, Piguet O, Kumfor F. Cognitive and affective empathy disruption in non-fluent primary progressive aphasia syndromes. Brain Impair. 2017;18(1):117–29.
49. Rankin KP, Kramer JH, Miller BL. Patterns of cognitive and emotional empathy in frontotemporal lobar degeneration. Cogn Behav Neurol. 2005;18(1):28–36.
50. Shamay-Tsoory SG, Aharon-Peretz J, Perry D. Two systems for empathy: a double dissociation between emotional and cognitive empathy in inferior frontal gyrus versus ventromedial prefrontal lesions. Brain. 2009;132(3):617–27.
51. Adjeroud N, Besnard J, El Massioui N, Verny C, Prudean A, Scherer C, et al. Theory of mind and empathy in preclinical and clinical Huntington's disease. Soc Cogn Affect Neurosci. 2015;11(1):89–99.
52. Sparks A, McDonald S, Lino B, O'Donnell M, Green MJ. Social cognition, empathy and functional outcome in schizophrenia. Schizophr Res. 2010;122(1-3):172–8.
53. Dermody N, Wong S, Ahmed R, Piguet O, Hodges JR, Irish M. Uncovering the neural bases of cognitive and affective empathy deficits in Alzheimer's disease and the behavioral-variant of frontotemporal dementia. J Alzheimers Dis. 2016;53(3):1–16.
54. Eslinger PJ, Moore P, Anderson C, Grossman M. Social cognition, executive functioning, and neuroimaging correlates of empathic deficits in frontotemporal dementia. J Neuropsychiatry Clin Neurosci. 2011;23(1):74–82.
55. Rankin KP, Gorno-Tempini ML, Allison S, Stanley CM, Glenn S, Weiner MW, et al. Structural anatomy of empathy in neurodegenerative disease. Brain. 2006;129(11):2945–56.
56. Baron-Cohen S, Wheelwright S. The Empathy Quotient - an investigation of adults with Asperger syndrome or high functioning autism, and normal sex differences. J Autism Dev Disord. 2004;34(2):163–75.
57. Jolliffe D, Farrington DP. Development and validation of the Basic Empathy Scale. J Adolesc. 2006;29(4):589–611.

58. Mehrabian A. Manual for the Balanced Emotional Empathy Scale (BEES). 1996. Available from Albert Mehrabian, 1130 Alta Mesa Road, Monterey, CA, USA 93940.
59. Hsieh S, Irish M, Daveson N, Hodges JR, Piguet O. When one loses empathy: its effect on carers of patients with dementia. J Geriatr Psychiatry Neurol. 2013;26(3):174–84.
60. Hutchings R, Hodges JR, Piguet O, Kumfor F. Why should I care? Dimensions of socio-emotional cognition in younger-onset dementia. J Alzheimers Dis. 2015;48(1):135–47.
61. Dziobek I, Rogers K, Fleck S, Bahnemann M, Heekeren HR, Wolf OT, et al. Dissociation of cognitive and emotional empathy in adults with Asperger syndrome using the Multifaceted Empathy Test (MET). J Autism Dev Disord. 2008;38(3):464–73.
62. Decety J, Michalska KJ, Kinzler KD. The contribution of emotion and cognition to moral sensitivity: a neurodevelopmental study. Cereb Cortex. 2012;22(1):209–20.
63. Bernhardt BC, Singer T. The neural basis of empathy. Annu Rev Neurosci. 2012;35:1–23.
64. Lamm C, Decety J, Singer T. Meta-analytic evidence for common and distinct neural networks associated with directly experienced pain and empathy for pain. NeuroImage. 2011;54(3):2492–502.
65. Baez S, Manes F, Huepe D, Torralva T, Fiorentino N, Richter F, et al. Primary empathy deficits in frontotemporal dementia. Front Aging Neurosci. 2014;6(262):1–11.
66. Ratnavalli E, Brayne C, Dawson K, Hodges JR. The prevalence of frontotemporal dementia. Neurology. 2002;58(11):1615–21.
67. Coyle-Gilchrist ITS, Dick KM, Patterson K, Rodríquez PV, Wehmann E, Wilcox A, et al. Prevalence, characteristics, and survival of frontotemporal lobar degeneration syndromes. Neurology. 2016;86(18):1736–43.
68. Piguet O, Hornberger M, Mioshi E, Hodges JR. Behavioral-variant frontotemporal dementia: diagnosis, clinical staging, and management. Lancet Neurol. 2011;10(2):162–72.
69. Rascovsky K, Hodges JR, Knopman D, Mendez MF, Kramer JH, Neuhaus J, et al. Sensitivity of revised diagnostic criteria for the behavioral variant of frontotemporal dementia. Brain. 2011;134(9):2456–77.
70. Hornberger M, Piguet O, Graham AJ, Nestor PJ, Hodges JR. How preserved is episodic memory in behavioral variant frontotemporal dementia. Neurology. 2010;74(6):473–9.
71. Seeley WW, Crawford R, Rascovsky K, Kramer JH, Weiner M, Miller BL, et al. Frontal paralimbic network atrophy in very mild behavioral variant frontotemporal dementia. Arch Neurol. 2008;65(2):249–55.
72. Whitwell JL, Przybelski SA, Weigand SD, Ivnik RJ, Vemuri P, Gunter JL, et al. Distinct anatomical subtypes of the behavioral variant of frontotemporal dementia: a cluster analysis study. Brain. 2009;132(11):2932–46.
73. Rosen HJ, Pace-Savitsky K, Perry RJ, Kramer JH, Miller BL, Levenson RL. Recognition of emotion in the frontal and temporal variants of frontotemporal dementia. Dement Geriatr Cogn Disord. 2004;17(4):277–81.
74. Keane J, Calder AJ, Hodges JR, Young AW. Face and emotion processing in frontal variant frontotemporal dementia. Neuropsychologia. 2002;40(6):655–65.
75. Snowden JS, Austin NA, Sembi S, Thompson JC, Craufurd D, Neary D. Emotion recognition in Huntington's disease and frontotemporal dementia. Neuropsychologia. 2008;46(11):2638–49.
76. Fernandez-Duque D, Black S. Impaired recognition of negative facial emotions in patients with frontotemporal dementia. Neuropsychologia. 2005;43(11):1673–87.
77. Miller L, Hsieh S, Lah S, Savage S, Hodges JR, Piguet O. One size does not fit all: face emotion processing impairments in semantic dementia, behavioral-variant frontotemporal dementia and Alzheimer's disease are mediated by distinct cognitive deficits. Behav Neurol. 2012;25(1):53–60.
78. Kumfor F, Hutchings R, Irish M, Hodges JR, Rhodes G, Palermo R, et al. Do I know you? Examining face and object memory in frontotemporal dementia. Neuropsychologia. 2015;71:101–11.

79. Haxby JV, Hoffman EA, Gobbini MI. The distributed human neural system for face perception. Trends Cogn Sci. 2000;4(6):223–33.
80. Kumfor F, Piguet O. Disturbance of emotion processing in frontotemporal dementia: a synthesis of cognitive and neuroimaging findings. Neuropsychol Rev. 2012;22(3):280–97.
81. Lavenu I, Pasquier F, Lebert F, Petit H, Van der Linden M. Perception of emotion in frontotemporal dementia and Alzheimer disease. Alzheimer Dis Assoc Disord. 1999;13(2):96–101.
82. Diehl-Schmid J, Pohl C, Ruprecht C, Wagenpfeil S, Foerstl H, Kurz A. The Ekman 60 faces test as a diagnostic instrument in frontotemporal dementia. Arch Clin Neuropsychol. 2007;22(4):459–64.
83. Bediou B, Ryff I, Mercier B, Milliery M, Henaff M-A, D'Amato T, et al. Impaired social cognition in mild Alzheimer disease. J Geriatr Psychiatry Neurol. 2009;22(2):130–40.
84. Omar R, Rohrer JD, Hailstone JC, Warren J. Structural neuroanatomy of face processing in frontotemporal dementia. J Neurol Neurosurg Psychiatry. 2011;82(12):1341–3.
85. Hsieh S, Hornberger M, Piguet O, Hodges JR. Brain correlates of musical and facial emotion recognition: evidence from the dementias. Neuropsychologia. 2012;50(8):1814–22.
86. Omar R, Henley S, Bartlett JW, Hailstone JC, Gordon E, Sauter DA, et al. The structural neuroanatomy of music emotion recognition: evidence from frontotemporal lobar degeneration. NeuroImage. 2011;56(3):1814–21.
87. Kumfor F, Irish M, Hodges JR, Piguet O. Discrete neural correlates for the recognition of negative emotions: insights from frontotemporal dementia. PLoS One. 2013;8(6):e67457.
88. Jastorff J, De Winter FL, Van den Stock J, Vandenberghe R, Giese MA, Vandenbulcke M. Functional dissociation between anterior temporal lobe and inferior frontal gyrus in the processing of dynamic body expressions: insights from behavioral variant frontotemporal dementia. Hum Brain Mapp. 2016;37(12):4472–86.
89. Virani K, Jesso S, Kertesz A, Mitchell D, Finger E. Functional neural correlates of emotional expression processing deficits in behavioral variant frontotemporal dementia. J Psychiatry Neurosci. 2013;38(3):174–82.
90. Lough S, Kipps CM, Treise C, Watson P, Blair JR, Hodges JR. Social reasoning, emotion and empathy in frontotemporal dementia. Neuropsychologia. 2006;44(6):950–8.
91. Torralva T, Kipps CM, Hodges JR, Clark L, Bekinschtein T, Roca M, et al. The relationship between affective decision-making and theory of mind in the frontal variant of frontotemporal dementia. Neuropsychologia. 2007;45(2):342–9.
92. Shany-Ur T, Poorzand P, Grossman S, Growden ME, Jang J, Ketelle RS, et al. Comprehension of insincere communication in neurodegenerative disease: lies, sarcasm and theory of mind. Cortex. 2012;48(10):1329–41.
93. Henry JD, Phillips LH, von Hippel C. A meta-analytic review of theory of mind difficulties in behavioral-variant frontotemporal dementia. Neuropsychologia. 2014;56:53–62.
94. Bora E, Walterfang M, Velakoulis D. Theory of mind in behavioral-variant frontotemporal dementia and Alzheimer's disease: a meta-analysis. J Neurol Neurosurg Psychiatry. 2015;86(7):714–9.
95. Snowden JS, Gibbons ZC, Blackshaw A, Doubleday E, Thompson J, Craufurd D, et al. Social cognition in frontotemporal dementia and Huntington's disease. Neuropsychologia. 2003;41(6):688–701.
96. Bertoux M, O'Callaghan C, Dubois B, Hornberger M. In two minds: executive functioning versus theory of mind in behavioral variant frontotemporal dementia. J Neurol Neurosurg Psychiatry. 2016;87(3):231–4.
97. Le Bouc R, Lenfant P, Delbeuck X, Ravasi L, Lebert F, Semah F, et al. My belief or yours? Differential theory of mind deficits in frontotemporal dementia and Alzheimer's disease. Brain. 2012;135(10):3026–38.
98. Lough S, Gregory C, Hodges JR. Dissociation of social cognition and executive dysfunction in frontal variant frontotemporal dementia. Neurocase. 2001;7(2):123–30.

99. Lough S, Hodges JR. Measuring and modifying abnormal social cognition in frontal variant frontotemporal dementia. J Psychosom Res. 2002;53(2):639–46.
100. Eslinger PJ, Moore P, Troiani V, Antani S, Cross K, Kwok S, et al. Oops! Resolving social dilemmas in frontotemporal dementia. J Neurol Neurosurg Psychiatry. 2007;78(5):457–60.
101. Irish M, Hodges JR, Piguet O. Right anterior temporal lobe dysfunction underlies theory of mind impairments in semantic dementia. Brain. 2014;137(4):1241–53.
102. Baez S, Morales JP, Slachevsky A, Torralva T, Matus C, Manes F, et al. Orbitofrontal and limbic signatures of empathic concern and intentional harm in the behavioral variant frontotemporal dementia. Cortex. 2016;75:20–32.
103. Oliver LD, Mitchell DG, Dziobek I, MacKinley J, Coleman K, Rankin KP, et al. Parsing cognitive and emotional empathy deficits for negative and positive stimuli in frontotemporal dementia. Neuropsychologia. 2015;67:14–26.
104. Gorno-Tempini ML, Hillis AE, Weintraub S, Kertesz A, Mendez M, Cappa SF, et al. Classification of primary progressive aphasia and its variants. Neurology. 2011;76(11):1006–14.
105. Landin-Romero R, Tan R, Hodges JR, Kumfor F. An update on semanic dementia: genetics, imaging and pathology. Alzheimers Res Ther. 2016;8(1):1–9.
106. Hodges JR, Mitchell J, Dawson K, Spillantini MG, Xuereb JH, McMonagle P, et al. Semantic dementia: demography, familial factors and survival in a consecutive series of 100 cases. Brain. 2009;133(1):300–6.
107. Thompson SA, Patterson K, Hodges JR. Left/right asymmetry of atrophy in semantic dementia: behavioral-cognitive implications. Neurology. 2003;61(9):1196–203.
108. Rosen HJ, Wilson MR, Schauer GF, Allison S, Gorno-Tempini ML, Pace-Savitsky C, et al. Neuroanatomical correlates of impaired recognition of emotion in dementia. Neuropsychologia. 2006;44(3):365–73.
109. Chan D, Anderson V, Pijnenburg Y, Whitwell J, Barnes J, Scahill R, et al. The clinical profile of right temporal lobe atrophy. Brain. 2009;132(5):1287–98.
110. Brambati SM, Amici S, Racine CA, Neuhaus J, Miller Z, Ogar J, et al. Longitudinal gray matter contraction in three variants of primary progressive aphasia: a tenser-based morphometry study. Neuroimage Clin. 2015;8:345–55.
111. Lam BY, Halliday GM, Irish M, Hodges JR, Piguet O. Longitudinal white matter changes in frontotemporal dementia subtypes. Hum Brain Mapp. 2014;35(7):3547–57.
112. Kumfor F, Landin-Romero R, Devenney E, Hutchings R, Grasso R, Hodges JR, et al. On the right side? A longitudinal study of left- versus right-lateralized semantic dementia. Brain. 2016;139(3):986–98.
113. Binney RJ, Henry ML, Babiak M, Pressman PS, Santos-Santos MA, Narvid J, et al. Reading words and other people: a comparison of exception word, familiar face and affect processing in the left and right temporal variants of primary progressive aphasia. Cortex. 2016;82:147–63.
114. Snowden J, Thompson J, Neary D. Knowledge of famous faces and names in semantic dementia. Brain. 2004;127(4):860–72.
115. Josephs KA, Whitwell JL, Vemuri P, Senjem ML, Boeve BF, Knopman DS, et al. The anatomic correlate of prosopagnosia in semantic dementia. Neurology. 2008;71(20):1628–33.
116. de Gelder B, Van den Stock J. Prosopagnosia. In: Wright JD, editor. International encyclopedia of the social & behavioral sciences. 2nd ed. Oxford: Elsevier; 2015. p. 250–5.
117. Simons JS, Graham KS, Galton CJ, Patterson K, Hodges JR. Semantic knowledge and episodic memory for faces in semantic dementia. Neuropsychology. 2001;15(1):101–14.
118. Péron JA, Piolino P, Moal-Boursiquot SL, Biseul I, Leray E, Bon L, et al. Preservation of person-specific semantic knowledge in semantic dementia: does direct personal experience have a specific role? Front Hum Neurosci. 2015;9(625):1–12.
119. Rosen HJ, Perry RJ, Murphy J, Kramer JH, Mychack P, Schuff N, et al. Emotion comprehension in the temporal variant of frontotemporal dementia. Brain. 2002;125(10):2286–95.
120. Kumfor F, Miller L, Lah S, Hsieh S, Savage S, Hodges JR, et al. Are you really angry? The effect of intensity on emotion recognition in frontotemporal dementia. Soc Neurosci. 2011;6(5-6):502–14.

121. Hsieh S, Hodges JR, Piguet O. Recognition of positive vocalizations is impaired in behavioral-variant frontotemporal dementia. J Int Neuropsychol Soc. 2013;19(4):483–7.
122. Hsieh S, Foxe D, Leslie F, Savage S, Piguet O, Hodges JR. Grief and joy: emotion word comprehension in the dementias. Neuropsychology. 2012;26(5):624–30.
123. Cohen MH, Carton AM, Hardy CJ, Golden HL, Clark CN, Fletcher PD, et al. Processing emotion from abstract art in frontotemporal lobar degeneration. Neuropsychologia. 2016;81:245–54.
124. Kumfor F, Irish M, Hodges JR, Piguet O. The orbitofrontal cortex is involved in emotional enhancement of memory: evidence from the dementias. Brain. 2013;136:2992–3003.
125. Rankin KP, Salazar A, Gorno-Tempini ML, Sollberger M, Wilson SM, Pavlic D, et al. Detecting sarcasm from paralinguistic cues: anatomic and cognitive correlates in neurodegenerative disease. NeuroImage. 2009;47(4):2005–15.
126. Irish M, Kumfor F, Hodges JR, Piguet O. A tale of two hemispheres: contrasting patterns of socioemotional dysfunction in left- versus right-lateralised semantic dementia. Dement Neuropsychol. 2013;7(1):88–95.
127. Duval C, Bejanin A, Piolino P, Laisney M, de La Sayette V, Belliard S, et al. Theory of mind impairments in patients with semantic dementia. Brain. 2012;135(1):228–41.
128. Bejanin A, Chételat G, Laisney M, Pélerin A, Landeau B, Merck C, et al. Distinct neural substrates of affective and cognitive theory of mind impairment in semantic dementia. Soc Neurosci. 2017;12:287–302.
129. Roos RA. Huntington's disease: a clinical review. Orphanet J Rare Dis. 2010;5(1):40–8.
130. Lee JM, Galkina EI, Levantovsky RM, Fossale E, Anne Anderson M, Gillis T, et al. Dominant effects of the Huntington's disease HTT CAG repeat length are captured in gene-expression data sets by a continuous analysis mathematical modeling strategy. Hum Mol Genet. 2013;22(16):3227–38.
131. Snowden JS, Craufurd D, Thompson J, Neary D. Psychomotor, executive, and memory function in preclinical Huntington's disease. J Clin Exp Neuropsychol. 2002;24(2):133–45.
132. Dumas EM, van den Bogaard SJ, Middelkoop HA, Roos RA. A review of cognition in Huntington's disease. Front Biosci (Schol Ed). 2013;5:1–18.
133. Tabrizi SJ, Reilmann R, Roos RA, Durr A, Leavitt B, Owen G, et al. Potential endpoints for clinical trials in premanifest and early Huntington's disease in the TRACK-HD study: analysis of 24 month observational data. Lancet Neurol. 2012;11(1):42–53.
134. Dogan I, Eickhoff SB, Schulz JB, Shah NJ, Laird AR, Fox PT, et al. Consistent neurodegeneration and its association with clinical progression in Huntington's disease: a coordinate-based meta-analysis. Neurodegener Dis. 2012;12(1):23–35.
135. Thieben MJ, Duggins AJ, Good CD, Gomes L, Mahant N, Richards F, et al. The distribution of structural neuropathology in pre-clinical Huntington's disease. Brain. 2002;125(8):1815–28.
136. Dogan I, Sass C, Mirzazade S, Kleiman A, Werner CJ, Pohl A, et al. Neural correlates of impaired emotion processing in manifest Huntington's disease. Soc Cogn Affect Neurosci. 2014;9(5):671–80.
137. Gray JM, Young AW, Barker WA, Curtis A, Gibson D. Impaired recognition of disgust in Huntington's disease gene carriers. Brain. 1997;120(11):2029–38.
138. Henley SM, Novak MJ, Frost C, King J, Tabrizi SJ, Warren JD. Emotion recognition in Huntington's disease: a systematic review. Neurosci Biobehav Rev. 2008;36(1):237–53.
139. Johnson SA, Stout JC, Solomon AC, Langbehn DR, Aylward EH, Cruce CB, et al. Beyond disgust: impaired recognition of negative emotions prior to diagnosis in Huntington's disease. Brain. 2007;130(7):1732–44.
140. Sprengelmeyer R, Young AW, Calder AJ, Karnat A, Lange H, Homberg V, et al. Loss of disgust. Perception of faces and emotions in Huntington's disease. Brain. 1996;119(5):1647–65.
141. Hayes CJ, Stevenson RJ, Coltheart M. The processing of emotion in patients with Huntington's disease: variability and differential deficits in disgust. Cogn Behav Neurol. 2009;22(4):249–57.

142. Wang K, Hoosain R, Yang RM, Meng Y, Wang CQ. Impairment of recognition of disgust in Chinese with Huntington's or Wilson's disease. Neuropsychologia. 2003;41(5):527–37.
143. Aviezer H, Bentin S, Hassin RR, Meschino WS, Kennedy J, Grewal S, et al. Not on the face alone: perception of contextualized face expressions in Huntington's disease. Brain. 2009;132(6):1633–44.
144. Croft RJ, McKernan F, Gray M, Churchyard A, Georgiou-Karistianis N. Emotion perception and electrophysiological correlates in Huntington's disease. Clin Neurophysiol. 2014;125(8):1618–25.
145. Lawrence AD, Watkins LH, Sahakian BJ, Hodges JR, Robbins TW. Visual object and visuospatial cognition in Huntington's disease: implications for information processing in corticostriatal circuits. Brain. 2000;123(7):1349–64.
146. Tabrizi SJ, Scahill RI, Owen G, Durr A, Leavitt BR, Roos RA, et al. Predictors of phenotypic progression and disease onset in premanifest and early-stage Huntington's disease in the TRACK-HD study: analysis of 36-month observational data. Lancet Neurol. 2013;12(7):637–49.
147. Labuschagne I, Jones R, Callaghan J, Whitehead D, Dumas EM, Say MJ, et al. Emotional face recognition deficits and medication effects in pre-manifest through stage-II Huntington's disease. Psychiatry Res. 2013;207(1-2):118–26.
148. Trinkler I, Cleret de Langavant L, Bachoud-Levi AC. Joint recognition-expression impairment of facial emotions in Huntington's disease despite intact understanding of feelings. Cortex. 2013;49(2):549–58.
149. Bora E, Velakoulis D, Walterfang M. Social cognition in Huntington's disease: a meta-analysis. Behav Brain Res. 2016;297:131–40.
150. Larsen IU, Vinther-Jensen T, Gade A, Nielsen JE, Vogel A. Do I misconstrue? Sarcasm detection, emotion recognition, and theory of mind in Huntington disease. Neuropsychology. 2016;30(2):181.
151. Baez S, Herrera E, Gershanik O, Garcia AM, Bocanegra Y, Kargieman L, et al. Impairments in negative emotion recognition and empathy for pain in Huntington's disease families. Neuropsychologia. 2015;68:158–67.
152. Robotham L, Sauter DA, Bachoud-Lévi A-C, Trinkler I. The impairment of emotion recognition in Huntington's disease extends to positive emotions. Cortex. 2011;47(7):880–4.
153. de Gelder B, Van den Stock J, Balaguer Rde D, Bachoud-Levi AC. Huntington's Disease impairs recognition of angry and instrumental body language. Neuropsychologia. 2008;46(1):369–73.
154. Novak MJ, Warren JD, Henley SM, Draganski B, Frackowiak RS, Tabrizi SJ. Altered brain mechanisms of emotion processing in pre-manifest Huntington's disease. Brain. 2012;135(4):1165–79.
155. Allain P, Havet-Thomassin V, Verny C, Gohier B, Lancelot C, Besnard J, et al. Evidence for deficits on different components of theory of mind in Huntington's disease. Neuropsychology. 2011;25(6):741–51.
156. Brüne M, Blank K, Witthaus H, Saft C. "Theory of mind" is impaired in Huntington's disease. Mov Disord. 2011;26(4):671–8.
157. Eddy CM, Sira Mahalingappa S, Rickards HE. Putting things into perspective: the nature and impact of theory of mind impairment in Huntington's disease. Eur Arch Psychiatry Clin Neurosci. 2014;264(8):697–705.
158. Eddy CM, Rickards HE. Theory of mind can be impaired prior to motor onset in Huntington's disease. Neuropsychology. 2015;29(5):792–8.
159. Mason SL, Zhang J, Begeti F, Guzman NV, Lazar AS, Rowe JB, et al. The role of the amygdala during emotional processing in Huntington's disease: from pre-manifest to late stage disease. Neuropsychologia. 2015;70:80–9.
160. Maurage P, Lahaye M, Grynberg D, Jeanjean A, Guettat L, Verellen-Dumoulin C, et al. Dissociating emotional and cognitive empathy in pre-clinical and clinical Huntington's disease. Psychiatry Res. 2016;237:103–8.

161. Alzheimer's Assosiation. 2015 Alzheimer's disease facts and figures. Alzheimers Dement. 2015;11:332–84.
162. Hebert LE, Weuve J, Scherr PA, Evans DA. Alzheimer disease in the United States (2010–2050) estimated using the 2010 census. Neurology. 2013;80(19):1778–83.
163. Tu S, Wong S, Hodges JR, Irish M, Piguet O, Hornberger M. Lost in spatial translation–a novel tool to objectively assess spatial disorientation in Alzheimer's disease and frontotemporal dementia. Cortex. 2015;67:83–94.
164. Galton CJ, Patterson K, Xuereb JH, Hodges JR. Atypical and typical presentations of Alzheimer's disease: a clinical, neuropsychological, neuroimaging and pathological study of 13 cases. Brain. 2000;123(3):484–98.
165. McKhann GM, Knopman DS, Chertkow H, Hyman BT, Jack CR Jr, Kawas CH, et al. The diagnosis of dementia due to Alzheimer's disease: recommendations from the National Institute on Aging-Alzheimer's association workgroups on diagnostic guidelines for Alzheimer's disease. Alzheimers Dement. 2011;7(3):263–9.
166. Mega MS, Cummings JL, Fiorello T, Gornbein J. The spectrum of behavioral changes in Alzheimer's disease. Neurology. 1996;46(1):130–5.
167. Pengas G, Hodges JR, Watson P, Nestor PJ. Focal posterior cingulate atrophy in incipient Alzheimer's disease. Neurobiol Aging. 2010;31(1):25–33.
168. Scheltens P, Leys D, Barkhof F, Huglo D, Weinstein HC, Vermersch P, et al. Atrophy of medial temporal lobes on MRI in "probable" Alzheimer's disease and normal ageing: diagnostic value and neuropsychological correlates. J Neurol Neurosurg Psychiatry. 1992;55(10):967–72.
169. Landin-Romero R, Kumfor F, Leyton CE, Irish M, Hodges JR, Piguet O. Disease-specific patterns of cortical and subcortical degeneration in a longitudinal study of Alzheimer's disease and behavioral-variant frontotemporal dementia. NeuroImage. 2017;151:72.
170. Mendez MF, Martin RJ, Smyth KA, Whitehouse PJ. Disturbances of person identification in Alzheimer's disease: a retrospective study. J Nerv Ment Dis. 1992;180(2):94–6.
171. Della Sala S, Muggia S, Spinnler H, Zuffi M. Cognitive modelling of face processing: evidence from Alzheimer patients. Neuropsychologia. 1995;33(6):675–87.
172. Tippett LJ, Blackwood K, Farah MJ. Visual object and face processing in mild-to-moderate Alzheimer's disease: from segmentation to imagination. Neuropsychologia. 2003;41(4):453–68.
173. Becker JT, Lopez OL, Boller F. Understanding impaired analysis of faces by patients with probable Alzheimer's disease. Cortex. 1995;31(1):129–37.
174. Werheid K, Clare L. Are faces special in Alzheimer's disease? Cognitive conceptualisation, neural correlates, and diagnostic relevance of impaired memory for faces and names. Cortex. 2007;43(7):898–906.
175. Lavallée MM, Gandini D, Rouleau I, Vallet GT, Joannette M, Kergoat M-J, et al. A qualitative impairment in face perception in Alzheimer's disease: evidence from a reduced face inversion effect. J Alzheimers Dis. 2016;51:1225–36.
176. Cheng P-J, Pai M-C. Dissociation between recognition of familiar scenes and of faces in patients with very mild Alzheimer disease: an event-related potential study. Clin Neurophysiol. 2010;121(9):1519–25.
177. Saavedra C, Olivares EI, Iglesias J. Cognitive decline effects at an early stage: evidence from N170 and VPP. Neurosci Lett. 2012;518(2):149–53.
178. Schefter M, Werheid K, Almkvist O, Lönnqvist-Akenine U, Kathmann N, Winblad B. Recognition memory for emotional faces in amnestic mild cognitive impairment: an event-related potential study. Aging Neuropsychol Cogn. 2013;20(1):49–79.
179. Phillips LH, Scott C, Henry JD, Mowat D, Bell JS. Emotion perception in Alzheimer's disease and mood disorder in old age. Psychol Aging. 2010;25(1):38–47.
180. Bucks RS, Radford SA. Emotion processing in Alzheimer's disease. Aging Ment Health. 2004;8(3):222–32.

181. Kumfor F, Sapey-Triomphe L-A, Leyton CE, Burrell JR, Hodges JR, Piguet O. Degradation of emotion processing ability in corticobasal syndrome and Alzheimer's disease. Brain. 2014;137(11):3061–72.
182. Bertoux M, de Souza LC, Sarazin M, Funkiewiez A, Dubois B, Hornberger M. How preserved is emotion recognition in Alzheimer disease compared with behavioral variant frontotemporal dementia? Alzheimer Dis Assoc Disord. 2015;29(2):154–7.
183. Klein-Koerkamp Y, Beaudoin M, Baciu M, Hot P. Emotional decoding abilities in Alzheimer's disease: a meta-analysis. J Alzheimers Dis. 2012;32(1):109–25.
184. Kumfor F, Irish M, Leyton CE, Miller L, Lah S, Devenney E, et al. Tracking the progression of social cognition in neurodegenerative disorders. J Neurol Neurosurg Psychiatry. 2014;85(10):1076–83.
185. Cadieux NL, Greve KW. Emotion processing in Alzheimer's disease. J Int Neuropsychol Soc. 1997;3(05):411–9.
186. Henry JD, Ruffman T, McDonald S, O'Leary M-AP, Phillips LH, Brodaty H, et al. Recognition of disgust is selectively preserved in Alzheimer's disease. Neuropsychologia. 2008;46(5):1363–70.
187. Gregory C, Lough S, Stone V, Erzinclioglu S, Martin L, Baron-Cohen S, et al. Theory of mind in patients with frontal variant frontotemporal dementia and Alzheimer's disease: theoretical and practical implications. Brain. 2002;125(4):752–64.
188. Fernandez-Duque D, Baird JA, Black SE. False-belief understanding in frontotemporal dementia and Alzheimer's disease. J Clin Exp Neuropsychol. 2009;31(4):489–97.
189. Leslie AM, Friedman O, German TP. Core mechanisms in 'theory of mind'. Trends Cogn Sci. 2004;8(12):528–33.
190. Dodich A, Cerami C, Crespi C, Canessa N, Lettieri G, Iannaccone S, et al. Differential impairment of cognitive and affective mentalizing abilities in neurodegenerative dementias: evidence from behavioral variant of frontotemporal dementia, Alzheimer's disease, and mild cognitive impairment. J Alzheimers Dis. 2016;50(4):1011–22.
191. Nash S, Henry JD, Mcdonald S, Martin I, Brodaty H, Peek-O'Leary MA. Cognitive disinhibition and socioemotional functioning in Alzheimer's disease. J Int Neuropsychol Soc. 2007;13(6):1060–4.
192. Sturm VE, Yokoyama JS, Seeley WW, Kramer JH, Miller BL, Rankin KP. Heightened emotional contagion in mild cognitive impairment and Alzheimer's disease is associated with temporal lobe degeneration. Proc Natl Acad Sci U S A. 2013;110(24):9944–9.
193. Midorikawa A, Leyton CE, Foxe D, Landin-Romero R, Hodges JR, Piguet O. All is not lost: positive behaviors in Alzheimer's disease and behavioral-variant frontotemporal dementia with disease severity. J Alzheimers Dis. 2016;54:549–58.
194. Sturm VE, McCarthy ME, Yun I, Madan A, Yuan JW, Holley SR, et al. Mutual gaze in Alzheimer's disease, frontotemporal and semantic dementia couples. Soc Cogn Affect Neurosci. 2011;6(3):359–67.
195. Hargrave R, Maddock RJ, Stone V. Impaired recognition of facial expressions of emotion in Alzheimer's disease. J Neuropsychiatry Clin Neurosci. 2002;14(1):64–71.
196. Cerami C, Dodich A, Canessa N, Crespi C, Marcone A, Cortese F, et al. Neural correlates of empathic impairment in the behavioral variant of frontotemporal dementia. Alzheimers Dement. 2014;10(6):827–34.
197. Hornberger M, Piguet O. Episodic memory in frontotemporal dementia: a critical review. Brain. 2012;135(3):678–92.
198. Collette F, Van der Linden M, Salmon E. Executive dysfunction in Alzheimer's disease. Cortex. 1999;35(1):57–72.
199. Kipps CM, Nestor PJ, Acosta-Cabronero J, Arnold R, Hodges JR. Understanding social dysfunction in the behavioral variant of frontotemporal dementia. The role of emotion and sarcasm processing. Brain. 2009;132(3):592–603.
200. Bertoux M, de Souza L, Cruz L, O'Callaghan C, Greve A, Sarazin M, et al. Social cognition deficits: the key to discriminate behavioral variant frontotemporal dementia from Alzheimer's disease regardless of amnesia? J Alzheimers Dis. 2015;49(4):1065–74.

201. Kipps CM, Hodges JR, Hornberger M. Nonprogressive behavioral frontotemporal dementia: recent developments and clinical implications of the 'bvFTD phenocopy syndrome'. Curr Opin Neurol. 2010;23(6):628–32.
202. Kipps CM, Nestor PJ, Fryer TD, Hodges JR. Behavioral variant frontotemporal dementia: not all it seems? Neurocase. 2007;13(4):237–47.
203. Devenney E, Forrest SL, Xuereb J, Kril JJ, Hodges JR. The bvFTD phenocopy syndrome: a clinicopathological report. J Neurol Neurosurg Psychiatry. 2016;87:1155–6.
204. Diehl-Schmid J, Schmidt EM, Nunnemann S, Riedl L, Kurz A, Förstl H, et al. Caregiver burden and needs in frontotemporal dementia. J Geriatr Psychiatry Neurol. 2013;26:221–9.
205. Kumfor F, Hodges JR, Piguet O. Ecologically valid assessment of emotional enhancement of memory in progressive nonfluent aphasia and Alzheimer's disease. J Alzheimers Dis. 2014;42(1):201–10.
206. Yang C, Zhang T, Li Z, Heeramun-Aubeeluck A, Liu N, Huang N, et al. The relationship between facial emotion recognition and executive functions in first-episode patients with schizophrenia and their siblings. BMC Psychiatry. 2015;15(241):1–8.
207. Ekman P. An argument for basic emotions. Cognit Emot. 1992;6(3-4):169–200.
208. Russell JA. A circumplex model of affect. J Pers Soc Psychol. 1980;39:1161–78.
209. Aviezer H, Hassin RR, Ryan J, Grady C, Susskind J, Anderson A, et al. Angry, disgusted, or afraid?: studies on the malleability of emotion perception. Psychol Sci. 2008;19(7):724–32.
210. Van den Stock J, Vandenbulcke M, Sinke CBA, Goebel R, de Gelder B. How affective information from faces and scenes interacts in the brain. Soc Cogn Affect Neurosci. 2014;9(10):1481–8.
211. Ibañez A, Manes F. Contextual social cognition and the behavioral variant of frontotemporal dementia. Neurology. 2012;78(17):1354–62.
212. Mesulam M. Frontal cortex and behavior. Ann Neurol. 1986;19(4):320–5.
213. Melloni M, Lopez V, Ibanez A. Empathy and contextual social cognition. Cogn Affect Behav Neurosci. 2014;14(1):407–25.
214. Baez S, García A, Ibanez A. The social context network model in psychiatric and neurological diseases, Current Topics in Behavioral Neurosciences. Berlin: Springer; 2016. p. 1–18.
215. Schilbach L, Timmermans B, Reddy V, Costall A, Bente G, Schlicht T, et al. Toward a second-person neuroscience. Behav Brain Sci. 2013;36(4):393–414.
216. Gleichgerrcht E, Torralva T, Roca M, Pose M, Manes F. The role of social cognition in moral judgement in frontotemporal dementia. Soc Neurosci. 2010;6(2):113–22.
217. Baez S, Kanske P, Matallana D, Montañes P, Reyes P, Slachevsky A, et al. Integration of intention and outcome for moral judgement in frontotemporal dementia: brain structures and signatures. Neurodegener Dis. 2016;16(3-4):207–17.
218. Bertoux M, Cova F, Pessiglione M, Hsu M, Dubois B, Bourgeois-Gironde S. Behavioral variant frontotemporal dementia patients fo not succumb to the Allais paradox. Front Neurosci. 2014;8(287):1–8.
219. Sommer T, Peters J, Gläscher J, Büchel C. Structure-function relationships in the processing of regret in the orbitofrontal cortex. Brain Struct Funct. 2009;213(6):535–51.
220. Liljegren M, Naasan G, Temlett J, Perry DC, Rankin KP, Merrilees J, et al. Criminal behavior in frontotemporal dementia and Alzheimer disease. JAMA Neurol. 2015;72(3):295–300.
221. Jensen P, Fenger K, Bolwig TG, Sørensen SA. Crime in Huntington's disease: a study of registered offences among patients, relatives and controls. J Neurol Neurosurg Psychiatry. 1998;65(4):467–71.
222. Manes F, Torralva T, Ibáñez A, Roca M, Bekinschtein T, Gleichgerrcht E. Decision-making in frontotemporal dementia: clinical, theoretical and legal implications. Dement Geriatr Cogn Disord. 2011;32(1):11–7.
223. Levy BR, Ferrucci L, Zonderman AB, Slade MD, Troncoso J, Resnick SM. A culture-brain link: negative age stereotypes predict Alzheimer's disease biomarkers. Psychol Aging. 2016;31(1):82–8.
224. Johnson R, Harkins K, Cary M, Sankar P, Karlawish J. The relative contributions of disease label and disease prognosis to Alzheimer's stigma: a vignette-based experiment. Soc Sci Med. 2015;143:117–27.

225. Williams JK, Erwin C, Juhl AR, Mengeling M, Bombard Y, Hayden MR, et al. In their own words: reports of stigma and genetic discrimination by people at risk for Huntington disease in the international RESPOND-HD study. Am J Med Genet B Neuropsychiatr Genet. 2010;153B(6):1150–9.
226. O'Callaghan C, Bertoux M, Irish M, Shine JM, Wong S, Spiliopoulos L, et al. Fair play: social norm compliance failures in behavioral variant frontotemporal dementia. Brain. 2016;139(1):204–16.
227. Miltiades HB, Thatcher WG. Social engagement during game play in persons with Alzheimer's 'Innovative practice'. Dementia. Jan 2017:1–6. doi:10.1177/1471301216687920
228. Kindell J, Keady J, Sage K, Wilkinson R. Everyday conversation in dementia: a review of the literature to inform research and practice. Int J Lang Commun Disord. 2017;52:392–406.
229. Young JA, Lind C, Steenbrugge W. A conversation analytic study of patterns of overlapping talk in conversations between individuals with dementia and their frequent communication partners. Int J Lang Commun Disord. 2016;51(6):745–56.
230. Kennedy DP, Adolphs R. The social brain in psychiatric and neurological disorders. Trends Cogn Sci. 2012;16(11):559–72.
231. Desgranges B, Laisney M, Bon L, Duval C, Mondou A, Bejanin A, et al. TOM-15: une épreuve de fausses croyances pour évaluer la théorie de l'esprit cognitive. Rev Neuropsychol. 2012;4(3):216–20.
232. Bowers D, Blonder LX, Heilman KM. The Florida affect battery. Gainsville, Florida: Center for Neuropsycholgical Studies; 1999.
233. Benton AL, Hamsher K, Varney NR, Spreen O. Test of facial recognition: Form SL. New York: Oxford University Press; 1983.
234. Rollins J, Flanagan S, McDonald S. The Awareness of Social Inference Test (TASIT). Oxford, UK: Pearson; 2002.
235. Happé FG. An advanced test of theory of mind: Understanding of story characters' thoughts and feelings by able autistic, mentally handicapped, and normal children and adults. J Autism Dev Disord. 1994;24(2):129–54.
236. White SJ, Coniston D, Rogers R, Frith U. Developing the Frith-Happé animations: A quick and objective test of theory of mind for adults with autism. Autism Res. 2011;4(2):149–54.

Psychotherapy and Social Neuroscience: Forging Links Together

Andrés Roussos, Malena Braun, Saskia Aufenacker, and Julieta Olivera

Abstract Psychotherapy drew on social science to forge the epistemological and methodological approach for its development and validation. Although its emphasis on psychosocial premises isolated it from neurobiological processes, psychotherapy never resigned such concerns. Against this background, here we review studies integrating psychotherapy and neuroscience, focusing on studies that show the path for establishing joint research programs. We describe strategies and instruments that have been used in the literature and identify relevant methodological challenges. In addition, we consider empathy and interpersonal relationships as concepts that can bridge social neuroscience with concepts from psychotherapy, such as therapeutic alliance and emotional regulation. Finally, we discuss the extent to which the integration of these two fields promotes practice-oriented research with valid information to empower practitioners (psychotherapists, psychiatrists, and other mental health professionals) in their work with patients.

A. Roussos (✉)
Equipo de Investigación en Psicología Clínica, Universidad de Buenos Aires,
Buenos Aires, Argentina

Consejo Nacional de Investigaciones Científicas y Técnicas (CONICET),
Buenos Aires, Argentina
e-mail: andres.roussos@comunidad.ub.edu.ar

M. Braun
Equipo de Investigación en Psicología Clínica, Universidad de Belgrano,
Buenos Aires, Argentina
e-mail: malenabraun@gmail.com

S. Aufenacker
Universidad del Salvador, Buenos Aires, Argentina
e-mail: auferackeri@hotmail.com

J. Olivera
Equipo de Investigación en Psicología Clínica, Universidad de Buenos Aires,
Buenos Aires, Argentina

Consejo Nacional de Investigaciones Científicas y Técnicas (CONICET),
Buenos Aires, Argentina
e-mail: joliveraryberg@gmail.com

© Springer International Publishing Switzerland 2017

A. Ibáñez et al. (eds.), *Neuroscience and Social Science*,
DOI 10.1007/978-3-319-68421-5_13

Keywords Psychotherapy • Psychotherapy research • Social neuroscience • Clinical significance • Empathy • Interpersonal relationships • Methodological challenges

> Many persons nowadays seem to think that any conclusion must be very scientific if the arguments in favor of it are derived from twitching of frogs' legs—especially if the frogs are decapitated—and that—on the other hand—any doctrine chiefly vouched for by the feelings of human beings—with heads on their shoulders must be benighted and superstitious. William James, Pragmatism: A New Name for Old Ways of Thinking (1907).

> We must recollect that all of our provisional ideas in psychology will presumably one day be based on organic substructure. Sigmund Freud, On narcissism: An introduction. Standard Edition, (1914).

1 Introduction

As we were drafting the Introduction to this chapter, we came across an article published in *Molecular Psychiatry*, one of Nature's journals. The results showed that rats respond differently to antidepressants depending on whether they were exposed to enriched or stressful conditions [1]. One of us, as a psychotherapist, found this article fascinating but naïve, since it presented the idea that context influences the effects of antidepressant medication as something new and innovative. How could it be that the statement: "drugs do not drive changes in mood per se but, by increasing brain plasticity" was presented as something new? ([1], p. 552).

Context, as a moderator variable, had been studied with monkeys [2], where no clear pattern emerged following the administration of amphetamines, until the primate's position in social hierarchy was considered. We also recalled the classic "Rat Park study", where the environment was crucial in determining which rats would become addicts [3]. And then we asked ourselves: "what about studies showing that the effect of oxytocin depends on context?" [4]. Are environmental and social factors still considered as a side effect instead of a key aspect to be taken into account in research? Haven't the models of mediators and moderators been incorporated as a standard means to comprehend brain activity? Why is the influence of contextual factors still presented as an innovative idea? Authors such as Clark-Polner and Clark [5], have already stated that the relational context is crucial for social neuroscience. Furthermore, isn't the whole field of social neuroscience, since the early 90s, promoting the necessity of taking multilevel analysis into account? [6].

Yet, we then have the article a closer look. It was clear that the authors were looking at events from very different standpoints: while their outlook was biological, ours was psychological. We then realized that this chapter will benefit from our diverse backgrounds: psychotherapy, psychotherapy research, and neuroscience. Our diversity, with the common goal of writing this chapter, generated a space for convergence. Neuroscience and psychotherapy research have been interacting for more than 20 years, but never merged. These disciplines are still not relating in a real interdisciplinary practice, and this is a pending debt that limits the utility and the dissemination of their work. Psychotherapists' day-to-day encounters with their patients have not been significantly affected by knowledge from neuroscience.

We need to create a new space for dialogue, where both disciplines can learn from each other, creating a different and new kind of knowledge. We do not need to learn to look through the other's stained glass; we need to construct a new one. Using the metaphor of the missing link, the aim is not to find one, since that would imply that it is already somewhere, waiting to be found. The purpose of the interaction is to forge new links from scratch. We think that the interaction of social neuroscience and psychotherapy can play a key role in the creation and refinement of psychological and neuroscientific theories.

To achieve this, we must link the methods and techniques of each participating discipline, and contrast them with relevant theories. This task is quite complex, since each of these disciplines is characterized by an unequal relationship between their theories and their techniques. On the one hand, psychotherapy has a highly sophisticated and conceptually comprehensive theoretical background while social neuroscience stands out for an increasing number of methods and techniques of great technical sophistication, and high descriptive and predictive accuracy.

Associated with this unbalanced situation is the belief that neuroscience will straightforwardly provide the empirical evidence that validates the theories of psychotherapy, and that, in turn, these theories will give coherence and amalgamate the findings of neuroscience. We do not believe that this will happen in a guileless manner. Rather, this merger is more likely to occur via what Laudan [7] describes as a new "research tradition", that is, a set of general presuppositions about the entities, processes, involved in field-specific research. This new tradition would be the common, interdisciplinary ground for researchers to test the explanatory, verifiable, and predictive quality of their theories and methods, while generating new integrative theories and tools.

Perhaps one of the most defying aspects in the preparation of this chapter was the organization of the existing bibliographical information. There is a copious amount of literature on neuroscience or psychotherapy. However, this is not the case when searching for literature on the interaction of neuroscience and psychotherapy. Additionally, when the search is refined to those empirical articles about social neuroscience and psychotherapy, the literature dwindles dramatically.

This scarcity is reflected in both disciplines: journals of psychotherapy have few empirical articles that include social neuroscience concepts, and vice versa. A further example of this paucity is that the *Handbook of Psychotherapy and Behavioral Change* [8], a fundamental guide in psychotherapy research, has not dedicated a single chapter to neuroscience in any of its six editions.

Since the 1990s ("the decade of the brain", as declared by the US Congress), psychotherapy has developed theories about its intersection with neuroscience and various authors, throughout these 20 years, have said that "the time is right now". However, there are currently few research programs dedicated entirely to this convergence.

Below we evaluate the possible reasons why, despite predictions and "good intentions", this blending has not met expectations.

Throughout this chapter, we will identify different types of obstacles in the integration of psychotherapeutic and neuroscientific knowledge. What we can confirm beforehand is that the task is not easy, though it is quite fulfilling. Psychotherapy research has been studying processes and results in psychotherapy for many years. One of the limitations towards the integration is that many times it has adopted a dualistic stance, leaving out information about the biological substrates that make us human. Most (but not all) of the research has focused on the "mind", without considering the "body".

Psychotherapy can be defined in different ways according to different levels of analysis. In Table 1 we provide definitions of psychotherapy and neuroscience from different perspectives in order to illustrate this point. Throughout the chapter, we will emphasize on the importance of multilevel analysis for understanding the complexities of reality. The definitions presented are neither the best nor the only ones available.

Psychotherapy, without a doubt, is a social phenomenon. By its very nature it involves the interaction of at least two human beings (or more than two in family therapy, couple therapy, and group therapy). Psychotherapy is also a psychological phenomenon, since it aims to modify cognition, attitudes, feelings, behaviors, and

Table 1 Multilevel definitions of psychotherapy and neuroscience

Level of analysis	Possible definitions of psychotherapy	Possible definitions of neuroscience
Social	Psychotherapy aims at the nullification of deviance that is not tolerable the individual and those about him or her. A person becomes deviant by breaking normative rules, values and symbolism. Therapists help this person recombine with their home folk in some new way by applying a form of symbolic behavior that is of a different order [9]	The study of the brain as a powerful means of providing new ways for understanding individuals and societies. Neuroscientific research… contributes to longstanding debates concerning free will, morality and madness [10]
Psychological	"The informed and intentional application of clinical methods and interpersonal stances derived from established psychological principles for the purpose of assisting individuals to modify their behaviors, cognitions, emotions, and/or other personal characteristics in directions that the participants deem desirable" ([11], p. 218)	A discipline that enables the finding of biological markers to help in diagnosis, and measure patient's change due to psychotherapeutic interventions
Biological	"… a specific kind of enriched environment designed to enhance the growth of neurons and the integration of neural networks" ([11], p. 27)	A discipline that includes different approaches used to study the molecular, cellular, developmental, structural, functional, evolutionary, computational, and medical aspects of the nervous system

other psychological constructs. Last but not least, psychotherapy is a biological phenomenon since human beings exist within a body.

The same could be said of neuroscience if one looks at available definitions of the field. None of the definitions presented in Table 1 excludes the other. In order to generate an interaction between these different levels and disciplines, it is necessary to incorporate knowledge from other levels. Psychoanalysis and neurology, born in the 1890s, are young disciplines. Both had a common origin and 3 years after the discovery of the synapse Freud was attempting to develop a neurologically based psychology [13]. Despites this early attempt for integration, both disciplines grew disconnected. Psychoanalysis was the steppingstone for the emergence of psychotherapy, and neuroscience branched out from neurology. Therefore, psychotherapy and neuroscience are even younger disciplines and their lack of integration is expected.

Before ending this introduction we would like to mention that several endeavors are being made in favor of this integration and are generating knowledge relevant to clinicians. For example, The Neurospychotherapist (http://www.neuropsychotherapist.com) gives news and information to psychotherapists about the ways of integrating neuroscientific research with psychotherapy. Likewise, the neuropsychoanalytic movement, which started in the 1990s, fosters the integration of psychoanalysis and neuroscience (https://npsa-association.org).

Within the psychotherapeutic world there was a moment in which any attempt of uniting psychotherapy and neuroscience was seen as a threat. Research groups and practitioners who see psychotherapy and neuroscience as an antagonist phenomenon are declining in number and strength, yet there are still difficulties for their integration. The first one is the belief that "someone else will do it, not us". This is a fierce obstacle because psychotherapy has not yet incorporated itself in the mainstream neuroscientific elements or interdisciplinary work. Another risky posture is to think that this interdisciplinary integration requires no extra work. Researching psychotherapy and neuroscience is not just pushing boundaries; it is confronting prejudice and, above all, creating new conceptual spaces and methodologies.

Historically, social sciences gave psychotherapy the epistemological and methodological approach for its development and validation. The fundamental core of its theoretical and clinical developments was psychosocial, isolating psychotherapy, for several decades, from the neurobiological processes. However, psychotherapy never resigned its neurobiological origins. Attempts at integrating neuroscience with psychotherapy have always being marked by a prevalence of one over the other. Such an interaction has been more like a fight for power than a real collaboration. Social neuroscience can act as a space in which new hypotheses can emerge, and old ones can be tested. This space should create new terminology to unify each of these fields without imposing one's terminology over the other. Integrative multilevel analysis allows us to think in a merged perspective rather than taking an antagonistic stance.

In this chapter, we first review studies integrating psychotherapy and neuroscience. Subsequently we focus on empathy and interpersonal relationships, as these concepts can bridge social neuroscience with constructs such as therapeutic alliance and emotional regulation, which were traditionally used by psychotherapy. Finally, we discuss whether the integration of these two worlds allow us to generate a truly practice-oriented research with valid information to empower practitioners (psychotherapists, psychiatrists and other mental health professionals), enlighten society, and, more importantly, help individuals by preventing suffering. We believe that an integration of neuroscientific and psychotherapeutic investigations will enable more precise methods of validation of our knowledge of psychotherapy, contributing to the efficacy and effectiveness of psychotherapy.

2 Results from the Literature Review

Our initial search identified a total of 434 articles. It was done using PubMed, PsycINFO, and Google scholar. At first we used the following keywords: psychotherapy AND neuroscience. In addition, reference lists of the identified articles were inspected for additional relevant studies. Also, we used Researchgate (researchgate.com) as an additional source. Articles were divided into four broad categories, or axes (Fig. 1), namely: (1) empirical studies on psychotherapy, using neuroscientific

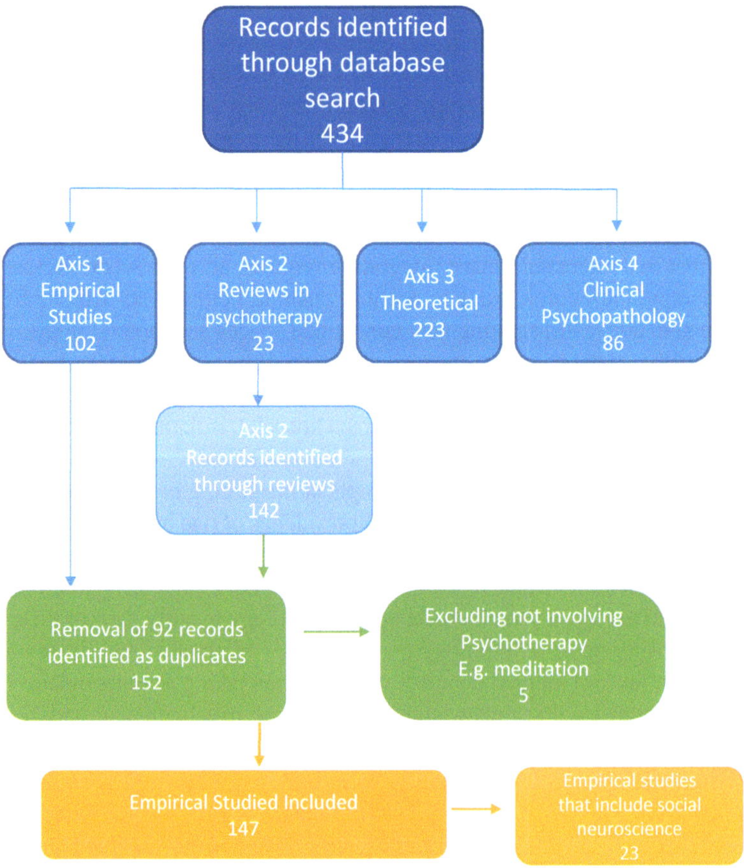

Fig. 1 Records findings

methods; (2) reviews of empirical studies on psychotherapy and neuroscience; (3) theoretical articles about psychotherapy and neuroscience; and (4) empirical articles about psychopathology, using neuroscientific methods.

It is important to consider that at first we decided to perform a non-systematic search, and this decision was based on different reasons. We wanted to adopt an exploratory stance in which defining a closed list of key words a priori could have limited potential results. We chose a more "qualitative" search, where new words were included in the search when they appeared in the articles found. For example, our first broad search using "neuroscience" and "psychotherapy" yielded articles about oxytocin and psychotherapy, and from that we performed a new search using "oxytocin" as a key word. Our search resembled a "snowball" sampling. As an effect of this type of search, different bias can be observed in our results. For example, there is an over-representation of neuroimaging studies compared to hormonal studies. This could be correctly representative of the actual distribution, or a search bias.

In a second instance we specifically searched the empirical articles cited in the reviews that we had not found before, arriving at a total of 147 empirical studies on psychotherapy and neuroscience (axis 2). The information incorporated to axes 3 and 4 was used to clarify concepts and the understanding of the material in axes 1 and 2.

3 Description of the Empirical Studies

The most frequent designs were randomized clinical trials (RCTs), featuring an experimental group that receives psychotherapy or a psychotherapeutic intervention and a control group that receives a proven pharmacological treatment, a placebo or remains in a waiting list until the end of the trial. Both groups had the same diagnosis, based on the Diagnostic Statistic Manual (DSM) [14] and/or the ICD [15]. Psychological assessment and a neuroscientific measure were administered pre- and post-intervention. In some studies, participants had to perform a task as they completed a neuroimaging protocol. Other designs compared a group of patients diagnosed with a specific disorder with a healthy matched sample, or, in rare occasions, two different psychotherapeutic treatments. Outcomes were assessed via traditional rating scales, such as the Beck Depression Inventory (BDI-II) [16] or the Hamilton Depression Scale (Ham-D) [17] for major depressive disorder, or the Yale-Brown OCD scale for OCD [18]—these being the most frequent disorders assessed. Other global treatment response scales were also used, such as the Clinical Global Impressions Severity Scale (CGI-S) and the Global Assessment of Functioning (GAF) [19], or the SCL-90 [20].

Most of the research reviewed used neuroimaging, such as fMRI, MRI, PET scan, and SPECT. Other tools employed measures of EEG, DNA, oxytocin level, cortisol level, MRS, and facial behavior (Fig. 2).

As previously mentioned, the most frequent disorders studied were OCD and MDD, followed by a range of anxiety disorders –PTSD, panic disorder, panic disorder with agoraphobia, specific phobia (mostly spider phobia), and social anxiety dis-

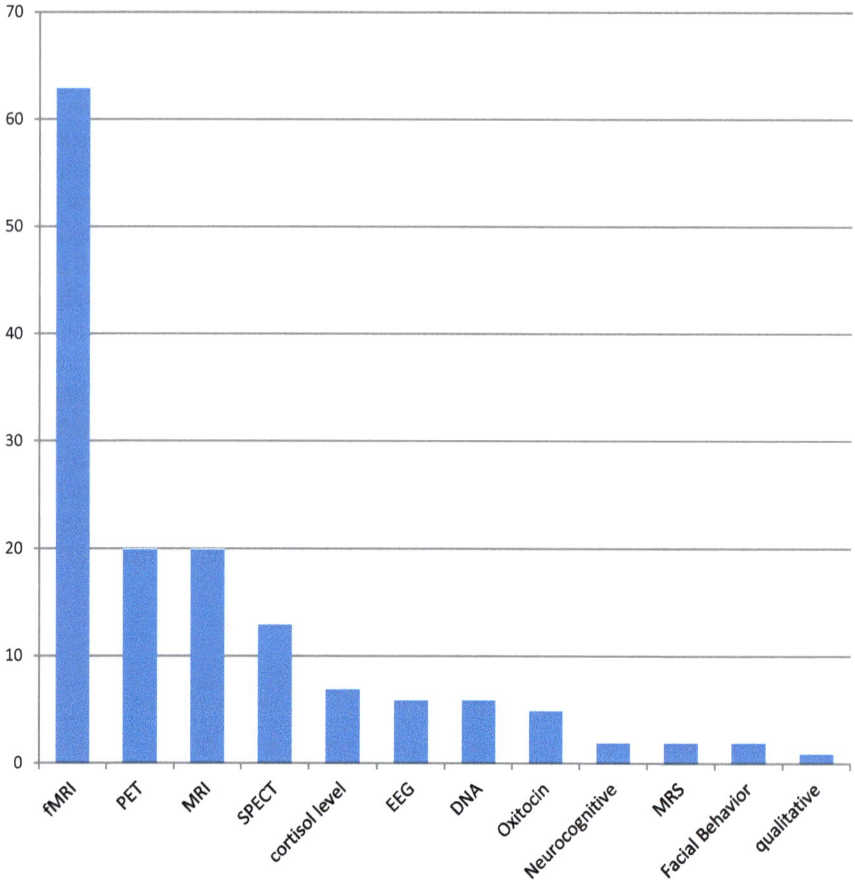

Fig. 2 Biological measures (N = 147). Note: Please see list of abbreviations

order–, and schizophrenia. Less frequently, with one or two studies each, we found research studies on substance dependence, dysthymia, game addiction, fibromyalgia, personality disorders, somatoform disorder, brain damage, chronic fatigue syndrome, Alzheimer's disease, clinical burnout, and post-natal depression (Fig. 3).

A variety of therapies and therapeutic interventions were included among the studies. Cognitive behavioral therapies (CBT) were the most widely employed, followed by psychodynamic approaches, exposure therapies, and behavioral therapies. More specific treatments, such as other cognitive therapies (including cognitive remediation therapy, cognitive rehabilitation therapy, and cognitive restructuring therapy), mindfulness and MBCT, interpersonal psychotherapy, Internet-delivered CBT and behavioral activation therapy were also applied. A large "other therapies" category was also created with those therapies present in only one or two articles. This category included family therapy, problem-solving therapy, brief eclectic therapy, and the Rosen method (Fig. 4).

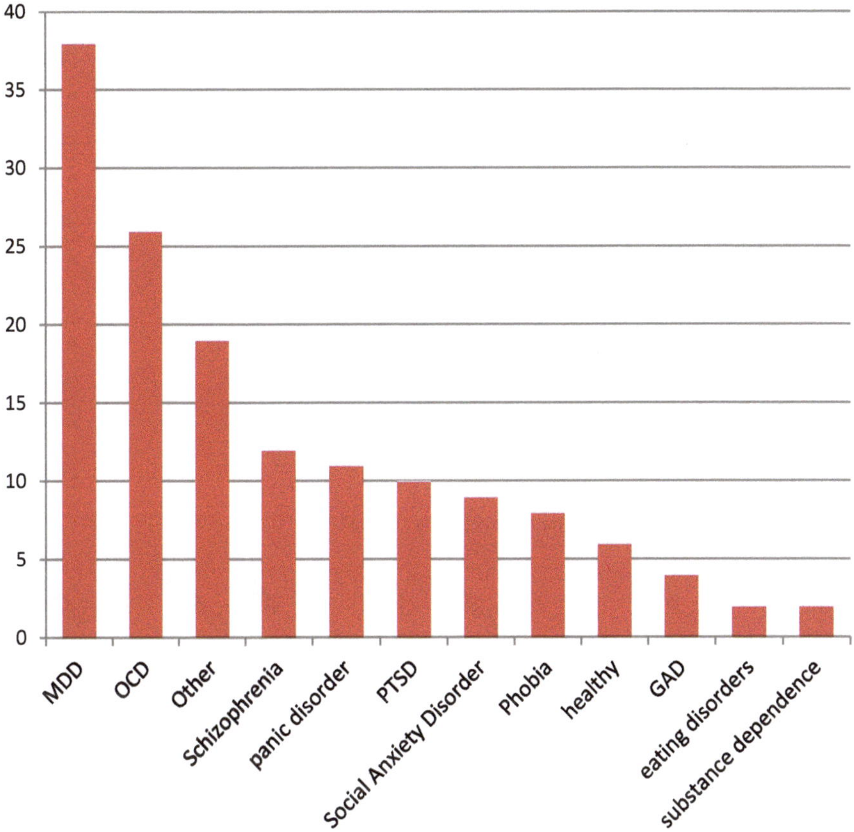

Fig. 3 Client's diagnosis ($N = 147$). Note: Please see list of abbreviations

The reviews reflected the same tendencies (Table 2). Not surprisingly, the most widely reviewed disorders were "emotional disorders". Neuroimaging studies were also the most utilized, showing the difficulty of finding alternative methods, such as hormonal and genetic studies, when the search used neuroscience as key word.

As it was observed, there is a great disproportion favoring outcome studies over those that study process or process-outcome. Frewen et al. [21] could not find in their review studies that included measures of psychological mechanisms of change, such as measures of alliance, interpersonal relations. Even though this trend is currently changing [22], there is still a great potential for interdisciplinary research in this area.

When looking at the reviews by year of publication, a change in trend of research can be inferred. Reviews from 2005 to 2015 aimed at linking psychotherapy interventions with observable changes in the brain. All the reviews from 2016 and 2017 have the word "prediction" in their title (and, of course, in their aims). This reflects a shift in the research agenda, from the existence of outcomes towards finding biomarkers that could predict results.

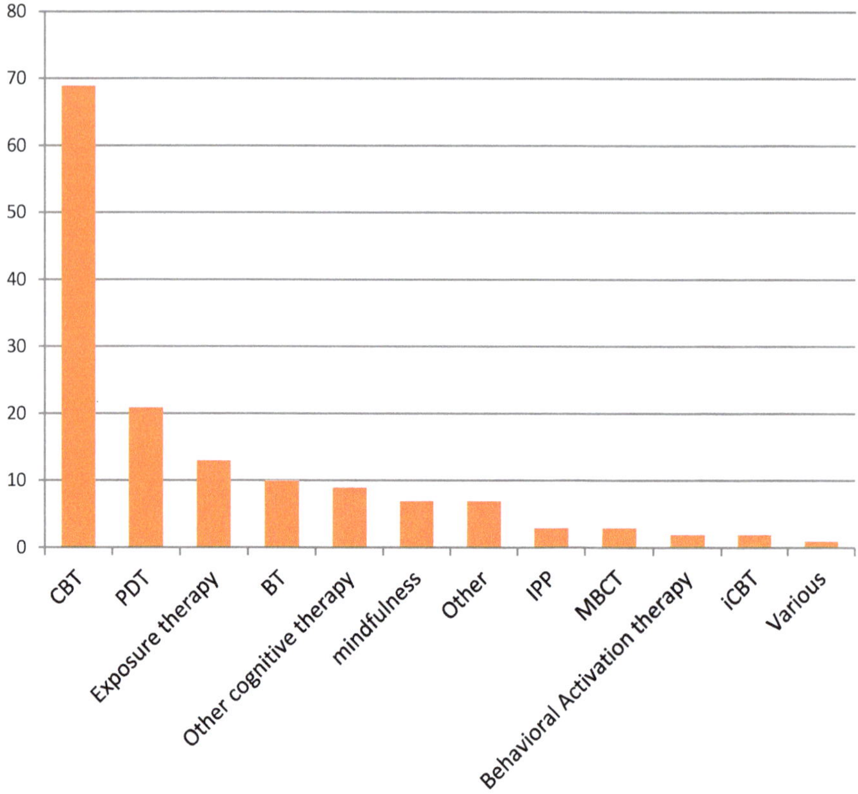

Fig. 4 Types of psychotherapy ($N = 147$). Note: Please see list of abbreviations

Both the analysis of the original investigations and the reviews show that there is still a lot of space for fruitful research. For example, using Cozolino's words, patients that are "somewhat less ill" have not been included in neuroscientific research [12], despite their importance for psychotherapy. Other groups that have still not been included in research are those with comorbidities and with "unspecified" disorders (a DSM category that is frequently used in clinical practice).

Finally, we found that neuroscience still needs to integrate within itself. There are almost no studies using more than one biological measure combined with behavioral measures. For example, it would be interesting to include hormonal measures, neuroimaging techniques, self reports and qualitative information. The usefulness of qualitative information in interdisciplinary research will be discussed later.

Social neuroscience and psychotherapy have been independently studying common concepts, such as interpersonal relations, empathy, mentalization, theory of mind, attachment, and attunement [50]. As shown in Fig. 5, the two most widely studied common concepts between social neuroscience and psychotherapy were empathy and interpersonal relations. Below we discuss these concepts analyzing their meaning for psychotherapy research and their potential for social neuroscience.

Table 2 Reviews on psychotherapy and neuroscience

Study	*n*	Techniques	Disorder/condition	Psychotherapy type
Abbass et al. [23]	11	Neuroimaging	Depression (atypical and typical), mixed depression, borderline personality disorder, panic disorder and somatoform disorder	PDT
Barsaglini et al. [24]	42	Neuroimaging	Not specified	Not specified
Beauregard [25]	7	Neuroimaging	Ocd panic depression spider phobia	Not specified
Beutel and Huber [26]	*	Neuroimaging	Not specified	PDT
Brooks and Stein [27]	19	fMRI	Anxiety and related disorders	CBT
Chakrabarty et al. [28]	40	Neuroimaging	MDD and anxiety	Not specified
Coloven et al. [29]	20	Biomarkers	PTSD	Evidence-based trauma-focused psychotherapies (i.e., PE, CPT, EMDR, CBT)
Fischer and Cleare [30]	6	Cortisol	Anxiety	Not specified
Fournier and Price [31]	*	Neuroimaging	Anxiety and depression	Not specified
Frewen et al. [21]	11	Neuroimaging	Mood and anxiety	CBT, IPT
Gonçalves et al. [32]	12	Biomarkers	PTSD	CBT
Jeon and Kim [33]	9	Neuroimaging	Depression	Not specified
Karlsson [34]	19	Neuroimaging	Depression, anxiety disorders, and borderline personality disorder	Not specified
Kumari [35]	8	Neuroimaging (PET, SPECT, fMRI)	Not specified	Not specified
Linden [36]	11	fMRI	Not specified	Not specified
Lueken and Hahn [37]	26	fMRI	Anxiety and depression	CBT and PDT
Lueken et al. [38]	60	Neurobiological markers	Anxiety	Not specified
Marano et al. [39]	16	Neuroimaging	Not specified	Not specified
Mason et al. [40]	15	Neuroimaging	Not specified	CBT

(continued)

Table 2 (continued)

Study	*n*	Techniques	Disorder/condition	Psychotherapy type
Messina et al. [41]	16	Neuroimaging	Depression and anxiety (except OCD)	Not specified
Peres and Nasello [42]	21	Neuroimaging	Not specified	Not specified
Porto et al. [43]	10	Neuroimaging	Anxiety	CBT
Quidé et al. [44]	63	Neuroimaging	Anxiety and MDD	Not specified
Roffman et al. [45]	14	Neuroimaging	Anxiety and depression	Not specified
Sharpley [46]	*	Neurobiological	Depression	Not specified
Sözeri-Varma and Karadağ [47]	11	Neuroimaging	MDD	Not specified (CBT; interpersonal, psychodynamic)
Thorsen et al. [48]	16	Neuroimaging	OCD	CBT and PDT
Weingartem and Strauman [49]	90	Neuroimaging	Not specified	Not specified

Notes: *Number of articles reviewed not specified. "Not specified" means that parameters were not fixed in the search for articles, but described in the results

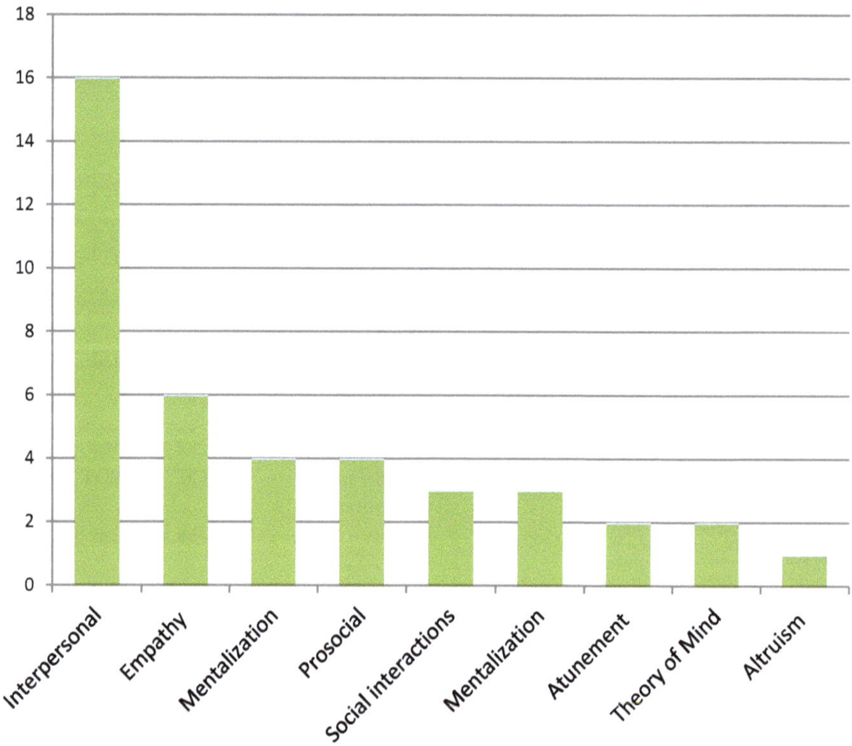

Fig. 5 Social neuroscience concepts (*N* = 23). Note: Please see list of abbreviations

4 Empathy and Interpersonal Relations: A Common Ground for Research

When the other person is hurting, confused, troubled, anxious, alienated, terrified; or when he or she is doubtful of self-worth, uncertain as to identity, then understanding is called for. The gentle and sensitive companionship of an empathic stance… provides illumination and healing. In such situations deep understanding is, I believe, the most precious gift one can give to another. Carl R. Rogers

4.1 Empathy

Ever since Titchener [51] coined it at the beginning of the twentieth century, the word "empathy" has been frequently used and everyone thinks they know what it means (see Kumfor et al., and Felisberti, this volume). The scientific literature shows this agreement, but in an opposite way, since there is only one thing every paper about empathy agrees on: there is no consensual definition of empathy. The disrupting fact is that, at the same time, we all have a similar representation of what empathy is, so why is it so difficult to find a universally accepted definition?

Among its many definitions, empathy is regarded as the ability to accurately infer another person's thoughts and feelings [52]. Metaphors such as "feeling another's shoes", "resonating" or "moving towards" try to expand this definition by indicating that empathy not only means comprehending but also feeling at least partially as the other person. In clinical psychology, Bohart and colleagues [53] state that empathy in the psychotherapy session is a cooperative dialogical process vividly grounded in the body.

Across different therapeutic approaches, empathy has always been considered an important aspect of psychotherapy. Since Carl Rogers emphasized the importance of therapist's empathy for patient-centered therapy [54], this concept has been studied and several measures were developed to assess it. Measures such as the Barret-Lennard Relationship Inventory (BLRI), which has an empathy subscale and was developed specifically to measure the therapeutic relationship, have been systematically used to evaluate therapist's empathy at different stages of psychotherapy [55]. As a result of these studies, therapist's empathy has been shown to be related to treatment outcome. A meta-analysis conducted in 2002 established a mean r of 0.32 while correlating therapist's empathy and outcome, concluding that therapist's empathy explains approximately 10% of variance in outcome, this number being larger than the specific intervention used [53]. However, among the large number of empirical studies regarding therapist's empathy, it is observed that the patient's experience of their therapists' empathy is a better predictor of patient's change than, for example, any specific intervention. This finding indicates that empathy is a moderately strong predictor of therapy outcome [56, 57].

Additionally, authors argued about the importance of patient's empathy [58]. There's a shift from considering that only therapist's empathy is important in psychotherapy to considering it as fundamentally interpersonal, thus including both therapists and patients. These authors sustain that when patients lack empathy, they may have a difficult time engaging in an effective working relationship [58].

Up to date, most research on empathy and psychotherapy is based on self-rating scales and measures. However, efforts are being made to establish empathy's bio-markers [59]. The first association between "psychological" empathy and "neural" empathy were the so-called "mirror neurons". These neurons would activate when someone observes an activity, in a similar way than when performing it [60]. Skin conductance [59] and oxytocin levels [61] have also been related to empathy. High levels of the latter, for example, have shown to enhance empathy in the "other perspective" condition–imagining another person in pain [61]. In a sample of pseudo-patients and therapists, electrodermal response was measured, and it correlated to an observer rated measure of empathy, adding evidence to a somatic underpinning of empathy [41]. In terms of brain location of empathy, Farrow and colleagues [62], for example, conducted an RCT with PTSD patients and found that activation of the middle temporal gyrus changes after modified CBT treatment. Other studies on empathy in patients usually relate to impairment, showing, for example, a distinction in brain areas of affective empathy—temporal structures—and cognitive empathy—frontal structures [50].

The relevance of empathy for both psychotherapy and social neuroscience is well established (for an example see [52]). We can think of concepts like empathy as models for the foundation of a new research tradition, which might help to answer both old and new questions.

4.2 Interpersonal Relationships

Interpersonal relationships play an essential role in psychotherapy, and are taken into account by most theoretical frameworks. Psychoanalysis emphasizes the importance of early relationships, how caregivers can interpret an infant's needs, and how the child lives its psychosexual evolution. During treatment, "transference" represents how interpersonal relationships affect the patient's life reflected in the specific relationship with the therapist [63]. Cognitive therapy postulates that our interpersonal relations are part of our core beliefs, and guide the way we perceive others and ourselves [64, 65]. In existential and humanistic psychotherapy, we see that individuals are constituted in a social atmosphere, which is pre-existent, and all of one's actions are done in this same atmosphere, so much so that at time one can get lost in the masses and loose one's individuality [66].

Interpersonal relations are also important in the psychotherapeutic process. We will analyze interpersonal relationships in two fundamental aspects: (a) the quality of the liaison between therapist and patients, named "therapeutic alliance"; and (b) attachment theory, one of the main psychological theories about how human interpersonal relationships are forged.

4.2.1 Therapeutic Alliance

A positive relationship with the therapist is a key factor for change in psychotherapy [67]. The importance of this relationship has been studied in different fields of psychotherapy, under such labels as therapeutic alliance, working alliance, therapeutic relationship, or helping alliance. It is a specific interpersonal relationship established between a therapist and a patient, oriented to help engagement with each other aimed at generating a beneficial change in the patient [68].

Bordin [69] defines therapeutic alliance differentiating three components: bonds, goals, and tasks. A positive working alliance is created by mutual acceptance of goals of the therapist and patient. A bond has to be made which is a network of positive attachment between both participating parties, hinging on tools such as trust, acceptance, and confidence. Tasks are the behaviors and cognitions that take place in the therapeutic process. For a positive outcome the therapist and patient must accept these conditions [70].

4.2.2 Attachment Theory

The importance of interpersonal relationships is explained in attachment theory, a well-known psychoanalytic theory that states that a patient's childhood interpersonal relations have a constituting effect on their personality. Attachment was first discussed by Freud and then developed by John Bowlby, and is one of the most empirically prolific domains of psychoanalytic theory. In "The Mind-brain relationship", Pally particularly develops the neuroscientific aspects of attachment, viewing it not only as a psychological phenomenon but also as a biological one. Attachment can regulate minds and bodies through non-verbal communication. Non-verbal communication carries information about bioemotional states, and regulates biological functions [71]. Pally does not use neuroscience as a way of validating theories, but rather as a way of understanding patients.

Attachment theory has surpassed psychoanalytic boundaries and has been one of the theories used to explain the natural interaction between neuroscience and psychotherapy, especially in the works of Allan Schore and Daniel Siegel [65, 72, 73]. In *The Developing Mind*, Siegel [65] presents a framework to show how interpersonal experiences shape the development of the mind and foster wellbeing. He explores how the mind is created in the interaction between biological processes and interpersonal experiences. These experiences are encoded in memory, creating, as Bowlby stated, a "secure base". The internal activation of the attachment system is also associated with how one reacts to external situations, perceiving them as threatening or not. This conditioning can be associated to the stimulation of the amygdala, which alters the sympathetic and parasympathetic systems, leading to whether or not one perceives a situation as stressful or threatening.

Emotion regulation as an aspect of attachment was predominately championed by Schore who has focused his research in the study of the importance of early experiences in personality formation [72–74]. His work shows how attachment influences the maturation of the orbitofrontal cortex, and how the social environment, mediated by attachment figures, influences brain development [72–74].

5 Methodological Challenges

After reviewing the research, we differentiated studies in two waves. The first wave, starting with Baxter in 1992 [75], represents those that tried to find whether psychotherapy outcome was related to biological changes. Genetic, hormonal, and brain mechanisms have been associated to change in psychotherapy [75–83]. This association is accepted in the mainstream, and the study of psychotherapy outcome and process via neuroimaging is now well established [49].

In the present review, we place Baxter in 1992 as the starting point [75]. However, several studies linking psychotherapy concepts and neuroscience were performed much earlier. For example, Kaplan and Bloom's [84] review on the use of sociological concepts in physiological research includes studies from the 1950s. One of the reviewed studies, titled "Physiological correlates of tension and antagonism in during psychotherapy: a study of 'interpersonal physiology'" [85], measured both patient's and therapist's heart rate during interviews. That review also includes studies on physiological aspects of personal interactions, and the physiological responses of empathy [86]. However, the previous examples are isolated events, and psychotherapy research has gone a long way since 1950.

A second wave is represented by research that tries to understand multiple determinism, generating answers for a key practitioners' question: What should I do to generate a stable and clinically significant change in my patients? This implies that researchers have to position themselves in an interdisciplinary context using concepts and tools from neuroscience and psychotherapy research, and using social neuroscience as a cornerstone.

The three principles of social neuroscience described by Cacciopo and Berntson [87] are also a characteristic of this second wave. These principles can be straightforwardly incorporated in psychotherapy research (Table 3).

Below we will present some useful concepts traditionally used in psychotherapy research that could accomplish that aim. In particular, we will address the issues of (1) their clinical significance, (2) practice-oriented research, (3) psychotherapy instruments for process studies, and (4) qualitative research.

6 Clinical Significance

The majority of the articles reviewed in the present study relied statistical significance to establish their results. Such a criterion aims to determine whether there is a significant difference between two groups, but does not account for the magnitude of this difference or for the usefulness, for patients and clinicians, of the results

Table 3 Principles of the doctrine multilevel analysis and their examples from psychotherapy research

Principles of the doctrine of multilevel analysis [87]	Examples from psychotherapy research
Multiple determinism: behaviors can have multiple antecedents within or across levels of organization	A psychotherapist generating a case conceptualization of a patient's potential diagnosis of depression will take into account medical studies, family history, employment, financial status, interpersonal relationships, social situation (e.g., economic crisis, a natural catastrophe, etc.)
Non-additive determinism: properties of the whole are not always readily predictable from the simple sum of the (initially recognized) properties of the parts	Research on mediators and moderatos of contextual elements (e.g., alliance, therapist characteristics, etc.)
Reciprocal determinism: there can be mutual influences among biological and social factors in determining behavior	Research on stress and psychological interventions (e.g., mindfulness-based stress reduction)

obtained [88]. The presence of a statistically significant effect does not guarantee the achievement of valuable and significant effects from the clinical perspective [89]. Clinical significance is the criterion used to establish whether the treatment was able to reach the efficacy parameters established by patients and psychotherapists [88, 89]. It identifies the extent to which treatments generate real and tangible effects in the lives of patients, achieving the results that patients and therapists seek when conducting a psychological consultation.

In our review, we found two studies [90, 91] that go beyond statistical significance and incorporate the concept of clinical significance, which is fundamental for understanding the quality of the change obtained by an intervention [90, 91]. The results found in studies that are based purely on statistical significance have limited conclusions to be drawn from the clinical point of view. As Jacobson and colleagues ([92], p. 306) argue, "patients who consult do not seek to achieve a statistically significant change but to achieve a reduction in their suffering." Psychotherapy researchers do not consider statistical significance as a sufficiently useful criterion when studying the effectiveness of treatments in psychotherapy [93]. Rather, they aim to generate studies that investigate the benefits of the changes produced by psychotherapy, along with their magnitude and impact on patients' daily lives [89].

Contrary to what is intuitively assumed, the evaluation of clinical significance does not imply a departure from statistics or mathematical models, but rather seeks to complement these efforts by estimating the differences between the values that represent health and disease. Clinical significance is based on fundamental mathematical resources for the measurement of its criteria. It uses quantitative parameters to account for the qualitative clinical criterion that is employed. It does not imply a break with the concept of statistical significance, since it incorporates some of the basic concepts, such as the *p*-value of probability. What clearly differentiates the notion of clinical significance is the inclusion of clinical concepts, and new statistical parameters to determine the effectiveness or efficacy of treatments, among other evaluations of clinical psychology. It is only through the interaction between clinical qualitative criteria, and sophisticated statistical-mathematical models that it is possible to generate a scientific model for evaluating the effects of an intervention in psychotherapy.

Efficacy and efficiency studies considering clinical significance are not usually contradicted by the results of other criteria, but tend to add more information [89]. Clinical significance is not intended to eradicate statistical significance or effect size. Instead, it seeks to coexist with them and relies on its methodologies: it uses the mathematical models of traditional criteria, and conjugates them with qualitative evaluations of clinical results. Therefore, clinical significance contributes with new data for a more complex understanding of the phenomenon of effectiveness in psychotherapy.

In order to assess the clinical significance of a change in psychotherapy, we must first define what should be considered a clinically significant change. In other words, what changes do clinicians and patients consider important to obtain from the process.

Ever since the first article by Jacobson et al. [94] in which this concept was introduced, different operational definitions have been enumerated to calculate clinical significance. Jacobson and Truax [89] assume, as a parameter of effectiveness, the patient's change towards normal functioning. From this theorization, patients who demand therapy are seen as part of a dysfunctional population. An effective or efficient treatment would be one that promotes change of a given patient until the patient is closer to the average performance of the functional population, instead of that of the pathological population. Some years later, the authors who created this definition [92, 95] stated that this parameter, for some disorders in which achieving normal functioning is not feasible, may be a little restrictive and excessively conservative, and should be modified and adapted to each particular situation studied. Other authors [96–99] criticize this bimodal way of perceiving the population (a functional population vs. a dysfunctional population), arguing that the distribution of the population is characterized by being part of a continuum that ranges from pathological to non-pathological. That is, they consider that there is no categorical point of demarcation that divides health and pathology. For these authors, the difference between health and disease is quantitative and non-qualitative. Therefore, individuals, should not be classified as people with or without disorders, but should be located in this continuum according to their degree of pathology or health. According to this point, Tingey et al. [97] argued that a population with the same diagnosis could be composed of people with different degrees of symptom severity and social impact of their symptoms.

Despite the criticisms of Jacobson and Truax's [89] definition, it has been the most used in the studying clinical significance [100–102]. However, as this definition of clinical significance has not satisfied many researchers in the area [88], some alternatives have been proposed, namely: (1) changes that reduce the risk of health problems [88, 89], (2) a level of change recognized by significant people [103], (3) elimination of the problem that leads to consultation [88], (4) improvements in the quality of life [99], (5) substantial changes in relationships with others [99], (6) modifications in the significant others of the patient from the therapy [99], (7) and reduction of functional disabilities [99, 104].Each of these definitions can be used for interdisciplinary research in order to include clinical significance. The information for their operationalization can be found in Kraemer et al. [88].

Kazdin [99] argues that even a treatment whose effect is the absence of changes, can generate results interpretable as clinically significant. In conditions characterized by progressive deterioration, such as dementia, the goal of therapy is not to reduce symptoms, but to maintain or delay the loss of the various functions affected. In these cases, the absence of change would be the appropriate parameter to assess the clinical significance of the process [99].

Tingey et al. [97], in the face of the already mentioned criticisms of the operational definition of Jacobson and Truax [89], they developed their own. Instead of dividing the population into two strata, a functional and a dysfunctional population, they divide it in several groups depending on the degree of social impact of the symptoms, and their possible intensity. Incorporating clinical significance—whatever the definition used—into neuroscience studies would allow greater precision in relation to patient improvement, and make studies much more attractive to practitioners.

7 Practice-Oriented Research

The gap between psychotherapy researchers' work and practitioners' interests has been debated for many years, and attempts have been made to bridge it. In recent years, a movement within psychotherapy research is gaining impetus. Practice-oriented research (POR) moves past the bridging the gap problem to the generation of a common space. POR is integrated in clinical routine. It is oriented towards collecting information which will be used in the clinical field, and involves the therapist in the processes of construction of the methodological mechanisms, as well as in the implementation and dissemination of the investigation [105]. POR allows clinicians to contribute in scientific development as well as influencing the future lines of investigation in the clinical field [106].

POR operates in a natural context, which evaluates the clinical practice and how it is normally conducted [105]. Since there are no imposed control laboratory clauses, the results obtained have more ecological validity. It is complementary with other investigation paradigms, like evidence based models, and both approaches can be mutually enriched [107, 108].

One of the most common types of POR is the monitoring of change during a session using standardized instruments [105]. Monitoring allows the study of patterns of change and the analysis of process variables associated. It has allowed the study of the effects of giving feedback about patients to therapists throughout the process.

Using this form of feedback involves a strategy to improve and motivate clinicians to participate in research. This is because the device used to gather information is at the same time a clinical resource that can help therapists with their patient's treatment [105]. The models of clinical monitoring and giving feedback to the therapist permit us to surpass potential benefit of the results of an investigation, and simultaneously improve clinical work [105].

At the same time, as stated by Fernández Alvarez et al. [109], monitoring sessions can be a valuable clinical resource that generates pieces of information, which can also be integrated in the training and supervision of therapists. Training and supervision are areas in which the integration with neuroscience is still missing.

8 Psychotherapy Instruments for Process Studies

In order to incorporate process elements to the study on psychotherapy and neuroscience, instruments traditionally used in psychotherapy research could be used. For example, the Core Conflictual Relationship Theme (CCRT) [110] is a psychodynamic process tool oriented to examine core patterns of relationship, initially using relationship anecdotes to establish them, and typically involves an exploration of early family transactions, as manifested through psychological projection and projective identification in outside life, as well as in the transference.

CCRT focuses on three aspects of a patient's central relationship: (1) conflict – their core desire or wish (W)–; (2) the response to it, elicited from other people—response of the other (RO); and their reaction in turn to that response—response of the self (RS). This technique is particularly useful for the study of social neuroscience, since it positions the subject in relation to a pattern of interpersonal and bidirectional character. It has been used by Loughead et al. [111] to evaluate brain activation during autobiographical relationship episode narratives. Although this approach is not proper of psychotherapy positions, the technique seems useful for studies of this type, as also shown by Roffmann et al. [112], who recently evaluated neural predictors of successful brief psychodynamic psychotherapy for persistent depression. These are examples of how a classic tool of psychotherapeutic process research, with more than 25 years of existence, makes it possible to carry out joint studies. There are other psychotherapy tools to evaluate process that can fit with the interdisciplinary approach.

9 Qualitative Research

Maybe stories are data with a soul. Brené Brown

On the first page of this chapter we presented a quote from William James, founder of pragmatism, calling for the scientific value of humans talking about their feelings. He warns against the illusion of science without human meaning. Qualitative research is the via *regia* to access meaning and human sense.

Qualitative researchers study individuals in their natural settings, attempting to interpret experiences in terms of the meanings people bring to them [113]. The incorporation of qualitative methods in interdisciplinary research is valuable for several reasons. The first one is that these methods enable the emergence of results that go beyond what was foreseen [56]. Qualitative data is useful for the generation of new hypothesis. For example, it can trigger new and different ideas to understand multideterminism. The second reason is that it is still the only method that enables the study of subjective meaning. The classical example of the British philosopher Gilbert Ryle illustrates in a fantastic way the necessity to inquire into subjectivity to understand reality. Here is the story: "Two boys fairly swiftly contract the eyelids of their right eyes. In the first boy this is only an involuntary twitch; but the other is winking conspiratorially to an accomplice. At the lowest or the thinnest level of description the two contractions of the eyelids may be exactly alike. From a cinematograph-film of the two faces there might be no telling which contraction, if either, was a wink, or which, if either, were a mere twitch" [114].

To understand the meaning of the wink we need to know the shared social conventions, the type of bond among the "winker," the "winkee," and the other people who are physically present (or not). If we ask the boy whether he winked or blinked we could avoid fruitless interpretations and, even if he lies, that lie could be studied as well. Gergen [115], based on Ryle's story, asserts that "the same form of biological activity may serve many different cultural functions." There are multiple levels to study winks, and blinks: biological, psychological and social. Qualitative research enables the study of meaning that transcends the three levels mentioned.

A third reason is the possibility to access perceptions of research participants. In the literature reviewed, we found an article in which Taubner et al. [116] inquires into the effect that participating in a fMRI and EEG study has both for patients and their psychotherapists. This study is part of the Hanse Neuropsychoanalysis Study (HNPS), which investigates neural correlates of change in chronically depressed patients before and after 8 and 16 months of psychoanalytic treatment. The research team consisted of psychoanalytic researchers who developed individually tailored stimuli for the fMRI and EEG. In the study groups in which the psychotherapists participated, they found that sometimes patients talked about the fMRI experience in treatment: One psychotherapist reported that the patient "... had difficulties engaging emotionally in the sentences that were presented during scanning" ([116], p. 280). While preparing this chapter, one of the authors (MB) participated as a control subject in an fMRI study, and reported the same difficulties. She recalls that while performing the Reading-the-Mind-in-the-Eyes Test, her mind used to wonder and sometimes was thinking about other things instead of what researchers were asking. Everyday concerns, such as what she wanted to eat that night, appeared in her head, conflating with researchers' instructions. She was surprised that at the end of the experiment no one asked her opinion about her performance in the tasks, and that this possible "errors" would not be taken into account. By including a post-participation interview, valuable information could be gathered which could inform further research.

The recollection of information from participants should not be restricted to what researchers define a priori, but should permit the incorporation of novel elements of what participants think and feel about their involvement in the studies. Human beings recall their experiences in the form of stories, a fundamental aspect of qualitative research.

9.1 Stories

"Although psychotherapy deals in stories, it turns out that they emerged from brain evolution to serve the purposes of increasing complexity, coordination, and connectivity between us. This is one of the many connections between interpersonal relationships and brain functioning that make psychotherapy a neuroscientific intervention" ([12], p. 463). Cozolino's challenging hypothesis about the stories and brain function could be a useful starting point for interdisciplinary research. Stories are both an effect of the brains' complexity, and also a triggering factor of complex brain activity. Incorporating stories as part of the material representing the complexity allows us to evaluate it, and to use narratives to trigger cognitive processes in another. The understanding of such dynamics requires, among other things, the capture of the meaning of these narratives and their associated complexity. Does the brain change when a human being changes the way his or her story is told? This could be a testable hypothesis in the future.

Corrective experience (CEs) is another concept from psychotherapy that can be an excellent candidate for interdisciplinary research. According to Castonguay and Hill [117], CEs in psychotherapy are "ones in which a person comes to understand or experience affectively an event or relationship in a different and unexpected way". Consider the following patient's narrative of a CEs which was obtained using qualitative methods [118]: "She told me that I had to understand that every morning lasted at least one year…When I listened to my therapist, her posture, her voice and calmness, gave me a peaceful sensation that had almost never happened to me".

The possibilities of interdisciplinary research of this material are numerous. For example, if we were able to look at the video from the session, it would be possible to double check whether the posture and voice of the therapist corresponds with a calm communication. Also, we could observe face movements patient's arousal. We could even include different biological measures (heart rate, EEG, or others), and every new piece of information could give us different parts of the puzzle. But only the patient is able to identify that for him or her it was a key moment when she recalls: "For me it was a key moment of therapy" [118]. External markers, including different sources, like patients self reports or biological information, can bring information to identify, clarify their qualities and maybe promote those CEs.

It seems that when CEs produces change, the patient is able to incorporate that change, condensing different kind of experiences, emotions, body sensations, and thoughts into one specific narrative [118]. Following an idea by Lane et al. [119], there are possible neurobiological explanations for the role of narratives and CEs in producing change in psychotherapy. Extracting meaning out of narratives requires intensive training and can be intimidating for researchers with a more experimental background. This kind of interdisciplinary research is not easy but its rewards are fulfilling.

10 Discussion

The aim of this chapter was to organize current information on the intersection of psychotherapy and social neuroscience. The main idea was to see whether this intersection really happened, and how can be enhanced. Figure 6 shows some of the fundamental components of the psychotherapeutic process, and concepts from social neuroscience in order to present a preliminary framework for interdisciplinary research.

Context, represented by the grey square, is the environment that surrounds psychotherapy. It includes the social networks of both patient and therapist, patient's family, institutions were patients and therapists belong, etc. Context also incorporates aspects of the country and city where therapy takes places, economic conditions, climate and every other imaginable aspect that could affect human behavior. Context has an effect before treatment, throughout treatment, and after the therapeutic intervention has ended.

Patient and therapist (represented in red) are each individually affected with their biological, sociological and psychological individual factors. It would be impossible to enumerate all these factors, but as an example, factors that have been studied in psychotherapy research are: religion, marital status, theoretical framework (therapist), age disparity, etc. Once therapist and patient (or patients) are working together, we need to analyze them as a dyad (green color). Theoretical concepts of this dyad studied in psychotherapy are, for example, therapeutic alliance, matching, corrective experiences, therapeutic interventions, etc. We mentioned earlier that psychotherapy research (represented in blue) and neuroscience (represented in yellow) have their own set of concepts and instruments to evaluate psychotherapeutic interventions. Concepts and techniques can be used by both disciplines (e.g. attachment comes from psychoanalytic theories), what really makes a difference is the generation of a common ground to work with. The key space of Fig. 6 is shared by those common concepts (also in green), this space must grow by incorporating terminology from different disciplines and, if possible, by creating new concepts in interaction.

All the central (non-methodological) components that make up Fig. 6 can (and should) be studied, both synchronously and diachronic. For example, evaluations of the figure and role of the psychotherapist should be studied in relation to the figure and role of the therapist in that specific dyad. This therapist maintains common factors in his/her treatments with other patients, but nevertheless that patient and therapist dyad (or multiple dyads in the same study) is a specific situation of study.

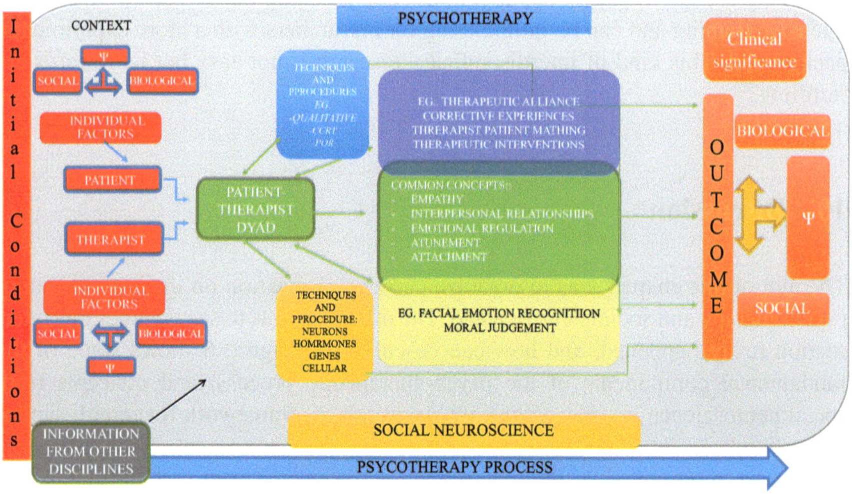

Fig. 6 Psychotherapeutic process and social neuroscience components

Outcomes (in orange) can be analyzed from different perspectives, considering behavioral measures, physiological factors, neuroimages, psychological constructs, social relationships, etc. Clinical significance could be a useful tool to evaluate the results of the psychotherapeutic treatment. Another characteristic of Fig. 6 is that it must be considered in a temporal sense as a continuum, this means that the conditions of onset are related to the whole psychotherapeutic process, as well as to the results of the same and the potential follow-up evaluations.

To date only a small proportion of studies on psychotherapy and neuroscience do so from a social perspective. Psychotherapy is an interpersonal process based on a special relation between a therapist and a patient (or a group of them). This interpersonal relationship implies common emotions (not necessary the same emotions), a common aim (explicit or implicit), and therefore a common task with a unique sense and meaning (or maybe a common misunderstanding).

Social neuroscience re unifies those aspects shifting from individual brains to social brains, analyzing the mutual space of creation and interaction [50, 120]. Psychotherapy changes whole system of emotions and cognitions and, associated with that, the involved brain activities.

Social neuroscience encourages studies with ecological validity and therefore greater use of their data to provide answers to clinicians. It is necessary to develop interdisciplinary work teams where each member is willing to both share their knowledge and learn form the others. A true interaction requires that researchers and clinicians accept that no one can explain the complexity of the brain and of the human beings by themselves. A very good example of an interdisciplinary team with research design as well as a high degree of ecological validity is the study of Martinez et al., where they study depression and anxiety symptoms in sessions with both therapist and patient connected to an EEG sensor [121].

In recent years, interdisciplinary studies are being promoted by the scientific community and have become almost mandatory. For example, in 2008 the National Institute of Mental Health (NIMH) presented a proposal denominated Research Domain Criteria (RDoC). This project, originally led by Thomas Insel and supported by the NIMH, has been perhaps one of the most complex and systematized proposals aimed at putting together an interdisciplinary model on psychopathology.

This framework points to the generation of a new classification scheme for mental health, which should include both biological and behavioral components. This should be achieved by integrating multiple levels of information (from genetic to self-reports) to achieve an understanding of the basic dimensions of the underlying functioning of the broad range of human behaviors, whether normal or abnormal.

The RDoC project is not a perfect model, but it represents a group of well-consolidated hypotheses to be tested. It also fits comfortably with the principles of the doctrine of multilevel analysis. It is a challenge for any research group, since it requires abandoning the comfort zone of our specific knowledge. However, the need for interdisciplinary work could be so great that our teams should have sociologists, anthropologists, psychologists, psychiatrists, biologists, physicists, and philosophers, among others. This is impossible, and may even be detrimental. Applying the metaphor of stained glass, if every discipline represents a color, using all of them at the same time would result in black and nothing could be seen. What we are proposing is a space for interdisciplinary research, in which researchers should be aware of the knowledge that is not present in our laboratories and complete those absences by reading the papers corresponding to those areas.

Bott et al. [122] warn psychotherapists against the incorporation of unwarranted neuroscientific language in their practice. The same could be said for social neuroscience researches about the incorporation of psychotherapeutic language. To grab theories without understanding their coherence or internal logic can be a risky business.

This chapter was written by psychotherapists and psychotherapy researchers, and it would have been different if it had been written by neuroscientists. Even though we claim for interdisciplinary research, we also believe the differences of disciplines are useful. We can interact, create common spaces and then go back, enriched, to our own spaces. As psychotherapists we always have to remember that "patients do not seek treatment for changes in blood flow or brain metabolism, but for subjective difficulties, suffering, and so forth" [26]. What we are talking about in this chapter is not just a fad or the conquest of one scientific field over another. What we are talking about is a true transformation of neuroscience and psychotherapy as disciplines, and the future looks bright for this interdisciplinary work aimed at forging new links.

References

1. Alboni A, van Dijk RM, Poggini S, Milior G, Perrotta M, Drenth T, Brunell N, Wolfer DP, Limatola C, Amrein I, Cirulli F. Fluoxetine effects on molecular, cellular and behavioral endophenotypes of depression are driven by the living environment. Mol Psychiatry. 2017;22:552–61.
2. Haber SN, Barchas PR. The regulatory effect of social rank on behavior after amphetamine administration. In: Barchas PR, editor. Social hierarchies: essays toward a sociophysiological perspective. Westport: Greenwood Press; 1983.
3. Alexander BK, Beyerstein BL, Hadaway PF, Coambs RB. Effect of early and later colony housing on oral ingestion of morphine in rats. Pharmacol Biochem Behav. 1981;15:571–6.
4. Bartz JA, Zaki J, Bolger N, Ochsner KN. Social effects of oxytocin in humans: context and person matter. Trends Cogn Sci. 2011;15:301–9.
5. Clark-Polner E, Clark MS. Understanding and accounting for relational context is critical for social neuroscience. Front Hum Neurosci. 2014;8:1–14.
6. Cacioppo JT, Decety J. Social neuroscience: challenges and opportunities in the study of complex behavior. Ann N Y Acad Sci. 2011;1224(1):162–73. https://doi.org/10.1111/j.1749-6632.2010.05858.x
7. Laudan L. Progress and its problems: towards a theory of scientific growth. Berkeley: University of California Press; 1977.
8. Lambert M. Bergin Garfield's handbook of psychotherapy and behavioral change. 6th ed. New York: Wiley; 2013.
9. Owen IR. Towards a sociology of psychotherapy. Couns Psychol Rev. 1993;8:6–9.
10. Pickersgill M. The social life of the brain: neuroscience in society. Curr Sociol. 2013;61:322–40.
11. Norcross JC. An eclectic definition of psychotherapy. In: Zeig JK, Munion WM, editors. What is psychotherapy? Contemporary perspectives. San Francisco: Jossey-Bass; 1990. p. 218–20.
12. Cozolino L. The neuroscience of psychotherapy. 2nd ed. New York: W. W Norton & Company; 2010.
13. Freud S. On narcissism: an introduction. In: Strachey J, et al., editors. The standard edition of the complete psychological works of Sigmund Freud, vol. XIV. London: Hogarth Press; 1914.
14. American Psychiatric Association. Diagnostic and statistical manual of mental disorders. 5th ed. Arlington: American Psychiatric Publishing; 2013.
15. World Health Organization. International statistical classification of diseases and related health problems, 10th revision (ICD-10). 1992.
16. Beck AT, Steer RA, Brown GK. Manual for the Beck depression inventory–II. San Antonio: Psychological Corporation; 1996.
17. Hamilton M. Development of a rating scale for primary depressive illness. Br J Soc Clin Psychol. 1967;6(4):278–96.
18. Goodman WK, Price LH, Rasmussen SA, Mazure C, Fleischmann RL, Hill CL, Henninger GR, Charney DS. The Yale-Brown obsessive compulsive scale: part I. Development, use, and reliability. Arch Gen Psychiatry. 1989;46:1006–12.
19. Endicott J, Spitzer RL, Fleiss JL, Cohen J. The global assessment scale; a procedure for measuring overall severity of psychiatric disturbance. Arch Gen Psychiatry. 1976;33:766–71.
20. Derogatis LR. SCL–90–R: administration, scoring, and procedures manual II. Towson: Clinical Psychometric Research; 1983.
21. Frewen PA, Dozois DJA, Lanius RA. Neuroimaging studies of psychological interventions for mood and anxiety disorders: empirical and methodological review. Clin Psychol Rev. 2008;28(2):229–47.

22. Buchheim A, Labek K, Walter S, Viviani R. A clinical case study of a psychoanalytic psychotherapy monitored with functional neuroimaging. Front Hum Neurosci. 2013;7:677. https://doi.org/10.3389/fnhum.2013.00677.
23. Abbass AA, Nowoweiski SJ, Bernier D, Tarzwell R, Beutel ME. Review of psychodynamic psychotherapy neuroimaging studies. Psychother Psychosom. 2014;83:142–7.
24. Barsaglini A, Sartori G, Benetti S, Pettersson-yeo W, Mechelli A. The effects of psychotherapy on brain function: a systematic and critical review. Prog Neurobiol. 2013;114:1–14.
25. Beauregard M. Mind does really matter: evidence from neuroimaging studies of emotional self-regulation, psychotherapy, and placebo effect. Prog Neurobiol. 2007;81:218–36.
26. Beutel ME, Huber M, Beutel ME, Mainz MH. Functional neuroimaging—can it contribute to our understanding of processes of change? Neuropsychoanalysis. 2008;10:5–16.
27. Brooks SJ, Stein DJ. A systematic review of the neural bases of psychotherapy for anxiety and related disorders. Dialogues Clin Neurosci. 2015;17:261–79.
28. Chakrabarty T, Ogrodniczuk J, Hadjipavlou G. Predictive neuroimaging markers of psychotherapy response. Harv Rev Psychiatry. 2016;24:396–405.
29. Colvonen APJ, Glassman LH, Crocker L, Buttner MM, Orff H, Schiehser DM, Norman SB, Afari N. Pretreatment biomarkers predicting PTSD psychotherapy outcomes: a systematic review. Neurosci Biobehav Rev. 2017;75:140–56.
30. Fischer S, Cleare AJ. Cortisol as a predictor of psychological therapy response in anxiety disorders−systematic review and meta-analysis. J Anxiety Disord. 2017;47:60–8. https://doi.org/10.1016/j.janxdis.2017.02.007.
31. Fournier JC, Price RB. Psychotherapy and neuroimaging. Focus. 2014;12(3):290–8.
32. Gonçalves R, Lages AC, Rodrigues H, Pedrozo AL, Silva E, Coutinho F, Neylan T, Figueira I, Ventura P. Potential biomarkers of cognitive behavior-therapy for post-traumatic stress disorder: a systematic review. Rev Psiq Clin. 2011;38:155–60.
33. Jeon SW, Kim YK. The effects of psychotherapy on brain function—major depressive disorder. In: Major depressive disorder-cognitive and neurobiological mechanisms; 2015. 10.13140/RG.2.1.4498.6088.
34. Karlsson H. How psychotherapy changes the brain. Psychiatr Times. 2011;28:1–5.
35. Kumari V. Do psychotherapies produce neurobiological effects ? Do psychotherapies produce neurobiological effects? Acta Neuropsychiatr. 2006;18:61–70.
36. Linden DEJ. How psychotherapy changes the brain–the contribution of functional neuroimaging. Mol Psychiatry. 2006;11:528–38.
37. Lueken U, Hahn T. Functional neuroimaging of psychotherapeutic processes in anxiety and depression: from mechanisms to predictions. Curr Opin Psychiatry. 2016;29:25–31.
38. Lueken U, Zierhurt K, Hahn T, Starbe B, Kircher T, Reif A, Richter J, Hamm A, Witterchen H-U, Domscke K. Neurobiological markers predicting treatment response in anxiety disorders: a systematic review and implications for clinical application. Neurosci Biobehav Rev. 2016;66:143–62.
39. Marano G, Traversi G, Nannarelli C, Pitrelli S, Mazza S, Mazza M. Functional neuroimaging: points of intersection between biology and psychotherapy. Clin Ter. 2012;163:445–56.
40. Mason L, Peters E, Kumari V. Functional connectivity predictors and mechanisms of cognitive behavioural therapies: a systematic review with recommendations. Aus New Zeal J Psychiatry. 2016;50:1–11.
41. Messina I, Palmieri A, Sambin M, Kleinbub JR, Voci A, Calvo V. Somatic underpinnings of perceived empathy: the importance of psychotherapy training. Psychother Res. 2012;23(2):169–77.
42. Peres J, Nasello AG. Psychotherapy and neuroscience: towards closer integration. Int J Psychol. 2007;43:1–15.
43. Porto PR, Oliveira L, Mari J, Volchan E, Figueira I, Ventura P. Does cognitive behavioral therapy change the brain? A systematic review of neuroimaging in anxiety disorders. J Neuropsychiatr Clin Neurosci. 2009;21:114–25.

44. Quidé Y, Witteveen AB, El-Hage W, Veltman DJ, Olff M. Differences between effects of psychological versus pharmacological treatments on functional and morphological brain alterations in anxiety disorders and major depressive disorder: a systematic review. Neurosci Biobehav Rev. 2012;36(1):626–44. Available from: http://www.sciencedirect.com/science/article/pii/S0149763411001710.
45. Roffman J, Macri C, Glick D, Dougherty DD, Rauch S. Neuroimaging and the functional neuroanatomy of psychotherapy. Psychol Med. 2017;35:1385–98.
46. Sharpley CF. A review of the neurobiological effects of psychotherapy for depression. Psychother Theor Res Pract Train. 2010;4:603–15.
47. Sözeri-varma G, Karada F. The biological effects of psychotherapy in major depressive disorders: a review of neuroimaging studies. Sci Res. 2012;3:857–63.
48. Thorsen AL, van den Heuvel OA, Hansen B, Kvale G. Neuroimaging of psychotherapy for obsessive compulsive disorder: a systematic review. Psychiatry Res Neuroimaging. 2017;233(3):306–13. https://doi.org/10.1016/j.pscychresns.2015.05.004.
49. Weingarten CP, Strauman TJ. Neuroimaging for psychotherapy research: current trends. Psychother Res. 2015;12:1–29.
50. Ibáñez A, García AM, Esteves S, Yoris A, Muñoz E, Reynaldo L, et al. Social neuroscience: undoing the schism between neurology and psychiatry. Soc Neurosci. 2016:1–39. Available from: https://doi.org/10.1080/17470919.2016.1245214.
51. Titchener BE. Lectures on the experimental psychology of the thought processes. Philos Rev. 1909;19:341–4.
52. Ickes W. Empathic accuracy: its links to clinical, cognitive, developmental, social, and physiological psychology. In: Decety J, Ickes I, editors. The social neuroscience of empathy. Cambridge: MIT Press; 2009. p. 57–70.
53. Bohart AC, Elliott R, Greenberg LS, Watson JC. Empathy. In: Norcross J, editor. Psychotherapy relationships that work; 2002. p. 89–108.
54. Rogers CR. The necessary and sufficient conditions of therapeutic personality change. J Cons Psychol. 1957;21:95–103.
55. Olivera J, Braun M, Roussos AJ. Instrumentos para la evaluación de la empatía en psicoterapia. Rev Argentina Clin Psicol. 2011;20:121–32.
56. Elliott R. Psychotherapy change process research: realizing the promise. Psychother Res. 2010;20(2):123–35.
57. Watson JC, Steckley PL, McMullen EJ. The role of empathy in promoting change. Psychother Res. 2014;24:286–98.
58. Dekeseyer M, Elliot R, Leijssen M. Empathy in psychotherapy: dialogue and embodied understanding. In: Decety J, Ickes W, editors. The social neuroscience of empathy. Cambridge: MIT Press; 2009. p. 113–24.
59. Riess H. Biomarkers in the psychotherapeutic relationship: the role of physiology, neurobiology, and biological correlates of E.M.P.A.T.H.Y. Harv Rev Psychiatry. 2011;19(3):162–74. Available from: http://www.ncbi.nlm.nih.gov/pubmed/21631162
60. Gallese V, Fadiga L, Fogassi L, Rizzolatti G. Action recognition in the premotor cortex. Brain. 1996;119:593–609.
61. Abu-Akel A, et al. The role of oxytocin in empathy to the pain of conflictual out-group members among patients with schizophrenia. Psychol Med. 2014;44(16):3523–32.
62. Farrow TF, Whitford TJ, Williams LM, Gomes L, Harris AW. Diagnosis-related regional gray matter loss over two years in first episode schizophrenia and bipolar disorder. Biol Psychiatry. 2005;58:713–23.
63. Høglend P. Exploration of the patient-therapist relationship in psychotherapy. Am J Psychiatry. 2014;17(10):1056–66. https://doi.org/10.1176/appi.ajp.2014.14010121. Review
64. Safran J. Interpersonal process in cognitive therapy. New York: Rowman & Littlefield Publisher Inc; 1994.
65. Siegel D. The developing mind, how relationships and the brain interact to shape who we are. New York, London: The Guilford Press; 1999.

66. Imbernón C, García-Valdecasas CJ. Psicoterapias humanístico-existenciales: fundamentos filosóficos y metodológicos. Rev Asoc Esp Neuropsiq. 2009;29:437–53.
67. Horvath AO. The therapeutic alliance: concepts, research and training. Aust Psychol. 2001;36: 170–6. https://doi.org/10.1080/00050060108259650.
68. Dinger U, Zimmermann J, Masuhr O, Spitzer C. Therapist effects on outcome and alliance in inpatient psychotherapy: the contribution of patients' symptom severity. Psychotherapy. 2016;54(2):167. https://doi.org/10.1037/pst0000059.
69. Bordin ES. The generalizability of the psychoanalytic concept of the working alliance. Psychother Theor Res Pract Train. 1979;16(3):252–60. https://doi.org/10.1037/h0085885.
70. Horvath AO, Luborsky L. The role of the therapeutic alliance in psychotherapy. J Consult Clin Psychol. 1993;61:561–73.
71. Pally R. The mind-brain relationship. New York: Karnac; 2000.
72. Schore AN. The effects of early relational trauma on right brain development, affect regulation, and infant mental health. Infant Ment Health J. 2001;22:201–69.
73. Schore AN. Advances in neuropsychoanalysis, attachment theory, and trauma research: implications for self psychology. Psychoanal Inq. 2002;22(3):433–85.
74. Schore AN. Affect regulation and the repair of the self. New York: Norton and Company; 2003.
75. Baxter LR, Schwartz JM, Bergman KS, Szuba MP, Guze BH, Mazziotta JC, Alazraki A, Sclin CE, Ferng H, Munford P, Phelps ME. Caudate glucose metabolic rate changes with both drug and behavior therapy for obsessive-compulsive disorder. Arch Gen Psychiatry. 1992;49(9):681–9. https://doi.org/10.1001/archpsyc.1992.01820090009002.
76. Brody AL, Saxena S, Stoessel P, et al. Regional brain metabolic changes in patients with major depression treated with either paroxetine or interpersonal therapy. Arch Gen Psychiatry. 2001;58:631–40.
77. Buchheim A, Viviani R, Kessler H, Kächele H, Cierpka M, Roth G, et al. Changes in prefrontal-limbic function in major depression after 15 months of long-term psychotherapy. PLoS One. 2012;7(3):e33745.
78. Dichter GS, Felder JN, Smoski MJ. Affective context interferes with cognitive control in unipolar depression: an fMRI investigation. J Affect Disord. 2009;114(1–3):131.
79. Dichter GS, Felder JN, Petty C, Bizzell J, Ernst M, Smoski MJ. The effects of psychotherapy on neural responses to rewards in major depression. Biol Psychiatry. 2010;66(9):886–97.
80. Fu CH, Williams SCR, Cleare AJ, Scott J, Mitterschiffthaler MT, Walsh ND, et al. Neural responses to sad facial expressions in major depression following cognitive behavioral therapy. Biol Psychiatry. 2008;64:505–12.
81. Furmark T, Tillfors M, Marteinsdottir I, Fischer H, Pissiota A, Langstrom B, Fredrikson M. Common changes in cerebral blood flow in patients with social phobia treated with citalopram or cognitive-behavioral therapy. Arch Gen Psychiatry. 2002;59:425–33.
82. Goldapple K, Segal Z, Garson C, Lau M. Modulation of cortical-limbic pathways in major depression: treatment-specific effects of cognitive behavior therapy. Arch Gen Psychiatry. 2004;61:34–41.
83. Kennedy SH, Konarski JZ, Segal Z, Lau MA, Bieling PJ, Mcintyre RS, Mayberg HS. Differences in brain glucose metabolism between responders to cbt and venlafaxine in a 16-week randomized controlled trial sidney. Am J Psychiatry. 2007;164:778–88.
84. Kaplan HB, Bloom SW. The use of sociological and social-psychological concepts in physiological research: a review of selected experimental studies. J Nerv Ment Dis. 1960;131: 128–34.
85. Dimascio A, Boyd RW, Greenblatt M. Physiological correlates of tension and antagonism during psychotherapy; a study of interpersonal physiology. Psychosom Med. 1957;19(2):99–104.
86. Malmo RB, Boag TJ, Smith AA. Physiological study of personal interaction. Psychosom Med. 1957;19(2):105–19.

87. Cacioppo JT, Berntson GG. Social psychological contributions to the decade of the brain. Doctrine of multilevel analysis. Am Psychol. 1992;47(8):1019–28. https://doi.org/10.1037/0003-066X.47.8.1019.
88. Kraemer BC, Zhang B, Leverenz JB, Thomas JH, Trojanowski JQ, Schellenberg GD. Neurodegeneration and defective neurotransmission in a Caenorhabditis Elegans model of tauopathy. Proc Natl Acad Sci U S A. 2003;100:9980–5.
89. Jacobson NS, Truax P. Clinical significance: a statistical approach to defining meaningful change in psychotherapy research. J Consult Clin Psychol. 1991;59:12–9.
90. Crowther A, Smoski MJ, Minkel J, Moore T, Gibbs D, Petty C, et al. Resting-state connectivity predictors of response to psychotherapy in major depressive disorder. Neuropsychopharmacology. 2015;40(7):1659–73.
91. Voderholzer U, Schwartz C, Freyer T, Zurowski B, Thiel N, Herbst N, Wahl K, Kordon A, Hohagen F, Kuelz AK. Cognitive functioning in medication-free obsessive-compulsive patients treated with cognitive-behavioural therapy. J Obsessive Compuls Relat Disord. 2013;2:241–8.
92. Jacobson NS, Roberts LJ, Berns SB, McGlinchey JB. Methods for defining and determining the clinical significance of treatment effects: description, application, and alternatives. J Consult Clin Psychol. 1999;67:300–7. https://doi.org/10.1037/0022-006X.67.3.300.
93. Thompson B. What future quantitative social science research could look like: confidence intervals for effect sizes. Educ Res. 2002;31:25–32.
94. Jacobson NS, Follette WC, Revenstorf D, Baucom DH, Hahlweg K, Margolin G. Variability in outcome and clinical significance of behavioral marital therapy: a reanalysis of outcome data. J Consult Clin Psychol. 1984;52:497–504.
95. Follette W, Callaghan G. The importance of the principle of clinical significance-defining significant to whom and for what purpose: a response to Tingey, Lambert, Burlingame, and Hansen. Psychother Res. 1996;6(2):133–43. https://doi.org/10.1080/10503309612331331658.
96. Wampold BE, Jensen WR. Clinical significance revisited. Behav Ther. 1986;17:302–5.
97. Tingey R, Lambert MJ, Burlingame G, Hansen N. Clinically significant change: practical indicators for evaluating psycho- therapy outcome. Psychother Res. 1996;6:144–53.
98. Martinovich Z, Saunders S, Howard K. Some comments on "assessing clinical change". Psychother Res. 1996;6:124–32.
99. Kazdin AE. The meanings and measurement of clinical significance. J Consult Clin Psychol. 1999;67:332–9.
100. Beckstead DJ, Hatch AL, Lambert MJ, Eggett DL, Vermeersch DA, Goates MK. Clinical significance of the outcome questionnaire(OQ–45.2). Behav Anal Today. 2003;4:74–90.
101. Lambert JJ, Ogles BM. The efficacy and effectiveness of psychotherapy. In: Lambert MJ, editor. Bergin and Garfield's handbook of psychotherapy and behavior change. 5th ed. New York: Wiley; 2004. p. 139–93.
102. Bauer S, Lambert MJ, Nielsen SL. Clinical significance methods: a comparison of statistical techniques. J Pers Assess. 2004;82(1):60–70.
103. Wolf M. Social validity: the case for subjective measurement or hoe applied behavior analysis is finding its heart. J Appl Behav Anal. 1978;11(2):203–14.
104. Karpenko V, Owens JS, Evangelista NM, Dodds C. Clinically significant symptom change in children with attention-deficit/hyperactivity disorder: does it correspond with reliable improvement in functioning? J Clin Psychol. 2009;65:76–93.
105. Castonguay LG, Youn SJ, Xiao H, Muran JC, Barber JP. Building clinicians-researchers partnerships: lessons from diverse natural settings and practice-oriented initiatives. Psychother Res. 2015;25:166–84.
106. Zarin DA, Pincus HA, West JC, McIntyre JS. Practice-based research in psychiatry. Am J Psychiatry. 1997;154:1199–208.
107. Barkham M, Margison F. Practice-based evidence as a complement to evidence-based practice: from dichotomy to chiasmus. In: Freeman C, Power M, editors. Handbook of evidence-based psychotherapies: a guide for research and practice. Chichester: Wiley; 2007. p. 443–76.

108. Barkham M, Hardy E, Mellor-Clark J. Developing and delivering practice-based evidence: a guide for the psychological therapies. 1st ed. Chichester: Wiley-Blackwell; 2010.
109. Fernández Alvarez H. Reflections on supervision in psychotherapy. Psychother Res. 2016; 1(26):1–10.
110. Luborsky L, Crits-Christoph P. Understanding transference: the core conflictual relationship theme method. New York: Basic Books; 1990.
111. Loughead J, Wileyto EP, Ruparel K, et al. Working memory-related neural activity predicts future smoking relapse. Neuropsychopharmacology. 2015;40(6):1311–20. https://doi.org/10.1038/npp.2014.318.
112. Roffman JL, Witte JM, Tanner AS, Ghaznavi S, Abernethy RS, Crain LD, Giulino PU, Lable I, Levy RA, Dougherty DD, Evans KC, Fava M. Successful brief psychodynamic psychotherapy for persistent depression. Psychother Psychosom. 2014;83:364–70. https://doi.org/10.1159/000364906.
113. Denzin NK, Lincoln YS. The handbook of qualitative research. 2nd ed. Thousand Oaks: Sage; 1994.
114. Ryle G. Thinking and re acting. In: Collected papers, Collected essays, vol. 2. New York: Barnes and Noble; 1929–1968. p. 465–79.
115. Gergen K. The acculturated brain. Theory Psychol. 2010;20(6):795–816. https://doi.org/10.1177/0959354310370906.
116. Taubner S, Buchheim A, Rudyk R, Kächele H, Bruns G. How does neurobiological research influence psychoanalytic treatments? Clinical observations and reflections from a study on the interface of clinical psychoanalysis and neuroscience. Am J Psychoanal. 2012;72:269–86.
117. Castonguay LG, Hill CE, editors. Transformation in psychotherapy: corrective experiences across cognitive behavioral, humanistic, and psychodynamic approaches. Washington, DC: American Psychological Association; 2012.
118. Roussos A, Braun M, Olivera J. "For me it was a key moment of therapy": corrective experience from the client's perspective. J Clin Psychol In Session. 2017;73(2):153–67.
119. Lane RD, Ryan L, Nadel L, Greenberg L. Memory reconsolidation, emotional arousal, and the process of change in psychotherapy: new insights from brain science. Behav Brain Sci. 2015;38:1–64.
120. Cacioppo JT, Cacioppo S, Dulawa S, Palmer AA. Social neuroscience and its potential contribution to psychiatry. World Psychiatry. 2014;13(2):131–9. https://doi.org/10.1002/wps.20118.
121. Martinez C, Tomicic A, Rodriguez E, Valdez C. The embodied nature of therapist-patient regulation: an EEG study of in-session neurodynamic. 2015. Available at: https://www.researchgate.net/publication/279940625_The_embodied_nature_of_therapist-patient_regulation_An_EEG_study_of_in-session_neurodynamic.
122. Bott NT, Radke AE, Kiely T. Ethical issues surrounding psychologists' use of neuroscience in the promotion and practice of psychotherapy. Prof Psychol Res Pract. 2016;47(5):321–9. https://doi.org/10.1037/pro0000103.

The Brain in the Public Space: Social Neuroscience and the Media

María Jimena Mantilla, Martín H. Di Marco, and Diego A. Golombek

Abstract Here we analyze public communication of neuroscience, in general, and social neuroscience, in particular, as well as the circulation of its particular discourse in mass media. We discuss particular issues of neuroscience communications in the context of science popularization. As an example, we offer an analysis of neuroscience coverage in a national newspaper of widespread distribution and conclude that even though news articles on social neuroscience do not represent a significant proportion of scientific reports in the press, they are important platforms to disseminate neuroscientific accounts of social processes. This is especially so as regards the topics of interpersonal ties and emotional mechanisms, two concepts traditionally dominated by the social sciences. Finally, we offer some recommendations for bridging the gap between academic research in the field and its popularization.

Keywords Brain • Neuroscience • Communication • Science • Print media • Health

M.J. Mantilla
Instituto de Investigaciones Gino Germani, Buenos Aires, Argentina

Consejo Nacional de Investigaciones Científicas y Técnicas (CONICET), Buenos Aires, Argentina
e-mail: mantillamariajimena@gmail.com

M.H. Di Marco
Consejo Nacional de Investigaciones Científicas y Técnicas (CONICET), Buenos Aires, Argentina

Instituto de Salud Colectiva, ISCo-UNLa, Buenos Aires, Argentina
e-mail: mh.dimarco@gmail.com

D.A. Golombek (✉)
Departamento de Ciencia y Tecnología, Universidad Nacional de Quilmes, Buenos Aires, Argentina

Consejo Nacional de Investigaciones Científicas y Técnicas (CONICET), Buenos Aires, Argentina
e-mail: dgolombek@unq.edu.ar

© Springer International Publishing Switzerland 2017

A. Ibáñez et al. (eds.), *Neuroscience and Social Science*,
DOI 10.1007/978-3-319-68421-5_14

1 Introduction

Expert knowledge on the brain has expanded significantly in recent decades and now circulates outside traditional academic spheres. Moreover, it has been established as a valid framework to understand everyday phenomena—in particular, those related to human behavior. The presence of neuroscientists in communication media, the growing appearance of journalism articles, and the emergence of popularization books (sometimes in the boundary with self-help literature), theater plays, social networks, and webpages are just some examples of this outbreak. The wealth of cultural spaces in which neuroscience is present shows the importance of the penetration of brain-centered discourses in the social arena.

Indeed, this is not a completely new phenomenon. The so-called decade of the brain in the 1990s identified cognitive psychology—one of the components on which social neuroscience is based—as one of its pillars, and, indeed, neuroscience has become a communicational tool for clinical psychologists [1, 2]. The guidelines for such a decade emphasized the need for studying the brain also in its sociocultural context, taking advantage of all the then-emerging technologies available for research (e.g., [3]). Indeed, cognitive psychology has also received considerable attention in the media, as well as other disciplines which also complement the general social neuroscience scheme, notably behavioral genetics [4]. However, neuroscience, in general, and social neuroscience, in particular, have experimented a tremendous growth in terms of their popular visibility, a fact that demands a specific analysis, both of its causes and its consequences. This translation from the lab and the clinic into the media also comprises an ethical dimension and, moreover, should also reflect an intention derived from public policies in science communication [5–8].

It is possible that neuroscientific explanations are somewhat more appealing to the general public. Indeed, there is evidence that, when provided, neuroscience information generates significant interest and might even interfere with the ability of critical analysis of judging the information. When neuroscientific terms were present, subjects judged explanations as better and more satisfying than those without brain jargon [9]. There is something special about neuroscience in current public communication, and this is obviously pervading the analysis and popularization of social research.

This chapter aims to analyze neuroscience circulation in the media, with a particular example based on the graphic press in Argentina. Will data analysis help us to cope with questions such as "when did this so-called neuro-boom start"? What kinds of themes are covered in the media? Have they changed in recent years? How are social processes described from a neuroscientific perspective? We focus on the rhetoric of the social neuroscience in the media, since this dimension is key in terms of public opinion, persuasion, and circulation [10], and, as such, the tone of journalism is as important as its content.

2 From Science Communication to Neuroscience Communication

The communication of science has turned into an autonomous research area, focused on the translational processes between the scientific field and the general public. According to Schäfer [11], mass media has become the most important public forum in contemporary society, including scientific information, providing a framework of societal self-observation and public opinion formation, among other aspects. This research area has developed in the context of increased lay publication of scientific information in newspapers, television, blogs, forums, among other outlets. Several authors (e.g., [12]) have shown, in different contexts, the growing publication trends of science-related content in newspapers, and how this boosts the circulation of scientific ideas in the public sphere.

In addition, in the current context of an increasing prevalence of online environments, traditional formats are being redefined, as revealed by the decrease in scientific sections in newspapers and the increase in blogs, forums, and webpages administered by both journalists and scientists [13]. These new communication environments have changed not only how information is disseminated—and its potential audience—but also a more frequent and extensive entailment of the audience with scientific information as reported by the media.

In his classic work about the political uses of science communication, Hilgartner [14] has stated that the culturally dominant view of the popularization of science suffers from conceptual problems which result in oversimplifying this process. While it is assumed that communication is based on a two-stage process—first the production of knowledge and then its dissemination—the actual diffusion of scientific ideas evidences the existence of ingrained beliefs in the purity of science and the potential pollution of knowledge by outsiders. In this context, the key question is what "appropriate simplification" is and who can draw the boundaries between oversimplified scientific information and insufficient translation of knowledge.

In particular, the communication of neuroscience-related news has been closely scrutinized by this field. Discussions arise regarding the sources of information of neuroscientists or the difficulties in the interaction between journalists and scientists [15], the possibility of spreading mistakes and polarized beliefs [16], and the effects of the so-called neuro-realism in the representations of the general audience [17]. As expected for most disciplines, the visibility of neuroscience-related news are closely related to their timing (i.e., whether the concept being communicated has become fashionable) and the specific media (e.g., newspaper type) [16].

In recent years, several studies have shown a growing interest in brain research and in cognitive and social neuroscience in particular [18–22]. One of the main focuses is the interest in a better public communication or areas related to clinical developments of neuroscience investigations. As we shall see, data derived from such studies are sometimes hard to interpret and can result in erroneous conclusions in the mass media [23], which stresses the need for a more informed and critical press that ensures a more precise communication [21, 24]. It is not uncommon to raise false or exaggerated expectations in the public, including the possibility of thought-reading [7, 22, 24]. Moreover, popular communication of neuroscience tends to argue for the *value* of the research and not necessarily for its content [25]. Among these values, the novelty and relevance of knowledge, together with its applicability, are usually emphasized. This is not qualitatively different from what happens in public communication of science in general, where the most frequently employed category of "value" is that representing technoscience as an activity extending the frontiers of knowledge [12].

3 Images in the Brain and in the Media

As stated above, neuroscience is particularly susceptible to false or exaggerated information in the media, giving raise to inaccurate perceptions of its real strength and limits. *Neurologisms* are a vivid example of this, with the popular use of terms such as neuromarketing, neuroeducation, neurogym, and many others. Illes et al. [26] suggest that the main challenges faced by neuroscience communication are (a) the complexity of the brain; (b) the personal, philosophical, and religious salience of the field; and (c) the burden of central nervous disease together with the stigma of neurological and mental disorders.

Indeed, an additional source for this complicated state of affairs is the intrinsic complexity (and, in some cases, inscrutability) of the most recent technologies used in order to understand the neural basis of individual and, in some cases, social behavior. Among these, neuroimaging techniques have certainly played an important role in the current expansion of neuroscience research and its mingling with other, more social, disciplines. In particular, functional imaging technologies have provided strong candidates for neural correlates of behavior and cognition. However, there is a frequent confusion between the actual realities of the experimental conclusions and its promises—and even perils—a confusion that is also spread to popularization products and media. Recent data suggests that media reporting of the results of functional imaging studies are "mostly positive and framed in terms of healthcare progress (…) (Without a) balance between technology opportunities and applications (…) and seems to favour oversimplification" [6]. Another analysis of press coverage of functional imaging studies concluded that the media "largely provided no explanation of the capabilities and limitations of fMRI (…) (and) had a mostly optimistic tone [20]." Moreover, even if the news coverage of neuroscientific research is relatively accurate, this does not guarantee and adequate reception by the general population. An extreme example is provided by the analysis of the media coverage of a single article ("Does bilingualism influence cognitive aging?", published by Thomas Bak in the *Annals of Neurology*) which, according to the authors, received a fair coverage from the international press but, nonetheless, the comments of readers throughout the world indicate that the public understanding of the main concepts of the study was, at the most, far from precise [27].

This increasing coverage, and sometimes misrepresentation, of neuroscience research seems to be a worldwide phenomenon. We shall now present some local data and analysis of press reports in this particular field.

3.1 An Example of Neuroscience Coverage in Mass Media

The analysis of news allows observing the emergence and evolution of scientific discourse on the brain in this particular framework of an increasing interest in the field [10, 11, 28]. Indeed, two of the authors of the present chapter (MJM and MDM) have recently reported a marked increase in news related to neuroscience in the last 15 years in Argentina [29]. Even considering the widespread of the areas being covered, they found that health- and disease-related news were prioritized, with content tending to construct a narrative of a healthy way of life. The combination between expert knowledge and medical recommendations, which characterizes some of these reports, certainly aids in the social legitimation of the ideas about the brain.

Here we present an initial approximation to the installation of social neuroscience in a popular context, by means of analyzing some of the main products of science popularization in the field. We have analyzed how new ideas about the brain and social neurosciences have emerged in the public agenda in Argentina and how they have been disseminated in print media. In order to achieve this objective, the publication trends and main topics of newspaper articles about neurosciences were analyzed. Furthermore, in-depth qualitative analysis of newspaper articles about social neuroscience was carried out, emphasizing new conceptions about the relation between the individual and society and the transformations of traditional views of social phenomena.

In order to build the *corpus* of analysis, all digital articles from *La Nación* (one of the main national newspapers in Argentina which, in addition, has more coverage on scientific news than other mainstream media; www.lanacion.com.ar) were compiled from 1996 to 2016. Articles were identified using the online browser of *La Nación*'s webpage, with the term "neurosciences" (*neurociencias,* in Spanish). Indeed, these two decades coincide with a significant historical moment when the neuroscientific field gained importance in Argentina, while images and ideas related to the brain and its study spread across the media, among other aspects.

For the selection of the cases, headlines of the newspaper articles were read and, in case of ambiguity, the entire article was read in order to evaluate its pertinence. The final *corpus* was composed of 754 articles, which were analyzed using qualitative and quantitative methods. All types of newspaper articles were gathered, including interviews, feature articles, editorials, columns, and opinion pieces.

In order to identify the newspaper articles about social neurosciences within the broader *corpus*, the headlines and lead paragraphs of all articles were read. No specific keywords were used to identify these articles in *La Nación*'s online browser, as social neuroscience is a subdiscipline with a particular approach on social phenomena that tend to tackle a wide range of topics. Hence pertinent articles were identified manually considering that they tackled social issues (such as interactions, morality, organizational aspects of society, relationships, etc.) from a neuroscientific perspective (i.e., relating the specific topic to the brain, its functioning, neurobiology, or neurochemistry).

For the quantitative analysis, a structured database was designed using SPSS (v. 19), and all the data was added manually. Variables included information about the article's approach and topics (type of newspaper article, publication year, source mention, type of source mentioned, main topic, origin, and section in the newspaper), although not all of them are analyzed in this chapter. These variables were selected to characterize key features of the dissemination of neuroscientific news and its evolution in time. For the qualitative analysis, the entire text of the articles were codified and analyzed with thematic content analysis. Preestablished, as well as emergent, dimensions of analysis were used to codify and organize the description of the articles.

The first finding is that the local trend resembled that from the rest of the world, where the "decade of the brain" was giving a boost to the rise of popular neuroculture [30], which implied the diffusion of representations of contemporary brain science in the means of communication—e.g., television, newspapers and magazines, blogs, and webpages [31].

Figure 1 shows the evolution of neuroscience-related newspaper articles in *La Nación*, from 1996 until 2016, supporting the hypothesis that Argentina—along other countries—also witnessed the rise and dissemination of neuroscientific discourses in the general public [32].

The two-decade period witnessed an increase in the number of articles published, with fluctuations between 2000 and 2010 that did not modify the overall growing tendency in the lay dissemination of neuroscientific ideas and discourses. From

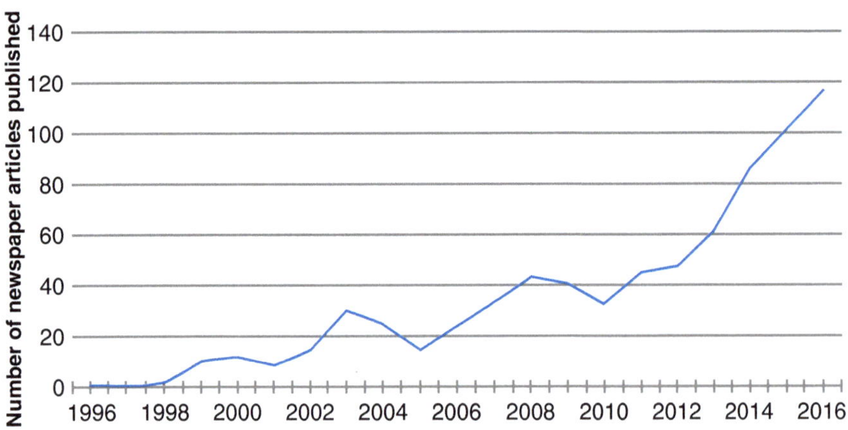

Fig. 1 Publication trend of newspaper articles about neurosciences. La Nación, 1996–2016. Source: prepared with information available in *La Nación*'s webpage (www.lanacion.com.ar). Total number of articles: 754

2010 onward, a steady rise in the number of articles can be seen, comprising the 65% of the entire number of articles (i.e., 491).

This growing tendency in the diffusion of neuroscientific ideas and news in the media confirms previous studies that have found a substantial rise in the communication of general scientific ideas [11], as well as in neuroscientific information in particular [21, 22, 26, 33, 34] to the lay public.

Table 1 illustrates the main neuroscience-related topics of the newspaper articles in *La Nación* during the period 1996–2016. Thirty two percent of these articles tackle health-disease issues. The majority of these articles are related to diseases (a 44% of health-related articles and a 14% of the overall articles about neurosciences in the 21-year period). Articles about diseases include neurodegenerative conditions (namely, Alzheimer's syndrome, Parkinson's disease, and multiple sclerosis), as well as a wide range of mental illnesses (e.g., depression, phobias, schizophrenia, autism), and other neurologically based disorders (e.g., epilepsy, migraine, stroke). Furthermore, a significant number of the articles tackle health-disease issues related to aging, including current theories to understand the aging process and treatments to specific health problems. While health-disease and brain topography (the latter representing 9% of the articles throughout this period) are among the most commonly and traditionally associated topics related to the brain and the emergence of neuroscience [29], the table shows that several other topics were covered by print media. The heterogeneity of topics ranges from education (10%) and emotions (11%) to decision-making (4%) and child development (3%).

Table 1 Main topics in newspaper articles about neurosciences. *La Nación*, 1996–2016

Topics	*N*	%
Health-disease	244	32
Emotions	80	11
Education	76	10
Brain topography	64	9
Others	38	5
Neuroscience and communication	34	5
Memory	33	4
Decision-making	32	4
Creativity	30	4
Technology	26	3
Child development	19	3
Psychoanalysis	17	2
Spirituality	15	2
Language	13	2
Cultural expressions	13	2
Gender	10	1
Economy	10	1
	754	100

Table prepared with information available in *La Nación*'s webpage (www.lanacion.com.ar)

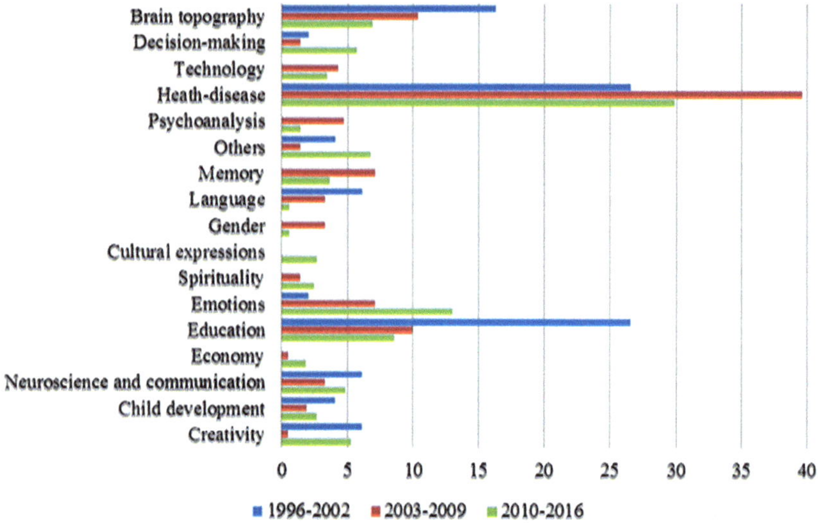

Fig. 2 Percentage of topics in newspaper articles: comparison between periods 1996–2009, 2003–2009, 2010–2016. La Nación newspaper, 1996–2016. Source: prepared with information available in *La Nación*'s webpage (www.lanacion.com.ar). Total number of articles: 754

Moreover, articles about communication of neuroscientific knowledge to the lay public represent 5% of the articles in this period. Despite the fact that this percentage is comparatively small, its presence indicates that articles which examine the nature of expertise, the communication of science and technology among professionals and to the public, and the scientific-lay translation barriers and strategies have contributed to the emergence of neuroscientific ideas among the general public. Furthermore, the fact that this category had a steady decrease in the number of articles throughout the period (as can be seen in Fig. 2) would support the hypothesis that the 1990s witnessed the emergence of neurosciences in the Argentine public agenda and, therefore, that the dissemination of scientific ideas was focused on communicating what the neurosciences are, the different subfields that have been developed, and its current and potential applications [22].

Figure 2 shows how much each topic was tackled comparatively in news articles during three periods: 1996–2002, 2003–2009 and 2010–2016. The chart shows that, despite the fact that the main patterns of topics dealt remain fairly constant throughout these 21 years, several significant transformations took place.

While health-disease issues remain the most recurrent in the newspaper articles during the period of analysis, other topics gained more visibility and dissemination. For instance, emotions, decision-making, and creativity were more represented during the third period, showing a subtle shift in the lay communication of scientific ideas. At the same time, news related to education gradually lost representation.

Moreover, during the second and third periods of analysis, previously unmentioned topics emerged, namely, economy and neuromarketing and cultural expressions. Both of these groups of articles are closely related to social neurosciences, since neurobiological explanations are given to understand fields that have been traditionally studied by social sciences and humanities. On the one hand, articles about economic and business-related news approach this social phenomenon from neurobiological standpoints. Whether it is to understand how companies are organized and what institutional structures could be used to encourage individual productivity, or how the neurochemistry of consumers conditions their decision-making process, articles about neuromarketing provide a new insight on old topics. On the other hand, articles about cultural expressions and activities provide explanations about the functions and changes in music, television, and literature. For example, the popularity of television series characterized by violence is explained by the neurochemical response of fear, which is associated with the release of dopamine and a consequential adrenaline state in the viewers.

Furthermore, the fact that the number of articles included in the category "Others" also increased during the last period can be seen as an indicator of the emergence of new specific topics that are being tackled by neurosciences and spread by popular means of communication.

As Racine, Waldman, and Rosenberg [22] have stated, the public interest in neurosciences and the brain and the expectations of the general population on this discipline have raised concerns and discussions about their potential implications. The growing interest that neurosciences have gained for social phenomena and the following dissemination of ideas from social neuroscience in the media (which can be seen in the rising number of articles about gender, cultural expressions, economy, and neuromarketing in Argentina) have created a new field where human and social sciences meet neurosciences [35, 36]. Fundamental dilemmas about human interactions and social organization—traditionally faced by social sciences and philosophy—are now being tackled by both scientific areas, bringing up new questions about the boundaries of scientific disciplines, the diffusion of current theories, and the discourses raised by the shifts in science.

3.2 *The Social Brain in the Print Media*

The previous section described the media coverage of neuroscience news in general, emphasizing the broader features of media dissemination of scientific ideas about the brain. The focus of this section, however, is to describe the newspaper articles that specifically dealt with social neurosciences.

Nonetheless, a clarification regarding the categorization of articles is needed, due to the complexity of dividing neurosciences and social neuroscience. The main reason why social neuroscience was not included as a category per se was that this specific academic area covers a wide range of the previously mentioned topics, such as emotions. In this case, emotions and attitudes are not exclusively researched from the "social perspective" of neurosciences. Therefore, the entire corpus of articles was recategorized in order to identify the newspaper articles that specifically tackled social neuroscience, namely, articles that dealt with social phenomena (morality, interactions, relationships, social organizations) from a neuroscientific perspective. The key criterion used to incorporate articles was that they linked neurobiology and neurochemistry to social issues and, in most cases, discussed the potentiality of this new approach on social topics.

Before exploring in-depth the ideas about the brain, individuals and society conveyed in these articles, several questions are relevant to understand some of the basic aspects of lay diffusion in the media: How quantitatively important is the social neuroscience in relation to neurosciences in general? Which are the main topics explored from this specific perspective?

Figure 3 shows the publication trend where the evolution of articles about neurosciences in general and social neuroscience in particular can be compared and analyzed. As the line chart depicts, the total number of newspaper articles published about social neuroscience is considerably small in comparison with the broad field of neurosciences. While the first article related to social neuroscience was published in 2001, most of the articles were published from 2006 onward. The fact that news dealing with social neuroscience have gained more dissemination indicates that, while social neuroscience is not the most quantitatively significant area in the lay communication of science, it may be a current growing research field.

Figure 4 illustrates the main topics tackled by the newspaper articles about social neuroscience. The majority of the articles (57%) focus on issues related to emotions and attitudes, ranging from scientific debates about the biological causes of violence and aggression to the neurochemical basis of morality and empathy. The second and third most common topics are economy and neuromarketing (9%) and brain topography (7%). The fact that most of the articles are concentrated in just one category, and that the other topics sum up comparatively small percentages, shows that the articles about social neuroscience have a similar publication pattern as the articles about neuroscience in general. Therefore, there is a high heterogeneity of topics in the newspaper articles.

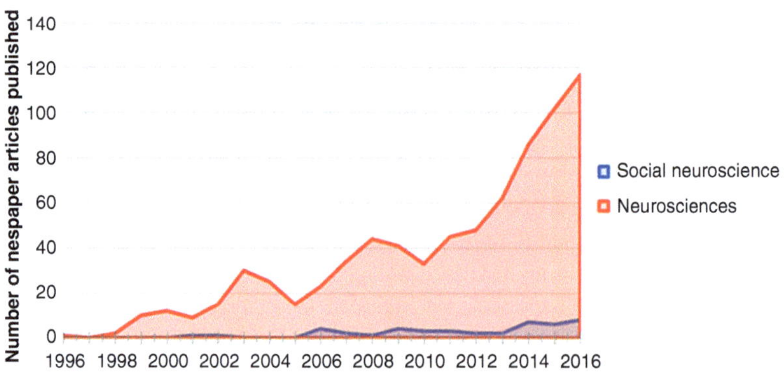

Fig. 3 Comparative publication trends of newspaper articles about neuroscience and social neuroscience. La Nación, 1996–2016. Source: prepared with information available in *La Nación*'s webpage (www.lanacion.com.ar). Total number of articles: 754

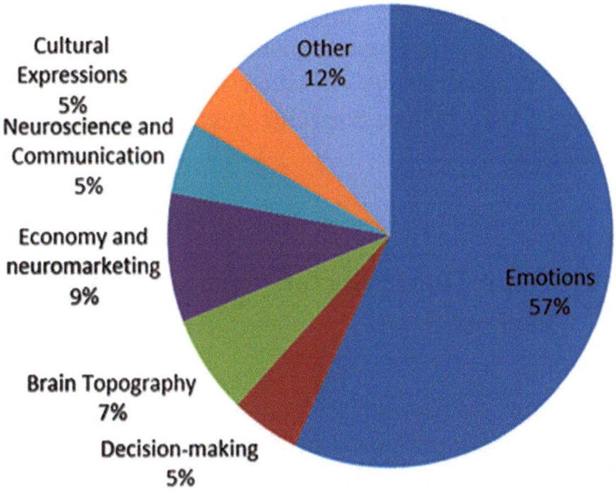

Fig. 4 Main topics of newspaper articles about social neuroscience. La Nación, 1996–2016. Source: prepared with information available in *La Nación*'s webpage (www.lanacion.com.ar). Total number of articles: 44

The fact that newspaper articles about social neurosciences are almost marginal in relation to the overall publication trend of articles about neurosciences sets a number of crucial questions that can only be tackled from a qualitative perspective. In this sense, it is tempting to conclude that the interest in social neuroscience for a popular audience resides in the fact that they integrate science to everyday experiences and is able to explain—at least partially—our personal problems, social ties, and emotional reactions.

Furthermore, it should be noted that media coverage of social neurosciences might be influenced by political interests as well as moral values circulating in the media. Several social studies have shown the influence of editorial policies on news coverage—e.g., the political and ideological affinity between media and right-wing parties in Latin American countries [37]. Moreover, other studies indicate that the ideological bias of editorial policies influence which news are published and how they are reported, including news related to violence [38, 39].

Indeed, it is clear that most, if not all, major newspapers are strongly politically biased, selecting both the type of topics covered and the particular point of view conveyed in the news. Social neuroscience is particularly prone to such kind of biases, as we have already shown in this chapter. An additional example relates to adolescence violence, which is being debated in social and political forums, including the possibility of decreasing the legal age for imprisonment. Indeed, several neuroscientific studies argue against lowering this age limit, considering the neurodevelopmental events taking place during this stage, including major modifications of cortical circuits during adolescence [40, 41], which should have important consequences from the neuroethical point of view [42]. However, these neuroscientific arguments are lacking in the newspaper reports and debates, which in some cases could reflect the political view of the editors.

Another example could be the growing evidence on the effects of poverty and malnutrition on brain development in children (e.g., [43–45]). Social communication of the scourge of poverty on children and youth does not usually consider scientific findings and might be adding, consciously or not, to the considerable stigmatization of its consequences.

A complete analysis of editorial policies and media coverage of social neurosciences would require a specific research study and therefore exceeds this chapter's objective.

4 What Does "Social" Mean From a Neuroscientific Perspective?

News report scientific information from areas such as the neural basis of racial prejudice, the rules for social behavior, the role of mirror neurons in social interaction, brain correlates of decision-making, moral judgment and theory of mind, etc. Indeed, the conception of what is "social" derived from the analysis of printed news is centered around interpersonal relationships [46], an area that is traditionally in the realm of sociology. In most, if not all, reports, the link between subjects is analyzed from a neuroscientific perspective that illuminates the neural basis of interpersonal actions.

News reports usually provide information by two mechanisms: first, by providing a summarized story of the scientific experiment and, second, by quoting the authors of such experiments. There is also another kind of report which does not convey a certain scientific finding but introduces the opinion of experts who judge the specific social problems from a neuroscientific point of view (e.g., violence at school, the rise of crime, etc.). This kind of opinion columns aids in the generation of consent regarding the legitimacy of neuroscience as a perspective to intervene in social problems. For instance, this type of news report is clearly illustrated in the following article about morality:

> Neurosciences have shown interest in aspects of human life that certain traditions considered distant and separate from science. One of these aspects is morality. The so-called 'values' translate into concrete facts that can be studied and understood scientifically. (…) A more detailed understanding of moral issues allows us to distinguish between different ways to live in society, and gives us a possibility to judge actions as better or worse, more or less ethical (Can morality be understood by science?, February 24, 2016).

Neuroscientific ideas that circulate in the graphic press explain, metaphorically speaking, how society is inscribed in the brain, by providing information about two-way processes: (1) how the brain mediates social interactions and (2) how social processes shape brain function. In other words, news validate the notion that human behavior results from neural activity. Two examples will be useful to interpret such processes, which have in common biological explanations of social experience.

Let's first consider the media analysis of cerebral architecture, stating that it is particularly adapted for social interactions. Recalling neuroscientific theory and experiments that aim to determine the precise localization of brain regions underlying specific behaviors, news convey the idea that human beings are especially (and anatomically) gifted for such interpersonal interactions. Quoting a report from the *La Nación* newspaper:

> For example, we can identify specific areas of the brain that act to inhibit violent and anti-social responses; other areas intervene in the moral process of socialization and in the capacity of responding to others' needs and not only to our own. (The importance of a happy brain, March 17, 2002)

Other than communicating the results of specific experiments, some of the news reports convey positive expectations about the promising character of neuroscience to explain social conduct. In this sense, sometimes the strategy is to put the scientific findings in an imprecise background, thus constructing a level of universality that favors the construction of an unrestricted realm for neuroscience. This strategy is common to journalists and scientists when writing for the general public. The notions of "correlation," "intervention," "cause," and "responsible" are loosely defined, without providing enough explanation about their reach.

On the other hand, the interpretation of social neuroscience experiments in the media is usually extremely general, without the proper context and specificity with which they are reported in the academic world. Indeed, the hypothetical nature of scientific results turns into certainties when depicted in mass media, probably due to a certain "cultural reputation of certainty" which, by translating academic discourses into popular texts, risks losing the necessary "nuances of science" that allow an ample interpretation of results [25].

A second example regards the invocation of "healthy behaviors," i.e., the reconceptualization of social links as a source for health or disease. In our data, most reports of social neuroscience mention the fact that emotions are closely related to the processes of social interaction, and the latter can become patterns for a healthy way of life. For example, the piece "Friendship has a surprising healing effect" (*La Nación*, October 15, 2006) states that subjects with a large network of social ties recover more quickly from disease and, indeed, it is neuroscience the area to study how do brains relate to each other and affect health.

Yet another example is a report on moral attitudes, stating that "neuroscience has proven that resentment and the difficulties to forgive potentiate chronic stress, cardiac injury, increases in blood pressure and even a higher alcohol or drug intake ("To forgive is always healthy," *La Nación* May 4, 2016). As we have pointed out in previous reports [29], the information regarding the relation between cerebral processes and health becomes even more relevant in a social context where a healthy lifestyle has become a moral imperative [47]. In this sense, considering social interactions as potential foes or friends of good health involves the inclusion of social life in health issues—a view that was not traditional in medicine and generates a myriad of novel metaphors and imagery about a biology molded by social context.

In summary, we have provided evidence about the typical way in which the graphic press depicts social neuroscience, which departs from the academic perspective of a neuroscientific view of society. This study of the rhetoric of neuroscience popularization is quite relevant taking into account that it is the main channel through which a general audience receives scientific information and therefore helps to construct cultural representations and social appropriation of science.

Moreover, it is remarkable that news about neuroscience also cover other spaces in the media. This can be seen in women's magazines, in weekend newspaper supplements, or in popular TV shows where neuroscientists are invited to express their views about a diversity of areas apparently unrelated to scientific scrutiny. Social neuroscience popularization certainly favors this kind of transmission, and journalism aids in their appropriation of scientific explanations about love, infidelity, maternity, and other themes in which hormones, the brain, and neurotransmitters become protagonists.

5 Concluding Remarks

We have shown the evolution of news about neuroscience in recent years, as well as the emergence of social neuroscience as a theme in the media. Although the latter has not become mainstream so far, their relevance relies in the kind of ideas it convey about the usefulness of brain science to understand interpersonal and emotional ties in society.

The analysis of the media coverage shows several key aspects of the communication of neurosciences. First, the number of articles related to this discipline has been steadily increasing in the last two decades, indicating the rise of neuroscientific ideas and discourses in the public space in Argentina. Second, these newspaper articles tackle a wide range of topics. Despite the fact that the majority focus on health-disease issues, the heterogeneity of topics illustrate the thematic diversification of neurosciences and, at the same time, the spectrum of aspects to which the public might relate to. Lastly, the specific analysis of articles about social neuroscience would indicate that, while it was not initially a popular topic in media coverage (the first articles were published in 2001), it is now gaining visibility, particularly with articles about emotions. Essential dilemmas about society and social interactions traditionally studied by social sciences—such as violence, morality, and empathy—are now faced by neurosciences.

The analysis suggests that public communication of social neuroscience generates new imagery, fantasies, and beliefs under the light of new findings of the social brain, by means of constructing a linear—and somewhat ambiguous—narrative of the relation between social and cerebral processes and mechanisms. The typical characteristics of science popularization (i.e., simplification and generalization of scientific results) do collaborate in this conception of a linear link between social and biological explanations. On the contrary, the journalistic language oscillates between causal and correlational explanations that link the brain to the social processes under study. Indeed, the use of undefined terms such as "correlates," "basis," "foundations," or "substrates" favors this ambiguous perspective about the nature of the link between social and biological operations. Moreover, the social representations arising from neuroscience popularization tend to reduce social relations to those of interactions between individuals, discarding other social dimensions which are traditionally studied by social sciences.

In summary, the novelty of neuroscience comprises not only the new and expanding areas of research but also novel ways of describing the social experiences to a lay audience. In this sense, the legitimacy of neuroscience relies, at least in part, outside the scientific expertise, since it encompasses a diversity of explanations that are absorbed by society as alternative interpretations of social experiences. The analysis of news in mass media contributes to unveil one of the circuits through which these ideas circulate in society. In other words, spreading of social neuroscience by the media collaborates in the hierarchy awarded to the brain in social behavior, as well as brings some legitimacy to the role of neuroscience as the most suitable area to study social processes that are traditionally the subject matter of other disciplines.

Having analyzed and somewhat diagnosed the current state of affairs of social neuroscience in the media as accessed by the general public, we should end with specific recommendations in order to shorten the gap between contemporary research in the area and its public communication. Universities and neuroscience schools certainly have a say in the process, since this is one of the fields where the gap begins. Among other proposals, social neuroscience courses could take advantage of social media as a tool for sharing up-to-date information on the subject and thus provide pathways for interactions between experts and the lay public [48].

On the other hand, although science communication has been professionalized in recent years, there is much to be done in terms of specialized training of both neuroscientists and journalists in neuroscience communication, a field that could also benefit from specific research which is currently quite scarce and fragmented [26]. In addition, social neuroscience deserves to be part of an "open science agenda" in which appropriately informed citizens can deliberate and discuss the reach and application derived from academic investigations. Citizens need (and demand) a realistic understanding of the dynamic and sometimes controversial nature of scientific authority. For this, it is necessary that scientists and specialized journalists describe the main features of experimental methods, the process of interpretation of scientific results, and their link with putative debatable issues, all of which show science as a social space of changing definitions (sometimes even competing), not as a closed an indisputable activity which does not represent the true everyday work of researchers. After all, neuroscience, much like social neuroscience in particular, refers to us, what we do, what we are, and what we feel, both as individuals and as a society.

References

1. Falk EB. Communication neuroscience as a tool for health psychologists. Health Psychol. 2010;29(4):355–7.
2. Roussos A, et al. Psychotherapy and social neuroscience: forging links together. In: Ibáñez A, Sedeño L, García AM, editors. Neuroscience and social science. New York: Springer; 2017.
3. Cacioppo JT, Berntson GG. Social psychological contributions to the decade of the brain. Doctrine of multilevel analysis. Am Psychol. 1992;47(8):1019–28.
4. Dingel MJ, Ostergren J, McCormick JB, Hammer R, Koenig BA. The media and behavioral genetics: alternatives coexisting with addiction genetics. Sci Technol Hum Values. 2015;40(4):459–86.

5. Zimmerman E, Racine E. Ethical issues in the translation of social neuroscience: a policy analysis of current guidelines for public dialogue in human research. Account Res. 2012;19(1):27–46.
6. de Jong IM, Kupper F, Arentshorst M, Broerse J. Responsible reporting: neuroimaging news in the age of responsible research and innovation. Sci Eng Ethics. 2016;22(4):1107–30.
7. Racine E, Illes J. Emerging ethical challenges in advanced neuroimaging research: review, recommendations and research agenda. J Empir Res Hum Res Ethics. 2007;2(2):1–10.
8. Zigmond MJ. Implementing ethics in the professions: preparing guidelines on scientific communication for the society for neuroscience. Commentary on 'implementing ethics in the professions: examples from environmental epidemiology' (Soskolne and Sieswerda). Sci Eng Ethics. 2003;9(2):191–200.
9. Weisberg DS, Keil FC, Goodstein J, Rawson E, Gray JR. The seductive allure of neuroscience explanations. J Cogn Neurosci. 2008;20(3):470–7.
10. Luz M, Sabino C, Mattos R, Ferla AA, Andres B, Alba RD. Contribution towards studying the contemporary social imaginary: rhetoric and images of biosciences in popular scientific periodicals. Interface. 2013;10(1):84–106.
11. Schäfer M. The media in the labs, and the labs in the media. What we know about the mediatization of science. In: Lundby K, editor. Mediatization of communication. Berlin: De Gruyter; 2014. p. 571–93.
12. Christidou V, Dimopoulos K, Kouladis V. Constructing social representations of science and technology: the role of metaphors in the press and the popular scientific magazines. J Sci Commun. 2004;13:347–62.
13. Brossard D. New media landscapes and the science information consumer. Proc Natl Acad Sci U S A. 2013;110(Suppl 3):14096–101.
14. Hilgartner S. The dominant view of popularization: conceptual problems, political uses. Public Underst Sci. 1990;20:519–39.
15. Allgaier J, Dunwoody S, Brossard D, Lo YY, Peters HP. Journalism and social media as means of observing the contexts of science. Bioscience. 2013;63(4):284–7.
16. van Atteveldt NM, van Aalderen-Smeets SI, Jacobi C, Ruigrok N. Media reporting of neuroscience depends on timing, topic and newspaper type. PLoS One. 2014;9(8):e104780.
17. Popescu M, Thompson RB, Gayton WF, Markowski V. A reexamination of the neurorealism effect: the role of context. J Sci Commun. 2016;15(6):A01.
18. Beck DM. The appeal of the brain in the popular press. Perspect Psychol Sci. 2010;5(6):762–6.
19. O'Connor C, Rees G, Joffe H. Neuroscience in the public sphere. Neuron. 2012;74(2):220–6.
20. Racine E, Bar-Ilan O, Illes J. fMRI in the public eye. Nat Rev Neurosci. 2005;6(2):159–64.
21. Racine E, Bar-Ilan O, Illes J. Brain imaging: a decade of coverage in the print media. Sci Commun. 2006;28(1):122–42.
22. Racine E, Waldman S, Rosenberg J, Illes J. Contemporary neuroscience in the media. Soc Sci Med. 2010;71(4):725–33.
23. Gonon F, Bezard E, Boraud T. Misrepresentation of neuroscience data might give rise to misleading conclusions in the media: the case of attention deficit hyperactivity disorder. PLoS One. 2011;6(1):e14618.
24. Racine E. Identifying challenges and conditions for the use of neuroscience in bioethics. Am J Bioeth. 2007;7(1):74–6. discussion W1–4
25. Jonhson MJ, Littlefield M. Lost and found in translation: popular neuroscience in the emerging neurodisciplines. In: Pickersgill M, Van Keulen I, editors. Solciological reflections on the neurosciences. Bingley: Emerald; 2011. p. 279–99.
26. Illes J, Moser MA, McCormick JB, Racine E, Blakeslee S, Caplan A, et al. Neurotalk: improving the communication of neuroscience research. Nat Rev Neurosci. 2010;11(1):61–9.
27. Broer T, Pickersgill M, Deary IJ. The movement of research from the laboratory to the living room: a case study of public engagement with cognitive science. Neuroethics. 2016;9:159–71.
28. Palma H. Infidelidad genética y hormigas corruptas. Una crítica al periodismo científico. Buenos Aires: Teseo; 2012. p. 245.

29. Mantilla MJ, Di Marco MH. La emergencia del cerebro en el espacio público: las noticias periodísticas sobre las neurociencias y el cerebro en la prensa gráfica en Argentina (2000–2012). Phys Rev Saúde Coletiva. 2016;26(1):177–200.
30. Ortega F, Vidal F. Neurocultures: glimpses into an expanding universe. Frankfurt am Main. New York: Peter Lang; 2011. p. 359.
31. Pitts-Taylor V. The plastic brain: neoliberalism and the neuronal self. Health. 2010;14(6):635–52.
32. Mantilla MJ. Educating 'cerebral subjects': the emergence of brain talk in the Argentinean society. BioSocieties. 2014;10(1):84–106.
33. Blakemore C. Neuroscience and the media: the need for communication. Neuroscience. 1993;57(1):217–26.
34. Miller G. Neuroscience. Neural communication breaks down as consciousness fades and sleep sets in. Science. 2005;309(5744):2148–9.
35. Callard F, Fitzgerald D. Rethinking interdisciplinarity across the social sciences and neurosciences. Basingstoke: Palgrave; 2015.
36. Meloni M, Williams S, Martin P. The biosocial: sociological themes and issues. Sociol Rev Monogr. 2016;64:7–25.
37. Goldstein AA. Qué afinidades políticas hay entre los principales diarios y partidos de "derecha" en Brasil, Chile y Argentina a inicios del siglo XXI? In: Bohoslavsky E, Echeverría O, editors. Las derechas en el Cono Sur, Siglo XX. Los Polvorines: Unive. Nac. de Gral. Sarmiento; 2013.
38. Federico L. Homicidios diarios. Análisis del discuros periodístico sibre homicidios por armas de fuego. Buenos Aires (Argentina) 2001–2002. Salud Colectiva. 2010;6(3):295–312.
39. Njaine K, De Souza Minayo MC. Análise do discurso da imprensa sobre rebelioes de jovens infratores em regime de privacao de liberdade. Ciencia Saude Coetiva. 2002;7(2):285–97.
40. Klein D, Rotarska-Jagiela A, Genc E, Sritharan S, Mohr H, Roux F, et al. Adolescent brain maturation and cortical folding: evidence for reductions in gyrification. PLoS One. 2014;9(1):e84914.
41. Uhlhaas PJ, Roux F, Singer W, Haenschel C, Sireteanu R, Rodriguez E. The development of neural synchrony reflects late maturation and restructuring of functional networks in humans. Proc Natl Acad Sci U S A. 2009;106(24):9866–71.
42. Evers K. Can we be epigenetically proactive. In: Metzinger T, Windt JM, editors. Open mind: philosophy and the mind sciences in the 21st century. Cambridge: MIT Press; 2016. p. 497–518.
43. Lipina SJ, Posner MI. The impact of poverty on the development of brain networks. Front Hum Neurosci. 2012;6:238.
44. Lipina SJ, Segretin MS. Strengths and weakness of neuroscientific investigations of childhood poverty: future directions. Front Hum Neurosci. 2015;9:53.
45. Segretin MS, Hermida MJ, Prats LM, Fracchia CS, Ruetti E, Lipina SJ. Childhood poverty and cognitive development in latin America in the 21st century. New Dir Child Adolesc Dev. 2016;2016(152):9–29.
46. Rose N, Abi-Rached J. Neuro: the new brain sciences and the management of the mind. Princeton: Princeton University Press; 2013. p. 352.
47. Clarke A, Shim J, Mamo L, Fosket R, Fishman J. Biomedicalization: technoscientific transformations of health, illness and US biomedicine. Am Sociol Rev. 2003;68(2):161–94.
48. Valentine A, Kurczek J. "Social" neuroscience: leveraging social media to increase student engagement and public understanding of neuroscience. J Undergrad Neurosci Educ. 2016;15(1):A94–A103.

Part III
Integration of Social and Neuroscientific Insights

Electrophysiological Approaches in the Study of the Influence of Childhood Poverty on Cognition

Marcos Luis Pietto, Juan E. Kamienkowski, and Sebastián J. Lipina

Abstract The influence of adverse environmental conditions on the organization and reorganization of the brain structure and function involves distinct neural systems at different levels of organization. Electroencephalographic (EEG) measures provide precise evidence on the temporal sequence in which relevant cognitive processes occur. Here, we offer a systematic review of EEG studies on the influence of childhood poverty on cognitive development. The paradigms used focused primarily on correlates of inhibitory control, selective attention, and unrelated task-event activity. Eighteen studies reported differences related to socioeconomic disparities, including (a) discrepancies in neural markers of interference control and early auditory sensory processing and (b) delays in the maturation of brain oscillations in frontal regions. Overall, EEG techniques appear to have predictive power over cognitive and academic performance of children. Therefore, EEG markers may be useful to evaluate the efficacy of interventions aimed to enhance cognitive development in children facing unfavorable social conditions.

Keywords EEG • ERP • Socioeconomic status • Childhood poverty • Cognitive development

---

M.L. Pietto (✉)
Unidad de Neurobiología Aplicada (UNA, CEMIC-CONICET), Buenos Aires, Argentina

Laboratorio de Inteligencia Artificial Aplicada (Departamento de Computación, FCEyN-UBA, CONICET), Buenos Aires, Argentina
e-mail: marcos.pietto@gmail.com

J.E. Kamienkowski
Laboratorio de Inteligencia Artificial Aplicada (Departamento de Computación, FCEyN-UBA, CONICET), Buenos Aires, Argentina

Departamento de Física (FCEyN-UBA, CONICET), Buenos Aires, Argentina
e-mail: juank@dc.uba.ar.com

S.J. Lipina
Unidad de Neurobiología Aplicada (UNA, CEMIC-CONICET), Buenos Aires, Argentina
e-mail: lipina@gmail.com.com

© Springer International Publishing Switzerland 2017
A. Ibáñez et al. (eds.), *Neuroscience and Social Science*,
DOI 10.1007/978-3-319-68421-5_15

1 Introduction

Early experiences influence emotional, cognitive, social, and learning-related developmental processes, which play an important role in children's educational and social integration opportunities during their first two decades of life [1–3]. Accruing evidence in the fields of developmental psychology and developmental cognitive neuroscience indicates that during such a period, adverse environmental experiences, associated with poverty, are related to changes in the development of different aspects of cognition at different levels of organization [4–18].

Although its complexity is not always considered adequately, poverty is a highly multidimensional, relational, and dynamic phenomenon. Its influences on cognitive development are given by a set of mediation and moderator factors that are part of the daily experience (see Kwon, this volume). Mediators and moderators involve both individual and contextual factors at different levels of organization. Some of the most important mediators that are postulated in the contemporary literature are (1) prenatal and perinatal health factors; (2) housing conditions; (3) neighborhood characteristics; (4) quality of home and school environment; (5) opportunities for cognitive and learning stimulation at home; (6) parenting and care styles; (7) parental mental health; (8) family, social, and cultural expectations about child development and learning; (9) access to social support networks; and (10) material and symbolic resources of families [13, 19–25]. In particular, the experience of poverty is associated with a set of potential cumulative and interacting risk factors [20, 26], which increase the likelihood of developing negative outcomes later in life [22, 23, 27, 28].

In addition, the impact of these risk factors on cognitive development may vary according to the individual's susceptibility and to the type, number, co-occurrence, and timing of exposure to deprivations [21, 22, 25, 26, 29–32]. Consequently, identifying factors of childhood poverty is a very complex task, comprising various theoretical, methodological, and logistical difficulties which make it difficult to generalize individual experiences. In turn, it is important to implement adequate research designs that can specify what aspects of experience of poverty contribute to individual differences in cognitive development and the efficiency of different neural networks [31], because the evidence suggests that different types of adverse experiences generate different influences on brain development [33, 34]. However, the measures of poverty that are commonly used in current studies on childhood poverty and cognition do not necessarily capture the complexity of the multiple adverse experiences. Moreover, no clear consensus has emerged on what indicators should be used to categorize an individual as poor. Thus, the present work focuses on the ways in which poverty is measured, highlighting the importance of improving our comprehension of childhood poverty as a multidimensional phenomenon in terms of individual experiences. From this perspective, we expect that this approach will contribute to the design of interventions to improve cognitive development.

2 The Neuroscientific Approach

There is an increasing body of neuroimaging evidence on the association between brain structure/functioning and childhood poverty, which indicates that the experience of childhood poverty is related to the activity and anatomical development of distinct brain networks. This evidence points regions implicated in cognitive domains such as language (e.g., left inferior frontal and fusiform gyrus), memory and learning (e.g., hippocampus), executive functioning (e.g., prefrontal cortex), and social-emotional processing (e.g., amygdala) [15, 35–39].

Here, we focus on electroencephalographic (EEG) studies that examine links between neural activity and measures of childhood poverty. These methodologies have the advantage of directly measuring neural activity and capturing cognitive dynamics in the time frame in which cognition occurs [40]. Their high temporal resolution allows tapping into the neural mechanisms engaged at each stage of information processing. For instance, examining the neural systems that underlie a particular cognitive ability can reveal subtle differences along information-processing streams, even in the absence of significant behavioral manifestations (e.g., [41]). This suggests that EEG methods may be helpful for elucidating fine-grained differences in brain processing associated with poverty. In addition, cognitive electrophysiological techniques are noninvasive, robust, fast to compute, applicable to large-scale screening, and much less expensive than other techniques. Such methodological attributes have important implications in building knowledge of cognitive development and the contextual modulation of poverty-related risks. In this sense, cognitive electrophysiology might offer an affordable, massive, and temporally precise approach to reveal cognitive indicators of negative and positive influences related to adverse (e.g., social inequality) and favorable (e.g., intervention programs) contextual experiences.

3 A Systematic Review of the Literature

The present study aimed to analyze the literature about the influences of childhood poverty on cognitive development from the perspective of cognitive electrophysiological explorations and to shed light on how poverty shapes brain function and impacts on cognitive components of behavior. In particular, we address the mechanisms supporting these processes and their association with children's poverty or low socioeconomic status (SES) experience.

After applying the *Preferred Reporting Items for Systematic Review and Meta-Analysis Protocols* (PRISMA-P) methodology for systematic reviews,[1] we identified a total of 18 studies from 12 articles from 5 countries, published between 1990 and 2016—most of them (67%) appearing over the last decade (Table 1).

3.1 Poverty Measures

In general, the studies explored the influence of poverty or low socioeconomic status on neural activity through three primary indicators: income, parental education, and occupation. Either combined or in isolation, these measures are commonly used to index SES. Importantly, the indicators varied among the studies. Some measures estimated low SES using a single variable such as *maternal education* [17, 42, 43] or *family income* [44], although others used both measures [45], or focused on variables based on family income, family income-to-needs ratio, parental occupation, or parental education [46–48]. Others used composite variables combining indexes of parental occupation, parental education, and family income, which were assessed by standardized questionnaires [5, 41, 48–56] (Table 1).

Most of the studies implemented discrete categories with different criteria to divide the measures into separate groups. For instance, when maternal education was used, if the mother had only completed high school education she was generally considered to have a low level of instruction [17, 42, 43]. When family income was used, it was considered either as gross family income [47] or as the percentage of the minimum monthly wage [45]. Finally, parental occupation was determined by one study [47] that used a category scale [57] to make three occupational groups (higher managerial or professional, intermediate and routine/semi-routine, and unemployed over the last 6 months). In turn, other studies implemented singular continuous estimates to explore the relationship between poverty and low SES vari-

[1] Our systematic review is based on the PRISMA-P standard protocol [113] to examine the association between poverty indicators and EEG activity in developmental cognitive studies. The search criteria contemplated: (a) articles published in English without restrictions on the range of the publication dates; (b) studies with an age range between birth and adolescence; and (c) experimental research reporting factors that were related to childhood poverty, EEG measures, and their relationship with cognitive development. Studies were identified by searching electronic databases and inspecting reference lists of articles. This search was applied to the National Library of Medicine's MEDLINE and EBSCO databases, considering the following terms: "SES," "income," "education," "occupation," "poverty," "social vulnerability," "ERP," "EEG," "children," "preschool," "kindergarten," and "school." Three reviewers selected the studies, and any disagreements were solved by consensus. We selected those articles in which the primary purpose was to measure the impact of poverty-related factors on brain and cognitive functioning. Conversely, the ones that were aimed mainly at addressing factors not necessarily associated to poverty (e.g., parental mental health or air pollution), or that were focused on extreme deprivation of these aspects (e.g., undernutrition, maltreatment), were not selected, even though they showed a certain relevance in assessing the impact of childhood poverty. The information that was extracted from each study included (1) sociodemographic characteristics of participants; (2) poverty measures (type, method of measurement, quantity and quality of considered factors); and (3) EEG and cognitive paradigms (amplitude, latency, power spectra of activity through scalp sites, accuracy, and reaction time of behavioral performances).

Table 1 Studies on the relationship between SES and EEG/ERP measures

Study	Participants	Poverty measure	Technique	Paradigm	Findings
Conejero et al. [48]	16–18 mos ($n = 52$)	SES[d] family income-to-need ratio	ERP/ freq. analysis	Error detection task	Large differences in frontal ERN (450–750 ms) and in theta power (300–600 ms after stimuli) among correct and incorrect configurations were, respectively, related to higher family SES and higher family education
		Parental occupation and education			SES in general, and parental education in particular, contributed to individual differences in the amplitude of ERN and associated theta power
Isbell et al. [51]	3–5 yrs ($n = 124$)	SES[b]	ERP	Auditory selective attention task	Early (100–200 ms) differential activation (attended-unattended story) at fronto-central sites was positively correlated to nonverbal IQ scores
Neville et al. [52]	3–5 yrs ($n = 141$)	SES[b]	ERP	Auditory selective attention task	Parent-based training showed more changes in the neural response (100–200 ms) underlying selective attentional processes
Ruberry et al. [44]	3–6 yrs ($n = 118$)	Family income	ERP	Go/no-go task and flanker task	Absence of significant correlations between ERP and income. Significant correlations between ERP and cognitive performance (executive control). ERP was associated with differential activity (N2, go minus no-go; P3, congruent minus incongruent) underlying the performance of go/no-go and flanker tasks

(continued)

Table 1 (continued)

Study	Participants	Poverty measure	Technique	Paradigm	Findings
Stevens et al. [43]	3–8 yrs ($n = 30$)	Maternal education	ERP	Auditory selective attention task	The refractory effect was faster in the attended story in children with higher maternal education, while lower maternal education children had similar refractory effects to attended and unattended stimuli
Stevens et al. [17]	3–8 yrs ($n = 32$)	Maternal education	ERP	Auditory selective attention task	Lower maternal-education children showed responses of greater amplitude in the 100–200 ms time-window to task-irrelevant stimuli at fronto-central scalp regions
Kishiyama et al. [41]	7–12 yrs ($n = 26$)	SES[a]	ERP	Visual detection task/novelty oddball paradigm	Lower SES children showed reduced P1 and N1 components to task-irrelevant stimuli at parieto-occipital leads and reduced N2 to novel stimuli at central scalp regions
D'Angiulli et al. [49, 50]	11–14 yrs ($n = 28$)	SES[b]	ERP/ freq. analysis	Auditory selective attention task	Early (100–400 ms) and late (600–800 ms) differential activation (attended-unattended auditory stimuli) was greater in higher SES children over mid-frontal cortical regions. However lower SES children had more mid-frontal and frontal theta power to the unattended than attended tones between 200 ms and 700 ms

(continued)

Table 1 (continued)

Study	Participants	Poverty measure	Technique	Paradigm	Findings
D'Angiulli et al. [49, 50]	13 yrs ($n = 28$)	SES[b]	ERP/ freq. analysis	Auditory selective attention task	Lower SES children showed an increase in selectivity of attention (Nd amplitude) concomitant to an increase in post ERP cortisol levels, whereas no such relationship was observed in higher SES children
D'angiulli et al. [5]	13 yrs ($n = 28$)	SES[b]	ERP/ freq. analysis	Auditory selective attention task	Children from lower SES backgrounds showed a right activation asymmetry at the mid-frontal scalp site in theta band, whereas higher SES showed the opposite pattern
					Individual mid-frontal right attentional activation was associated with individual differences across SES rank, task-dependent cortisol reactivity, and increase in boredom at the start of the task
Skoe et al. [42]	14–15 yrs old ($n = 66$)	Maternal education	ABR	Passive listening paradigm	ABRs from lower maternal education adolescents showed a lower consistency of response, a weaker encoding of speech and greater noisier activity
Brito et al. [46]	At birth ($n = 66$)	Parental education, family income, family income-to-needs	freq. analysis	Sleep (~10 min)	EEG spectrum was not correlated to SES. However, it was associated with cognitive performance (memory and language) at 15 months

(continued)

Table 1 (continued)

Study	Participants	Poverty measure	Technique	Paradigm	Findings
Tomalski et al. [47]	7–8 mos ($n = 55$)	Gross family income or maternal occupation	freq. analysis	Watching videos	Infants from lower-income families and mother occupation had lower frontal gamma AP
Otero [55]	20–30 mos ($n = 50$)	SES[c]	freq. analysis	Sleep (~30 min)	Lower SES children showed significantly higher delta power in all scalp regions, lower alpha power in frontal, central and occipital regions
Otero [54]	4 yrs ($n = 42$)	SES[c]	freq. analysis	Eyes closed (~10 min)	Children from lower SES showed significantly higher total power over anterior sites, higher power in lower delta and theta bands over frontal leads, and lower alpha power over frontal, occipital and temporal sites
Otero et al. [53]	5–6 yrs ($n = 42$)	SES[c]	freq. analysis	Eyes closed (~10 min)	Lower SES children showed at 5 years higher power in theta and delta bands over frontal areas and lower power in alpha band, especially over posterior areas. At 6 years of age, differences remained the same for theta and alpha, respectively, at frontal regions and temporal-occipital scalp regions
Harmony et al. [45]	6–13 yrs ($n = 118$)	Maternal education and income per head	freq. analysis	Eyes closed	Lower SES children had higher power in delta, theta and beta bands over frontal regions. Moreover, lower power in alpha band over frontal, temporal and occipital regions were observed in lower SES children

(continued)

Table 1 (continued)

Study	Participants	Poverty measure	Technique	Paradigm	Findings
Tomarken et al. [56]	12–14 yrs ($n = 39$)	SES[b]	freq. analysis	Counterbalanced eyes open and eyes closed	High-risk children had higher power in alpha band in left relative to right frontal areas. SES, but not risk status, significantly predicted asymmetry measures

ABR auditory brainstem response, freq. analysis band frequency analysis, *ERP* event-related potential, *SES* socioeconomic status
[a]MacArthur sociodemographic questionnaire
[b]Four-factor index of social status [115])
[c]Questionnaires from [114])
[d]Z-transformed scores based on parental occupation, parental education, and family income-to-need ratio

ables and EEG measures, instead of collapsing the information into discrete categories [42, 44, 46, 48, 56].

The methods used to measure poverty and SES differ among studies in terms of how scores are calculated and the quantity and quality of involved factors. It is thus unclear whether implemented poverty measures across studies capture similar underlying factors and how this impacts on their comparability. Importantly, each poverty indicator is related to the presence or absence of resources that may influence brain structure and functioning in different ways [31, 33, 34, 58]. For example, it has been shown that distinct socioeconomic factors are associated with specific features of neuroanatomical development, such as surface area [37]. In particular, parental education and family income seem to be associated in different ways with brain areas that are considered critical for language, memory, and cognitive development. Noble and colleagues [37] found that family income was logarithmically related to brain surface areas, but parental education had a linear association with those areas.

These findings highlight the need for the implementation of experimental designs that allow us to explore the specific influence of poverty and low SES indicators on brain structure and function separately. Most EEG/ERP studies tend to underestimate the fact that these poverty indicators are based on different conceptual frameworks related to cognitive outcomes. However, the few studies that examined poverty indicators separately found null correlations or similar associations between EEG/ERP patterns and each indicator [46–48]. Moreover, the use of one poverty or low SES indicator, or a set of poverty or low SES indicators, does not contemplate the temporal dynamics in the experience of childhood poverty. Adverse experiences related to poverty and their influence on brain development are no stable across the first two decades of life [8, 59]. Furthermore, the correlation between poverty and low

SES indicators and EEG/ERP outcomes could also be the result of the combination of other individual differences in temperament and environmental susceptibility [34], which in general are not considered in the reviewed studies.

In sum, electrophysiological approaches to study the influences of poverty on brain functioning apply classic unidimensional indicators not considering the variability of different aspects of the adversity experiences (as shown by distinct indicators), and the dynamic nature of changes during development as well. This creates a partial characterization of the individual experience of poverty or low SES, and overlooks the complex scenario comprised of mediation mechanisms that support the correlation between poverty constructs and EEG/ERP outcomes [33, 34]. These issues constitute a fertile field for the interdisciplinary exploration between neuroscience and the social sciences to contribute to the design of childhood poverty and low SES indicators that could help deepen the knowledge of their associations and mechanisms.

3.2 Electrophysiological Paradigms

Two major measures were implemented across the 18 EEG studies reviewed: (a) frequency analysis of baseline EEG activity and (b) ERPs. In seven articles, baseline EEG activity was recorded to assess overall differences in the patterns of EEG between SES groups through a broadband frequency analysis [45–47, 53–56]. This unrelated task-event activity is generally utilized to infer overall characteristics of neural architecture. Broadband frequency analysis allows quantifying oscillatory electrical activity at different frequencies. Although baseline EEG recording intends to represent a general unrelated task-event activity, it could be acquired using different paradigms and experimental conditions (e.g., resting state, ERP). This is inevitable when performing experiments at different developmental stages, but it poses an additional difficulty when comparing through them. In many studies presented in this chapter, the children remained awake with their eyes closed [45, 53, 54], and this could be counterbalanced with "open-eyes" trials [56]. In the other two studies, resting state was acquired during sleep [46, 55]. Finally, in one study, the children watched videos of toys and interacting faces [47] (Table 1).

In the remaining 11 articles, electrical activity that was associated with a perceptual or cognitive task was recorded (ERPs) [5, 17, 41–44, 48–52] (Table 1). The activity that was related to the tasks was also used to perform a spectral power analysis in each trial before averaging them [5, 48–50].

The paradigms implemented in the reviewed ERP studies were mainly aimed to explore executive control processes (Table 2). First, three different tasks were implemented to examine neural mechanisms of selective attention: (1) a *nonspatial auditory attention task*, in which participants had to give an overt response [5, 49, 50]; (2) a *spatial auditory attention task*, in which no active response was required [17, 43, 51, 52]; and (3) a *novelty oddball paradigm* that was implemented to assess visual selective attention [41]. Second, the neural mechanisms that underlay different inhibitory control processes were evaluated in two separate tasks [44]: a novel *go/no-go task* was designed to assess mainly the response inhibition, and a modified *flanker task* was administered to evaluate fundamental control functions interference. Third, brain mechanisms that were involved in error detection processes were investigated through a passive paradigm [48]. Finally, a passive listening task was used to measure auditory brainstem responses (ABRs) [42]. Different aspects of the ABRs were examined collectively under the term "auditory neural acuity." Authors defined this term as "the nervous system's ability to resolve and reliably transmit fine-grained information about acoustic signals within the environment" [42] (Table 2).

3.3 ERP Studies on Socioeconomic Status Throughout Development

To investigate the effect of different developmental environments on brain functioning, investigators have examined prefrontal-dependent functions and auditory brainstem processing using ERP. Conejero et al. [48] conducted a study in toddlers (16–18 months) aimed to investigate whether neural mechanisms involved in error detection were related to SES variables. Electrophysiological responses (ERP and oscillatory neural activity in theta band) from different conditions (correct, position error, conceptual error) of an error detection paradigm were measured. Briefly, the results showed a significant increase in the amplitude of the error-related negativity (ERN; 450–750 ms poststimulus onset) and in theta power, within 300–600 ms after stimulus onset, and over the fronto-central scalp regions for incorrect trials in all groups. Correlational analysis showed that these electrophysiological measures were also associated with SES. Specifically, a decrease in expected differences in ERN between correct and incorrect configurations were related to lower family SES and lower family education, and a decrease in differences in theta power between correct and incorrect configurations was related to lower family education. The authors reasoned that adverse environmental conditions related to low SES might affect the executive attention network in early stages of cognitive development. This argument is supported by evidence that shows that both ERN and frontal theta oscillations were associated with other executive attentional-related tasks [60–62] and the activity of the anterior cingulate cortex [63, 64], which is an important node of executive network that is involved in regulation of conflict [65]. Nevertheless, a reduced response of error-related signals in children from lower SES families may indicate a poorer activation of the executive attention network that is related to conflict detection, or a debilitated representation of stimulus configurations, or both.

Executive attentional processes that are related to inhibitory control were explored with a large sample that included children aged 3–6 [44]. The researchers evaluated whether differences observed in executive control tests that were related to family income could be accounted for by differences in the underlying neural processes. Specific ERPs were calculated at frontal (N2) and parietal (P3) scalp sites in two inhibitory control tasks (*flanker task* and *go/no-go task*). Both income and ERP measures were associated separately with behavioral performance on an executive control battery. On the one hand, lower income was correlated with poorer performance. These results are in line with prior behavioral findings that show that children from poor homes present a lower performance in executive control task [28, 66, 67]. On the other hand, better performance on the executive control battery was correlated with (1) larger differences in activity on N2 for go minus no-go trials (*go/no-go task*), (2) larger differences in activity on P3 for congruent minus incongruent conditions (*flanker task*), and (3) smaller positive P3 amplitude for incongruent trials. Importantly, nonsignificant correlations were found between the amplitude of ERPs on these inhibitory control tasks and family income [44]. One possible explanation is related to the design of the ERP tasks. For instance, the performance measured in computerized ERP paradigms often has a high level of accuracy, because these tasks are programed intentionally not to be over-demanding to keep the underlying electrophysiological activity reliable. Therefore, SES effects might not be observed at the neural level because the task was not sensitive enough (had less power) to capture the predicted association with the neural mechanisms that underlie ERP. Moreover, executive control performance was associated separately with ERP and income. These significant associations might be noticeable because a specific test battery collects great amounts of single tasks assessing different dimensions of the complex evaluated function; hence, it is more reliable in capturing individual differences in the entire sample. Another explanation, suggested by Ruberry et al. [44], is that the observed income disparities in executive control performance might be related to other mechanisms than executive attention and inhibitory control that were assessed by *go/no-go* and *flanker tasks*.

Several studies have reported differences in ERP measures of selective attention between children from poor and nonpoor families [5, 17, 41, 43, 49, 50]. Kishiyama et al. [41] examined neural signatures of visual selective attention and performance on executive function tests in relationship to SES, in children between 7 and 12 years of age. During the selective visual attention task, the children were asked to respond upon detection of the low-probability targets that were embedded in streams of the task-irrelevant stimuli (novel or high-probability standard stimuli). Although both SES groups had similar amplitude to target stimuli, lower SES children had a decreased amplitude of parietal P1 and N1 to standard stimuli, and a decreased amplitude of fronto-central N2 to novel stimuli than the higher SES counterparts (see Table 2 for details). These results indicated that electrophysiological measures of attention were reduced in lower SES children to task-irrelevant and novelty stimuli.

Stevens et al. [17, 43] examined the effects of maternal education level (HME, higher maternal education; LME, lower maternal education) on a *selective auditory attention task* in children 3–8 years old. The ERPs were calculated in relation to the probe stimuli that were superimposed to both attended and unattended channels (i.e., attended and unattended narratives that were administered in the right and left ear) (Table 2). Although children remembered both stories equally well, brain activity differed between groups over central and frontal scalp sites. Specifically, both

Table 2 ERP paradigms

Paradigm	Studies	Experimental design	ERP components
Nonspatial auditory attention task	D'Angiulli et al. [5, 49, 50]	*Instructions*: Respond as fast and accurately as possible to one of four tones presented binaurally. The relevant tone was indicated at the beginning of the experimental session	Subtraction of the maximum negative deflection, between attended nontarget duration tones and unattended nontarget duration tones
		Stimuli: Tone, {800 Hz, 1200 Hz}; duration {100 ms, 250 ms}	
		Interstimulus interval: 1 second	*Latencies*: {100–400 ms and 600–800 ms}
		Conditions: Target tones, 10%; unattended target tones, 10%; attended nontarget tones, 40%; and unattended nontarget tones, 40%	*Scalp sites*: {Fronto-central}
Spatial auditory attention task	Isbell et al. [51], Stevens et al. [15, 43], Neville et al. [52]	*Instructions*: Attend to a story presented from either the left or the right speaker, while ignoring the other story -presented on the other side. The two stories always differed in story content and narrator voice (male/female). Small images from the attended story together with small arrow pointing toward attended channel were displayed on a monitor	Mean amplitudes were compared between probe stimuli presented on the attended and unattended channels
			Latencies: {100–200 ms}
		Stimuli: Linguistic and nonlinguistic probe stimuli {70 dB} superimposed on both narratives; duration, {100 ms}	*Scalp sites*: {Fronto-central}
		Interstimulus interval: {200 ms, 500 ms, 1000 ms}	
		Condition: Attended vs. unattended	
Selective visual attention task	Kishiyama et al. [41]	*Instructions*: Detect the low-probability targets embedded in streams of task-irrelevant stimuli (novel and standard stimuli)	For standard stimuli P1 and N1 components were quantified
			Latencies: {50–150 ms, 100–250 ms}
		Stimuli: Black triangles {target, standard} and digitized color images {novel}. The target triangles were tilted to the right relative to upright standard triangles	For target and novel stimuli P2 and N2 were computed
			Latencies: {50–250 ms, 100–350 ms}
			Scalp sites: {Parieto-central}
		Duration: {250 ms}	
		Interstimulus interval: {1000 ms}	
		Condition: Target, 10%; novel, 15%; standard, 75%	

(continued)

Table 2 (continued)

Paradigm	Studies	Experimental design	ERP components
Go/no-go task	Ruberry et al. [44]	*Instructions*: Press a button when the target changed their original color to blue	For each task condition N2 and P3 components were quantified
		Stimuli: Frog and fish displayed randomly on the screen {flickered at 3 Hz and 5 Hz}; duration, 1200 ms	*Latencies*: {250–400 ms, 400–700 ms}
		Conditions: 25% Were "go trials" in which target stimuli changed their color and children had to press the button, 25% "no-go trials" in which distractor stimuli changed their color and was not required to respond, 50% were "standard trials" in which neither stimuli changed their color	*Scalp sites*: {Frontal, parietal}
Flanker task	Ruberry et al. [44]	*Instructions*: Pay attention to the center target fish and to press the button that matched its direction	For each task condition N2 and P3 components were computed
		Stimuli: Row of five fish centered in the middle of screen; duration, 5000 ms	*Latencies*: {200–400 ms, 400–700 ms}
		Conditions: Congruent: 50%, The flanker fish faced in the same direction as the center fish	*Scalp sites*: {Frontal, parietal}
		Incongruent: 50%, The flanker fish faced the opposite direction of the center target fish	
Error detection task	Rueda et al. [76]	*Instructions*: Pay attention to the progressive completion of puzzles presented on a computer screen	Errors vs. correct contrasts of mean amplitude of ERN component were computed. Further, time-frequency analysis was conducted (theta power)
		Stimuli: Three-piece puzzles of cartoon animals	
			Latencies: ERN {120–160 ms, for adults; 459–750 ms, for toddlers}
		Conditions: Correct completion: 33.3%	*Scalp sites*: Mid-frontal
		Incorrect completion (position error): 33.3%	
		Incorrect completion (conceptual error): 33.3%	

(continued)

Table 2 (continued)

Paradigm	Studies	Experimental design	ERP components
Passive listening task	Skoe et al. [42]	*Instruction*: Attend to a movie and ignore the stimulus that was presented at a rapid rate to the right ear	ABRs were passively collected from the stimuli presentations. The consistency along the experimental session, the extent on which the stimulus is represented in the response, and the noise level in the response were examined from ABRs
		Stimuli: Syllable "da" {80 dB}; duration, 63 ms	
		Rate of presentation: {10.9/s}	

groups had larger positivity within 100–200 ms of the probe onset in the attended versus unattended channel, but HME had a smaller amplitude of response to probes in the unattended channel than LME [17]. In other words, there were no group differences in the ERP response in the attended channel, but the LME group exhibited a higher amplitude response to the probes in the unattended one. Authors interpreted this pattern of activity as indicative of a reduced ability to filter irrelevant information (i.e., to suppress the response to ignored sounds) in the LME group. Moreover, between-group discrepancy in selective attentional processing was also evident when stimuli were presented at fast rates that caused an auditory refractory effect [68]. Specifically, LME had a similar refractory period effect to both attended and unattended stimuli. The difference in the amplitude of the neural response for stimuli that was presented at inter-stimulus intervals of 500 versus 1000 ms was not significant under either task condition, which suggested full recovery regardless of the direction of selective attention. In contrast, children with HME exhibited the same pattern only in the attended channel, which suggested that full recovery was affected by the direction of selective attention. In other words, auditory refractory effects between children with HME and LME differed specifically for the unattended stimuli [43].

Similar attentional differences related to SES, both in ERPs and spectral analysis, have been found by D'Anguilli et al. in a series of studies using a *nonspatial auditory attention task* [5, 49, 50] (Table 1). Adolescents who were 11–14 years old were instructed to attend and respond to a specific pitch tone (attended channel) and to ignore tones with the other pitch (unattended channel). Whereas higher SES children showed greater ERP differentiation between attended and unattended auditory stimuli, this differentiation was small or absent in lower SES children. This pattern was found over mid-frontal cortical regions at early (100–400 ms) and late (600–800 ms) stages of processing. Consistent with the study by Stevens et al. [17], these results suggested that low SES children may process the irrelevant information differently, paying equally attention to the distracting and target stimuli. Moreover, in the spectral analyses from auditory selective attention task, they showed that a lower SES background was associated with right activation asymmetry for the theta band over mid-frontal sites, and higher theta power was associated with unattended (irrelevant) stimuli compared to attended (relevant) stimuli, but the opposite pattern was related to higher SES environments [5, 49, 50]. Importantly, low and high SES children performed behaviorally similarly, despite the fact that they exhibited different neural responses. Thus, the authors suggested that lower SES children have a differential processing "preference." In other words, they suggested that the last may also attend to distractors that allocate additional attentional resources to task-irrelevant information (higher theta power to unattended stimuli) and, thus, they perform attentional tasks like their higher SES counterparts exert more effortful control (i.e., higher right theta over mid-frontal sites).

Combining the results of these selective attention studies, it appears that differential activation patterns are involved in control attentional processes, especially in early stages of information processing between children with different SES. These findings highlight the need to design more specific types of paradigms to elucidate which attentional control mechanisms might explain these findings. In fact, undifferentiated activity between relevant and irrelevant information could be due to a greater susceptibility to attention, capture by irrelevant items, and a slower attentional disengagement from distractors [69]. Moreover, research efforts should focus on identifying effects and intervening mechanisms that contribute to the association between poverty measures and these attentional patterns. It is plausible that children from poor homes may adopt alternative strategies due to an adaptive response toward the stressful environmental settings that characterize poor homes and neighborhoods, to anticipate potentially challenging, negative, or threatening situations [20, 70]. Poor children could have learned to maintain greater sensitivity toward what surrounds them (general sustained attentional response), which may be associated with the processing of a broad set of information in their environment independently of current goals [50].

At this point, several studies present electrophysiological differences between groups of children from low and high SES families, but an interesting question is how these differences are distributed among individuals. Using the same *auditory selective attention task*, Isbell et al. [51] found that ERP modulations related to selective attention accounted for individual variability in nonverbal cognitive skills in a group of preschool children from low SES families. Larger frontal and central mean amplitude differences between ERPs to probes, which were embedded in attended versus unattended stories during the selective auditory attention task, were associated with higher nonverbal IQ scores based on multiple regression analysis. These findings extend previous results showing similar links between electrophysiological measures of attentional control system and higher order functions of young children from poor families [69, 71]. Beyond the design limitations to support causal relationships, the importance of these findings resides in the fact that they provide initial evidence about individual relationships between measures according to two levels of organization (i.e., neural activation and cognitive performance).

All the reviewed studies focused on cognitive-related neural activity, and they did not consider neural activity at a lower level of information processing. Sensory neural activity is directly susceptible to exposure to environmental inputs, and these inputs influence higher level processes. Skoe et al. [42] demonstrated neural discrepancies of more basic underlying mechanisms in adolescents with different years of maternal education. They found that the LME level was related to less efficient auditory processing in the brainstem during the passive listening paradigm. In addition, the latter was also associated with a lower performance on working memory and language processes. Specifically, adolescents who had mothers with LME showed less consistency in their response, a weaker encoding of speech, and greater noisier activity in the auditory brainstem responses (ABRs), which reflected lower auditory neural acuity. Furthermore, correlational analyses between the actual years of maternal education and each of the neural measures revealed that the number of years of maternal education was positively associated with a greater consistency of the response, and more robust speech encoding.

Studies on the effects of sensory enrichment, such as musical training and bilingualism, have shown that expertise could be associated with enhanced auditory neural acuity in the brainstem [72, 73]. This implies that improvement in auditory neural acuity could be associated with the level of exposure to specific sound characteristics. Thus, the current state of the nervous system that was provided by the individual's life experience with sound will be reflected in the auditory brainstem response. In turn, it is known that early experiences of the basic sensory system influence the development of higher level functions [74]. In the context of poverty, it has been documented that children from poor families live in backgrounds with lower levels of language exposure, quantitatively and qualitatively, and that these experiences are associated with children's language development [18].

Future research would benefit from a design that allows us to elucidate how brainstem response mediates or accounts for the relationship between poverty-related variables, such as early language exposure and children's receptive and expressive language skills. Because brainstem responses do not require motor or cognitive engagement, these measures could be useful for examining the relationship between lower sensory and cognitive neural networks in children from poor homes. For example, present findings suggest that there may be more basic underlying mechanisms that account for the influence of neural circuitry that subserves attention allocation. That is, an impoverished perceptual representation might be responsible for the degree to which executive attentional network is recruited during cognitive processing.

All these studies pointed out several differences in the neural mechanisms of attention skills and sensory encoding on a variety of tasks. During development, particularly in the first years of life, the nervous system is highly plastic so that important gains in the efficiency of brain functioning may occur because of individual experiences. Thus, an open question that could have a large applied impact is when and how we can implement interventions to take advantage of this neural plasticity to change those initial differences that are related to different developmental contexts. In this sense, only one study evaluated brain activation patterns before and after an intervention (an attentional program training) in lower SES preschoolers [52]. More than 100 children, who were enrolled in a Head Start program, were randomly assigned to the Training Program (TP), Head Start (HS) alone, or to an Active Control Group (ACG). The TP was the only one combining intervention sessions for parents with attention training exercises for children. Although the ACG only performed classroom training for children, the HS group did not receive supplemental activities. The results showed that children who performed the family-based TP had more self-regulatory gains than children who had participated in the other two groups. Specifically, children not only showed higher scores in both nonverbal intelligence and receptive language tasks, but they also showed an increase in the neural response that was reflected in the early attentional modulation (100–200 ms) to attended stimuli, in the spatial auditory attention task (Table 2). In addition, parental reports on children's behavior expressed greater social skills, fewer problematic behaviors, and less parental perceived stress. Finally, the TP group also showed favorable changes in objective laboratory observations of language and interaction patterns. From a neural functioning perspective, these results indicated that the SES disparities in brain activation during development are not necessarily fixed.

The importance of intervention programs resides in the possibility of identifying activities that are able to induce changes in brain development and to determine what aspect of the efficiency of different neural networks could be influenced by different mediating mechanisms. On the one hand, the study by Neville et al. [52] provided important evidence about how activities oriented to parents could improve the brain activity that was related to attentional processes. On the other hand, it did not include direct measures of child and parent stress, or measures of parent-child interactions, such as language exposure or maternal interaction style. Thus, as reasoned by the authors, it is not possible to assess trajectory models that evaluate the mechanism of change, or to establish whether neural attentional changes were mediated by parental changes and/or decreases in child stress regulation.

ERP studies have mainly examined aspects of selective and executive attention, which involve processes of conflict resolution, inhibitory control, and error detection [75]. These processes are associated with a neural network that involves medial frontal cortex, anterior cingulate, lateral prefrontal, and parietal cortices [65, 76]. In addition, ERP evidence indicates associations between poverty and neural processing even when behavioral differences do not emerge [e.g., [41]]. In sum, these studies provide convergent evidence for the association between of poverty on executive and selective attention mechanisms [5, 17, 41, 43, 48–50].

3.4 Frequency Analysis of EEG Baseline Activity and Socioeconomic Status Throughout Development

A number of studies have used frequency analysis of EEG baseline activity to assess how specific power oscillations were associated with different developmental contexts. Brito et al. [46] tested infants at birth using resting-state EEG activity during sleep. They found that frontal and parietal power in gamma bands were associated with memory and language skills at 15 months of age. However, results also showed a nonsignificant correlation between neonatal EEG power and SES variables (i.e., parental education, family income). These null findings suggested that EEG disparities that were associated with SES-related variables, such as education and income, may arise during postnatal experience. Nevertheless, longitudinal designs that include mediation analysis are needed to test whether the EEG differences are explained by different prenatal and postnatal experiences related to poverty.

Baseline brain activity was recorded as early as 6–9-month-olds while viewing video clips [47]. The infants from lower SES homes (measured by gross family income and maternal occupation) showed significantly lower gamma power over frontal regions than those from higher SES homes. Particularly, when infants were compared merely according to gross family income, authors found differences in the power of lower gamma bands (21–30 Hz), whereas differences in high gamma band power (31–45 Hz) were found when groups were compared based on maternal occupation [47]. Based on previous studies [77–80], reduced gamma band activity over frontal areas in infants from low SES backgrounds was interpreted by the authors as a possible early indicator of potential developmental difficulties in attentional control processes and language. Accordingly, differences in resting EEG gamma power correlated with language and cognitive abilities during infancy [46, 79, 81]. For instance, frontal gamma power measured at birth and during the first 3 years of age has been associated positively with individual differences in language and cognitive skills at 1 [46] and 4–5 years of age [81].

In another study, resting-state recordings of adolescents whose mothers had a history of depression manifested greater relative left versus right alpha-band power on alpha band over left mid-frontal scalp areas. This was not predicted by the risk of depression, but rather by SES-related variables such as lower occupation, fewer years of education of the parents, and a smaller probability of being married [56]. These differences were interpreted as indicating a left frontal hypo-activity in lower SES adolescents.

A 6-year prospective study of preschool children made by Otero et al. [53] found differences in EEG power spectra at specific frequencies (Table 1). In the first session, baseline activity of 20–30-month-old infants was recorded while they were sleeping. The findings showed that infants from low SES homes had higher delta and lower alpha power during sleep [55]. The second session was implemented when children were 4 years of age, and in this case, the resting-state activity was recorded in children that were awake and with their eyes closed. The results showed that low SES children had higher power in lower bands (delta and theta) over frontal leads, and lower alpha power, especially over occipital and temporal sites [54]. Interestingly, EEG pattern differences continued during the third session when children were 5 years old. For example, lower SES children showed higher power values in lower bands over frontal areas, but they also showed lower power in alpha band over posterior areas. Finally, although the differences between low and high SES samples diminished with age, these remained at 6 years in frontal theta and occipital-temporal alpha bands [53].

The relevance of these findings resides in the analysis of contextual effects on maturation-related EEG activity changes at different developmental stages. Poverty experienced at 2–6 years of age was associated with different patterns of neural maturation, as assessed by EEG. In addition, the study showed that disparities in neural maturation between groups decreased during the course of development. On the assumption that adverse experiences during the investigated period remained fixed, it could be argued that early disparities were likely to grow during the course of development if these were caused by the accumulation of adversities or stress factors. Otherwise, it could be argued that differences decreased if schooling experience partially counteracted the impact of adverse experiences, which allowed children from poor backgrounds to overcome virtual developmental gaps. Thus, these longitudinal electrophysiological patterns could be partially accounted for by changes in the susceptibility of children to the type of adverse experience during development [82]. Future investigations should focus on how the link between poverty variables and brain signatures is influenced by changes in susceptibility and type of poverty experiences during development.

Another study investigated spontaneous EEG activity patterns in school-age children (6–13 years of age) while having their eyes closed [45]. Consistent with Otero et al. [54], results indicated that children from low SES homes had greater power values than children from high SES backgrounds in delta and theta bands over frontal areas and lower power values in alpha band over temporal and occipital sites. Alpha power was lower, and beta power was greater over frontal areas in low SES children, when compared to the other group. In addition, absolute power decreased with age, whereas relative power increased for higher bands and decreased for lower bands. The authors interpreted these data as showing that children from low SES backgrounds had the EEG characteristics of younger children. In effect, it is known that during infancy and early childhood, there is a decrease in the power of lower frequencies linked to a concomitant increase in the power of higher frequencies [83–86]. Beyond the fact that these spectral trends were found in the studies reviewed here [45, 53], children from poor backgrounds showed a higher prevalence of lower bands that was combined frequently with a lower prevalence of higher frequencies compared to their counterparts at every age [45, 47, 53–55].

Despite the correlational and cross-sectional nature of the great majority of the studies reviewed here, the findings supported the notion of a possible maturational lag, which is in line with MRI findings that show slower rates of brain growth in low SES children between 5 months and 4 years of age [87]. Yet, it is important to note that these findings represent an initial line of evidence, although more longitudinal data are necessary to support that children from poverty context present a maturational lag. In addition, mediation analysis and adjustments for confounding factors are necessary to elucidate how specific poverty experiences explain differences on EEG maturation. Moreover, mediation analysis would help to test whether EEG power differences help to explain the influences of distinct developmental contexts on cognition. These efforts would result in the possibility to use disparities in developmental trajectories of EEG power as cognitive markers that reflect differences in the general cognitive development between children from different SES backgrounds. However, it remains uncertain to which extent these neurophysiological differences are associated with behavioral outcomes. Despite important evidence showing that EEG power and behavior are associated with poverty experience, little is known about how EEG mediates the link between poverty experience and behavioral outcomes in, for example, the acquisition of cognitive skills during infancy.

Taken together, these studies suggest that poverty context may influence a wide frequency range of resting EEG during development. Studies reviewed here showed that children from low SES backgrounds have an increase in the power of low frequencies over anterior sites, and often a decrease in the power of alpha and higher frequencies over the anterior or posterior scalp sites, compared to higher SES samples. These findings that are derived from baseline EEG activity are also consistent with behavioral [28, 67, 88], MRI evidence [37], and ERP studies [5, 17, 41, 48–50]. They suggest that poor environments might exert its influence over brain networks that are related to executive processes, episodic memory, and learning skills.

3.5 Mediation Mechanisms

Almost all EEG studies on socioeconomic disparities lack evidence about mechanisms that could mediate the relationship between childhood poverty experience and brain functioning. Conversely, the literature on the impact of childhood poverty on brain development has proposed two main conceptual hypotheses that could partially explain this link: the *experience of stress* and *early language exposure* [18]. Although the action of these mechanisms is not likely to be independent, specific brain networks would be affected by each of them.

The *experience of stress* in low SES children is likely to be caused by both family and broader environmental characteristics. For instance, children growing up in poverty are more likely to experience bad parenting, family conflict, separation, and to live in chaotic, noisier, crowded, and more dangerous environments [20], all of which can contribute to increase the stress regulation response. Previous evidence suggested the existence of a deregulation of the hypothalamic-pituitary-adrenal (HPA) axis, which usually controls the secretion of cortisol hormone, among others, which contributes to the physiological stress response. Although several studies have agreed on this deregulation hypothesis, some of them have shown a pattern of hypercortisolism [89–92], although others found hypocortisolism [93–95] associated with impoverished backgrounds. The explanations for these discrepancies have focused on participants' characteristics, such as gender, age, and the diversity of adverse experiences [18]. Importantly, at the neurobiological level, a deregulation in stress physiology could have consequences for brain networks with high concentrations of corticosteroid receptors, such as the amygdala, the hippocampus, and the prefrontal cortex (PFC). These areas are sensitive to the effect of stress hormone exposure, and high levels of stress could alter their functioning [18, 90, 96, 97]. On the one hand, the hippocampus and the PFC are involved in the feedback that downregulates the functioning of the HPA axis, while the amygdala plays a facilitating role in the activation of HPA. Sustained exposure of stress hormones, such as cortisol, can produce cellular death, which can damage the functioning and structure of the hippocampus and promote the reactivity of the amygdala. On the other hand, in response to stress, the amygdala evokes the release of high levels of catecholamines and glucocorticoids, which can alter PFC functioning and increases amygdala reactivity [96]. Thus, because higher levels of stress can alter PFC and hippocampal functioning, it increases the functioning of the amygdala that leads to information processing and behavior switches from slow, thoughtful, and "top-down" regulation to a rapid, reflexive, and "bottom-up" regulation.

Previous studies showed that exposure to chronic stressors during childhood mediates the relationship between lower family income in childhood and reduced PFC activity during the regulation of emotions in adulthood [98]. In addition, Blair and colleagues [89] studied a large population that was predominantly low-income, and they found that children who had experienced fewer positive parenting behaviors had higher basal cortisol levels, which was associated with lower performance of executive functions [89]. Inconsistent, unpredictable, and less responsive parenting practices could be stressful for children, because they may feel a lack of control over their physical, social, and emotional needs. In this sense, it was hypothesized that a sustained exposure to stress in unpredictable living environments, and a lower sense of control, could lead children to exhibiting a general "alarm" state [99].

At the neural level, this involves a greater recruitment of networks that are involved in automatic and vigilance processing [18]. These adaptive responses toward a more automatic processing of information could help children to anticipate potentially challenging, negative, or threatening situations. However, it could also have consequences on the self-regulation of behaviors, thoughts, and emotions. Consistent with the *experience of stress* hypothesis of mediation, children with lower maternal education and SES showed comparable frontal activity to relevant and irrelevant information for task goals [5, 17, 49, 50]. Thus, low SES children may be more prone to process the broad set of information available and to have more difficulties inhibiting irrelevant information.

Although the *experience of stress* hypothesis of mediation is gaining more influence on the field of developmental neuroscience [98], it still remains little tested by EEG approaches. D'Angiulli et al. [5] used a direct measure of stress and found that low SES children had marginally higher levels of cortisol than high SES ones. In addition, only the low SES children showed an increase in the electrophysiological response of selective attention, which corresponded to an increase in post-task cortisol levels. Thus, it seems to be the case that low SES children became more stressed by exerting more effortful control to perform the task adequately. In addition, Neville et al. [52] using non-direct stress measure (i.e., self-reports of parenting stress) found a significant large decrease in parenting perceived-stress, after a training program with intervention sessions for parents and attentional exercises for children, relative to either attentional exercises for children alone or normal development of the HS program alone [52].

The *language exposure* hypothesis of mediation is supported by an extensive body of literature that has shown that growing up in low SES backgrounds is associated with poor quantity and quality of language exposure at home. First, it has been shown that parents of children from high SES families read more to their children than parents in low SES families. Second, it has been shown that mothers from low

SES backgrounds use fewer words, less complicated syntax, talk less frequently with their children, and, when they do talk, are more likely to be directing their children's behavior than simply eliciting conversation [100]. Third, the activities that parents choose for interacting with their children are likely to differ according to the SES, and this can influence concrete language-learning opportunities [100, 101]. For instance, some studies have shown that when mothers look for books with their preschool children, they use a more complex and richer speech during this selection process than in other activities [100].

These distinct language-learning experiences across SES were associated in different studies with differences in children's language skills, including vocabulary, phonological awareness, and syntax [36, 100, 102–104]. However, the neural mechanisms through which language exposure may influence child development of language- relevant brain networks are still unclear [101].

Following the *language exposure* hypothesis of mediation, it has been hypothesized that poor language environment could affect brain areas that are related to language processing [105], such as auditory (perisylvian) regions, the visual word form area, and the anterior inferior frontal cortex [101]. Moreover, it has been suggested that both conceptual models, *language exposure* and *experience of stress* hypotheses, are not likely to be completely independent, and they could be supported by overlapping neural mechanisms [82]. On the one hand, stress exposure is likely to interfere with language acquisition. For instance, because a deregulation of stress response could lead to a dysfunction in higher-order cognitive processes, children that are affected by high stress exposure probably have greater difficulty processing complex syntactic structures and concentrating in educational settings [101]. Alternatively, fewer and poorer language-learning experiences could reduce the opportunities to receive rich and complex language stimulation, experiences that help children to develop new skills taxing working memory resources [8, 18].

Up to now, EEG studies have not explored children exposure to language in association with SES disparities. Commonly, composite variables of SES were used to test directly their link with electrophysiological markers or language competencies. For instance, Tomalski et al. [47] found that infants from low SES backgrounds have reduced frontal gamma power, a pattern related to lower language skills in toddlers [79]. In turn, Kishiyama et al. [41] documented that children from low SES families had lower performance on a vocabulary test, but they did not investigate how this was related to the reduced activity found in early EEG components during a visual attention task.

From the perspective of developmental neuroscience, the accumulated evidence suggests that children are specially susceptible to the influence of adverse experiences [59]. For instance, during childhood, there are rapid and important changes in brain functioning, and early exposure to adverse experiences could alter the development more easily and more profoundly than adverse experiences that occur later on. Importantly, specific early alterations can influence the development of other functioning domains later in childhood. Electrophysiological approaches provide a direct measure of that neural functioning and, thus, these techniques are critical for studying how early experience of poverty influences development. Very often, at an early age, differences at the neural level of organization are more evident than differences at the cognitive and behavioral levels [59].

In any case, more research is needed to understand the mediating mechanisms by which the experience of poverty may impact the efficiency of different neural networks during development. Future research should include direct measures of parent and child stress physiology, linguistic environment, and other related poverty experiences that could be used to assess and analyze mechanisms of change. These approaches would help to elucidate the pathways and mechanisms through which distinct experiences of adversities, which are related to poverty, operate at different levels of organization.

4 Future Directions

The exposure to material and sociocultural deprivations is associated with a complex range of influences on neural organization and reorganization at different levels. A key question regarding the influences of poverty on neural and cognitive development is whether these disparities can be overcome by interventions, and what levels of analysis (e.g., molecular, neural, cognitive, behavioral) can support and guide these possible changes. Recent studies indicate that distinct types of interventions were effective in improving the performance levels in cognitive tasks in preschool children who lived in conditions of social vulnerability due to poverty [24, 106–111]. The evidence from the neural level—assessed through EEG—indicates that educational programs promoting parenting skills and cognitive stimulation in children positively influences cognitive performance and neural activity in low SES children. In some cases, these types of gains were achieved in a relatively short time [52]. This preliminary evidence allows both the identification of potential targets and time frames for the design of interventions to generating changes in neural and cognitive development. Furthermore, interventions including EEG measures could help to determine both the underlying mechanisms of gains and the extent of mutability of impacts that are generated by verifiable deprivations in distinct developmental contexts.

Evaluations that consider multiple levels of organization are not applied generally in the context of cognitive interventions beyond the laboratory settings, such as in schools or homes. The inclusion of neural analysis often imposes limitations for use outside the laboratory, because of the added burden of noise, logistics, and transportation. Therefore, it is important to broaden the efforts to extend the design and the implementation of these approaches. Currently, novel methodologies are being developed to improve the signal quality of portable EEG equipment, both in terms of hardware and signal processing, such as artifacts and single-trial analysis techniques [40, 112]. In this regard, efforts that are aimed at transferring laboratory methodologies to different developmental contexts creates the possibility of extending their inclusion to studies with greater ecological value.

In addition, future studies should focus on innovative efforts to include a wider range of EEG paradigms that can be used to test suitable hypotheses about how early adverse experience is related to different patterns of brain development. The findings about effects of poverty or low SES on brain functioning were achieved by using unidimensional measures that could not explain the mechanisms through which poverty impacts brain circuits. Thus, the measures that have been implemented up to now only captured the status of each child indirectly and partially, but they have not considered individual factors that could better characterize the child's experience due to poverty. Conceptual advancements should thus generate new definitions of poverty specifically considering the dynamics of the adversity on children's experiences. This could be achieved by using analyses that have been applied commonly in recent studies of childhood poverty and cognition, such as mixed models [24] or multiple mediation models [8]. These methodological approaches allow the identification of those socio-environmental risks or protective factors that explain the variance of poverty measures on cognitive outcomes [8]. This is especially important because it would provide an ecological and dynamic perspective for each developmental context, which would enable both to capture the effects of specific contextual deprivations on several cognitive systems during development and to fine-tune targets for improvement the design of innovative intervention programs. Thus, future research would benefit from thinking about a definition of poverty in terms of a continuum of effects with several possible outcomes, which depend on the interaction of several crucial factors that are defined by the type, number, and accumulation of risk factors to which children are exposed, the co-occurrence of deprivations, the timing of exposure, and the individual susceptibility to each one. Moreover, electrophysiological approaches would help to elucidate the predictive role of adverse experience on the development of brain functioning. In an interdisciplinary context, electrophysiological approaches would also help to generate information at different interconnected levels of analysis, and they would contribute to building a concept of poverty as a complex phenomenon.

5 Conclusions

ERP studies in relation to poverty have focused mainly on the assessment of attentional mechanisms. The verified associations between poverty and attentional processes might be related to a domain-general effect, which could complement the findings within the social sciences and neuroscience regarding the associations of SES, language, and executive functions. However, resting-state EEG studies have suggested that poverty contexts may influence a wide range of frequencies during development, indicating that a poor environment might influence prefrontal brain areas and their related cognitive processes. Also, some of the studies suggested that there is a maturational lag between children from low and high SES families but, up to now, there are no longitudinal studies that support this hypothesis to demonstrate how these differences evolve through development. Importantly, only one study has explored how these differences change with intervention programs that take advantage of the brain plasticity, especially at young ages. Finally, future studies must benefit from the large conceptual advances that have been made by developmental psychology about the mediators of the influence of poverty on cognitive development, and they should attempt to discern between the two main conceptual theories that have been proposed: *the experience of stress and language exposure*.

The available evidence of influences of childhood poverty on brain functioning supports the notion that improving our understanding about what aspects of deprivations would influence the cognitive development requires (a) the building of an ecological and dynamic approach considering the variability of cognitive outcomes, which depend on the mediating mechanisms associated with the specific adverse experiences that have a dynamic nature and change during development, (b) the design of more elaborate conceptual paradigms to integrate current neuroscientific evidence on indicators of adverse experiences with patterns of brain structure and function, and (c) the assessment of the impact of interventions outside a laboratory setting to incorporate greater ecological measures of children's functioning.

Acknowledgments Authors are supported by CONICET, FONCYT, CEMIC, Fundación Conectar, and University of Buenos Aires. Authors thank Thomas A. Gavin, Professor Emeritus, Cornell University, for help with editing the English in this chapter. The authors declare that there is no conflict of interest.

References

1. Allan NP, Hume LE, Allan DM, Farrington AL, Lonigan CJ. Relations between inhibitory control and the development of academic skills in preschool and kindergarten: a meta-analysis. Dev Psychol. 2014;50(10):2368–79. https://doi.org/10.1037/a0037493.
2. Bull R, Lee K. Executive functioning and mathematics achievement. Child Dev Perspect. 2014;8(1):36–41. https://doi.org/10.1111/cdep.12059.
3. Shonkoff JP. Leveraging the biology of adversity to address the roots of disparities in health and development. Proc Natl Acad Sci U S A. 2012;109(Suppl 2):17302–7. Available from http://www.pnas.org/content/109/Supplement_2/17302.abstract.
4. Blair C, Raver CC. Poverty, stress, and brain development: new directions for prevention and intervention. Acad Pediatr. 2016;16(3):S30–6. Available from http://www.sciencedirect.com/science/article/pii/S1876285916000267.
5. D'Angiulli A, Van Roon PM, Weinberg J, Oberlander TF, Grunau RE, Hertzman C, et al. Frontal EEG/ERP correlates of attentional processes, cortisol and motivational states in adolescents from lower and higher socioeconomic status. Front Hum Neurosci. 2012;6:306. Available from http://www.pubmedcentral.nih.gov/articlerender.fcgi?artid=3500742&tool=pmcentrez&rendertype=abstract.
6. Gianaros PJ, Hackman D. Contributions of neuroscience to the study of socioeconomic health disparities. Psychosom Med. 2013;75(7):610–5. Available from http://europepmc.org/articles/PMC3816088.
7. Hackman DA, Farah MJ, Meaney MJ. Socioeconomic status and the brain: mechanistic insights from human and animal research. Nat Rev Neurosci. 2010;11(9):651–9. Available from http://www.ncbi.nlm.nih.gov/pubmed/20725096.

8. Hackman DA, Gallop R, Evans GW, Farah MJ. Socioeconomic status and executive function: developmental trajectories and mediation. Dev Sci. 2015;18(5):686–702. https://doi.org/10.1111/desc.12246.
9. Hackman DA, Farah MJ. Socioeconomic status and the developing brain. Trends Cogn Sci. 2009;13(2):65–73. Available from http://www.sciencedirect.com/science/article/pii/S1364661308002635.
10. Lipina SJ, Segretin MS. Strengths and weakness of neuroscientific investigations of childhood poverty: future directions. Front Hum Neurosci. 2015;9(53):1–5. Available from http://www.frontiersin.org/human_neuroscience/10.3389/fnhum.2015.00053/abstract.
11. Lipina SJ, Colombo JA. Poverty and brain development during childhood: an approach from cognitive psychology and neuroscience. Washington: American Psychological Association; 2009. 172 p. Available from http://ovidsp.ovid.com/ovidweb.cgi?T=JS&PAGE=reference&D=psyc6&NEWS=N&AN=2009-08043-000.
12. Lipina SJ, Posner MI. The impact of poverty on the development of brain networks. Front Hum Neurosci. 2012;6:1–12. Available from http://journal.frontiersin.org/article/10.3389/fnhum.2012.00238/abstract.
13. Lipina SJ. Biological and sociocultural determinants of neurocognitive development: central aspects of the current scientific agenda. In: Vaticana LE, editor. Bread and brain, education and poverty. Vatican City: Pontifical Academy of Sciences; 2014. p. 1–30.
14. Moffitt TE, Arseneault L, Belsky D, Dickson N, Hancox RJ, Harrington H, et al. A gradient of childhood self-control predicts health, wealth, and public safety. Proc Natl Acad Sci. 2011;108(7):2693–8. Available from http://europepmc.org/articles/PMC3041102.
15. Pavlakis AE, Noble K, Pavlakis SG, Ali N, Frank Y. Brain imaging and electrophysiology biomarkers: is there a role in poverty and education outcome research? Pediatr Neurol. 2015;52(4):383–8. https://doi.org/10.1016/j.pediatrneurol.2014.11.005.
16. Raizada RDS, Kishiyama MM. Effects of socioeconomic status on brain development, and how cognitive neuroscience may contribute to levelling the playing field. Front Hum Neurosci. 2010;4:3. Available from http://europepmc.org/articles/PMC2820392.
17. Stevens C, Lauinger B, Neville H. Differences in the neural mechanisms of selective attention in children from different socioeconomic backgrounds: an event-related brain potential study. Dev Sci. 2009;12(4):634–46. https://doi.org/10.1111/j.1467-7687.2009.00807.x.
18. Ursache A, Noble KG. Neurocognitive development in socioeconomic context: multiple mechanisms and implications for measuring socioeconomic status. Psychophysiology. 2016;53(1):71–82. https://doi.org/10.1111/psyp.12547.
19. Bradley RH, Corwyn RF. Socioeconomic status and child development. Annu Rev Psychol. 2002;53(1):371–99. https://doi.org/10.1146/annurev.psych.53.100901.135233.
20. Evans GW. The environment of childhood poverty. Am Psychol. 2004;59(2):77–92. https://doi.org/10.1037/0003-066X.59.2.77.
21. Gassman-Pines A, Yoshikawa H. The effects of antipoverty programs on children's cumulative level of poverty-related risk. Dev Psychol. 2006;42(6):981–99. https://doi.org/10.1037/0012-1649.42.6.981.
22. Rhoades BL, Greenberg MT, Lanza ST, Blair C. Demographic and familial predictors of early executive function development: contribution of a person-centered perspective. J Exp Child Psychol. 2011;108(3):638–62. Available from http://www.sciencedirect.com/science/article/pii/S0022096510001633.
23. Sarsour K, Sheridan M, Jutte D, Nuru-Jeter A, Hinshaw S, Boyce WT. Family socioeconomic status and child executive functions: the roles of language, home environment, and single parenthood. J Int Neuropsychol Soc. 2011;17(1):120–32. Available from http://journals.cambridge.org/article_S1355617710001335.
24. Segretin MS, Lipina SJ, Hermida MJ, Sheffield TD, Nelson JM, Espy KA, et al. Predictors of cognitive enhancement after training in preschoolers from diverse socioeconomic backgrounds. Front Psychol. 2014;5:205. Available from http://europepmc.org/articles/PMC3952047.

25. Walker SP, Wachs TD, Meeks Gardner J, Lozoff B, Wasserman GA, Pollitt E, et al. Child development: risk factors for adverse outcomes in developing countries. Lancet. 2007;369(9556):145–57. Available from http://www.sciencedirect.com/science/article/pii/S0140673607600762.
26. Evans GW, Li D, Whipple SS. Cumulative risk and child development. Psychol Bull. 2013;139(6):1342–96. https://doi.org/10.1037/a0031808.
27. Cadima J, McWilliam RA, Leal T. Environmental risk factors and children's literacy skills during the transition to elementary school. Int J Behav Dev. 2010;34(1):24–33. Available from https://www.scopus.com/inward/record.uri?eid=2-s2.0-74049089341&partnerID=40&md5=1c7587681e58f468561e7669efa506c0.
28. Lipina S, Segretin S, Hermida J, Prats L, Fracchia C, Camelo JL, et al. Linking childhood poverty and cognition: environmental mediators of non-verbal executive control in an argentine sample. Dev Sci. 2013;16(5):697–707. https://doi.org/10.1111/desc.12080.
29. Belsky J, Bakermans-Kranenburg MJ, van IJzendoorn MH. For better and for worse differential susceptibility to environmental influences. Curr Dir Psychol Sci. 2007;16(6):300–4. Available from http://www.ingentaconnect.com/content/bpl/cdir/2007/00000016/00000006/art00003.
30. Najman JM, Aird R, Bor W, O'Callaghan M, Williams GM, Shuttlewood GJ. The generational transmission of socioeconomic inequalities in child cognitive development and emotional health. Soc Sci Med. 2004;58(6):1147–58. Available from http://www.sciencedirect.com/science/article/pii/S0277953603002867.
31. Sheridan MA, KA ML. Dimensions of early experience and neural development: deprivation and threat. Trends Cogn Sci. 2014;18(11):580–5. Available from http://www.ncbi.nlm.nih.gov/pubmed/25305194.
32. Stanton-Chapman TL, Chapman DA, Kaiser AP, Hancock TB. Cumulative risk and low-income children's language development. Top Early Child Spec Educ. 2004;24(4):227–37. Available from http://tec.sagepub.com/content/24/4/227.abstract.
33. Duncan GJ, Magnuson K. Socioeconomic status and cognitive functioning: moving from correlation to causation. Wiley Interdiscip Rev Cogn Sci. 2012;3(3):377–86. https://doi.org/10.1002/wcs.1176.
34. Lipina SJ. Critical considerations about the use of poverty measures in the study of cognitive development. Int J Psychol. 2017;52(3):241–50. https://doi.org/10.1002/ijop.12282.
35. Raizada RDS, Richards TL, Meltzoff A, Kuhl PK. Socioeconomic status predicts hemispheric specialisation of the left inferior frontal gyrus in young children. NeuroImage. 2008;40(3):1392–401. Available from http://www.sciencedirect.com/science/article/pii/S1053811908000475.
36. Noble KG, Wolmetz ME, Ochs LG, Farah MJ, McCandliss BD. Brain-behavior relationships in reading acquisition are modulated by socioeconomic factors. Dev Sci. 2006;9(6):642–54. https://doi.org/10.1111/j.1467-7687.2006.00542.x.
37. Noble KG, Houston SM, Brito NH, Bartsch H, Kan E, Kuperman JM, et al. Family income, parental education and brain structure in children and adolescents. Nat Neurosci. 2015;18(5):773–8. Available from http://europepmc.org/articles/PMC4414816.
38. Avants BB, Hackman DA, Betancourt LM, Lawson GM, Hurt H, Farah MJ. Relation of Childhood Home Environment to Cortical Thickness in Late Adolescence: Specificity of Experience and Timing. PLoS One. 2015;10(10):e0138217. https://doi.org/10.1371/journal.pone.0138217.
39. Hair NL, Hanson JL, Wolfe BL, Pollak SD. Association of child poverty, brain development, and academic achievement. JAMA Pediatr. 2015;169(9):822–9. Available from http://archpedi.jamanetwork.com/article.aspx?doi=10.1001/jamapediatrics.2015.1475.
40. Cohen MX. Analyzing neural time series data: theory and practice. London: MIT Press; 2014. 600 p. Available from https://mitpress.mit.edu/books/analyzing-neural-time-series-data.
41. Kishiyama MM, Boyce WT, Jimenez AM, Perry LM, Knight RT. Socioeconomic disparities affect prefrontal function in children. J Cogn Neurosci. 2009;21(6):1106–15. Available from http://www.ncbi.nlm.nih.gov/pubmed/18752394.

42. Skoe E, Krizman J, Kraus N. The impoverished brain: disparities in maternal education affect the neural response to sound. J Neurosci. 2013;33(44):17221–31. Available from http://www.ncbi.nlm.nih.gov/pubmed/24174656
43. Stevens C, Paulsen D, Yasen A, Neville H. Atypical auditory refractory periods in children from lower socio-economic status backgrounds: ERP evidence for a role of selective attention. Int J Psychophysiol. 2015;95(2):156–66. https://doi.org/10.1016/j.ijpsycho.2014.06.017.
44. Ruberry EJ, Lengua LJ, Crocker LH, Bruce J, Upshaw MB, Sommerville JA. Income, neural executive processes, and preschool children's executive control. Dev Psychopathol. 2017;29(1):143–54. Available from http://www.journals.cambridge.org/abstract_S095457941600002X.
45. Harmony T, Marosi E, Diaz de Leon AE, Becker J, Fernandez T. Effect of sex, psychosocial disadvantages and biological risk factors on EEG maturation. Electroencephalogr Clin Neurophysiol. 1990;75(6):482–91. Available from http://www.sciencedirect.com/science/article/pii/0013469490901357.
46. Brito NH, Fifer WP, Myers MM, Elliott AJ, Noble KG. Associations among family socioeconomic status, EEG power at birth, and cognitive skills during infancy. Dev Cogn Neurosci. 2016;19:144–51. Available from http://www.sciencedirect.com/science/article/pii/S1878929315301201.
47. Tomalski P, Moore DG, Ribeiro H, Axelsson EL, Murphy E, Karmiloff-Smith A, et al. Socioeconomic status and functional brain development – associations in early infancy. Dev Sci. 2013;16(5):676–87. https://doi.org/10.1111/desc.12079.
48. Conejero Á, Guerra S, Abundis-Gutiérrez A, Rueda MR. Frontal theta activation associated with error detection in toddlers: influence of familial socioeconomic status. Dev Sci. 2016. https://doi.org/10.1111/desc.12494.
49. D'Angiulli A, Herdman A, Stapells D, Hertzman C. Children's Event-related potentials of auditory selective attention vary with their socioeconomic status. Neuropsychology. 2008;22(3):293. https://doi.org/10.1037/0894-4105.22.3.293.
50. D'Angiulli A, Weinberg J, Grunau R, Hertzman C, Grebenkov P. Towards a cognitive science of social inequality: children's attention-related ERPs and salivary cortisol vary with their socioeconomic status. In: Proceedings of the 30th cognitive science society annual meeting. Washington, DC: Cognitive Science Society; 2008. p. 211–216.
51. Isbell E, Wray AH, Neville HJ. Individual differences in neural mechanisms of selective auditory attention in preschoolers from lower socioeconomic status backgrounds: an event-related potentials study. Dev Sci. 2016;19(6):865–80. https://doi.org/10.1111/desc.12334.
52. Neville HJ, Stevens C, Pakulak E, Bell TA, Fanning J, Klein S, et al. Family-based training program improves brain function, cognition, and behavior in lower socioeconomic status preschoolers. Proc Natl Acad Sci U S A. 2013;110(29):12138–43. https://doi.org/10.1073/pnas.1304437110.
53. Otero GA, Pliego-Rivero FB, Fernández T, Ricardo J. EEG development in children with sociocultural disadvantages: a follow-up study. Clin Neurophysiol. 2003;114(10):1918–25. Available from http://www.ncbi.nlm.nih.gov/pubmed/14499754.
54. Otero GA. Poverty, cultural disadvantage and brain development: a study of pre-school children in Mexico. Electroencephalogr Clin Neurophysiol. 1997;102(6):512–6. Available from http://www.sciencedirect.com/science/article/pii/S0013469497952139.
55. Otero GA. EEG spectral analysis in children with sociocultural handicaps. Int J Neurosci. 1994;79(3–4):213–20. Available from http://www.ncbi.nlm.nih.gov/pubmed/7744563.
56. Tomarken AJ, Dichter GS, Garber J, Simien C. Resting frontal brain activity: linkages to maternal depression and socio-economic status among adolescents. Biol Psychol. 2004;67(1–2):77–102. Available from http://www.sciencedirect.com/science/article/pii/S030105110400033X.
57. UK Office for National Statistics (2010). Standard Occupational Classification 2010. Volume 3: The National Statistics Socio-economic Classification User Manual. Basingstoke: Palgrave Macmillan.
58. Lipina SJ, Simonds J, Segretin MS. Recognizing the child in child poverty. Vulnerable Child Youth Stud. 2011;6(1):8–17. https://doi.org/10.1080/17450128.2010.521598.

59. Duncan GJ, Magnuson K, Votruba-Drzal E. Moving beyond correlations in assessing the consequences of poverty. Annu Rev Psychol. 2017;68(1):413–34. Available from http://www.annualreviews.org/doi/10.1146/annurev-psych-010416-044224.
60. Berger A, Tzur G, Posner MI. Infant brains detect arithmetic errors. Proc Natl Acad Sci U S A. 2006;103(33):12649–53. https://doi.org/10.1073/pnas.0605350103.
61. Reid VM, Hoehl S, Grigutsch M, Groendahl A, Parise E, Striano T. The neural correlates of infant and adult goal prediction: evidence for semantic processing systems. Dev Psychol. 2009;45(3):620–9. Available from http://www.ncbi.nlm.nih.gov/pubmed/19413420.
62. Ciavarro M, Ambrosini E, Tosoni A, Committeri G, Fattori P, Galletti C. rTMS of medial parieto-occipital cortex interferes with attentional reorienting during attention and reaching tasks. J Cogn Neurosci. 2013;25(9):1453–62. Available from http://www.ncbi.nlm.nih.gov/pubmed/23647519.
63. Luu P, Tucker DM, Derryberry D, Reed M, Poulsen C. Electrophysiological responses to errors and feedback in the process of action regulation. Psychol Sci. 2003;14(1):47–53. Available from http://www.ncbi.nlm.nih.gov/pubmed/12564753.
64. Tsujimoto T, Shimazu H, Isomura Y. Direct recording of theta oscillations in primate prefrontal and anterior cingulate cortices. J Neurophysiol. 2006;95(5):2987–3000. Available from http://www.ncbi.nlm.nih.gov/pubmed/16467430.
65. Petersen SE, Posner MI. The attention system of the human brain: 20 years after. Annu Rev Neurosci. 2012;35(1):73–89. Available from http://www.ncbi.nlm.nih.gov/pubmed/22524787.
66. Raver CC, Blair C, Willoughby M. Poverty as a predictor of 4-year-olds' executive function: new perspectives on models of differential susceptibility. Dev Psychol. 2013;49(2):292–304.
67. Lipina SJ, Martelli M, Vuelta B, Colombo JA. Performance on the a-not-b task of Argentinean infants from unsatisfied and satisfied basic needs homes. Int J Psychol. 2005;39(1):49–60.
68. Coch D, Skendzel W, Neville HJ. Auditory and visual refractory period effects in children and adults: an ERP study. Clin Neurophysiol. 2005;116(9):2184–203. Available from http://www.sciencedirect.com/science/article/pii/S1388245705002312.
69. Fukuda K, Vogel EK. Individual differences in recovery time from attentional capture. Psychol Sci. 2011;22(3):361–8. Available from http://pss.sagepub.com/content/22/3/361.abstract.
70. Evans GW, Gonnella C, Marcynyszyn LA, Gentile L, Salpekar N. The role of chaos in poverty and children's socioemotional adjustment. Psychol Sci. 2005;16(7):560–5. Available from http://www.ncbi.nlm.nih.gov/pubmed/16008790.
71. Gulbinaite R, Johnson A, de Jong R, Morey CC, van Rijn H. Dissociable mechanisms underlying individual differences in visual working memory capacity. Neuroimage. 2014;99:197–206. Available from http://www.ncbi.nlm.nih.gov/pubmed/24878830.
72. Kraus N, Chandrasekaran B. Music training for the development of auditory skills. Nat Rev Neurosci. 2010;11(8):599–605. Available from http://www.ncbi.nlm.nih.gov/pubmed/20648064.
73. Krizman J, Marian V, Shook A, Skoe E, Kraus N. Subcortical encoding of sound is enhanced in bilinguals and relates to executive function advantages. Proc Natl Acad Sci U S A. 2012;109(20):7877–81. Available from http://www.ncbi.nlm.nih.gov/pubmed/22547804.
74. Fox SE, Levitt P, Nelson CA. How the timing and quality of early experiences influence the development of brain architecture. Child Dev. 2010;81(1):28–40. https://doi.org/10.1111/j.1467-8624.2009.01380.x.
75. Rueda MR, Posner MI, Rothbart MK. The development of executive attention: contributions to the emergence of self-regulation. Dev Neuropsychol. 2005;28(2):573–94. https://doi.org/10.1207/s15326942dn2802_2.
76. Rueda MR, Pozuelos JP, Combita LM. Cognitive neuroscience of attention from brain mechanisms to individual differences in efficiency. AIMS Neurosci. 2015;2(4):183–202. https://doi.org/10.3934/Neuroscience.2015.4.183.
77. Crone NE, Hao L, Hart J, Boatman D, Lesser RP, Irizarry R, et al. Electrocorticographic gamma activity during word production in spoken and sign language. Neurology. 2001;57(11):2045–53. Available from http://www.ncbi.nlm.nih.gov/pubmed/11739824.

78. Gross DW, Gotman J. Correlation of high-frequency oscillations with the sleep-wake cycle and cognitive activity in humans. Neuroscience. 1999;94(4):1005–18. Available from http://www.ncbi.nlm.nih.gov/pubmed/10625043.
79. Benasich AA, Gou Z, Choudhury N, Harris KD. Early cognitive and language skills are linked to resting frontal gamma power across the first 3 years. Behav Brain Res. 2008;195(2):215–22. Available from http://linkinghub.elsevier.com/retrieve/pii/S0166432808004993.
80. Ray S, Niebur E, Hsiao SS, Sinai A, Crone NE. High-frequency gamma activity (80-150Hz) is increased in human cortex during selective attention. Clin Neurophysiol. 2008;119(1):116–33. Available from http://www.ncbi.nlm.nih.gov/pubmed/18037343.
81. Gou Z, Choudhury N, Benasich AA. Resting frontal gamma power at 16, 24 and 36 months predicts individual differences in language and cognition at 4 and 5 years. Behav Brain Res. 2011;220(2):263–70. Available from http://www.sciencedirect.com/science/article/pii/S016643281100088X.
82. McLaughlin KA, Sheridan MA. Beyond cumulative risk: a dimensional approach to childhood adversity. Curr Dir Psychol Sci. 2016;25(4):239–45. Available from http://www.ncbi.nlm.nih.gov/pubmed/27773969.
83. Clarke AR, Barry RJ, McCarthy R, Selikowitz M. Age and sex effects in the EEG: development of the normal child. Clin Neurophysiol. 2001;112(5):806–14. Available from http://www.sciencedirect.com/science/article/pii/S1388245701004886.
84. Marshall PJ, Bar-Haim Y, Fox NA. Development of the EEG from 5 months to 4 years of age. Clin Neurophysiol. 2002;113(8):1199–208. Available from http://www.sciencedirect.com/science/article/pii/S1388245702001633.
85. Takano T, Ogawa T. Characterization of developmental changes in EEG-gamma band activity during childhood using the autoregressive model. Pediatr Int. 1998;40(5):446–52. https://doi.org/10.1111/j.1442-200X.1998.tb01966.x.
86. Kondacs A, Szabó M. Long-term intra-individual variability of the background EEG in normals. Clin Neurophysiol. 1999;110(10):1708–16. Available from http://www.ncbi.nlm.nih.gov/pubmed/10574286.
87. Hanson JL, Hair N, Shen DG, Shi F, Gilmore JH, Wolfe BL, et al. Family poverty affects the rate of human infant brain growth. PLoS One. 2013;8(12):e80954. https://doi.org/10.1371/journal.pone.0080954.
88. Mezzacappa E. Alerting, orienting, and executive attention: developmental properties and sociodemographic correlates in an epidemiological sample of young, urban children. Child Dev. 2004;75(5):1373–86. https://doi.org/10.1111/j.1467-8624.2004.00746.x.
89. Blair C, Granger DA, Willoughby M, Mills-Koonce R, Cox M, Greenberg MT, et al. Salivary cortisol mediates effects of poverty and parenting on executive functions in early childhood. Child Dev. 2011;82(6):1970–84. Available from http://www.ncbi.nlm.nih.gov/pubmed/22026915.
90. Lupien SJ, King S, Meaney MJ, McEwen BS. Can poverty get under your skin? Basal cortisol levels and cognitive function in children from low and high socioeconomic status. Dev Psychopathol. 2001;13(3):653–76. Available from http://search.ebscohost.com/login.aspx?direct=true&db=psyh&AN=2001-18325-012&site=ehost-live&scope=site.
91. Lupien SJ, King S, Meaney MJ, BS ME. Child's stress hormone levels correlate with mother's socioeconomic status and depressive state. Biol Psychiatry. 2000;48(10):976–80. Available from http://www.ncbi.nlm.nih.gov/pubmed/11082471.
92. Chen E, Cohen S, Miller GE. How low socioeconomic status affects 2-year hormonal trajectories in children. Psychol Sci. 2010;21(1):31–7. Available from http://www.ncbi.nlm.nih.gov/pubmed/20424019.
93. Badanes LS, Watamura SE, Hankin BL. Hypocortisolism as a potential marker of allostatic load in children: associations with family risk and internalizing disorders. Dev Psychopathol. 2011;23(3):881–96. Available from http://www.ncbi.nlm.nih.gov/pubmed/21756439.
94. Chen E, Paterson LQ. Neighborhood, family, and subjective socioeconomic status: how do they relate to adolescent health? Health Psychol. 2006;25(6):704–14. Available from http://www.ncbi.nlm.nih.gov/pubmed/17100499.

95. Kliewer W, Reid-Quinones K, Shields BJ, Foutz L. Multiple risks, emotion regulation skill, and cortisol in low-income African American youth: a prospective study. J Black Psychol. 2008;35(1):24–43. https://doi.org/10.1177/0095798408323355.
96. Arnsten AFT. Stress signalling pathways that impair prefrontal cortex structure and function. Nat Rev Neurosci. 2009;10(6):410–22. https://doi.org/10.1038/nrn2648.
97. Lupien SJ, Lepage M. Stress, memory, and the hippocampus: can't live with it, can't live without it. Behav Brain Res. 2001;127(1-2):137–58. Available from http://linkinghub.elsevier.com/retrieve/pii/S0166432801003618.
98. Kim P, Evans GW, Angstadt M, Ho SS, Sripada CS, Swain JE, et al. Effects of childhood poverty and chronic stress on emotion regulatory brain function in adulthood. Proc Natl Acad Sci. 2013;110(46):18442–7. https://doi.org/10.1073/pnas.1308240110.
99. Ursin H, Eriksen HR. The cognitive activation theory of stress. Psychoneuroendocrinology. 2004;29(5):567–92. Available from http://www.sciencedirect.com/science/article/pii/S030645300300091X.
100. Hoff E. How social contexts support and shape language development? Dev Rev. 2006;26(1):55–88. Available from http://linkinghub.elsevier.com/retrieve/pii/S0273229705000316.
101. Perkins SC, Finegood ED, Swain JE. Poverty and language development: roles of parenting and stress. Innov Clin Neurosci. 2013;10(4):10–9. Available from http://www.ncbi.nlm.nih.gov/pubmed/23696954.
102. Hoff E. Causes and consequences of SES-related differences in parent-to-child speech. In: Socioeconomic status, parenting, and child development. Mahwah: Lawrence Erlbaum Associates; 2003. p. 147–60.
103. Huttenlocher J, Vasilyeva M, Cymerman E, Levine S. Language input and child syntax. Cogn Psychol. 2002;45(3):337–74. Available from http://www.ncbi.nlm.nih.gov/pubmed/12480478.
104. Pan BA, Rowe ML, Singer JD, Snow CE. Maternal correlates of growth in toddler vocabulary production in low-income families. Child Dev. 2005;76(4):763–82. https://doi.org/10.1111/j.1467-8624.2005.00876.x.
105. Brito NH, Noble KG. Socioeconomic status and structural brain development. Front Neurosci. 2014;8:276. Available from http://www.ncbi.nlm.nih.gov/pubmed/25249931.
106. Burger K. How does early childhood care and education affect cognitive development? An international review of the effects of early interventions for children from different social backgrounds. Early Child Res Q. 2010;25(2):140–65. Available from http://www.sciencedirect.com/science/article/pii/S0885200609000921.
107. Cybele Raver C, McCoy DC, Lowenstein AE, Pess R. Predicting individual differences in low-income children's executive control from early to middle childhood. Dev Sci. 2013;16(3):394–408. https://doi.org/10.1111/desc.12027.
108. Goldin AP, Hermida MJ, Shalom DE, Elias Costa M, Lopez-Rosenfeld M, Segretin MS, et al. Far transfer to language and math of a short software-based gaming intervention. Proc Natl Acad Sci. 2014;111(17):6443–8. Available from http://www.pnas.org/content/111/17/6443.abstract.
109. Jolles DD, Crone EA. Training the developing brain: a neurocognitive perspective. Front Hum Neurosci. 2012;6(76):76. Available from http://www.frontiersin.org/human_neuroscience/10.3389/fnhum.2012.00076/abstract.
110. Rueda MR, Rothbart MK, McCandliss BD, Saccomanno L, Posner MI. From the cover: training, maturation, and genetic influences on the development of executive attention. Proc Natl Acad Sci. 2005;102(41):14931–6. Available from http://www.pnas.org/content/102/41/14931.abstract.
111. Lipina SJ, Segretin MS, Hermida MJ, Colombo JA. Research on childhood poverty from a cognitive neuroscience perspective: examples of studies in Argentina. In: Handbook of mental health in children and adolescents. London: Sage; 2012. p. 256–74.

112. Pietto ML, Kamienkowski JE, Lipina SJ. Electrophysiological approaches in the study of cognitive development outside the lab. Buenos Aires: Latin American Brain Mapping Network; 2017.
113. Shamseer L, Moher D, Clarke M, Ghersi D, Liberati A, Petticrew M, et al. Preferred reporting items for systematic review and metaanalysis protocols (PRISMA-P) 2015: elaboration and explanation. BMJ. 2015;329:g7647. https://doi.org/10.1136/bmj.g7647.
114. Valdez, J.L., Campos S., & Ortega, M.A. Las condiciones de vida en familias de escasos recursos consideradas de "Alto y Bajo Riesgo Psicosocial". Paper presented at the International Seminary of Cerebral Damage. Toluca (Mexico); 1989.
115. Hollingshead, A. A. (1975). Four-factor index of social status. Unpublished manuscript, Yale University, New Haven, CT.

The Cultural Neuroscience of Socioeconomic Status

Jung Yul Kwon, Ryan S. Hampton, and Michael E.W. Varnum

Abstract In this chapter, we review an emerging body of research that has used neuroscientific techniques (EEG, ERP, fMRI) to examine how our socioeconomic status (SES) affects brain functioning. We focus on SES effects on neural responses reflecting (1) attunement to others, (2) vigilance, (3) trait inference, and (4) emotion regulation. We also address relevant findings regarding the effects of SES on (5) selective attention from a cultural neuroscience perspective. We end by outlining future directions for cultural neuroscience research on the impact of SES, including expanding the scope of inquiry to assess potential interactions between SES and broader cultural context.

Keywords Socioeconomic status • Cultural neuroscience • Culture • Empathy • Motor resonance • Emotion regulation • Vigilance • Trait inference

1 Socioeconomic Status

For centuries, socioeconomic status (SES) has been a topic of interest to social scientists and scholars in a broad variety of disciplines (see Pietto et al., this volume). SES is a complex, continuous construct that comprises objective indicators, such as educational attainment and occupational prestige, as well as subjective indicators, such as perception of one's own place in a larger hierarchical structure. Although sometimes considered distinct from social class, which is often conceptualized as a categorical rather than a continuous variable [1, 2], here we use the terms interchangeably as this is common in the literature we will review, and as arguably both notions tap into the same underlying constructs.

SES constitutes a powerful and pervasive context throughout one's life, influencing various aspects of psychology including cognitive tendencies [3], agency and choice [4], health [5], and subjective well-being [6] (see Kemp et al., this volume). Recently, the application of neuroscientific methods in SES research has proved

J.Y. Kwon (✉) • R.S. Hampton • M.E. Varnum
Department of Psychology, Arizona State University,
PO Box 871104, Tempe, AZ 85287-1104, USA
e-mail: jungyulkwon@asu.edu; ryan.hampton@asu.edu; mvarnum@asu.edu

© Springer International Publishing Switzerland 2017

A. Ibáñez et al. (eds.), *Neuroscience and Social Science*,
DOI 10.1007/978-3-319-68421-5_16

fruitful, revealing various ways that SES affects neural functioning. The present chapter aims to illustrate how the intersection of cultural neuroscience and SES has unveiled exciting possibilities for SES research.

2 SES and Cultural Neuroscience

The fundamental assumptions in cultural neuroscience are that the brain is plastic, and that frequent and systematic experience accounts for substantial amounts of variation in neural function and organization [7–10]. Culture comprises ideas, values, and beliefs shared by a group of people. These shared meanings are embedded in behavioral scripts or practices, as well as the organizational and behavioral institutions that help transmit such values and beliefs across generations. Various components of culture are mutually reinforcing—for example, while specific values may be emphasized in environments with certain consistent constraints on behavior, norms and practices may be produced and transmitted in order to promote those values [11, 12]. Culture shapes our everyday realities in concrete and tangible ways, and, as different cultural contexts lead a group of people to repeatedly engage in certain behavioral patterns and construct traditions that reinforce these behaviors, it is reasonable to predict that the accumulation of these experiences within and across generations would manifest in reliable changes to the brain and resultant psychology [9, 11, 13].

It has been posited that SES should also be regarded as a type of cultural context, as divergent social class contexts provide systematically different environments which promote different values, beliefs, and behaviors [1, 2, 14, 15]. Individuals with low SES tend to have fewer and less stable resources, and experience greater threat from a variety of sources. This type of ecology would engender behaviors, norms, values, and beliefs that are decidedly dissimilar from those in high-SES contexts who do not face similar challenges. Further, like other forms of culture, those associated with SES may be transmitted across individuals and generations in addition to being direct evoked responses to ecological conditions. Accordingly, one would expect to see these differences reflected in the social cognitive processes and neural functioning as a response to these different ecologies [7].

3 Benefits of Neural Measures

Although the psychological consequences of SES have typically been studied using common social psychological methods (i.e., self-report, implicit measures, and measures of behavior), the use of neural measures in studying the effects of SES on social cognition offers several advantages. Because neuroscientific techniques, such as EEG, ERP, and fMRI, provide more direct measures of cognitive activity without having to rely on downstream behavioral consequences, they allow researchers to circumvent common problems with self-report including group differences in social desirability [16] or response styles [17]. Further, these measures are often precognitive, occurring less than 400 ms after the onset of stimuli and thus provide more direct access to mental activity rather than overt social behavior [18]. Previous studies in cultural neuroscience have revealed group differences in neural signals in the absence of differences in downstream behavior [19] and in the presence of contradictory self-report [20], as well as absence of neural differentiation in the presence of self-report differences [21], helping disentangle competing behavioral and self-report effects [22].

Thus, combining neural measures with more traditional ones can have a number of advantages. On the one hand, when consistent with the results obtained via traditional methods, neuroscientific results can bolster our confidence in preexisting findings. On the other hand, when the results are inconsistent, the location of the discrepancy can give us a more nuanced understanding of how SES affects psychological processes. For example, if similar patterns of behavioral performance are observed between people who differ in SES, but differences in neural activity are observed, this may suggest that SES affects automatic processes required to achieve a given outcome or the degree of effort required, while a more controlled behavioral response may compensate, leading to similar downstream results. Also, the neuroscientific approach is more sensitive than other traditional approaches, enabling us to detect cultural differences even when no overt behavioral differences exist or when observable behavioral output is absent. Examples and implications are detailed later in this chapter. However, neuroscientific approaches are not without limitations [18, 23], and there is great benefit to combining these methods with more traditional behavioral and self-report measurements. Furthermore, scientists may often be interested in behavior itself, or conscious subjective experiences, phenomena which neural paradigms may not be well positioned to capture. Hence we do not argue that other methods are obsolete or that neural approaches are always best suited to answer any given question about how SES affects the ways people think, feel, and behave.

In what follows, we first summarize recent neuroscience findings on the effects of SES on social cognitive processes. Then, we discuss theories that account for why these effects might occur. We conclude by suggesting some possible future directions in the cultural neuroscience approach to studying SES.

4 SES and Attunement to Others

Individuals with lower SES have been found to have more interdependent views of the self, and to be more sensitive to social cues [24, 25]. Cultural neuroscience research finds further support for this idea, suggesting that there is a neural basis for the tendency of working class individuals to be more attuned to others' behaviors, intentions, and emotions. For example, in an EEG study examining the fronto-central P2, an ERP component thought to be a neural marker for empathy [26, 27],

Varnum and colleagues [20] found that those with lower SES showed stronger empathic neural responses to images of faces expressing pain. It should be noted here that the self-report trait empathy was actually higher for participants with high SES than those with low SES. This contradictory result highlights one of the advantages of neural measures, indicating that traditional approaches might not always tell the whole story.

Another EEG study suggests an association between SES and motor resonance [28]. In this study, low-SES participants showed stronger Mu suppression while they watched on the computer screen a stranger's hand repeatedly open and close. Mu suppression is thought to index the degree of activation of the mirror neuron system (MNS) [29, 30]. The MNS, which includes the premotor cortex, the supplementary motor area, and the primary somatosensory cortex, is a collection of neurons that fire both when one is performing a motor activity, and when one is watching and "mirroring" or simulating someone else's observed activity [31]. This provides some evidence that the MNS is more reactive for low-SES individuals.

Recent fMRI studies indicate specific brain regions that may be linked to SES and attunement to others. In one such study, Muscatell and colleagues [32] found that brain regions associated with mentalizing—the dorsomedial prefrontal cortex (dmPFC), the medial prefrontal cortex (mPFC), and the posterior cingulated cortex (PCC)/precuneus—showed stronger activation for low-SES individuals while they viewed images of other people with brief passages containing social information. Another study demonstrates how SES may moderate the relationship between neural responses to social events and downstream behavior [33]. In this study, participants first played Cyberball, a procedure in which participants are made to feel socially excluded in a rigged online interaction, inside the fMRI scanner, and later performed a simulated driving task with a passenger who expressed different norms regarding risk taking. Stronger activation in the brain regions associated with social pain and reward sensitivity while experiencing social exclusion predicted more conformity to peer norms but only for low-SES individuals. The relationship was reversed for high-SES individuals.

These studies using neural measures confirm that those who are lower in SES appear more attuned to others and to social feedback. This is in line with a contextualist perspective on social class in which diminished resources and lower rank are related to situations that constrain behavior and outcomes [1]. This heightened focus on others' emotions, actions, and intentions is likely adaptive in environments characterized by fewer resources and greater threat.

5 SES and Threat

In addition to being more attuned to others in general, those who are low in SES also appear to have a greater sensitivity to potential threats in the environment. Research examining neural responses to viewing facial expressions of anger has found that lower SES is related to stronger activation in the amygdala [32, 34]. This is consistent with previous findings linking low SES with negative emotional and health outcomes [35], providing further evidence for the idea that living in an environment characterized by frequent exposure to potential harm or threat would lead to a psychology that is systematically different from one that develops in a safe and secure environment. A recent finding suggests that, in addition to heightened vigilance to threat, low SES may be associated with greater capacity to consciously regulate affective responses to threatening and other negative events [36]. This EEG study examined LPP, an ERP component indexing affective arousal and linked with amygdala activation [37], while participants viewed high-arousal, negative-valence images depicting violence, gore, and natural threats. When instructed to enhance or suppress their affective responses to these stimuli, low-SES participants showed more suppression of central-parietal LPP compared to high-SES participants. Thus, it appears that the neural systems of individuals from low-SES contexts may be more attuned to potential threats and that people from these ecologies are consequentially better able to regulate affective responses to such threats.

6 SES and Trait Inference

Neural measures in juxtaposition to behavioral measures have also revealed new insight into the relationship between SES and trait inference. Varnum and colleagues [19] used the N400, an ERP component indexing semantic incongruity [38], to investigate whether people from different SES backgrounds would vary in their tendency to spontaneously infer traits when given only scant behavioral information about the targets. In this study, participants were first instructed to memorize pairings of faces and behavioral statements. Then, they performed a lexical decision task in which they saw each of the previously seen faces followed by a word describing a trait consistent with the behavioral statements, the antonym of that trait word, or a pseudoword. As expected, increased N400 modulations were observed when the antonyms of the previously implied trait were presented but only for the high-SES participants. This suggests that, during the memorization phase, high-SES participants spontaneously inferred traits from the behavioral statements, but low-SES participants did not. On the other hand, no differences were found between the two groups at the implicit behavioral level, and groups showed no differences in task performance or recall for face trait pairings. This evidence supports previous findings that spontaneous trait inference may be specific to certain cultural groups rather than a universal, automatic process [39]. While this conclusion is largely consistent with previous findings that used downstream behavioral measures to demonstrate greater dispositional bias in high-SES individuals [3, 24, 40], the finer temporal resolution provided by this ERP paradigm allows us to better locate at which stage of cognitive processing these differences might originate.

7 SES and Selective Attention

Several EEG studies using ERP paradigms have investigated the effect of SES on selective attention using auditory selective attention tasks [41–44]. In these, participants were instructed to either attend to or ignore particular auditory stimuli. The results of such studies show that children from low-SES backgrounds have stronger neural responses to the distracters when they were instructed to ignore them. Although these findings tend to be subsumed under a larger body of literature documenting the discrepancy between high- and low-SES individuals in cognitive performance [45], it is important to note that these studies found no differences in performance (i.e., reaction time and accuracy) or recall. Therefore, rather than implying an impairment in executive functioning, these results may indicate greater attentional breadth, consistent with ecological demands for attending to scarce resources and abundant threats [14]. This highlights the importance of understanding that not all neural differences between groups reflect deficits, and likely many if not most will reflect adaptations.

8 Self-Construal and SES

A key dimension in explaining cultural variation is self-construal, which varies along the independence-interdependence continuum [46]. An independent construal involves viewing the self as separate and distinct from others and emphasizing one's unique, internal attributes, whereas an interdependent construal involves viewing the self as less differentiated from, and fundamentally interconnected to, close others. Systematic differences in psychological phenomena between East Asian and Western societies have been well documented and understood through the lens of self-construal, as East Asians tend to have relatively more interdependent orientations, and Westerners tend to have more independent orientations [46, 47]. There is also evidence of comparable differences between low- and high-SES individuals [3, 4, 48], which suggest that the distinct social environments inhabited by these two groups give rise to different ways of understanding the self.

Moreover, the growing body of cultural neuroscience research comparing East Asians and Westerners indicates that variations in self-construal may also account for social class differences in neural processes. For example, when making personality trait judgments about oneself or one's mother, East Asians showed less differentiation in the activation of mPFC than Westerners for whom there was stronger activation for their own traits [49]. In the same vein, whereas European Americans showed a self-enhancement effect at self-report, behavioral, and neural levels using an N400 paradigm, Chinese participants showed the opposite trend at each level, enhancing both a close and unfamiliar other over the personal self [22]. In the realm of vicarious experience, Varnum and colleagues [50] found that, in a task resulting in monetary rewards for the self or one's friend, priming an interdependent self-construal led to less differentiation in the activation of the ventral striatum. Neural evidence also supports the idea that East Asians tend to be more sensitive to social evaluative threat [51] and have a more holistic cognitive style, displaying greater attention to contextual information [39, 47, 52–54]. To the extent that self-construal varies as a function of SES, one might then expect corresponding variations in neural functioning for low- and high-SES individuals. It should be noted that the similarities we see between East-West comparisons and SES effects also favor the notion that social class differences may be better viewed as adaptations than deficits.

9 Life History Theory and SES

Life history theory [55] posits that, given the limited nature of resources, an organism's decisions regarding the allocation of its resources always implicate trade-offs. In evolutionary terms, there are two broad domains in which an organism can invest, namely, somatic growth/maintenance and reproduction. How organisms solve this fundamental problem of managing trade-offs during their lifetime is referred to as life history strategy. While there is a wide range of variation across species in their life history strategies, within-species variation also results from immediate ecological pressures or cues that influence to which life domain resources should be invested and when [56]. For instance, an environment characterized by prevalence of infectious diseases, high mortality rate, resource scarcity, harshness, or unpredictability would cue a "faster" life history strategy, which includes earlier investment in reproduction, less parental investment, focus on short-term outcomes, and aggressive behavior [57–64]. Hence, we might expect social class differences in neural responses or circuitry involved in reward processing, and in inhibition of violent or sexual impulses.

As far as SES can be thought of as a proxy for distinct types of ecologies with differing levels of environmental pressures, life history theory is a useful perspective through which SES effects or adaptations can be interpreted. Since low-SES individuals are more likely to inhabit the kind of ecology illustrated above, we would expect faster life history strategies to be more prevalent [60, 61]. In addition, given the importance of early developmental context in the choice and persistence of life history strategy, it is reasonable to suspect neural bases for such variations. Although the suite of behaviors associated with a fast strategy is often viewed as a maladjustment by society, from a life history perspective these behaviors and ways of thinking are in fact adaptive [65].

10 Adaptive Responses to Ecology

In addition to promoting different life history strategies, the differing ecologies of low- and high-SES contexts may lead to other psychological adaptations. Given that people who are lower in SES come from an environment characterized by harshness and scarcity, at first glance, it might be tempting to explain neural differences as a function of SES, as deficits resulting from the challenges and disadvantages that low-SES individuals face in the context of development. Indeed, a large literature on performance discrepancies and health disparities between low- and high-SES individuals is consistent with this perspective. However, it may be useful to try to understand some of these results as successful adaptations to the environment [14]. For instance, while the effects of SES on responses to threat may be interpreted as unnecessary and perhaps detrimental over-activation of the amygdala, hypervigilance to potential physical harm would actually be adaptive in an environment in which violence is prevalent. In such contexts, the failure to detect danger could be much more costly than perceiving danger when there is none [66]. Similarly, greater breadth of attention would be beneficial for identifying danger in complex and unpredictable settings. That low-SES individuals appear to have enhanced ability to consciously regulate their affective responses to negative events is also consistent with the interpretation that these differences in neural responses are in fact adaptive.

As people's material resources and position in the social hierarchy place differential constraints on their behavior, it would make sense that people from opposite SES backgrounds learn to navigate their social worlds in dissimilar ways to achieve the same goal. We could speculate that a scarce and harsh environment demands low-SES individuals to be more oriented toward others as they would represent opportunities for cooperating or sharing risks en route to a common objective. Stronger mu suppression while watching others' actions, and increased P2 responses to images of painful facial expressions, respectively provide evidence for enhanced motor resonance and empathy, which would facilitate coordinating as well as building and maintaining relationships with others. The same reasoning extends to the findings on SES and the mentalizing regions of the brain and conformity. One might also argue that many of these findings could reflect adaptations to more dangerous environments as well.

It is important to note that these adaptations, while useful for the individual from an evolutionary perspective, imply trade-offs in other domains. For instance, behavioral traits associated with a fast life history strategy may be perceived as more or less favorable by others, depending on the particular cultural context. In addition to the direct negative outcomes that may result from one's focus on short-term outcomes, unrestricted sexual strategies, aggressive behaviors, etc., the social implications of this suite of behaviors may also lead to long-term negative psychological and physical health outcomes. Moreover, when these adaptations are transmitted across time and space, they can interact with the immediate environment in complex ways that make downstream consequences very difficult to predict.

11 Future Directions

Although the cultural neuroscience framework has already helped reveal many differences in the ways people from different SES backgrounds perceive and interact with the world, we have barely scratched the surface of what this approach offers.

In this section, we offer a number of avenues for future research using neuroscience to understand how SES affects the ways we think.

One promising area for future research is in the domain of vicarious reward. As noted earlier, people from low-SES backgrounds tend to have greater attunement to others than those from high-SES backgrounds [20, 24, 25, 28]. Due to limited resources and relatively low achievement, especially in academic settings [67], one might expect those from low-SES backgrounds to be more likely to engage in "Basking in Reflected Glory" [68]. Though empathizing with others' success may appear to be a helpful way to buffer self-esteem of low-SES individuals, this connection has yet to be made. However, similar work in the field of cultural neuroscience hints at the presence of such an effect. Varnum and colleagues [50] primed independent and interdependent self-construals in an fMRI study of reward. For those primed with an independent self, activation in the bilateral ventral striatum was greater for personal rewards compared to rewards for a friend; however, for those primed with an interdependent self, the ventral striatum activated equally for both. One might expect then that lower SES might be linked to stronger vicarious reward responses as well as potentially stronger reward responses when observing another's success.

Another area in which SES may affect neural responses is that of impulsivity/delay discounting. Consistent with a life history theory approach, Griskevicius and colleagues [60] have found that lower SES tends to be linked to greater temporal discounting under conditions of threat. More generally, lower SES is associated with greater impulsivity [69] and a suite of behaviors reflecting faster life history strategies [57, 61, 63]. We suspect that we might observe neural evidence of greater attunement to smaller, more immediate, rewards among those lower in SES and that such rewards may produce stronger activations in the brain's reward network than among those higher in SES.

Finally, recent advances in EEG methodologies may allow future researchers to test interactive relationships of SES and prosociality using hyperscanning and network analysis. Hyperscanning involves simultaneously recording EEG signals from pairs of participants in order to analyze neural event-related and oscillatory synchrony [70, 71]. The combination of the brain data from each participant results in a single network of activation between both participants based on a graph theory approach to analyzing brain data [72, 73]. Early applications of this methodology revealed increase phase synchrony during guitar improvisation [74]. More recent work directly examined the synchrony between partners' brains during an iterative prisoners dilemma game [75]. They found that partners who cooperated showed stronger interbrain synchrony, while those who were more prone to defection had very few of these types of connections. Taken together with neurological evidence suggesting people from low-social-status groups activate mentalizing regions of the brain more, EEG hyperscanning may allow researchers to delve into the complex interplay of status and prosociality. Such work may also help us to understand how social class shapes interpersonal interactions in a more ecologically meaningful way.

12 Conclusion

In this chapter, we have summarized evidence that SES affects the way our brains function. We have shown that this is the case for a variety of psychological phenomena, ranging from motor resonance to affective regulation. We have explored different theories for why such differences may exist, arguing that the latter are most likely adaptive responses to ecological conditions. Finally, we have also provided what we hope are some useful suggestions for researchers who wish to use neuroscience to understand the consequences of social class. Although this field is still in its infancy, we believe that it will continue to generate important insights that will inform our understanding of the profound consequences of SES.

References

1. Kraus MW, Piff PK, Mendoza-denton R, Rheinschmidt ML, Keltner D. Social class, solipsism, and contextualism: how the rich are different from the poor. Psychol Rev. 2012;119(3):546–72.
2. Santos HC, Grossmann I, Varnum MEW. Culture, cognition, and cultural change in social class. In: The Oxford handbook of cognitive sociology. Oxford: Oxford University Press. In press.
3. Grossmann I, Varnum MEW. Social class, culture, and cognition. Soc Psychol Personal Sci. 2011;2(1):81–9.
4. Stephens NM, Markus HR, Townsend SS. Choice as an act of meaning: the case of social class. J Pers Soc Psychol. 2007;93(5):814–30.
5. Gallo LC, de los Monteros KE, Shivpuri S. Socioeconomic status and health: what is the role of reserve capacity? Curr Dir Psychol Sci. 2009;18(5):269–74.
6. Diener E, Ng W, Harter J, Arora R. Wealth and happiness across the world: material prosperity predicts life evaluation, whereas psychosocial prosperity predicts positive feeling. J Pers Soc Psychol. 2010;99(1):52–61.
7. Chiao JY. Cultural neuroscience: a once and future discipline. In: Chiao JY, editor. Cultural neuroscience: cultural influences on brain function, Volume 178 of Progress in brain research. New York: Elsevier; 2009. p. 287–304.
8. Han S, Northoff G, Vogeley K, Wexler BE, Kitayama S, Varnum MEW. A cultural neuroscience approach to the biosocial nature of the human brain. Annu Rev Psychol. 2013;64:335–59.
9. Kitayama S, Uskul AK. Culture, mind, and the brain: current evidence and future directions. Annu Rev Psychol. 2011;62:419–49.
10. Kim HS, Sasaki JY. Cultural neuroscience: biology of the mind in cultural contexts. Annu Rev Psychol. 2014;65(1):487–514.
11. Cohen D. Cultural psychology. In: Mikulincer M, Shaver PR, Borgida E, Bargh JA, editors. APA handbook of personality and social psychology, volume 1: attitudes and social cognition, APA handbooks in psychology. Washington, DC: American Psychological Association; 2015. p. 415–56.
12. Kitayama S, Cohen D. Handbook of cultural psychology. New York: Guilford Press; 2010.
13. Kitayama S, Park J. Cultural neuroscience of the self: understanding the social grounding of the brain. Soc Cogn Affect Neurosci. 2010;5(2-3):111–29.
14. Varnum MEW. The emerging (social) neuroscience of SES. Soc Personal Psychol Compass. 2016;10(8):423–30.
15. Cohen AB. Many forms of culture. Am Psychol. 2009;64(3):194–204.

16. Dudley NM, McFarland LA, Goodman SA, Hunt ST, Sydell EJ. Racial differences in socially desirable responding in selection contexts: magnitude and consequences. J Pers Assess. 2005;85(1):50–64.
17. Mottus R, Allik J, Realo A, Rossier J, Zecca G, Ah-Kion J, et al. The effect of response style on self-reported conscientiousness across 20 countries. Personal Soc Psychol Bull. 2012;38(11):1423–36.
18. Luck SJ. An introduction to the event-related potential technique. 2nd ed. Cambridge: The MIT Press; 2014.
19. Varnum MEW, Na J, Murata A, Kitayama S. Social class differences in N400 indicate differences in spontaneous trait inference. J Exp Psychol Gen. 2012;141(3):518–26.
20. Varnum MEW, Blais C, Hampton RS, Brewer GA. Social class affects neural empathic responses. Cult Brain. 2015;3(2):122–30.
21. Varnum MEW, Hampton RS. Cultures differ in the ability to enhance affective neural responses. Soc Neurosci. 2017;12:594.
22. Hampton RS, Varnum MEW. Do cultures vary in self-enhancement? ERP, behavioral, and self-report evidence. 2017. In press.
23. Luck SJ, Gaspelin N. How to get statistically significant effects in any ERP experiment (and why you shouldn't). Psychophysiology. 2017;54:146–57.
24. Kraus MW, Cote S, Keltner D. Social class, contextualism, and empathic accuracy. Psychol Sci. 2010;21(11):1716–23.
25. Stellar JE, Manzo VM, Kraus MW, Keltner D. Class and compassion: socioeconomic factors predict responses to suffering. Emotion. 2012;12:449–59.
26. Sheng F, Han S. Manipulations of cognitive strategies and intergroup relationships reduce the racial bias in empathic neural responses. NeuroImage. 2012;61:786–97.
27. Sheng F, Liu Y, Zhou B, Zhou W, Han S. Oxytocin modulates the racial bias in neural responses to others' suffering. Biol Psychol. 2013;92(2):380–6.
28. Varnum MEW, Blais C, Brewer GA. Social class affects mu-suppression during action observation. Soc Neurosci. 2016;11(4):449–54.
29. Arnstein D, Cui F, Keysers C, Maurits NM, Gazzola V. μ-suppression during action observation and execution correlates with BOLD in dorsal premotor, inferior parietal, and SI cortices. J Neurosci. 2011;31(40):14243–9.
30. Braadbaart L, Williams JHG, Waiter GD. Do mirror neuron areas mediate mu rhythm suppression during imitation and action observation? Int J Psychophysiol. 2013;89(1):99–105.
31. Molenberghs P, Cunnington R, Mattingley JB. Is the mirror neuron system involved in imitation? A short review and meta-analysis. Neurosci Biobehav Rev. 2009;33(7):975–80.
32. Muscatell KA, Morelli SA, Falk EB, Way BM, Pfeifer JH, Galinsky AD, et al. Social status modulates neural activity in the mentalizing network. NeuroImage. 2012;60:1771–7.
33. Cascio CN, O'Donnell MB, Simons-Morton BG, Bingham CR, Falk EB. Cultural context moderates neural pathways to social influence. Cult Brain. 2017;5:50–70.
34. Gianaros PJ, Horenstein JA, Hariri AR, Sheu LK, Manuck SB, Matthews KA, et al. Potential neural embedding of parental social standing. Soc Cogn Affect Neurosci. 2008;3(2):91.
35. Gallo LC, Matthews KA. Understanding the association between socioeconomic status and physical health: do negative emotions play a role? Psychol Bull. 2003;129(1):10–51.
36. Kwon JY, Hampton RS, Varnum MEW. Social class differences in the ability to suppress affective neural responses. Poster presented at: 17th Annual Meeting of the Society for Personality and Social Psychology, Advances in Cultural Psychology Preconference; 2017 Jan 19; San Antonio, TX.
37. Liu Y, Huang H, McGinnis-Deweese M, Keil A, Ding M. Neural substrate of the late positive potential in emotional processing. J Neurosci. 2012;32(42):14563–72.
38. Kutas M, Federmeier KD. Thirty years and counting: finding meaning in the N400 component of the event-related brain potential (ERP). Annu Rev Psychol. 2011;62:621–47.
39. Na J, Kitayama S. Spontaneous trait inference is culture-specific: behavioral and neural evidence. Psychol Sci. 2011;22(8):1025–32.

40. Kraus MW, Piff PK, Keltner D. Social class, sense of control, and social explanation. J Pers Soc Psychol. 2009;97(6):992–1004.
41. D'angiulli A, Roon V, Maria P, Weinberg J, Oberlander T, Grunau R, et al. Frontal EEG/ERP correlates of attentional processes, cortisol and motivational states in adolescents from lower and higher socioeconomic status. Front Hum Neurosci. 2012;6:306.
42. D'Angiulli A, Herdman A, Stapells D, Hertzman C. Children's event-related potentials of auditory selective attention vary with their socioeconomic status. Neuropsychology. 2008;22(3):293–300.
43. Stevens C, Lauinger B, Neville H. Differences in the neural mechanisms of selective attention in children from different socioeconomic backgrounds: an event-related brain potential study. Dev Sci. 2009;12(4):634–46.
44. Stevens C, Paulsen D, Yasen A, Neville H. Atypical auditory refractory periods in children from lower socio-economic status backgrounds: ERP evidence for a role of selective attention. Int J Psychophysiol. 2015;95(2):156–66.
45. Hackman DA, Farah MJ. Socioeconomic status and the developing brain. Trends Cogn Sci. 2009;13(2):65–73.
46. Markus HR, Kitayama S. Culture and the self: implications for cognition, emotion, and motivation. Psychol Rev. 1991;98(2):224–53.
47. Varnum MEW, Grossmann I, Kitayama S, Nisbett RE. The origin of cultural differences in cognition: the social orientation hypothesis. Curr Dir Psychol Sci. 2010;19:9–13.
48. Na J, McDonough IM, Chan MY, Park DC. Social-class differences in consumer choices: working-class individuals are more sensitive to choices of others than middle-class individuals. Personal Soc Psychol Bull. 2016;42(4):430–43.
49. Zhu Y, Zhang L, Fan J, Han S. Neural basis of cultural influence on self-representation. NeuroImage. 2007;34(3):1310–6.
50. Varnum MEW, Shi Z, Chen A, Qiu J, Han S. When "Your" reward is the same as "My" reward: self-construal priming shifts neural responses to own vs. friends' rewards. NeuroImage. 2014;87:164–9.
51. Park J, Kitayama S. Interdependent selves show face-induced facilitation of error processing: cultural neuroscience of self-threat. Soc Cogn Affect Neurosci. 2014;9(2):201.
52. Hedden T, Ketay S, Aron A, Markus HR, Gabrieli JDE. Cultural influences on neural substrates of attentional control. Psychol Sci. 2008;19(1):12–7.
53. Lao J, Vizioli L, Caldara R. Culture modulates the temporal dynamics of global/local processing. Cult Brain. 2013;1(2-4):158–74.
54. Nisbett RE, Peng K, Choi I, Norenzayan A. Culture and systems of thought: holistic versus analytic cognition. Psychol Rev. 2001;108(2):291–310.
55. Del Giudice M, Gangestad SW, Kaplan HS. Life history theory and evolutionary psychology. In: Handbook of Evolutionary Psychology. Hoboken: Wiley; 2015.
56. Davies NB, Krebs JR, West SA. An introduction to behavioural ecology. Oxford: Wiley; 2012.
57. Brumbach BH, Figueredo AJ, Ellis BJ. Effects of harsh and unpredictable environments in adolescence on development of life history strategies. Hum Nat. 2009;20(1):25–51.
58. Figueredo AJ, Vásquez G, Brumbach BH, Schneider SMR, Sefcek JA, Tal IR, et al. Consilience and life history theory: from genes to brain to reproductive strategy. Dev Rev. 2006;26(2):243–75.
59. Figueredo AJ, Vásquez G, Brumbach BH, Sefcek JA, Kirsner BR, Jacobs WJ. The K-factor: individual differences in life history strategy. Personal Individ Differ. 2005;39(8):1349–60.
60. Griskevicius V, Tybur JM, Delton AW, Robertson TE. The influence of mortality and socioeconomic status on risk and delayed rewards: A Life History Theory approach. J Pers Soc Psychol. 2011;100(6):1015–26.
61. Griskevicius V, Delton AW, Robertson TE, Tybur JM. Environmental contingency in life history strategies: the influence of mortality and socioeconomic status on reproductive timing. J Pers Soc Psychol. 2011;100(2):241–54.

62. Hill SE, Boehm GW, Prokosch ML. Vulnerability to disease as a predictor of faster life history strategies. Adapt Hum Behav Physiol. 2016;2(2):116–33.
63. Hill SE, Prokosch ML, DelPriore DJ, Griskevicius V, Kramer A. Low childhood socioeconomic status promotes eating in the absence of energy need. Psychol Sci. 2016;27(3):354–64.
64. Simpson JA, Griskevicius V, Kuo SI-C, Sung S, Collins WA. Evolution, stress, and sensitive periods: the influence of unpredictability in early versus late childhood on sex and risky behavior. Dev Psychol. 2012;48(3):674–86.
65. Dishion TJ. Social influences on executive functions development in children and adolescents: steps toward a social neuroscience of predictive adaptive responses. J Abnorm Child Psychol. 2016;44(1):57–61.
66. Nesse RM. Natural selection and the regulation of defenses: a signal detection analysis of the smoke detector principle. Evol Hum Behav. 2005;26(1):88–105.
67. Reardon SF. The widening academic achievement gap between the rich and the poor: new evidence and possible explanations. In: Duncan GJ, Murnane RJ, editors. Whither opportunity? Rising inequality, schools, and children's life chances. New York: Russel Sage Foundation; 2011. p. 91–116.
68. Cialdini RB, Borden RJ, Thorne A, Walker MR, Freeman S, Sloan LR. Basking in reflected glory: three (football) field studies. J Pers Soc Psychol. 1976;34(3):366–75.
69. McLoyd VC. Socioeconomic disadvantage and child development. Am Psychol. 1998;53(2):185–204.
70. Konvalinka I, Roepstorff A. The two-brain approach: how can mutually interacting brains teach us something about social interaction? Front Hum Neurosci. 2012;6:215.
71. Dumas G, Lachat F, Martinerie J, Nadel J, George N. From social behaviour to brain synchronization: review and perspectives in hyperscanning. IRBM. 2011;32(1):48–53.
72. Bullmore E, Sporns O, et al. Nat Rev Neurosci. 2009;10(3):186–98.
73. Stam CJ, Reijneveld JC. Graph theoretical analysis of complex networks in the brain. Nonlinear Biomed Phys. 2007;1(1):3–22.
74. Lindenberger U, Li S-C, Gruber W, Müller V. Brains swinging in concert: cortical phase synchronization while playing guitar. BMC Neurosci. 2009;10:22.
75. Astolfi L, Toppi J, Fallani DVF, Vecchiato G, Cincotti F, Wilke CT, et al. Imaging the social brain by simultaneous hyperscanning during subject interaction. IEEE Intell Syst. 2011;26(5):38–45.

Social Ties, Health and Wellbeing: A Literature Review and Model

Andrew Haddon Kemp, Juan Antonio Arias, and Zoe Fisher

Abstract Humanity is facing an increasing burden of chronic disease and an ageing population that will lead to more years lived with disability. Dealing with these issues is difficult, especially if we consider the deterioration of social ties and the decline in social connectedness, which may also impact on health and wellbeing. However, research on the association between social ties and health outcomes has been characterized by conceptual difficulties, controversy and simplistic models. Here, we (1) review the literature on the associations between social ties and health outcomes, (2) identify various mechanisms through which these associations may arise and (3) propose a model on which future research activity could be based. We observe that social ties are an important contributor to health outcomes that may rival the effects of many traditional risk factors including smoking, alcohol consumption and physical activity. A complex network of behavioural, psychological and physiological mechanisms drives the health of individuals, and sociostructural factors will either facilitate or impede desired health outcomes within community ecosystems. The GENIAL [*g*enomics-*e*nvironment-vagus *n*erve-social *i*nteraction-*a*llostatic regulation-*l*ongevity] model is proposed, and important mediators and moderators are characterized along a pathway to wellbeing and longevity. A major regulatory role is given to the vagus nerve—indexed by heart rate variability—as it is responsible for a host of psychological and physiological processes that influence social ties, subsequent health and wellbeing. Future research needs to move beyond the disciplinary dilemma, initiate multidisciplinary exchange and facilitate new lines of interdisciplinary enquiry. We further argue that extending beyond the self by focusing on relationships with others and our connections to the environment will aid a much-needed transition to a more caring and understanding world.

A.H. Kemp (✉)
Department of Psychology and the Health and Wellbeing Academy, College of Human and Health Sciences, Swansea University, Singleton Campus, Swansea SA2 8PP, UK
e-mail: a.h.kemp@swansea.ac.uk

J.A. Arias
Department of Psychology, Swansea University, Swansea, UK
e-mail: jescarbaciones@gmail.com

Z. Fisher
Traumatic Brain Injury Service, Morriston Hospital, Swansea, UK
e-mail: Zoe.Fisher4@wales.nhs.uk

© Springer International Publishing Switzerland 2017
A. Ibáñez et al. (eds.), *Neuroscience and Social Science*,
DOI 10.1007/978-3-319-68421-5_17

Keywords Social ties • Social integration • Health and wellbeing • Vagal function • Heart rate variability • GENIAL model • Social isolation • Loneliness

Community, in a word,
is the beating heart of life,
and we neglect it at our peril.
Robin Dunbar (2010). How Many Friends Does One Need?:
Dunbar's Number and Other Evolutionary Quirks

1 Introduction

Social ties are linked to one's capacity to achieve and maintain health and wellbeing, driven by a fundamental human need to form social bonds. Yet, research on this topic has been contradictory, controversial and characterized by simplistic models. The construct of social ties is heterogeneous and multidimensional [1, 2], comprising objective measures of network size and degree of participation in social activities (social support, social integration), as well as subjective measures, including the perception of social connectedness and loneliness. Similarly, the assessment of wellbeing has focused on at least three different aspects including life satisfaction, positive emotions and human flourishing [3–6]. While it is not surprising that the literature is a minefield of contradictory findings and false leads, the scientific search for pathways to health and wellbeing is a noble endeavour and an important societal step forward. The goals for this chapter are (1) to consider the evidence for associations between social ties, health and wellbeing; (2) to examine what the mediating and moderating paths might be; and (3) to propose a simplified, yet sophisticated working model on which future research activity could be based.

Varied definitions of wellbeing have led to some (initially) counter-intuitive findings. For instance, researchers have warned that overvaluing the need for happiness—paradoxically—leads to compromised wellbeing, including increases in depressive symptoms and major depression [7]. The reason for this is that people are often disappointed with their level of happiness and may ultimately feel less happy [8]. Furthermore, emphasizing the importance of positive emotions over negative emotions is unproductive, as normal fluctuations in negative affect may have certain advantages, including improved memory performance, reduction in judgemental errors, enhanced motivation and more effective interpersonal strategies [9, 10]. These findings highlight the 'upside of your downside' [10] and a need for emotional agility and psychological flexibility [11], rather than the importance of positive over negative emotions, as has been argued [12, 13]—and criticized [14, 15]—previously. That been said, the advantages of negative affect are likely specific to normal fluctuations in mood states, rather than more extreme forms of negative affect characteristic of the affective disorders, which are associated with impairments in attention and memory recall in particular (e.g. [16, 17]). While affective components of wellbeing will colour our psychological moments, 'eudaimonia'—a Greek word translated to human flourishing [18]—may have stronger associations with health and longevity [19, 20] (see also [21–23]).

Much has been written about the potential effects of social ties on health outcomes and intermediate variables along this pathway. However, conceptual difficulties (e.g. [24]) and simplistic models (e.g. [25]) have led to considerable controversy (e.g. [26]; and see also [27]). Two generic pathways through which health outcomes may arise are the effects of social ties regardless of the experience of stress and stress buffering [1]. In this regard, social integration will promote positive psychological states leading to health-promoting physiological responses, while social support will help to buffer the effects of stressful experiences by promoting less threatening interpretations of adverse events and effective coping strategies [1]. Recent works in the field of positive psychology have focused on how social ties might impact on physical health. For instance, positive emotions have been associated with social connectedness and physical health [28, 29] in a self-sustaining upward spiral dynamic [25, 30]. Positive psychological attributes (optimism and hedonic wellbeing, in particular) have also been linked to increased engagement in positive health behaviours (e.g. healthy eating, physical activity) as well as improved cardiac health [28, 29]. While epidemiological findings from the UK Million Women Study observed no direct effects of happiness (in particular) on mortality over a 10-year period [20], analyses on the English Longitudinal Study of Ageing [31] came to different conclusions. This study, however, focused on eudaimonia, encompassing meaning, purpose and flourishing, and demonstrated that those in the highest quartile of wellbeing had a 58% reduced mortality risk, after adjustment for age and sex. After further adjustment for sociodemographic factors, this effect decreased to a still significant 30%. While it could be hypothesized that social personality traits such as sociability underpin these health outcomes, findings from the Terman life-cycle study [32] suggest that this may not be the case. The Terman study [32], conducted over more than eight decades, shows that childhood sociability promotes social ties and flourishing in midlife. However, sociability was also associated with alcohol abuse. Therefore, sociability exerts differential influences on health. Other findings from the same study further indicate that cheerful children may grow up to be more careless about their health leading to an increased risk for premature mortality, which may include a substantial number of unverified suicides [33]. These discrepancies and controversies in the field highlight a need for an up-to-date review of the literature and more sophisticated models of the associations between social ties, health and wellbeing, all goals of the present book chapter.

The aims of this chapter are threefold:

(1) To examine the epidemiological evidence for a link between social ties, health and wellbeing. While this is a useful starting point to examine whether associations may exist, the epidemiological literature is less helpful in establishing causal mechanisms that may underpin such associations, leading us to the following aim.

(2) To characterize potential mechanisms that might mediate or moderate associations between social ties, health and wellbeing and explore how various mechanisms might overlap.
(3) Based upon the reviewed evidence, we then sought to develop a model that might better explain the association between social ties and health outcomes. A model is proposed and considerations for future research on this topic are discussed.

Several points regarding our review should be noted. First, we discuss research published in a variety of distinct fields and disciplines; therefore, a comprehensive review of the literature is beyond the scope of our chapter. Instead, we draw upon relevant, published reviews and highlight recent studies that build upon this work. Second, the World Health Organization defines health as composed of physical, mental and social components. For the purpose of the current chapter, we emphasize the importance of physical and mental components, allowing us to consider various mechanisms that might drive and support associations between social ties and health outcomes, the single best measure of which is longevity [6]. Third, we have already noted that social ties are heterogeneous and multidimensional. Social ties may also be either positive or negative and the emotional tone of these relationships differentially impacts on health outcomes. Fourth, recent data suggest that social media use is associated with worse wellbeing [34–36]. While one needs to be careful about generalizing these findings to all online interactions (e.g. virtual support groups), real-world interactions may be an especially important contributor to positive health outcomes.

2 On the Association Between Social Ties, Health and Wellbeing

Investigating the effects of loneliness and social isolation on coronary heart disease (CHD) and stroke across 16 longitudinal datasets ($N = 181{,}006$), poor social ties were associated with a 29% increase in risk of incident CHD and a 32% increase risk in stroke [37] (two leading causes of morbidity globally) over 3–21 years. No differences were observed between the association of loneliness or social isolation with CHD, nor were there any differences between males and females. By contrast, outcomes from another meta-analysis [38] investigating the association between social ties and mortality risk—drawing conclusions from 148 studies comprising more than 300,000 participants—revealed a 50% increased likelihood of survival in those with stronger relationships over an average of 7.5 years follow-up. Importantly, this research excluded studies in which mortality was a result of suicide or injury, and results were consistent across age, sex, initial health status, follow-up period and cause of death. This study also reported that the influence of social ties on reduced risk for mortality was comparable to other well-known factors including smoking cessation, abstaining from alcohol, engaging in physical activity and having a lean body mass index (Fig. 1).

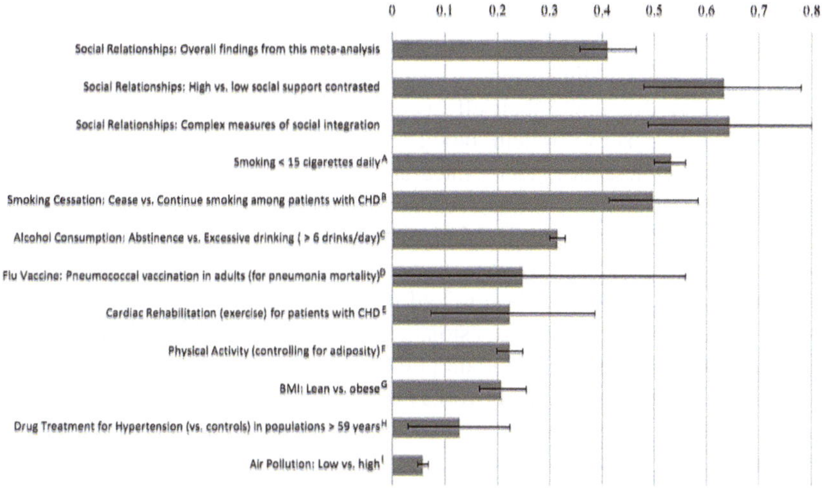

Holt-Lunstad J, Smith TB, Layton JB (2010) Social Relationships and Mortality Risk: A Meta-analytic Review. PLOS Medicine 7(7): e1000316. https://doi.org/10.1371/journal.pmed.1000316
http://journals.plos.org/plosmedicine/article?id=10.1371/journal.pmed.1000316

PLOS | MEDICINE

Fig. 1 Comparison of odds (lnOR) of decreased mortality across several conditions associated with mortality. Effect size of zero indicates no effect. The effect sizes were estimated from meta-analyses. Figure from Holt-Lunstad et al. [38], reprinted with permission under the terms of the Creative Commons Attribution License, which permits unrestricted use, distribution and reproduction in any medium, provided the original author and source are credited

This study [38] also reported that findings were strongest for complex measures of social integration comprising multiple components such as marital status, network size and participation (OR = 1.91; 95% CI 1.63–2.23) and lowest for binary indicators of residential status (living alone versus with others) (OR = 1.19; 95% CI 0.99–1.44). In a more recent work by these authors [39], social isolation, loneliness and living alone were examined as risk factors for mortality. A total of 70 investigations with 48,673 participants were identified for inclusion in analysis. Findings revealed that social isolation, loneliness and living alone contributed to a 29%, 26% and 32% increased risk for mortality over a 7-year follow-up period, respectively. No differences were observed between objective (social contact and living alone) and subjective (feelings of loneliness) social isolation. It is also interesting to note that the authors observed that adults *less than* 65 years of age were at a greater risk of mortality when they lived alone or were lonely (OR = 1.57 for adjusted data), compared with older individuals under the same conditions (OR = 1.25 for those aged between 65 and 75 and OR = 1.14 for those aged older than 75). They [39] suggested that the widespread belief that health risks are greater in older adults may be inaccurate. The possibility that social isolation may have adverse effects in younger people is consistent with another study [40] that drew on data from four nationally representative longitudinal studies spanning adolescence through to late adulthood. Social integration was associated with better physiological functioning in a dose-response fashion in both early and later life. The latter [40] investigated structural and functional dimensions of social ties (social integration, social support and social strain), examining their effects on objectively measured biomarkers of physical health (C-reactive protein, blood pressure, waist circumference and body mass index). In adolescence, social integration was associated with a 40% lower odds of elevated inflammation, while social isolation (OR = 1.27) raised the odds to a comparable degree with physical inactivity (OR = 1.21). These physiological effects were partly explained by socioeconomic status, negative health behaviours (smoking, physical inactivity, obesity) and prior chronic disease. In older age, the effects of social isolation on hypertension risk (OR = 2.42) even exceeded the effect of diabetes (OR = 1.49). The important point from this study is that social integration and embeddedness in social networks during adolescence may impact on metabolic and cardiovascular functioning and contribute to health risks even before symptoms of disease emerge.

Findings from another study [41] revealed that both social isolation and loneliness predict mortality over a 7-year follow-up period in 6500 older people from the English Longitudinal Study of Ageing. Absolute proportions of deaths ($N = 918$) were 21.9 vs. 12.3% for high- and low-average isolation groups, respectively, and 19.2 vs. 13.0% in the high- and low-average loneliness groups, respectively. However, the association with the emotional experience of loneliness was shown to be largely accounted for by baseline mental and physical health, and control for loneliness did not reduce the hazard ratio for social isolation, leaving the authors to conclude that loneliness may not be the primary mechanism through which social isolation transmits its effects. Similar non-significant findings had been reported in an earlier study when controlling for baseline health, functional limitations and depression [42]. In the more recent research [41], the hazard ratio for mortality was 1.50 in the high social isolation group, and adjustment for demographic factors, baseline health status and depression was associated with a reduced, although still significant, hazard ratio of 1.26. No sex differences were observed. The authors' [41] discussion of their findings is illuminating, highlighting the important role of lifestyle factors and need for additional help to those who experience social isolation to engage in positive health behaviours to help reduce risk of mortality.

It is noted that conflicting findings for the effects of social ties have also been reported. For instance, outcomes from the Melbourne Collaborative Cohort Study [43] reported that objective assessment of social connectedness—defined using questions on marital status, the number of people in the household, number of relatives visiting each month, number of friends that could be visited without invitation and number of hours of social activity per week—was not associated with successful ageing 12 years later. The authors concluded that while social connectedness may be related to the perception of ageing well, it does not help avoid common conditions associated with ageing. Successful ageing was defined as age 70 years or over and absence of diabetes, heart attack, coronary artery bypass graft surgery, angioplasty, stroke and cancer, impairment and perceived major difficulty with physical functioning and low risk of psychological distress. These conflicting findings may reflect methodological differences associated with how social ties are defined and variable outcomes of interest for each study. In this case, the Melbourne Collaborative Cohort Study [43] explored quantitative aspects, rather than perceptions of social connectedness or the quality of the interactions. Other research has also demonstrated that social networks can have negative [44] as well as positive effects. This study investigated the spread of obesity through social ties over a 32-year period in 12,067 people recruited as part of the Framingham Heart Study. It was reported that the chance of someone becoming obese increased by 37–56% if he or she had had a spouse, sibling, or friend who became obese, and these findings extended to three degrees of separation.

In summary, social ties appear to be an important contributor to health outcomes that may rival many of the traditional risk factors including smoking, alcohol consumption and physical activity. While the epidemiological literature sheds light on the existence of associations, it is difficult to draw causal conclusions from this data. Therefore, the next important step is to understand how (i.e. mediation) and when (moderation) these associations might arise. In the next section, we shift our focus to potential mediators and moderators of the link between social ties and health, working towards a sophisticated model that might provide a foundation for future research activities.

3 Potential Mechanisms

Social ties may impact on health outcomes through variety of tightly intertwined behavioural, psychological and physiological mechanisms (Fig. 2).

Social ties will influence whether individuals exercise, eat healthy food, smoke, consume alcohol and use illicit drugs through social control and influence. In 1897, Emile Durkheim proposed that social integration and widely held norms function to regulate behaviour including the tendency for suicide [54]. More recent social capital [55] and social cognitive [45] theories provide further theoretical context for understanding how individuals, their relationships with others and the communities in which they belong influence health behaviour. *Positive* social ties will promote healthy behaviours and reduce risky behaviours subsequently contributing to good health. In fact, it has been argued that if the benefits of positive health behaviours were able to be compressed into a single pill, this achievement would be declared a milestone in the field of medicine [45]. However, *negative* social ties may lead to risky behaviours (e.g. smoking, alcohol abuse, drug use) with subsequent impacts on wellbeing. (See [46] for a review on the complex interrelationships between social ties and health behaviours.) These *negative* social ties have even been shown to extend to the spread of obesity [44] such that weight gain in one person is associated with weight gain in friends, siblings, spouse and neighbours. The authors point to the social nature of these associations, concluding that the perception of social norms regarding the acceptability of obesity may have contributed to the findings. Intriguingly, the authors further suggest that network phenomena might be exploited

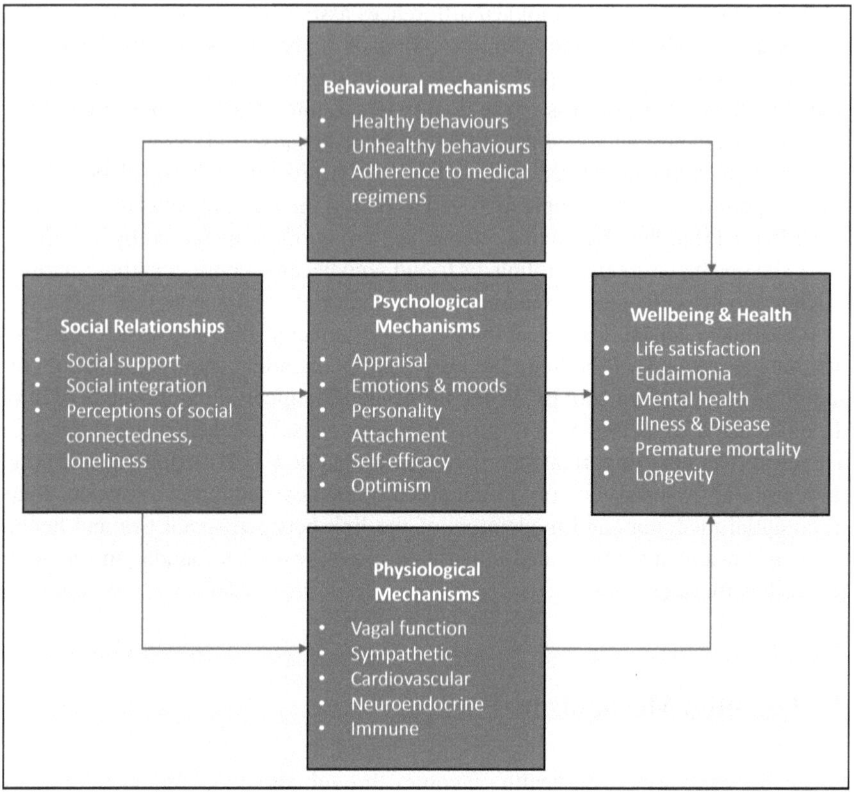

Fig. 2 Summary of pathways that have been proposed to mediate or moderate the link between social ties, health and wellbeing. The influence of sociostructural factors over these pathways is denoted by the *grey box overlay*, emphasizing that the health of individuals is not achieved within a vacuum, but within community ecosystems. This figure summarizes the mechanisms identified in previously published reviews on this topic [6, 45–53]. While the arrows are unidirectional, it is likely that pathways are bidirectional, reflecting the possibility that proposed mechanisms and some outcome measures of health and wellbeing will themselves impact on social ties (see Fig. 4). Past research has been limited by simplistic models (e.g. positive social ties → health), highlighting a need for more sophisticated models (Fig. 4) that consider the complexity of interrelationships between various proposed mechanisms (in this figure)

to spread positive health behaviours. While there has been significant debate over the extent to which individuals are able to achieve certain health outcomes, academics [45, 55] emphasize the need for a combination of individualist and structuralist approaches to health promotion, involving a focus on individual self-efficacy in combination with sociostructural factors that impact on one's capacity to achieve health goals.

Psychological factors such as personality and attachment styles may mould the social environment, which may, over time, have effects on health outcomes. Findings from the Terman life-cycle study [6, 47] have pinpointed conscientiousness as the most important personality trait linked to longevity, and individuals with low conscientiousness as well as high neuroticism might be at especially high risk due to impulsivity, disorganization, anxiety and high emotionality [6] (see also [56]). Personality factors have also been demonstrated to account for as much as 63% of the variance in subjective wellbeing [57]. Other research [58] has demonstrated that only conscientiousness is related to mortality risk across seven different cohorts including the British Household Panel Survey, 2006–2009; the German Socio-Economic Panel Study, 2005–2010; the Household, Income and Labour Dynamics in Australia Survey, 2006–2010; the US Health and Retirement Study, 2006–2010; the Midlife in the United States Study, 1995–2004; and the Wisconsin Longitudinal Study's graduate and sibling samples, 1993–2009. Individuals in the lowest tertile of conscientiousness were shown to have a 1.4 times higher risk of death compared to those in the top two tertiles, and this association was robust to adjustment for health behaviours, marital status and education.

Individual attachment styles (a dyadic characteristic) are also linked to health outcomes [48]: secure attachment is associated with wellbeing and mental health [59], while insecure attachment predicts inflammatory illnesses 30 years later [60]. Attachment theory [61–63] has made important contributions to understanding how early life experiences may shape the development of personality—an individual characteristic—and how adults perceive and react during various types of social encounters (see [64]). Attachment styles have been shown to influence close relationships as well as interactions with unknown people, perhaps underpinned by its effects on psychological moments [59, 65]. Attachment styles can be divided into two main categories: secure and insecure attachments. A secure attachment style develops when the primary caregiver provides a secure base for the infant and responds consistently to restore emotional balance in times of distress. In contrast, an insecure attachment style will emerge if attachment figures are repeatedly experienced as unresponsive or inconsistent in their responses in times of need and stress. Further subdivisions of unsecure attachments include anxious and avoidant attachment styles. Avoidant attachment can be experienced when proximity seeking to the primary caregiver is perceived as pointless or even dangerous because of the distress caused when proximity seeking fails. Anxious attachment styles occur when a perceived failure to handle threats independently encourages the infant to seek support despite the fact that attachment figures are experienced as inconsistent.

The impact of stress will be mediated by these psychological factors, consistent with the transactional model of stress and coping [66]. While perceptions of risk may lead to the initiation of various coping mechanisms, it may also lead to psychological distress and adverse health outcomes [67]. Research [68] demonstrates that attachment styles influence momentary affective states, cognitive appraisals and social functioning. Compared with securely attached individuals, individuals with anxious attachment reported higher negative affect, stress and perceived social rejection. Individuals with an avoidant attachment style reported decreased positive states and a decreased desire to be with others when alone. The appraisal of situations as stressful will have subsequent downstream effects that will then impact on health outcomes, increasing mortality risk two- to threefold among middle-aged men and women aged 36–52 years [69]. Similarly, other research [70] has shown that psychological distress increases risk for mortality from all causes, cardiovascular disease and external causes, even in those who do not come to the attention of mental health services. Furthermore, individuals who report a high degree of stress *and a belief that stress is harmful* have a 43% increased risk of premature death [67].

Physiological stress responses and the capacity to regulate these responses play a mediating role in pathways to health and wellbeing. Social isolation, low social support and *negative* social ties lead to chronic activation of immune, neuroendocrine and metabolic systems, increasing risk of cardiovascular disease [71, 72]. In rodents, social defeat—characterized by repeated physical attacks and declaration of subordination by the non-dominant animal—has considerable behavioural and physiological impacts. In the short term (minutes to hours), defeat produces vagal withdrawal, tachycardia, hypertension, elevated levels of glucocorticoids and catecholamines and reduced concentrations of testosterone [73, 74]. Over the longer term (days and weeks), lasting changes in behaviour (anxiety), hypothalamic-pituitary-adrenocortical axis activity (increased) and neurotransmitter systems are observed [74]. In fact, chronic psychosocial stress including threat of physical attack and daily episodes of aggression by a dominant male over a period of 2 weeks led to structural damage at the level of the heart (fibrotic tissue accumulation) [75]. While acute stress responses were observed to habituate over time, repeated episodes of social defeat led to a sixfold larger amount of reparative tissue, increasing susceptibility to cardiac arrhythmias. These findings suggest that despite short-term adaptations to stress, chronic psychosocial stress will lead to multisystemic overstimulation (or 'allostatic load'), contributing to permanent pathological alterations [74, 76]. In humans, lasting social conflict increases the chance of getting a cold after exposure to a common cold virus [77], and risk of inflammation caused by social isolation is of similar magnitude to the risk associated with physical inactivity [40]. By contrast, social connectedness and support are associated with positive affect, which might be associated with lower heart rate, higher rate variability, blood pressure and inflammatory markers, which benefit health [78]. Social integration has also been found to decrease risk of physiological dysregulation in a dose-response manner [40]. Further research indicates that the vagus nerve—indexed by heart rate variability (or HRV)—may play a causal regulatory role over our psychological moments (Figs. 3 and 4).

Two neurobiological models have been proposed that help to understand the link between psychological and physiological factors. These models include polyvagal theory [79–82] and the neurovisceral integration model [83–86], the major features of which are summarized in Fig. 3. Increased function within prefrontal-vagal pathways supports prosocial behaviour and positive emotions, while decreased function facilitates response to environmental change, fight-flight-or-freeze responses and negative emotions. Increased function along prefrontal-vagal pathways is driven by cortical inhibition of the central nucleus of the amygdala (CeA), which then activates the vagus nerve within the nucleus ambiguous (increasing HRV) and facilitates socially engaging facial expressions and positive social interactions. The nucleus tractus solitarius within the medulla oblongata receives vagal afferent feedback from the viscera and internal milieu, and this information is then directed to

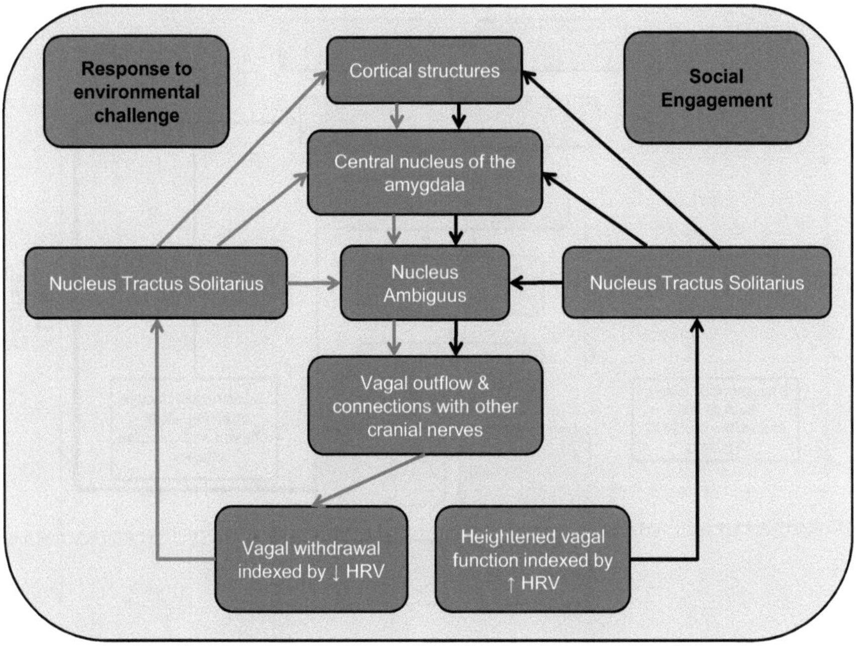

Fig. 3 Neurobiological components and associated behaviours that contribute to a psychological moment. A prominent role is given to the vagus nerve within an integrated brain-body network that either facilitates engagement with others (*black arrows*) or supports rapid whole-body response to environmental challenge (*grey arrows*). The arrows represent both efferent projections from the brain, contributing to rapid alterations in vagal function and related behavioural responses, as well as afferent feedback from peripheral end organs allowing for effective regulation of ongoing processing. These bidirectional pathways from and to the brain provide a psychophysiological framework through which psychological moments reciprocally and prospectively contribute to alterations in vagal function (e.g. mutual causation between emotional experience and vagal function). Afferent projections also provide a theoretical basis through which many behavioural interventions such as massage, exercise, meditation, yoga and HRV biofeedback may be understood

cortical structures responsible for the top-down, flexible regulation of psychological moments (Fig. 3, black arrows). By contrast, decreased function along prefrontal-vagal pathways is associated with responsiveness to environmental challenge (e.g. orienting) and withdrawal from the environment (e.g. fear, anxiety). Decreased function is driven by disinhibition of CeA (the major efferent source for modulation of cardiovascular, autonomic and endocrine responses) and vagal withdrawal, triggering fight-flight-or-freeze responses. Again, information relating to the status of the viscera and internal milieu are fed back to the nucleus of solitary tract and the cortex, allowing for subsequent regulation of psychological moments (Fig. 3, grey arrows).

Brain development is influenced by a child's early social environment. Limbic regions are in a critical period of growth in the first 2 years, and these same neurobiological structures will mediate stress-coping capacities for the rest of the life

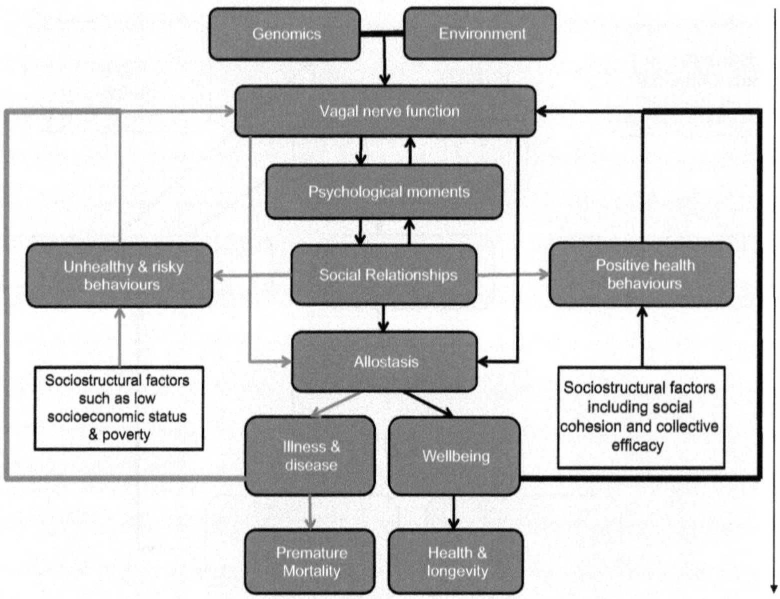

Fig. 4 The GENIAL model: *g*enomics-*e*nvironment-vagus *n*erve-social *i*nteraction-*a*llostatic regulation-*l*ongevity. The GENIAL model is a simplified yet sophisticated model for better understanding pathways to premature mortality or longevity, drawing on evidence from multiple disciplines, highlighting important mediating roles for vagal nerve function and social ties. In this model, vagal nerve function impacts on and is influenced by social ties and regulates a variety of allostatic mechanisms leading to either premature mortality or longevity. Vagal function provides a structural link between psychological moments and physiological processes. Illness and disease (and wellbeing) will further impact on vagal function in a downward (*upward*) self-sustaining spiral. Health behaviours and sociostructural factors represent important moderators of the pathways to health and wellbeing. Temperament (denoted by the *light grey overlay*) is a lens though which the world is viewed and a foundation on which psychological moments arise

span [87]. Recent neuroimaging work has explored neuroanatomical differences underpinning different attachment styles in adulthood and how this may influence social and emotional processing (see [88] for a review). Secure and insecure attachment styles may differentially recruit functional brain networks for interacting with others. According to a functional neuroanatomical model? of adult attachment style on social processes [88], two neural compartments mediate automatic affective evaluations versus more controlled cognitive processes. These include systems for (1) rapid, automatic affective appraisals (emotional mentalization), which is primarily involved in encoding basic dimensions of safety versus threat or approach versus aversion tendencies in social contexts, and (2) controlled social processing and regulation (cognitive mentalization), operating in a more conscious, voluntary mode, which is involved in representing the mental states of others and regulating one's own behaviour, thoughts and emotions. These two functional components rely on distinct brain networks [81, 89, 90], which involve limbic cortico-subcortical areas (e.g. amygdala, striatum, insula, cingulate, hippocampus) for affective evaluations and fronto-temporal areas (e.g. medial prefrontal cortex, orbitofrontal cortex, superior temporal sulcus, temporo-parietal junction) for cognitive mentalization and regulation. Importantly, these components may entertain a reciprocal dynamic balance between each other. The model argues that specific attachment styles, emotions, and behavioural responses are associated with the differential recruitment of these components.

Social engagement can only occur when the environment is perceived as safe and defensive circuits are inhibited. Therefore, an insecure attachment style—associated with impairments in vagal function [91]—will adversely impact the extent to which successful social engagement can occur. Vagal impairment will subsequently impact the degree to which downstream pathways are able to be regulated, having implications for health outcomes. There is increasing evidence for a regulatory role of the vagus nerve over downstream pathways (Fig. 4). Recent studies [92, 93] have demonstrated that chronically administered vagal nerve stimulation (VNS) may trigger beneficial effects, and these were even observed in obese insulin-resistant rats fed with a high-fat diet for 12 weeks. The authors reported that VNS decreased plasma insulin, insulin resistance, total cholesterol, triglyceride, LDL and visceral fat [93], relative to controls. VNS also decreased blood pressure, increased HRV and improved left ventricular function [93]. Finally, VNS exerted antioxidant, anti-apoptosis and anti-inflammation properties. In another study by these authors [92], VNS attenuated brain mitochondrial dysfunction, improved brain insulin sensitivity, decreased cell apoptosis and increased dendritic spine density, leading to improved cognitive function. These studies lend support to data from humans demonstrating that lower vagal function predicts elevated systemic inflammation—indexed by C-reactive protein—4 years later [94]. Other research has explored the pathways that might mediate the association between HRV and cognitive impairment [95], an important and relevant investigation considering that some researchers consider cognitive function to be an important component of wellbeing [6]. While past studies had suggested that the vagus nerve might regulate downstream pathways that could then impact on cognitive function, no prior study had investigated this possibility. Findings indicated that reduced vagal function was associated with increased insulin resistance—a feature of type II diabetes characterized by poor regulation of glucose in the body—leading to a thickening of the carotid arteries (higher intima-media thickness) and cognitive dysfunction [95]. These findings were further supported in five separate sensitivity and specificity analyses. It was concluded [95] that vagal function might provide a 'spark' that initiates a cascade of adverse downstream effects subsequently leading to cognitive impairment.

Common genetic variants have also been shown to impact on the social brain, which may have downstream effects on biological processes that contribute to disease and mortality. The gene coding for the oxytocin receptor (OXTR) has been shown to play an important role in contributing to individual differences in social behaviour and cognition [96, 97]. For example, the G allele of the single nucleotide polymorphism (*rs53576*) located in the OXTR has been shown to be associated with higher pro-sociality in nonverbal displays, as judged by outside observers' ratings of silent behaviour [98]. By contrast, carriers of the OXTR rs53576 A allele have been shown to display lower levels of sensitive responsiveness to their toddlers [99], empathy [100] and positive affect [101]. In fact, haplotypes constructed with three polymorphisms of the OXTR (rs53576, rs2254298 and rs2228485) are associated with positive affect, negative affect and loneliness [101]. Others [102] have reported that the rs53576 polymorphism is essential for the stress buffering effects of social connectedness, such that those participants with at least one copy of the *rs53576* G allele display higher HRV when social support is provided during the Trier Social Stress Test, a standardized, laboratory-based assessment of psychological stress. Recent research [103] has further demonstrated that while variation in three neuropeptide receptor genes (oxytocin, β-endorphin and dopamine) display important associations with sociality, endorphins and dopamine may have a much wider spectrum of effects than oxytocin. Furthermore, β-endorphin has been shown to operate effectively at dyadic and group levels [104–106] because its release can be triggered in others by touch, unlike oxytocin [103, 106].

Research in the field of epigenetics has further demonstrated how life experiences can be written into DNA. For instance, some mother rats spend considerable time licking and grooming (LG) their pups, while others do not. Offspring of mothers that show high levels of LG show differences in DNA methylation (one of several epigenetic mechanisms that cells use to control gene expression), as compared to offspring of low LG mothers [107]. As adults, these rats displayed stress responses that were dependent on amount of LG. Specifically, rats that received the most LG had an optimal response to stress, while those that had received less LG displayed an exaggerated stress response. This work demonstrated that the epigenomic state of a gene can be established through behavioural programming. A decade of research now shows that LG is translated into biochemical signals that enter the DNA and programme it differently, allowing the animal to equip itself for life and the environment. While caution is required over linking LG reactions in rats to high-quality mother-infant interactions in humans, evidence in humans [108, 109] supports the conclusion that maternal stress leads to lasting, broad and functionally organized DNA methylation signatures in offspring, which may be linked to internalizing and externalizing disorders [110], lower cognitive and language abilities [111] and increased risk for metabolic disease [112, 113]. These findings demonstrate the adverse effects that early life experience may have on health or disease.

In summary, a host of mechanisms have been proposed to influence the pathways to health outcomes, and community ecosystems may either facilitate or impede engagement in health behaviours. While a variety of behavioural, psychological and physiological mechanisms have been proposed, research is typically characterized by a restricted focus on simple models involving single mediators or moderators in isolation. Recognizing this issue, recent reviews [6, 49] have highlighted a need for sophisticated models that take into account the complex pathways between social ties and health outcomes. This is our goal for the next section.

4 A Model and Foundation for Future Research Activity

To find a solution, we need a new way
of understanding the problem.
Robert Maunder & Jonathan Hunter (2015),
Love, Fear and Health

Heeding calls for more sophisticated models [6, 49], this section of our book chapter makes an important contribution to the literature by bridging the gap from psychology through to epidemiology [114–116] and laying a foundation for future research activity. We propose the GENIAL model for pathways to wellbeing and longevity, a comprehensive model spanning *g*enomics and its interaction with the *e*nvironment through to health outcomes, highlighting a major regulatory role for the vagus *n*erve over social *i*nteraction and *a*llostatic regulation, subsequently leading to premature mortality or *l*ongevity (Fig. 4). Four key features of our model are worth emphasizing. First, vagal function (indexed by HRV) plays a key regulatory role over pathways leading to either premature mortality or longevity [114]. Second, vagal function will influence our psychological moments, cognitive functions, psychological flexibility to environmental change and capacity to engage with others [82, 84, 114, 117] and plays a critical regulatory role over allostatic systems [76], providing a structural link between mental wellbeing and physical health. We further propose that individual differences in resting vagal function will influence capacity for regulating psychological and physiological mechanisms (discussed further below). Third, social ties are supported by and impact on vagal function, consistent with polyvagal theory [82] and the proposal that positive (negative) emotions and vagal function influence one another in an upward (downward)-spiral dynamic [25, 30, 118]. Fourth, sociostructural factors within community ecosystems will either facilitate or impede health behaviour [45, 50, 51], subsequently impacting on vagal nerve function [25, 30, 82, 118]. Positive health behaviours, as well as unhealthy behaviours, will also be influenced by social ties directly [45, 54, 55].

At the top of the model (Fig. 4), genomics and interactions with the environment are proposed to play an important role in influencing individual variability in vagal function. Nurture and nature can no longer be regarded as discretely separate issues. Genetic susceptibilities are activated by environmental influences, a phenomenon labelled as the gene by environment interaction, and advances in epigenetics demonstrate how such interactions can shape health outcomes. Although research on the genomics of human wellbeing is in its infancy, there are several studies [22, 23, 119] that have laid a foundation for future research in this area. Extended periods of stress and threat may increase expression of inflammatory genes, preparing the immunity system for a potential wound-related bacterial infection derived from social conflict. By contrast, positive socialization is hypothesized to increase transcriptional levels of antiviral-related genes, protecting the immune system against potential viral infections derived from increases in social interaction with other members of the species [120]. While controversial [26, 121, 122] (but see [123]), this research lays a preliminary framework through which genomics will impact on downstream processes, including the vagus nerve (see Fig. 4), which plays a key role in regulating the immune system [124–126], amongst others. A recent genome-wide association study on data from 59 cohorts (n = 298,420) [119] identified three genetic loci that were associated with subjective wellbeing, defined by positive emotions and life satisfaction. Biological analyses revealed associations with anxiety disorders (ill-being) but only small genetic correlations with physical health phenotypes, including BMI, ever-smoker status, coronary artery disease and fasting glucose levels and triglyceride levels, further highlighting the need for more sophisticated models of pathways to health and wellbeing, as we present here. The construct of wellbeing in this study may have also contributed to the lack of associations with physical health.

Our model emphasizes a critical role for the vagus nerve because it is responsible for the regulation of a host of psychological and physiological processes that impact on social ties, health and wellbeing and vice versa. Several studies have demonstrated that nasal administration of oxytocin may augment vagal function [127, 128], reflecting an enhanced capacity for social approach and engagement [127]. These results together with genetic findings discussed above suggest that neuropeptides involved in social bonding—such as oxytocin, β-endorphin and dopamine—may drive individual differences in vagal function, which then allow (or restrict) individuals from engaging in and maintaining social ties. These ideas are supported by other research [25, 30, 118] demonstrating that loving-kindness meditation leads to increased positive emotions, an effect moderated by baseline vagal activity. Results further indicated that increases in positive emotion led to further increases in vagal activity, a finding that was mediated by the perception of greater social connections. A simple model was proposed through which positive emotions might build physical health, an idea further developed here. The model put forward by these authors [25, 30, 118] suggests that associations between positive emotions and social ties might both drive and be supported by vagal function, representing a self-sustaining upward-spiral dynamic. We suggest that there may also be a bidirectional relationship between vagal tone and negative emotion, and this mutual causation may contribute to a self-sustaining, downward spiral that leads to illness, disease and premature mortality. Reductions in vagal tone have been shown to precede affective disorder [129], and these reductions [130–133] are not ameliorated by antidepressant treatment [130, 131] or even transcranial direct current stimulation [132], despite amelioration of symptoms.

Typically, research findings are interpreted from the vantage point of one's own discipline, a phenomenon known as the disciplinary dilemma. In the current book chapter, we have sought to build on past work by bridging the gap between parallel lines of evidence from different fields of research. The vagus nerve is known to play an important role in maintaining homeostasis and achieving stability through change, a process known as 'allostasis'. The vagus plays an important role in regulating allostasis, yet past models tend to overlook this contribution. Multiple body systems are regulated by prefrontal-vagus pathways including the sympathetic nervous system and hypothalamic-pituitary-adrenal axis [82], inflammatory pathways [125, 126] and metabolism including glucose regulation [134, 135]. The vagus nerve may also stimulate neurogenesis by regulating the expression of brain-derived neurotrophic factor (BDNF) [136, 137], a key molecule involved in the regulation of metabolic efficiency, eating behaviour, synaptic plasticity and learning and memory [138]. These alterations may contribute to improvements in cognitive function and mood observed with vagal nerve stimulation [139, 140]. It has even been suggested that the vagus might lead to sustainable epigenetic modifications [141]. Diet and exercise have substantial anti-ageing effects, effects that may be mediated by afferent projections of the vagus ([115, 142], see [143]), and research is beginning to demonstrate that these positive health behaviours can alter the epigenome that may then stabilize and become inherited (see [144] for review). Vagal function may therefore regulate (or fail to regulate) allostasis through various, complementary psychological and physiological mechanisms. The first of these is a generalized inhibitory function of prefrontal-vagal pathways that serve to 'sculpt' goal-directed behaviour [84], enabling the individual to better respond to environmental change, facilitating effective regulation of allostasis [145]. Julian Thayer's neurovisceral integration model emphasizes a tightly integrated brain-body network regulated by prefrontal-vagal pathways [83, 84, 86]. The second potential mechanism is the role of vagus in stabilizing physiological arousal leading to improved allostatic regulation. According to Stephen Porges' polyvagal theory [82], the myelinated vagus nerve in combination with the cranial nerves support social engagement, providing a 'vagal brake' over the phylogenetically older sympathetic nervous system and unmyelinated vagus nerve. Further to this, Jos Brosschot recently published a psychological model of 'generalized unsafety' [116], in which he argues that stress is a default physiological response (characterized by low HRV) that must be turned off, rather than a physiological response elicited by some trigger. Being part of a cohesive social network is proposed to be a critical safety signal that turns off this default, physiological stress response. The third mechanism through which vagal function regulates allostasis is the cholinergic anti-inflammatory reflex [125, 126, 135, 146]. According to Kevin Tracey's model, the vagal nerve controls immune function and pro-inflammatory responses such that the afferent vagus nerve is involved in the detection of cytokines and pathogen-derived products, while the efferent vagus is responsible for the regulation and control of cytokine release. In summary, impairment in vagal function will lead to chronic activation of the stress response and overstimulation of allostatic systems ['allostatic load' [70]]. Dysregulation of allostatic systems will subsequently lead to ill health from a host of conditions and diseases including disability, prolonged infection, delayed wound healing, obesity, diabetes, atherosclerosis, osteoporosis, arthritis, frailty, Alzheimer's disease, periodontal disease and cancer [114, 115, 147, 148].

The model we propose here bridges a very large gap from psychology to epidemiology, illustrating an intimate link between psychological factors, health and wellbeing (see [149]). Pathways to health and wellbeing are dependent on genetic and environmental factors that directly influence vagal function, supporting the capacity for social engagement and promoting effective regulation of allostatic systems, leading to resilience, psychological wellbeing and longevity, or, if dysregulated, psychiatric illness, physical disease and premature mortality. The model is obviously a simplification of reality but provides a foundation on which future research on pathways to health and wellbeing could be developed. Bidirectional pathways that feedback on vagal function, psychological factors, and social relationships are also recognized. In this regard, those with chronic conditions will have lower vagal function that will impact on capacity for social engagement, which will then limit one's capacity for social integration, leading to further social isolation. It is possible, therefore, that those who would benefit the most from the effects of positive relationships will have fewer opportunities to experience them, leading to a downward spiral of negative emotions, social isolation, loneliness and ill health. These considerations highlight opportunities for improving the lives of people living with chronic conditions.

In summary, the GENIAL model is novel for at least five reasons:

First, vagal function is characterized as a major regulator and driver of health and wellbeing outcomes including longevity. By contrast, the field typically presents vagal function as one of many time-limited biomarkers that naturally fluctuate as the body maintains homeostasis (e.g. [6]).

Second, we build on prior research emphasizing single mechanisms such as genomics [23, 150], vagal tone [25, 30, 118] and personality [6] by accounting for the complexity and interactions between behaviour, psychological and physiological mechanisms and the influence of sociostructural factors.

Third, by adding in vagal function as an upstream regulator of pathways to wellbeing and longevity, we build on earlier systemic models of health and disease such as the allostasis model [76], immune dysregulation theory [151] and the causal network model linking depression and coronary heart disease [152].

Fourth, our model establishes a much-needed bridge between psychology and epidemiology, linking psychological factors to wellbeing (or illness and disease), contributing to longevity (or premature mortality).

Fifth, our model combines individualist and structuralist approaches [45] to understanding health and wellbeing over the life course, thereby placing the health of individuals within the context of community ecosystems. While our model is obviously a simplification of reality, the statistician George Box explains: 'Essentially, all models are wrong, but some are useful' [153].

5 Discussion and Conclusions

Humanity is facing major challenges and uncertainty, highlighting an urgent need for social harmony, unity and understanding. Nevertheless, social relations are increasingly strained, fractious and disconnected. Cultural shifts towards greater individualism and rapidly advancing technologies have led to an inflated sense of self and cultural narcissism [154]. Society has become isolating, homeowners distrusting, and people are dying lonely deaths that remain undiscovered for long periods of time, a phenomenon the Japanese call 'kodokushi'. In the United Kingdom, more than a quarter of all households in 2016 include people living alone, increasing by 51% in those aged 45–64 years between 1996 and 2016 [155]. Similar findings have been reported for the United States, according to the US Census Bureau [156]. Strikingly, more than a third of homeowners think that their immediate neighbours cannot be trusted [157]. Findings from the European Quality of Life Survey (2011/2012) indicated that around 1 in 10 people (11%) report feeling lonely all, most, or more than half of the time and that just over a third of people surveyed wished they could spend more time with family and have more social contacts [157, 158]. Astonishingly, other research indicates that around three in five teenagers (62%) report feeling lonely [159], increasing risk for mental ill health and other problems. Loneliness is associated with generalized distress, especially interpersonal sensitivity (low self-esteem) [160] and depression [161], and psychological distress has been shown to increase risk for all-cause mortality over an 8-year follow-up period [70]. In fact loneliness may increase risk for mortality to such an extent that it is equivalent to smoking 15 cigarettes a day [38]. Loneliness may lead to affective disorders (including major depression and anxiety disorders), which may reduce life expectancy by up to 18 years, an effect that is even greater than heavy smoking [162]. It is not surprising therefore that mental health problems represent the largest single source of world economic burden attributable to non-communicable disease [163, 164]. The burden of mental health problems cost the UK economy an estimated £70 billion (or 4.5% of the gross domestic product) through direct costs associated with health and social services and indirect costs associated with reductions in workforce productivity [165]. A contributor to this problem is the substantial mental health treatment gap and lag. The first refers to the numbers of people who need treatment that are not receiving it, which exceeds 50% in all countries of the world and reaches 90% in those with less resources [166]. The amount of time taken to receive care when it does exist—treatment lag—is estimated to be as long as 10 years [167]. These astonishing figures present significant challenges to health-care systems, and one wonders whether policy-makers and health-care providers are sufficiently prepared to cope with the predicted rise in the prevalence of disability and chronic disease associated with ageing of populations globally. So how might scientific research impact on this current state of affairs?

In 2010, the World Psychiatric Association carried out a systematic survey of leaders of psychiatry in nearly 60 countries on the strategies needed to reduce the treatment gap and lag in relation to mental health, neurological conditions and substance abuse disorders [166]. Three broad themes emerged: first, numbers of psychiatrists and other mental health professionals must increase; second, greater involvement by non-specialist providers is needed; and third, service users and their family members must be empowered to actively participate in service planning and delivery. The second and third of these themes may be facilitated through a strategy known as 'task shifting', involving the delegation of health care—where appropriate—to less specialized health workers. By reorganizing the workforce in this way, task shifting can make more efficient use of the limited numbers of mental health professionals that are available. Task shifting to non-specialist health workers has been shown to be a cost-effective way of improving outcomes for people with mental health disorders as long as there is supervision and support from health-care professionals [168]. Social prescribing is another innovative and complementary approach to managing and addressing treatment gap and lag. Social prescribing aims to promote the use of the voluntary sector within health-care settings by creating referral pathways that allow primary health-care patients with non-clinical needs to be directed to local voluntary services and community groups in addition to having their medical needs met. The substantial gap and lag in treatment are societal problems, and societal solutions including task shifting and social prescribing align well with mounting evidence (reviewed above) on the associations between social ties and health outcomes and the way in which these associations are mediated. For example, task shifting looks to local communities and its initiatives to support people with chronic conditions, facilitating positive health outcomes. Social prescribing involves health-care professionals prescribing community engagement in addition to lifestyle advice, psychological therapies and medication. However, for these solutions to be successful, better links need to be created between community organizations, the health-care sector and academia. This may allow the construction of appropriate social that will then facilitate evidence-based individual pathways to health and wellbeing. Such initiatives require vigorous empirical evaluation through strategic university partnerships.

Health and wellbeing do not automatically emerge once ‘the swamps of suffering are drained’ [169]. Therefore, the domain of positive psychology has much to offer those wondering how the treatment gap and lag might be managed. Positive psychology refers to the scientific study of human flourishing and an applied approach to enabling individuals, communities and organizations to thrive [170, 171]. We now know that the influence of life circumstances is far less than what might be expected, highlighting opportunities for targeted intervention. Wellbeing is determined by three factors [172]: a genetically determined set point, intentional activity and circumstantial factors according to a 50-40-10 formula. While genetics accounts for approximately 50% of the population variation in happiness, life circumstances only make a small contribution, around 10%. The important point here is that intentional activities contribute remaining variance—in the order of 40%—providing considerable opportunities for increasing wellbeing through structured activities such as engagement in positive interventions [173]. Social ties clearly make an important contribution to wellbeing, through social support, social integration as well as perceptions of social connectedness and loneliness. In the current chapter, we have emphasized how social ties and engagement with others might contribute to increased wellbeing, through a host of closely intertwined mechanisms within community ecosystems (see Fig. 4).

The focus of positive psychology on what makes life worth living is beginning to impact on community initiatives to improve the quality of life in neighbourhoods and cities. Here are some examples that emphasize a key role for social ties and positive relationships with others for transitioning to a better world. The first example—the Transition Network (http://www.transitionnetwork.org)—is a community-led movement that seeks to address some of the world’s biggest challenges (such as economic decline, social inequality and climate change) by connecting with ourselves, others and the natural world while developing and promoting positive possibilities such as urban food markets, community energy projects and local currencies such as the Brixton Pound. Another example is the 'Happy City Initiative' (http://www.happycity.org.uk), which aims to enhance community wellbeing through cultural and cooperative activities that extend beyond economic progress. Central to this initiative is the quality of relationships with others, in addition to supportive and active communities and the availability of opportunities for meaningful engagement with others. Yet another example is the 'Down to Earth Project' (http://www.downtoearthproject.org.uk), which aims to improve people's lives through outdoor communitarian activities and wellbeing programmes. Anecdotally, we have seen the lives of those with acquired brain injury transformed by engaging in outdoor activities such as the 'Building Community' project, which involves building sustainable facilities for future community activities. Further examples include the Action for Happiness initiative, which involves delivering training packages based on positive psychology and disseminating skills to communities (http://www.actionforhappiness.org) and social prescription of nature-based interventions [174, 175]. While it remains to be seen whether these community-based initiatives will succeed over the long term and the extent to which health and wellbeing is improved in the process, programmes like these have great potential for transformative change.

The epidemiological transition to non-communicable disease including mental ill health, diabetes, obesity and cardiovascular disease is now the dominant source of disease burden. Chronic disease has replaced acute (and communicable) disease as leading burdens of morbidity, mortality and health-care expenditures [168, 176–182]. However, our models of health care have not adapted to reflect these changes. The traditional approach to health care is the 'acute medical model', which involves the passive receipt of care. The assumptions underpinning this model is that illness is likely to be short-lived, and thus the aim of care is 'cure' and 'return to a pre-illness state of health' [183]. Although the acute care model is extremely important and is responsible for the saving of many lives, it does not provide an effective model for chronic care. Chronic disease is different [168] to acute disease as cure is usually not possible. Treatment outcomes for people with chronic conditions are contingent on an effective collaboration between clinician and patient, for example, encouraging patient adherence to treatment regimens and adoption of recommended lifestyle changes, among others. Thus, patients are no longer passive recipients of care but need to be active and equal partners in the management of their chronic disease. The acute medical model is further limited by a narrow definition of 'health' that is focused on illness or impairment, rather than flourishing.

Researchers [51] have argued that a complete understanding of social ties and their associated health outcomes will only be achieved through a more holistic approach to community health by defining communities as 'human ecological systems'. (Readers may be interested in looking at the work of Nicholas A. Christakis in particular, who focuses on how social networks impact on health and wellbeing.) The structure and organization of communities have important implications for community health, especially sociostructural factors (Fig. 4). An increasing body of work has examined the health of individuals and populations within the context of community ecosystems. For instance, a recent study [184] reporting on associations between psychological language that is used on the Twitter platform and county-level heart disease mortality argued that tweets of younger adults may disclose characteristics of their community, reflecting the physical and social environments that influence the health behaviours and stress experiences of their residents. This study [184] examined associations between 148 million county-mapped tweets across 1347 counties and county-level, age-adjusted mortality rates for atherosclerotic heart disease obtained from the Centers of Disease Control and Prevention. Patterns reflecting negative social ties, disengagement, and negative emotions emerged as risk factors, while positive emotions and psychological engagement emerged as protective factors. Intriguingly, the authors reported that their model based only on Twitter language predicted mortality better than a model combining ten common demographic, socioeconomic and health risk factors, including smoking, diabetes, hypertension and obesity. This focus on systems is also consistent with recent calls [185] for researchers to approach psychopathology as a system, drawing on developments in dynamical systems theory, network theory, chaos theory and other branches of the complexity sciences. Together with new insights offered by the GENIAL model, these methodological developments may help to transform our understanding of pathways to health and wellbeing within the context of the health of individuals and their community ecosystems.

In summary, we have reviewed the literature on associations between social ties and health outcomes, identified various mechanisms that might underpin the association and proposed the GENIAL model for pathways to wellbeing and longevity, emphasizing social ties as an important mediator along this pathway. We suggest that past research on the link between social ties and health outcomes has been held back for at least four reasons. First, the field has been characterized by multiple competing simplistic models leading to calls for more sophisticated models [6, 49] that take into account the complex pathways between social ties and health outcomes, as well as more robust statistical methods that could then be applied, drawing on, for example, developments in complex systems theory and dynamics [185]. Second, the field typically considers the vagal nerve as an epiphenomenon of the stress response, rather than a regulator of multisystemic, downstream pathways that can lead to either premature mortality or longevity. This is despite available theory [82–84, 124] and data [94, 95, 186] that provide a framework through which to understand how the vagus might regulate downstream systems. Third, research on vagal tone is typically conducted with a restricted focus on the individual, rather than their connections with others, social integration and community ecosystems, further highlighting a need for drawing on modern analytical methods. Fourth, research findings are typically interpreted from the vantage point of one's own discipline, a phenomenon known as the 'disciplinary dilemma'. While this is understandable, it is also unfortunate. It could be argued that the greatest insights in science arise through multidisciplinary exchange, allowing for new lines of interdisciplinary and even transdisciplinary enquiry to arise.

Future research will benefit from investigating more sophisticated models of pathways to health and wellbeing and drawing on recent developments in statistical methods (e.g. [187]) to large datasets. Multidisciplinary collaboration and interdisciplinary science are essential for a more complete understanding of the mechanisms underpinning the associations between social ties, health and wellbeing. Our working hypothesis underpinning the GENIAL model is that vagal function plays an important regulatory role over pathways to health and wellbeing, which include psychological factors, social ties and allostatic processes. Understanding that the health of individuals is not achieved within a vacuum, we recognize that physical and social environments will either impede or facilitate individual pathways to health and wellbeing. Within this framework, gene-environment interactions will contribute to individual differences in vagal function that will both influence and be influenced by psychological moments and social ties. Individual differences in vagal function will lead to variation in regulation of allostasis, and if regulation is optimal, one will be set along pathway to wellbeing, health and longevity; however, if dysregulated, the course will be set for illness, disease and premature mortality. The GENIAL model has important implications for (a) teaching service users about the importance of emotional flexibility and relationships with health outcomes and (b) creating social contexts that facilitate positive experiences and social ties especially for those with chronic conditions, who experience fewer opportunities for social integration and relationships with others. It is our hope that future research will proceed by combining individualist and structuralist approaches to health and converge into a transdisciplinary science that draws on developments in complex systems theory and dynamics, leading to a better understanding of how social ties influence human health and, ultimately, better care for managing patients with chronic mental disorders and physical disease.

Acknowledgements Contributions to an early draft of this manuscript from several students (Ingrid Bjerknes Røyne and Jasmine Marie Rollings), enrolled in the taught Masters (MSc) programme in Psychology at Swansea University, are gratefully acknowledged. Constructive criticism from two anonymous reviewers was also particularly helpful in the development of the ideas presented herein.

References

1. Cohen S. Social relationships and health. Am Psychol. 2004;59(8):676–84.
2. Berkman LF, Glass T, Brissette I, Seeman TE. From social integration to health: Durkheim in the new millennium. Soc Sci Med. 2000;51(6):843–57.
3. Diener E. Subjective well-being. Psychol Bull. 1984;95(3):542–75.
4. Seligman M. Flourish: a visionary new understanding of happiness and well-being. New York: Simon and Schuster; 2012.
5. Ryff CD. Psychological well-being revisited: advances in the science and practice of eudaimonia. Psychother Psychosom. 2014;83(1):10–28.
6. Friedman HS, Kern ML. Personality, well-being, and health. Annu Rev Psychol. 2014;65(1): 719–42.
7. Ford BQ, Shallcross AJ, Mauss IB, Floerke VA, Gruber J. Desperately seeking happiness: valuing happiness is associated with symptoms and diagnosis of depression. J Soc Clin Psychol. 2014;33(10):890–905.

8. Mauss IB, Tamir M, Anderson CL, Savino NS. Can seeking happiness make people unhappy? Paradoxical effects of valuing happiness. Emotion. 2011;11(4):807–15.
9. Forgas JP. Don't worry, be sad! On the cognitive, motivational, and interpersonal benefits of negative mood. Curr Dir Psychol Sci. 2013;22(3):225–32.
10. Kashdan T, Biswas-Diener R. The upside of your dark side: why being your whole self–not just your "good" self–drives success and fulfillment. New York: Plume Books; 2015. p. 1.
11. Kashdan T, Rottenberg J. Psychological flexibility as a fundamental aspect of health. Clin Psychol Rev. 2010;30(7):865–78.
12. Fredrickson BL, Losada MF. Positive affect and the complex dynamics of human flourishing. Am Psychol. 2005;60(7):678–86.
13. Fredrickson BL. Updated thinking on positivity ratios. Am Psychol. 2013;68(9):814–22.
14. Brown NJL, Sokal AD, Friedman HL. The complex dynamics of wishful thinking: the critical positivity ratio. Am Psychol. 2013;68(9):801–13.
15. Brown NJL, Sokal AD, Friedman HL. The persistence of wishful thinking. Am Psychol. 2014;69(6):629–32.
16. Quinn C, Harris A, Kemp AH. The interdependence of subtype and severity: contributions of clinical and neuropsychological features to melancholia and non-melancholia in an outpatient sample. J Int Neuropsychol Soc. 2012;18(2):361–9.
17. Quinn CR, Harris A, Felmingham K, Boyce P, Kemp AH. The impact of depression heterogeneity on cognitive control in major depressive disorder. Aust N Z J Psychiatry. 2012;46(11):1079–88.
18. Robinson D. Aristotle's psychology. New York: Columbia University Press; 1999.
19. Ryan RM, Deci EL. On happiness and human potentials: a review of research on hedonic and eudaimonic well-being. Annu Rev Psychol. 2001;52(1):141–66.
20. Liu B, Floud S, Pirie K, Green J, Peto R, Beral V, et al. Does happiness itself directly affect mortality? The prospective UK million women study. Lancet. 2016;387(10021):874–81.
21. Ryff CD, Singer BH, Dienberg Love G. Positive health: connecting well-being with biology. Philos Trans R Soc B. 2004;359(1449):1383–94.
22. Fredrickson BL, Grewen KM, Coffey KA, Algoe SB, Firestine AM, Arevalo JMG, et al. A functional genomic perspective on human well-being. Proc Natl Acad Sci. 2013;110(33):13684–9.
23. Fredrickson BL, Grewen KM, Algoe SB, Firestine AM, Arevalo JMG, Ma J, et al. Psychological well-being and the human conserved transcriptional response to adversity. PLoS ONE. 2015;10(3):e0121839.
24. Disabato DJ, Goodman FR, Kashdan TB, Short JL, Jarden A. Different types of well-being? A cross-cultural examination of hedonic and eudaimonic well-being. Psychol Assess. 2016; 28(5):471–82.
25. Kok BE, Coffey KA, Cohn MA, Catalino LI, Vacharkulksemsuk T, Algoe SB, et al. How positive emotions build physical health: perceived positive social connections account for the upward spiral between positive emotions and vagal tone. Psychol Sci. 2013;24(7):1123–32.
26. Coyne JC. Highly correlated hedonic and eudaimonic well-being thwart genomic analysis. Proc Natl Acad Sci U S A. 2013;110(45):E4183.
27. Brown N, Lomas T, Eiroá-Orosa FJ. The Routledge international handbook of critical positive psychology. London: Routledge; 2017.
28. Dubois CM, Beach SR, Kashdan TB, Nyer MB, Park ER, Celano CM, et al. Positive psychological attributes and cardiac outcomes: associations, mechanisms, and interventions. Psychosomatics. 2012;53(4):303–18.
29. Boehm JK, Kubzansky LD. The heart's content: the association between positive psychological well-being and cardiovascular health. Psychol Bull. 2012;138(4):655–91.
30. Kok BE, Fredrickson BL. Upward spirals of the heart: autonomic flexibility, as indexed by vagal tone, reciprocally and prospectively predicts positive emotions and social connectedness. Biol Psychol. 2010;85(3):432–6.
31. Steptoe A, Deaton A, Stone AA. Subjective wellbeing, health, and ageing. Lancet. 2015; 385(9968):640–8.

32. Kern ML, Porta Della SS, Friedman HS. Lifelong pathways to longevity: personality, relationships, flourishing, and health. J Pers. 2013;82(6):472–84.
33. Martin LR, Friedman HS, Tucker JS, Tomlinson-Keasey C, Criqui MH, Schwartz JE. A life course perspective on childhood cheerfulness and its relation to mortality risk. Personal Soc Psychol Bull. 2002;28(9):1155–65.
34. Shakya HB, Christakis NA. Association of facebook use with compromised well-being: a longitudinal study. Am J Epidemiol. 2017;185(3):203–11.
35. McDool E, Powell P, Roberts J, Taylor K. Social media use and children's wellbeing: IZA DP No. 10412. IZA Institute of Labor Economics; 2016.
36. Kross E, Verduyn P, Demiralp E, Park J, Lee DS, Lin N, et al. Facebook use predicts declines in subjective well-being in young adults. PLoS ONE. 2013;8(8):e69841.
37. Valtorta NK, Kanaan M, Gilbody S, Ronzi S, Hanratty B. Loneliness and social isolation as risk factors for coronary heart disease and stroke: systematic review and meta-analysis of longitudinal observational studies. Heart. 2016;102(13):1009–16.
38. Holt-Lunstad J, Smith TB, Layton JB. Social relationships and mortality risk: a meta-analytic review. PLoS Med. 2010;7(7):e1000316.
39. Holt-Lunstad J, Smith TB, Baker M, Harris T, Stephenson D. Loneliness and social isolation as risk factors for mortality: a meta-analytic review. Perspect Psychol Sci. 2015;10(2):227–37.
40. Yang YC, Boen C, Gerken K, Li T, Schorpp K, Harris KM. Social relationships and physiological determinants of longevity across the human life span. Proc Natl Acad Sci. 2016;113(3):578–83.
41. Steptoe A, Shankar A, Demakakos P, Wardle J. Social isolation, loneliness, and all-cause mortality in older men and women. Proc Natl Acad Sci. 2013;110(15):5797–801.
42. Luo Y, Hawkley LC, Waite LJ, Cacioppo JT. Loneliness, health, and mortality in old age: a national longitudinal study. Soc Sci Med. 2012;74(6):907–14.
43. Hodge AM, English DR, Giles GG, Flicker L. Social connectedness and predictors of successful ageing. Maturitas. 2013;75(4):361–6.
44. Christakis NA, Fowler JH. The spread of obesity in a large social network over 32 years. N Engl J Med. 2007;357(4):370–9.
45. Bandura A. Health promotion by social cognitive means. Health Educ Behav. 2004;31(2): 143–64.
46. Umberson D, Crosnoe R, Reczek C. Social relationships and health behavior across the life course. Annu Rev Sociol. 2010;36(1):139–57.
47. Friedman HS, Martin LR. The longevity project: surprising discoveries for health and long life from the landmark eight decade study. New York: Hudson Street Press; 2011.
48. Maunder R, Hunter J. Love, fear, and health: how our attachments to others shape health and health care. Toronto: University of Toronto Press; 2015.
49. Uchino BN, Bowen K, Carlisle M, Birmingham W. Psychological pathways linking social support to health outcomes: a visit with the "ghosts" of research past, present, and future. Soc Sci Med. 2012;74(7):949–57.
50. Hernandez LM, Blazer DG. The impact of social and cultural environment on health. In: Hernandez LM, Blazer DG, editors. Genes, behavior and the social environment: moving beyond the nature/nurture debate. Washington, DC: National Academies Press; 2006.
51. Wilson SM. An ecologic framework to study and address environmental justice and community health issues. Environ Justice. 2009;2(1):15–24.
52. Tay L, Tan K, Diener E, Gonzalez E. Social relations, health behaviors, and health outcomes: a survey and synthesis. Appl Psychol Health Well Being. 2012;5(1):28–78.
53. Uchino BN. Social support and health: a review of physiological processes potentially underlying links to disease outcomes. J Behav Med. 2006;29(4):377–87.
54. Durkheim E. Suicide: a study in sociology. Glencoe: Free Press; 1897. p. 1–427.
55. Eriksson M. Social capital and health – implications for health promotion. Glob Health Action. 2011;4(1):5611.
56. Bogg T, Roberts BW. Conscientiousness and health-related behaviors: a meta-analysis of the leading behavioral contributors to mortality. Psychol Bull. 2004;130(6):887–919.

57. Steel P, Schmidt J, Shultz J. Refining the relationship between personality and subjective well-being. Psychol Bull. 2008;134(1):138–61.
58. Jokela M, Batty GD, Nyberg ST, Virtanen M, Nabi H, Singh-Manoux A, et al. Personality and all-cause mortality: individual-participant meta-analysis of 3,947 deaths in 76,150 adults. Am J Epidemiol. 2013;178(5):667–75.
59. Mikulincer M, Shaver PR. Attachment in adulthood: structure, dynamics, and change. New York: Guilford Publications; 2007.
60. Puig J, Englund MM, Simpson JA, Collins WA. Predicting adult physical illness from infant attachment: a prospective longitudinal study. Health Psychol. 2013;32(4):409–17.
61. Bowlby J. Attachment and loss. 3. Loss. New York: Basic Books; 1980.
62. Bowlby J. Attachment and loss. 1. Attachment. New York: Basic Books; 1969.
63. Bowlby E. Attachment and Loss. 2. Separation: anger and anxiety. New York: Basic Books; 1998.
64. Vrtička P. The social neuroscience of attachment. In: Ibáñez A, Sedeño L, García AM, editors. Neuroscience and social science. New York: Springer; 2017.
65. Niedenthal PM, Brauer M, Robin L, Innes-Ker ÅH. Adult attachment and the perception of facial expression of emotion. J Pers Soc Psychol. 2002;82(3):419–33.
66. Lazarus R, Folkman S. Stress, appraisal, and coping. New York: Springer; 1984.
67. Keller A, Litzelman K, Wisk LE, Maddox T, Cheng ER, Creswell PD, et al. Does the perception that stress affects health matter? The association with health and mortality. Health Psychol. 2012;31(5):677–84.
68. Sheinbaum T, Kwapil TR, Ballespí S, Mitjavila M, Chun CA, Silvia PJ, et al. Attachment style predicts affect, cognitive appraisals, and social functioning in daily life. Front Psychol. 2015;6:296.
69. Lund R, Christensen U, Nilsson CJ, Kriegbaum M, Hulvej Rod N. Stressful social relations and mortality: a prospective cohort study. J Epidemiol Community Health. 2014;68(8):720–7.
70. Russ TC, Stamatakis E, Hamer M, Starr JM, Kivimaki M, Batty GD. Association between psychological distress and mortality: individual participant pooled analysis of 10 prospective cohort studies. BMJ. 2012;345:e4933.
71. Penwell LM, Larkin KT. Social support and risk for cardiovascular disease and cancer: a qualitative review examining the role of inflammatory processes. Health Psychol Rev. 2010;4(1):42–55.
72. Cacioppo JT, Hawkley LC. Social isolation and health, with an emphasis on underlying mechanisms. Perspect Biol Med. 2003;46(3):S39–52.
73. Sgoifo A, Koolhaas J, De Boer S, Musso E, Stilli D, Buwalda B, et al. Social stress, autonomic neural activation, and cardiac activity in rats. Neurosci Biobehav Rev. 1999;23(7):915–23.
74. Sgoifo A, Carnevali L, Grippo AJ. The socially stressed heart. Insights from studies in rodents. Neurosci Biobehav Rev. 2013;39:51–60.
75. Costoli T, Bartolomucci A, Graiani G, Stilli D, Laviola G, Sgoifo A. Effects of chronic psychosocial stress on cardiac autonomic responsiveness and myocardial structure in mice. Am J Physiol Heart Circ Physiol. 2004;286(6):H2133–40.
76. McEwen BS. Stress, adaptation, and disease: allostasis and allostatic load. Ann N Y Acad Sci. 1998;840(1):33–44.
77. Cohen S, Doyle WJ, Skoner DP, Rabin BS. Social ties and susceptibility to the common cold. JAMA. 1997;277:1940–4.
78. Steptoe A, Dockray S, Wardle J. Positive affect and psychobiological processes relevant to health. J Pers. 2009;77(6):1747–76.
79. Porges SW. Orienting in a defensive world: mammalian modifications of our evolutionary heritage. A polyvagal theory. Psychophysiology. 1995;32(4):301–18.
80. Porges SW. The polyvagal theory: phylogenetic substrates of a social nervous system. Int J Psychophysiol. 2001;42(2):123–46.
81. Porges SW. Social engagement and attachment: a phylogenetic perspective. Ann N Y Acad Sci. 2003;1008:31–47.
82. Porges SW. The polyvagal theory: neurophysiological foundations of emotions, attachment, communication, and self-regulation. New York: W.W. Norton & Company; 2011.

83. Thayer J, Lane RD. Claude Bernard and the heart-brain connection: further elaboration of a model of neurovisceral integration. Neurosci Biobehav Rev. 2009;33(2):81–8.
84. Thayer J, Hansen AL, Saus-Rose E, Johnsen BH. Heart rate variability, prefrontal neural function, and cognitive performance: the neurovisceral integration perspective on self-regulation, adaptation, and health. Ann Behav Med. 2009;37(2):141–53.
85. Thayer J, Ahs F, Fredrikson M, Sollers Iii JJ, Wager TD. A meta-analysis of heart rate variability and neuroimaging studies: implications for heart rate variability as a marker of stress and health. Neurosci Biobehav Rev. 2012;36(2):747–56.
86. Smith R, Thayer J, Khalsa SS, Lane RD. The hierarchical basis of neurovisceral integration. Neurosci Biobehav Rev. 2017;75:274–96.
87. Mesulam MM. From sensation to cognition. Brain. 1998;121(Pt 6):1013–52.
88. Vrtička P, Vuilleumier P. Neuroscience of human social interactions and adult attachment style. Front Hum Neurosci. 2012;6:212.
89. Lieberman MD. Social cognitive neuroscience: a review of core processes. Annu Rev Psychol. 2007;58(1):259–89.
90. Fonagy P, Luyten P, Strathearn L. Borderline personality disorder, mentalization, and the neurobiology of attachment. Infant Ment Health J. 2011;32(1):47–69.
91. Diamond LM. Attachment style, current relationship security, and negative emotions: the mediating role of physiological regulation. J Soc Pers Relat. 2005;22(4):499–518.
92. Chunchai T, Samniang B, Sripetchwandee J, Pintana H, Pongkan W, Kumfu S, et al. Vagus nerve stimulation exerts the neuroprotective effects in obese-insulin resistant rats, leading to the improvement of cognitive function. Sci Rep. 2016;6(1):26866.
93. Samniang B, Shinlapawittayatorn K, Chunchai T, Pongkan W, Kumfu S, Chattipakorn SC, et al. Vagus nerve stimulation improves cardiac function by preventing mitochondrial dysfunction in obese-insulin resistant rats. Sci Rep. 2016;6(1):19749.
94. Jarczok MN, Koenig J, Mauss D, Fischer JE, Thayer J. Lower heart rate variability predicts increased level of C-reactive protein 4 years later in healthy, nonsmoking adults. J Intern Med. 2014;276(6):667–71.
95. Kemp AH, López SR, Passos VMA, Bittencourt MS, Dantas EM, Mill JG, et al. Insulin resistance and carotid intima-media thickness mediate the association between resting-state heart rate variability and executive function: a path modelling study. Biol Psychol. 2016;117:216–24.
96. Meyer-Lindenberg A, Tost H. Neural mechanisms of social risk for psychiatric disorders. Nat Neurosci. 2012;15(5):663–8.
97. Kumsta R, Heinrichs M. Oxytocin, stress and social behavior: neurogenetics of the human oxytocin system. Curr Opin Neurobiol. 2013;23(1):11–6.
98. Kogan A, Saslow LR, Impett EA, Oveis C, Keltner D, Rodrigues Saturn S. Thin-slicing study of the oxytocin receptor (OXTR) gene and the evaluation and expression of the prosocial disposition. Proc Natl Acad Sci. 2011;108(48):19189–92.
99. Bakermans-Kranenburg MJ, van Ijzendoorn MH. Oxytocin receptor (OXTR) and serotonin transporter (5-HTT) genes associated with observed parenting. Soc Cogn Affect Neurosci. 2008;3(2):128–34.
100. Rodrigues SM, Saslow LR, Garcia N, John OP, Keltner D. Oxytocin receptor genetic variation relates to empathy and stress reactivity in humans. Proc Natl Acad Sci. 2009;106(50):21437–41.
101. Lucht MJ, Barnow S, Sonnenfeld C, Rosenberger A, Grabe HJ, Schroeder W, et al. Associations between the oxytocin receptor gene (OXTR) and affect, loneliness and intelligence in normal subjects. Prog Neuro-Psychopharmacol Biol Psychiatry. 2009;33(5):860–6.
102. Kanthak MK, Chen FS, Kumsta R, Hill LK, Thayer J, Heinrichs M. Oxytocin receptor gene polymorphism modulates the effects of social support on heart rate variability. Biol Psychol. 2016;117:43–9.
103. Pearce E, Wlodarski R, Machin A, Dunbar RIM. Variation in the β-endorphin, oxytocin, and dopamine receptor genes is associated with different dimensions of human sociality. Proc Natl Acad Sci U S A. 2017;8:201700712.
104. Dunbar RIM, Baron R, Frangou A, Pearce E, van Leeuwen EJC, Stow J, et al. Social laughter is correlated with an elevated pain threshold. Proc Biol Sci. 2012;279(1731):1161–7.

105. Weinstein D, Launay J, Pearce E, Dunbar RIM, Stewart L. Group music performance causes elevated pain thresholds and social bonding in small and large groups of singers. Evol Hum Behav. 2016;37(2):152–8.
106. Nummenmaa L, Tuominen L, Dunbar R, Hirvonen J, Manninen S, Arponen E, et al. Social touch modulates endogenous μ-opioid system activity in humans. NeuroImage. 2016;138:242–7.
107. Weaver ICG, Cervoni N, Champagne FA, D'alessio AC, Sharma S, Seckl JR, et al. Epigenetic programming by maternal behavior. Nat Neurosci. 2004;7(8):847–54.
108. King S, Dancause K, Turcotte-Tremblay A-M, Veru F, Laplante DP. Using natural disasters to study the effects of prenatal maternal stress on child health and development. Birth Defect Res C. 2013;96(4):273–88.
109. Cao-Lei L, Massart R, Suderman MJ, Machnes Z, Elgbeili G, Laplante DP, et al. DNA methylation signatures triggered by prenatal maternal stress exposure to a natural disaster: project ice storm. PLoS ONE. 2014;9(9):e107653.
110. Grizenko N, Fortier M-È, Gaudreau-Simard M, Jolicoeur C, Joober R. The effect of maternal stress during pregnancy on IQ and ADHD symptomatology. J Can Acad Child Adolesc Psychiatry. 2015;24(2):92–9.
111. Laplante DP, Brunet A, Schmitz N, Ciampi A, King S. Project ice storm: prenatal maternal stress affects cognitive and linguistic functioning in 5 1/2-year-old children. J Am Acad Child Adolesc Psychiatry. 2008;47(9):1063–72.
112. Dancause KN, Laplante DP, Fraser S, Brunet A, Ciampi A, Schmitz N, et al. Prenatal exposure to a natural disaster increases risk for obesity in 5½-year-old children. Pediatr Res. 2012;71(1):126–31.
113. Dancause KN, Veru F, Andersen RE, Laplante DP, King S. Prenatal stress due to a natural disaster predicts insulin secretion in adolescence. Early Hum Dev. 2013;89(9):773–6.
114. Kemp AH, Quintana DS. The relationship between mental and physical health: insights from the study of heart rate variability. Int J Psychophysiol. 2013;89(3):288–96.
115. Thayer J, Yamamoto SS, Brosschot JF. The relationship of autonomic imbalance, heart rate variability and cardiovascular disease risk factors. Int J Cardiol. 2010;141(2):122–31.
116. Brosschot JF, Verkuil B, Thayer J. Exposed to events that never happen: generalized unsafety, the default stress response, and prolonged autonomic activity. Neurosci Biobehav Rev. 2017; 74(Part B):287–96.
117. Kemp AH. Heart rate variability, affective disorders and health. In: Baune BT, Tully PJ, editors. Cardiovascular diseases and depression. Cham: Springer; 2016. p. 167–85.
118. Kok BE, Fredrickson BL. Evidence for the upward spiral stands steady: a response to Heathers, Brown, Coyne, and Friedman (2015). Psychol Sci. 2015;26(7):1144–6.
119. Okbay A, Baselmans BML, De Neve J-E, Turley P, Nivard MG, Fontana MA, et al. Genetic variants associated with subjective well-being, depressive symptoms, and neuroticism identified through genome-wide analyses. Nat Genet. 2016;48(6):624–33.
120. Cole SW, Hawkley LC, Arevalo JMG, Cacioppo JT. Transcript origin analysis identifies antigen-presenting cells as primary targets of socially regulated gene expression in leukocytes. Proc Natl Acad Sci. 2011;108(7):3080–5.
121. Brown NJL, MacDonald DA, Samanta MP, Friedman HL, Coyne JC. A critical reanalysis of the relationship between genomics and well-being. Proc Natl Acad Sci. 2014;111(35):12705–9.
122. Brown NJL, MacDonald DA, Samanta MP, Friedman HL, Coyne JC. More questions than answers: continued critical reanalysis of Fredrickson et al.'s studies of genomics and well-being. PLoS ONE. 2016;11(6):e0156415.
123. Fredrickson BL. Selective data analysis in Brown et al.'s continued critical reanalysis. PLoS ONE. 2016;11(8):e0160565.
124. Pavlov V, Tracey K. The vagus nerve and the inflammatory reflex-linking immunity and metabolism. Nat Rev Endocrinol. 2012;8(12):743–54.
125. Tracey KJ. Physiology and immunology of the cholinergic antiinflammatory pathway. J Clin Invest. 2007;117(2):289–96.
126. Tracey KJ. The inflammatory reflex. Nature. 2002;420(6917):853–9.

127. Kemp AH, Quintana DS, Kuhnert R-L, Griffiths K, Hickie IB, Guastella AJ. Oxytocin increases heart rate variability in humans at rest: implications for social approach-related motivation and capacity for social engagement. PLoS ONE. 2012;7(8):e44014.
128. Norman GJ, Berntson GG, Cacioppo JT, Morris JS, Malarkey WB, Devries AC. Oxytocin increases autonomic cardiac control: moderation by loneliness. Biol Psychol. 2011;86(3):174–80.
129. Jandackova VK, Britton A, Malik M, Steptoe A. Heart rate variability and depressive symptoms: a cross-lagged analysis over a 10-year period in the Whitehall II study. Psychol Med. 2016;46(10):2121–31.
130. Kemp AH, Quintana DS, Gray MA, Felmingham KL, Brown K, Gatt JM. Impact of depression and antidepressant treatment on heart rate variability: a review and meta-analysis. Biol Psychiatry. 2010;67(11):1067–74.
131. Kemp AH, Brunoni AR, Santos IS, Nunes MA, Dantas EM, Carvalho de Figueiredo R, et al. Effects of depression, anxiety, comorbidity, and antidepressants on resting-state heart rate and its variability: an ELSA-Brasil cohort baseline study. Am J Psychiatr. 2014;171(12):1328–34.
132. Brunoni AR, Kemp AH, Dantas EM, Goulart AC, Nunes MA, Boggio PS, et al. Heart rate variability is a trait marker of major depressive disorder: evidence from the sertraline vs. electric current therapy to treat depression clinical study. Int J Neuropsychopharmacol. 2013;16(9):1937–49.
133. Kemp AH, Quintana DS, Quinn CR, Hopkinson P, Harris AWF. Major depressive disorder with melancholia displays robust alterations in resting state heart rate and its variability: implications for future morbidity and mortality. Front Psychol. 2014;5:1387.
134. Berthoud H-R. The vagus nerve, food intake and obesity. Regul Pept. 2008;149(1–3):15–25.
135. Tracey KJ, Pavlov VA. The vagus nerve and the inflammatory reflex–linking immunity and metabolism. Nat Rev Endocrinol. 2012;8(12):743–54.
136. Follesa P, Biggio F, Gorini G, Caria S, Talani G, Dazzi L, et al. Vagus nerve stimulation increases norepinephrine concentration and the gene expression of BDNF and bFGF in the rat brain. Brain Res. 2007;1179:28–34.
137. Biggio F, Gorini G, Utzeri C, Olla P, Marrosu F, Mocchetti I, et al. Chronic vagus nerve stimulation induces neuronal plasticity in the rat hippocampus. Int J Neuropsychopharmacol. 2009;12(9):1209.
138. Gomez-Pinilla F. The influences of diet and exercise on mental health through hormesis. Ageing Res Rev. 2008;7(1):49–62.
139. Groves DA, Brown VJ. Vagal nerve stimulation: a review of its applications and potential mechanisms that mediate its clinical effects. Neurosci Biobehav Rev. 2005;29(3):493–500.
140. Vonck K, Raedt R, Naulaerts J, De Vogelaere F, Thiery E, Van Roost D, et al. Vagus nerve stimulation…25 years later! What do we know about the effects on cognition? Neurosci Biobehav Rev. 2014;45:63–71.
141. Stilling RM, Dinan TG, Cryan JF. Microbial genes, brain & behaviour-epigenetic regulation of the gut-brain axis. Genes Brain Behav. 2013;13(1):69–86.
142. Bravo JA, Forsythe P, Chew MV, Escaravage E, Savignac HM, Dinan TG, et al. Ingestion of lactobacillus strain regulates emotional behavior and central GABA receptor expression in a mouse via the vagus nerve. Proc Natl Acad Sci U S A. 2011;108(38):16050–5.
143. Carter JB, Banister EW, Blaber AP. Effect of endurance exercise on autonomic control of heart rate. Sports Med. 2003;33(1):33–46.
144. Ling C, Rönn T. Epigenetic adaptation to regular exercise in humans. Drug Discov Today. 2014;19(7):1015–8.
145. Thayer J, Sternberg E. Beyond heart rate variability: vagal regulation of allostatic systems. Ann N Y Acad Sci. 2006;1088(1):361–72.
146. Huston JM, Tracey KJ. The pulse of inflammation: heart rate variability, the cholinergic anti-inflammatory pathway and implications for therapy. J Intern Med. 2010;269(1):45–53.
147. Thayer J, Lane RD. The role of vagal function in the risk for cardiovascular disease and mortality. Biol Psychol. 2007;74(2):224–42.
148. Thayer J, Loerbroks A, Sternberg EM. Inflammation and cardiorespiratory control: the role of the vagus nerve. Respir Physiol Neurobiol. 2011;178(3):387–94.

149. Barberis SD, et al. A pluralist framework for the philosophy of social neuroscience. In: Ibáñez A, Sedeño L, García AM, editors. Neuroscience and social science. New York: Springer; 2017.
150. Fredrickson BL, Grewen KM. A functional genomic perspective on human well-being. Proc Natl Acad Sci U S A. 2013;110:13684–9.
151. Kiecolt-Glaser JK, McGuire L, Robles TF, Glaser R. Emotions, morbidity, and mortality: new perspectives from psychoneuroimmunology. Annu Rev Psychol. 2002;53(1):83–107.
152. Stapelberg NJ, Neumann DL, Shum DHK, McConnell H, Hamilton-Craig I. A topographical map of the causal network of mechanisms underlying the relationship between major depressive disorder and coronary heart disease. Aust N Z J Psychiatry. 2011;45(5):351–69.
153. Box G, Draper NR. Empirical model-building and response surfaces. New York: Wiley; 1987.
154. Twenge JM, Foster JD. Birth cohort increases in narcissistic personality traits among American college students, 1982–2009. Soc Psychol Personal Sci. 2010;1(1):99–106.
155. Statistical bulletin: families and households in the UK; 2016. www.ons.gov.uk.
156. Vespa J, Lewis JM, Kreider RM. America's families and living arrangements: 2012 [Internet]. census.gov; 2013 [cited 2017 Apr 25]. https://www.census.gov/prod/2013pubs/p20-570.pdf.
157. Siegler V. Measuring national well-being-an analysis of social capital in the UK. Office of National Statistics; 2015.
158. Beutel ME, Klein EM, Brähler E, Reiner I, Jünger C, Michal M, et al. Loneliness in the general population: prevalence, determinants and relations to mental health. BMC Psychiatry. 2017;17(1):97.
159. Lau J. Social intelligence and the next generation. London: National Service Citizen; 2016.
160. Jackson J, Cochran SD. Loneliness and psychological distress. J Psychol. 1991;125(3):257–62.
161. Cacioppo JT, Hughes ME, Waite LJ, Hawkley LC, Thisted RA. Loneliness as a specific risk factor for depressive symptoms: cross-sectional and longitudinal analyses. Psychol Aging. 2006;21(1):140–51.
162. Chesney E, Goodwin GM, Fazel S. Risks of all-cause and suicide mortality in mental disorders: a meta-review. World Psychiatry. 2014;13(2):153–60.
163. Mental Health Foundation. The fundamental facts about mental illness. London: Mental Health Foundation; 2015.
164. Bloom DE, Cafiero E, Jané-Llopis E, Abrahams-Gessel S, Bloom LR, Fathima S, et al. The global economic burden of noncommunicable diseases. PGDA working papers. Program on the global demography of aging; 2012.
165. OECD. Mental health and work: United Kingdom. Paris: OECD Publishing; 2014.
166. Patel V, Maj M, Flisher AJ, De Silva MJ, Koschorke M, Prince M, et al. Reducing the treatment gap for mental disorders: a WPA survey. World Psychiatry. 2010;9(3):169–76.
167. Wang PS, Berglund PA, Olfson M, Kessler RC. Delays in initial treatment contact after first onset of a mental disorder. Health Serv Res. 2004;39(2):393–416.
168. Institute of Medicine (US) Committee on Quality of Health Care in America. Crossing the quality chasm: a new health system for the 21st century. Washington, DC: National Academies Press; 2001.
169. Nesse RM. Natural selection and the elusiveness of happiness. Philos Trans R Soc Lond B. 2004;359(1449):1333–47. https://www.ncbi.nlm.nih.gov/pmc/articles/PMC1693419/pdf/15347525.pdf
170. Gable SL, Haidt J. What (and why) is positive psychology? Rev Gen Psychol. 2005;9(2):103–10.
171. Sheldon KM, King L. Why positive psychology is necessary. Am Psychol. 2001;56(3):216–7.
172. Lyubomirsky S, Sheldon K, Schkade D. Pursuing happiness: the architecture of sustainable change. Rev Gen. 2005;9:111–31.
173. Parks AC, Biswas-Diener R. Positive interventions: past, present and future. In: Mindfulness, acceptance, and positive psychology. Oakland: Context Press; 2013.
174. Bragg R, Leck C. Good practice in social prescribing for mental health: The role of nature-based interventions. Natural England Commissioned Reports; 2016.
175. Song C, Ikei H, Miyazaki Y. Physiological effects of nature therapy: a review of the research in Japan. Int J Environ Res Public Health. 2016;13(8):781.

176. Murray CJ, Lopez AD. Global mortality, disability, and the contribution of risk factors: global burden of disease study. Lancet. 1997;349(9063):1436–42.
177. Murray CJ, Lopez AD. Alternative projections of mortality and disability by cause 1990–2020: global burden of disease study. Lancet. 1997;349(9064):1498–504.
178. Lopez AD, Project DCP. Global burden of disease and risk factors. Oxford: Oxford University Press; 2006. p. 1.
179. Whiteford HA, Degenhardt L, Rehm J, Baxter AJ, Ferrari AJ, Erskine HE, et al. Global burden of disease attributable to mental and substance use disorders: findings from the global burden of disease study 2010. Lancet. 2013;382(9904):1575–86.
180. GBD 2013 Mortality and Causes of Death Collaborators. Global, regional, and national age-sex specific all-cause and cause-specific mortality for 240 causes of death, 1990–2013: a systematic analysis for the global burden of disease study 2013. Lancet. 2014;385(9963):117–71.
181. Vos T, Barber RM, Bell B, Bertozzi-Villa A. Global, regional, and national incidence, prevalence, and years lived with disability for 301 acute and chronic diseases and injuries in 188 countries, 1990–2013: a systematic analysis for the global burden of disease study 2013. Lancet. 2015;386:743–800.
182. GBD 2015 DALYs and HALE Collaborators. Global, regional, and national disability-adjusted life-years (DALYs) for 315 diseases and injuries and healthy life expectancy (HALE), 1990-2015: a systematic analysis for the global burden of disease study 2015. Lancet. 2016;388(10053):1603–58.
183. Shahady EJ. Barriers to care in chronic disease: how to bridge the treatment gap. Consultant. 2006;46:1149–52.
184. Eichstaedt JC, Schwartz HA, Kern ML, Park G, Labarthe DR, Merchant RM, et al. Psychological language on twitter predicts county-level heart disease mortality. Psychol Sci. 2015;26(2):159–69.
185. Nelson B, McGorry PD, Wichers M, Wigman JTW, Hartmann JA. Moving from static to dynamic models of the onset of mental disorder. JAMA Psychiat. 2017;74(5):528–34.
186. Wang H, Yu M, Ochani M, Amella CA, Tanovic M, Susarla S, et al. Nicotinic acetylcholine receptor alpha7 subunit is an essential regulator of inflammation. Nature. 2003;421(6921):384–8.
187. Christakis NA, Fowler JH. Social contagion theory: examining dynamic social networks and human behavior. Stat Med. 2013;32(4):556–77.

Part IV
Philosophical Contributions to Theoretical, Methodological, and Ethical Questions

The Self-Domesticated Animal and Its Study

Mario Bunge

Abstract The present chapter aims (a) to emphasize two points made abundantly clear by contemporary social cognitive and affective neuroscientists and (b) to note the philosophical nature of those points. These are (a) that all mental processes are brain processes and (b) that since all humans belong to several social systems, their mental life can only be understood by social psychology. Lastly, I propose that the main difference between the classical and the contemporary phases of that science is that, whereas the former sought to *describe* the psychosocial realm, nowadays we also wish to *understand* it—by unveiling its underlying mechanisms, such as the negative effect of social exclusion on neuroimmune processes. This more ambitious goal suggests merging biopsychology with the social sciences instead of either isolating the former or attempting to reduce it to either zoology or sociology. Such a call for merger should discourage all talk about neuropolitics and the like, for social science is about social systems, not isolated individuals, whereas psychology, whether individual or social, is about socially embedded individuals.

1 Introduction

Although the official name of our species is *Homo sapiens*, some people have preferred *Homo faber*, *loquens*, *adorans*, *bellator*, *ludens*, or *crudelis*. Still others opt for the *self-domesticated*, *problematizing*, *soul-owning*, *political animal*, *God's imitation*—or even, as televangelist Billy Graham once put it, "we are fallen creatures living in a fallen world."

That famous man of God did not mention the cooperatives of destitutes, even abandoned children, who organize themselves into self-managed cooperatives to eke out their livelihoods in big third-world cities. While there are plenty of feral cats, there have been only two well-documented feral children: Kaspar Hauser and Victor de Aveyron, both in the early nineteenth century. When abandoned, children tend to get together, help each other, and survive by scavenging and collecting, sorting, and selling refuse. They are marginal but refuse to be totally excluded from

M. Bunge (✉)
Department of Philosophy, McGill University, Montreal, Quebec, Canada, H3A 0G4
e-mail: marucho.bunge@gmail.com

© Springer International Publishing Switzerland 2017
A. Ibáñez et al. (eds.), *Neuroscience and Social Science*,
DOI 10.1007/978-3-319-68421-5_18

society. They are certainly downtrodden, but not fallen in the theological sense, for they can get up, climb mountains, and survive through hard work and solidarity.

I prefer to be known as a *social animal*—the title of a standard textbook in social psychology [1]—or, better yet, as the *self-domesticated* animal, for these names encompass some of the previously listed nicknames. In addition, they suggest that all the disciplines that study us are *biosociological* rather than biological (naturalism), spiritual (idealism), moral (Hume), or human—as if the other sciences were unhuman.

The qualifier *biosociological* invites us to fuse all the disciplines dealing with people, from zoology, neuroscience, demography, and epidemiology to anthropology, sociology, and historiography. The same qualifier warns us not to try and circumvent the social, by jumping from the individual brain to economic transactions, as if these were individual processes like digesting and navel gazing. And the qualifier *social* in "cognitive and affective neuroscience" focuses on the neural aspect of social behavior, from love and play to trade and war. The same qualifier in "biosocial evolution" reminds us that, because humans domesticate themselves, human evolution has been artifactual as well as natural from the moment the first tool and the earliest social norm were crafted (see [2, 3]).

Feeling love or hatred is personal, but marriage and trade are social. Actually, all human behavior (except scratching one's head) is social to some extent, because it happens in a social context, affects others, and leaves traces on the environment as well as on the actor's brain. That sociality greatly contributes to defining us, and other simians are obvious from the way social exclusion diminishes us and the revulsion sociopathy causes in most of us. Thus, autism is a serious disease, solitary confinement a harsh punishment and even a kind of torture, and deafness a painful shortcoming because it isolates individuals even more so than blindness or muteness. Nor is the deprivation of social contact damaging just to people. Puppies reared in cages develop into abnormal adults; and hungry capuchin monkeys prefer the company of conspecifics to food.

To humans, sociality is much more than gregariousness: being social involves crafting or maintaining social systems or "circles" of many kinds. And human social systems are more than shoals, flocks, herds, or other groups of conspecifics: it involves inventing, observing, or altering norms of behavior and institutions such as schooling, teamwork, mutual aid, and more. Such institutions emerged and evolved for defense, conflict resolution, production, and trade. We struggle for existence but cooperate for coexistence [4].

When such institutions fail, as it happens in cases of unprovoked aggression and the stealing of land or people, the aggressors are said to behave savagely, even though the idea that all primitives are violent has been just as discredited as the myth of the good savage. Kindness and wickedness are learned, not inborn. Only the capacity to learn to become either a good or a bad person is innate. Nativism is not a result of scientific research but an ideology designed to excuse racism, misogyny, and even school-tax evasion. In short, Aristotle's born slave, Lombroso's born criminal, Chomsky's born linguist, and Gopnik's born scientist are just fantasies. As Marx put it, we make society, and society makes us—but, *pace* Marx, society does not feel or think through us, because it is brainless. Hence, although we must distinguish individuals from their social hosts, we should not detach them.

Adverse social circumstances may sicken in various ways. For example, loneliness and forced social exclusion due to discrimination, arbitrary subordination, economic insecurity unemployment, or restricted access to public health facilities may cause anxiety, stress, depression, social phobia, eating disorders, and even heart disease and self-harm [5]. No wonder that emigration can be deeply unsettling and that many of the patients of clinical psychologists suffer from the "broken heart syndrome" following widowhood.

Uprooting harms people no less than trees. Just think of the refugees from persecution, ethnic cleansing, and war. They suffer not just from biological deprivations but also from abandoning their habitual social "circles" or systems. Much the same is also true of the unemployed, who lose not only their lifestyles and self-respect but also their reassuring contacts with their erstwhile colleagues. Cuts in social services, in particular in public health, have a similar effect. For example, the longevity of Britons decreased significantly under the so-called neoliberal rule of Margaret Thatcher [6].

The neuroscientist's job is to discover the neural systems and processes involved in feeling, planning, or controlling social processes, that is, strings of events that affect other people. For instance, they may wish to discover whether a particular action was free (spontaneous) or compelled by an external stimulus, as well as the brain subsystems activated or inhibited during that action. The result of such a study may be used to design and implement behavior norms and institutions aiming at either encouraging or discouraging actions of that kind. In general, we ought to learn before acting.

The most popular topics in recent social cognitive and affective neuroscience are self-recognition, self-reflection, self-knowledge, self-control, self-started processes, and the corresponding deficits [7]. The study of reflecting on one's current experience has led to a closer study of medial prefrontal cortex (BA10) [8]. This is the region of the prefrontal cortex that is disproportionately larger in humans than in other primates.

For this reason, a biological reductionist might propose calling ourselves *batens* or owners of the BA10 region. But sociological reductionists might then argue that our species deserves to be known as *Homo credulus*, as it takes humanness to worship cruel gods and trust deceitful politicians. Wild animals are not as easily duped because they are not trapped by ideology and advertising.

Besides, recent research has cast serious doubts on the existence of a particular part of the human brain in charge of sociality [9]. It seems that nearly all of our brain is social, even though one region specializes in feeling nociceptive pain (one's own), another in empathic pain, and so on. In the human brain, localization combines with coordination ([10]: 166 ff). This is why systemism, rather than either individualism or holism, is the ticket.

2 A Formula for the Types of Mental Activity

The anomalous size of the BA10 region in humans is related to the importance of internally focused processes versus externally focused ones. These differences may be compressed into the formula $M = S + B + SB + BS + BSB$, where M designates the intensity of mental activity, S that of the automatic response to external stimulus, and B that of the spontaneous controlled mental process.

The combinations of the two main kinds of process are SB (exo-endo) and BS (endo-exo). SB stands for environmentally biased mental constructions or moral deliberation, whereas BS represents action biased by intellectual or moral processes. Sensory deprivation is represented by $S = \varnothing$, whereas $B = \varnothing$ stands for the blank state. Typically, sociologically oriented social psychologists stress SB processes, whereas psychologically oriented social psychologists focus on BS processes. However, both trends tend to treat B as a black box: only those who are neuroscientifically inclined dare open the box and look for the specific neural circuits that perform the mental operations in question. The following section contains a few examples of each of the four categories.

Since neither of the three variables in question is well defined, the previous formula is so far only a mnemonic prop and heuristic device. Still, it also summarizes a whole research project: that of properly defining all three variables. In particular, B would presumably be defined in terms of such parameters as neuronal firing frequency and synaptic plasticity.

A further function of the same formula is that it encapsulates the two main classical alternatives to the current approach: $B = \varnothing$ or behaviorism and $S = \varnothing$ or the mind-over-matter (or downward causation) doctrine. (In cognitive neuroscience, downward causation means either cerebral cortex → rest of the body or society → individual.)

The former school is that of Thomas Aquinas, Hume, Condillac, Mill, Watson, Skinner, and Vygotsky, whereas the latter or internalist school is that of Plato, Augustine, Berkeley, Maine de Biran, Freud, Merleau-Ponty, Eccles, Popper, Chomsky, and Pinker. The philosophical concomitants of these trends are empiricism cum externalism and spiritualism cum internalism, respectively.

Regrettably, most historians of philosophy repeat the idea that the formula “There is nothing in the intellect that was not previously in the senses” was invented by the British empiricists, in particular Bacon, Locke, and Hume. To begin with, far from being new, said principle was held by all the schoolmen belonging to the Aristotelic-Thomistic school. Secondly, Bacon stated explicitly that, far from resembling ants, which only gather what they find, humans resemble honeybees, in that they transform into honey and wax the pollen they gather. Thirdly, Locke acknowledged that mathematical knowledge does not derive from sense impressions—which is why some Locke experts have called him a ratio-empiricist. Only Hume was a radical empiricist, as shown by his rejection of Newton’s theories, which went far beyond appearances. And, because of his monarchical and racist opinions, he lagged far behind the radical fringe of the French Enlightenment [11].

Clearly, social cognitive neuroscience fits neither of the traditional philosophical trends, for it places cognition and emotion in the brain and puts the brain in its social context. Thus, perception is sensitive to social pressure, but it occurs in the brain. Donald Hebb's classical experiments on sensory deprivation and Jean Piaget's on the constructive nature of memory, learning and thinking, support the current view of the brain as a *tabula rasa* (blank slate) at birth but thereafter as a creative organ, always ready to read, misread, or ignore external stimuli, as well as to imagine ideas of many degrees of abstraction. Anyone who has suffered hallucinations caused by a stroke will bear witness to the frightening inventiveness of a brain free from environmental controls.

It has been conjectured that each kind of mental process is performed by a neural circuitry of its own kind [12]. Automatic processes, such as unconditioned reflexes, proprioceptive sensations, tasting food, falling asleep, and waking up, would occur in neural systems whose cellular components are held together by "rigid," or rather elastic, synaptic connections, some of which are inborn. By contrast, plastic neural systems would be those where controlled processes occur—or, in the shifty parlance of the day, they would "mediate" the learned patterns.

Roughly, automatic would equal inborn, and controlled would equal learned. However, the automatic/controlled distinction is not a dichotomy, for some automatic processes are plastic. For instance, children can be trained to control their bowel movements—something that chimpanzees cannot do. Even the brainstem, a phylogenetically old brain structure, is plastic. Indeed, the optokinetic reflex, which stabilizes images in the retina as the animal navigates in its environment, can learn to adjust to drastic environmental changes, such as confinement into a cage [13].

It has become customary to say of brain structures that they *mediate* or *subserve* their specific functions, as in "brainstem neurons mediate (or subserve) innate motor behaviors." But if organ *A does B*, one should not say that *A* "mediates" *B* nor that *A* "subserves" *B*, for there are no intermediaries between structures and their functions, and the latter do not gratify their structures. Ordinarily we say that hands grasp, not that they "mediate" or "subserve" grasping. Talk of mediation or subserving in cognitive neuroscience is bad science and bad grammar, on top of an attempt to disguise dualism.

Another rather popular expression is "instantiate." Actually, the visual system sees, and the auditive one hears, just as legs do the walking and lungs the breathing. Likewise, it is wrong to say that the legs are the anatomical "correlates" of walking: legs just walk—though of course controlled by the motor centers, starting with the cerebellum. None of these parts of the body is a means to an end or goal, and none of them instantiates (exemplifies) a generalization. Straight talk is always preferable to circumlocution.

To understand a process, we must find out *what* it is and *where* it occurs. Brain imaging techniques help solve the latter problem, but to tackle the former problem, neuroscientists must deploy all the physiological and biochemical techniques elaborated since the scientific revolution, along with the biochemical ones invented since the beginning of the nineteenth century.

For example, to find out the mechanism of social isolation and thus that of social reinsertion, it has been found necessary to follow the trajectory of dopamine molecules in and out of the dorsal raphe nucleus in the brainstem—the cusp of the spinal cord [14]. In turn, the uncovering of that trajectory involves the electrophysiological techniques invented in mid-nineteenth century by Emil Du Bois-Reymond. This outspoken materialist and atheist started his scientific career studying electric fishes, a subject that most corporate-minded academic administrators would find unpromising and thus unworthy of support: they never heard about unanticipated events.

3 Random Sample of Findings

Let us list a few typical findings of social cognitive neuroscience—an exercise that should emphasize how much the *neuro* approach to the mental and behavior contributes to transforming the psycho black box into a translucid box allowing us to peek into its mechanisms.

3.1 Spontaneous Processes

Spontaneous or self-started processes are those that occur without any external inputs. Feeling a headache, dreaming, having a sudden idea, and exercising free will are familiar examples of such processes. Presumably, they are not localized, or, better, they may occur in different brain regions. Moreover, although these processes are not stimulus bound, most of them, in particular self-consciousness, pride, shame, and the wishes to succeed and to be well thought of, are likely to have been learned in the course of social interactions. In any event, they violate the stimulus-response schema, and they are not "computed" either, so they are counterexamples to both behaviorism and information-processing or computational psychology.

3.2 Automatic Processes

Raw perceptions and feelings, as well as conditioned reflexes, are the best-known examples of automatic processes. Pavlov's dogs (which salivated upon hearing a gong strike that used to accompany the delivery of food) and Skinner's study of the pigeons that danced when their seed containers were filled have been amply vulgarized but only as late as the first half of the twentieth century.

Only mechanist philosophers, from Descartes to La Mettrie, argued that all nonhuman animals are unfeeling automata. Margaret Mead claimed that Samoans do not feel any emotions, but nobody shared this extravagance. Some present-day philosophers, namely, the upholders of the computer or information-processing psychology, such as Putnam and Fodor, have adopted the same view, though replacing the mechanical automaton, such as Vaucanson's ingenious duck, with the electronic computer.

The computer view of the mental is at variance with the well-known facts that computers work only when fed algorithms, that emotions are notoriously unruly, and that social life evokes such emotions as empathy, fear, compassion, love, and hatred, all of which occur in neural networks involving the amygdala, a subcortical organ likely to have emerged much earlier than the neocortex and whose volume correlates with social network size and complexity [15]. In general, whereas the tendency for personal electronic gadgets is to miniaturize, neocortex size has been increasing with group size [16].

Novelty detection is another subject of contemporary research, which engages not only psychologists but also roboticians. Among lower animals, a novelty cue is a sign of danger, hence, a source of fear and a warning to flee or freeze. In contrast, among humans and other higher vertebrates, novelty provokes curiosity as well as caution, and sometimes it motivates investigation, which may garner new knowledge.

It is currently believed that the hippocampus is the main human novelty detector. In any case, we are getting close to learning why some animals are neophilic, whereas others are neophobic. The eventual impact of this research on political psychology should be obvious, as long as we do not forget that vested interests, which escape neuroscience, contribute powerfully to shaping political attitudes. This is why, it would be foolish to engage in neuropolitics.

3.3 Controlled Processes

Imitation is one of the best-studied mental processes since Gabriel Tarde's once-popular book on groupthink (1890). Imitation research received a sudden boost when Giacomo Rizzolatti et al. [17], at Parma University, discovered the mirror neurons in the macaque's inferior parietal cortex. Since then, similar neuronal systems have been located in humans and in some birds. These studies have confirmed the hypothesis that simians and other species possess a "theory of mind"—the name that David Premack and Guy Woodruff [18] gave the ability of imputing feelings and beliefs to others. A rough equivalent is the *Verstehen* or empathic understanding that Wilhelm Dilthey imputed to the students of social matters.

Of course, it is wrong to call *theory* the capacity to understand the minds of others. So far, it is only an ability waiting for a theory. And it is not obvious that a synonym for "empathy" is really needed. What is clear is that the ability in question is not mind reading but "reading," or rather interpreting, outward or behavioral *indicators* of mental processes. It is obvious that such studies have not only enriched animal psychology but have also enhanced our respect for monkeys and domestic dogs, as well as our admiration for Darwin's attribution of empathy to nonhuman animals.

Further investigations of the neural sources of empathy using functional magnetic resonance imaging have revealed the participation of much more than the mirror neuron system. Indeed, the spontaneous, intuitive, or preanalytic understanding minds of others are so important in social transactions that in humans it engages multiple brain systems [19]. Moreover, empathy is so strong in monkeys that they can go hungry to prevent electric shocks to conspecifics. Human infants, too, give signs of distress when they see or hear other babies in distress. The level of distress decreases with age, but as compensation, the readiness to help other children increases. The school bully inspires fear but has no friends.

Granted, Stanley Milgram's sensational experiments [20] seemed to show that we all enjoy watching others being tortured. However, this was not Milgram's point: what he showed is that fear of authority can trump fellow feeling. A similar point was made by the hugely successful German play and film "The Captain from Köpenick" (1931, 1956), where a humble shoemaker masquerades as a Prussian officer and, as he marches through a small town, gathers a growing following who take over City Hall and "confiscate" the money in its coffer. The idea is of course that the good German citizen at the beginning of the twentieth century was eager to obey military orders. Could anything similar happen today in Washington DC?

Finally, let us not forget that the $B \rightarrow S$ process can be exaggerated to the point of delusions of grandeur. Berkeley's formula "To be is to be perceived" is a case in point, for it enslaves the entire world to the perceiving subject. Social constructionism is a recent version of that delusion, for it views everything social as a product of the ego [21]. For instance, according to the "social model of disability," the latter is "wholly and exclusively social" [22].

Thus, even quadriplegia and Down syndrome would be only in our minds; and, being social constructions rather than medical conditions, the remedy for them would be a radical transformation of society—that is, waiting for that to happen rather than trying to help right now, for instance, through brain prostheses translating thoughts into actions or teaching manual skills to retarded youngsters. Fortunately, Anastasiou and Kauffman [23] have disabled that defeatist offshoot of social constructionism.

3.4 Exo-Endo Processes

Religiosity is a classical case of the *SB* kind. Lucien Lévy-Bruhl's 1922 bestseller [24] was the earliest if failed attempt to characterize what he called "primitive mentality" and to explain it as an adaptation to the imagined environment of our remote ancestors. The speculations of the recent evolutionary psychologists have been much more detailed and were ridiculed by Telmo Pievani [25].

Surprisingly, the first scientific investigation of the religiosity-societal health correlation in the prosperous democracies was published only recently [26]. It found that higher rates of religiosity "correlate with higher rates of homicide, juvenile and early adult mortality, STD infection rates, teen pregnancy, and abortion rates […]

the United States is almost always the most dysfunctional of the developed democracies, sometimes spectacularly so, and almost always scores poorly," while at the same time, it is the most religious and also the most inclined to reject evolutionary biology and other scientific achievements. In contrast, Japan, Scandinavia, and France are the most secular nations in the West and at the same time some of the least unequal.

Data-driven research is another instance of a thought process initiated by a striking observation of the environment—that is, one that clashes with received wisdom or just fills a gap in our body of knowledge. The end product of this process is also known as a chance discovery or lucky finding.

Actually there is an element of luck, good or bad, even in the most carefully designed observation or experiment, as we are always bound to miss some variable or other. In addition, it is well to keep in mind Louis Pasteur's wise remark that "Chance favors only the prepared mind." For example, the ancient Chinese astronomers-astrologers observed and admired Eta Carinae, this extremely brilliant variable star, but only recently has it been learned that it is actually composed of two stars with a total mass of about 100 solar masses and that its colossal explosions result from nuclear reactions in their interior.

3.5 Endo-Exo Processes

All free rational choices and decisions, as well as the resulting actions, are spontaneous or self-initiated processes in the prefrontal cortex. One of the most familiar experiences of this kind is free will, that is, volition not controlled by external stimuli, as when, after careful consideration, we follow a course of action congruent with our moral principles, even if we realize that it is likely to harm us.

Hypothesis-led scientific research belongs to this category. Indeed, the projects of this kind are backed not only by the usual philosophical presuppositions, such as realism and intelligibility, but also by specific guesses, such as the possible binding of the molecule being investigated with special receptors on the membrane of the target cell. Such assumed specificity guides the research, which is then anything but an erratic trial-and-error search.

Presumably, concussions, strokes, and other severe brain lesions, as well as deficits in neurotransmitters due to malnutrition or excessive alcohol consumption, translate into abnormal mental or behavioral symptoms, from apathy and recent memory deficits to poor scholastic achievement and disastrous political policies and actions. A pioneering investigation of the strong negative correlation between malnutrition and cortex thickness, and the corresponding poor scholastic achievement of Mexican children [27], was revealing yet still hardly known to the international developmental psychology community. It is recalled here mainly to emphasize the usefulness of science in the detection of social issues and the elaboration of social policies to resolve them [28].

3.6 Exo-Endo-Exo Processes

In addition to the unidirectional processes listed above, we have loops of the *SBS* type. An obvious case of this kind is the *hurrah* shout expressing the joy felt when watching a goal made by our soccer team. Its dual, *schadenfreude*, is socially and morally very different from healthy joy, but presumably it engages the same neural systems in addition to the vocal one.

Another familiar case of an *SBS* loop is the so-called Thomas theorem. This is summed up in the formula "People do not react to facts but to the way they perceive facts." For example, we often buy merchandise or vote for politicians, whose "image" has been manufactured by publicity agencies that have embellished the product in question. In other words, some of our actions are driven by false beliefs, and sometimes we guess that others, too, are fooled in like manner.

Until recently, it was thought that this ability to impute others' false beliefs is specific to humans and, moreover, one that emerges only after the fourth year. Recent work by Michael Tomasello's team [29] suggests that younger infants, as well as three species of great apes, can anticipate that conspecifics will act according to false beliefs. This finding also suggests that the distinction between truth and falsity is several million years old rather than a recent philosophical invention. So much for Nietzsche's brutal slogan "Let life be and truth perish."

Lastly, let us peek at the popular belief that "chronic raiding and feuding characterize life in a state of nature" [30]. This opinion, first voiced four centuries ago by Thomas Hobbes and popularized in recent years by armchair evolutionary psychologists, has been challenged by anthropologists such as Fry and Söderberg [31], who studied 21 mobile forager bands distributed among four continents. They conclude that "most incidents of lethal aggression can aptly be called homicides, a few others feud, and only a minority warfare."

4 Conclusion

In conclusion, the simian brain is highly social, and some regions of it are more susceptible than others to social stimuli. But, as Hebb's sensory deprivation experiments showed in the 1950s, the waking brain acts spontaneously all the time even in the absence of external stimulation, though it tends to hallucinate. The normal brain interacts with its immediate environment as well as with the rest of the organism. This finding suggests that psychology is a biosociological science rather than either a chapter of zoology, as biologism has it, or a purely social science, as sociologism imagined. In fact, the recent trend in psychology is toward merger or convergence rather than toward independence, let alone reduction [32].

Acknowledgments I thank Agustín Ibáñez (INECO) for his critical remarks, as well as Verónica Bunge Vivier (Ecología, Conaycit, México DF), Iris Mauss (Psychology, UCA, Berkeley), and Ignacio Morgado (Neurobiology, Universidad Autónoma de Barcelona), for helping with the references.

References

1. Aronson E. The social animal. New York: Worth/Freeman; 2011.
2. Trigger BG. Sociocultural evolution. Malden: Blackwell; 1998.
3. Richerson PJ, Boyd R. Not by genes alone: how culture transformed human evolution. Chicago: University of Chicago Press; 2005.
4. Bowles S, Gintis H. A cooperative species. Princeton University Press: Princeton; 2011.
5. Marmot MG, Rose G, Shipley MJ, Hamilton PJ. Employment grade and coronary heart disease in British civil servants. J Epidemiol Community Health. 1978;32:244–9.
6. Wilkinson R, Pickett K. The spirit level: why equality is better for everyone. London: Penguin Books; 2010.
7. Ibáñez A, Hesse E, Manes M, García AM. Freeing free will: a neuroscientific perspective. Appendix 1. In: Bunge M, editor. Doing science in the light of philosophy. Singapore: World Scientific Publications; 2017.
8. Lieberman MD. Social cognitive neuroscience: a review of core processes. Annu Rev Psychol. 2007;58:259–89.
9. Singer T. The past, present and future of social neuroscience: a European perspective. NeuroImage. 2012;61:437–49.
10. Bunge M. Matter and mind (Boston studies in the philosophy of science). New York: Springer; 2010.
11. Israel J. Revolutionary ideas: an intellectual history of the French revolution. Princeton: Princeton University Press; 2014.
12. Bunge M. The mind-body problem. Oxford: Pergamon; 1980.
13. Liu B, Huberman AD, Scanziani M. Cortico-fugal output from visual cortex promotes plasticity of innate motor behavior. Nature. 2017;538:383–7.
14. Matthews GA, Nieh EH, Vander Weele CM, Wildes CP, Ungless MA, Tye KM. Dorsal raphe dopamine neurons represent the experience of social isolation. Cell. 2016;164(4):617–31.
15. Bickart KC, Wright CI, Dautoff RJ, Dickerson BC, Feldman Barrett L. Amygdala volume and social network size in humans. Nat Neurosci. 2011;14:163–6.
16. Dunbar RI. The social brain hypothesis and its implications for social evolution. Ann Hum Biol. 2009;36:562–72.
17. Rizzolatti G, Craighero L. The mirror-neuron system. Annu Rev Neurosci. 2004;27:169–92.
18. Premack D, Woodruff G. Does the chimpanzee have a theory of mind? Behav Brain Sci. 1978; 1(4):515–26.
19. Preston SD, de Waal FBM. Empathy: its ultimate and proximate bases. Behav Brain Sci. 2002; 25:1–71.
20. Milgram S. Behavioral study of obedience. J Abnorm Soc Psychol. 1963;67:371–8.
21. Bunge M. The sociology-philosophy connection. Brunswick: Transaction Publishers; 1999. Reissued in paperback, 2012
22. Oliver M. Understanding disability. Basingstoke: Macmillan; 1996.
23. Anastasiou D, Kauffman JM. The social model of disability. J Med Philos. 2013;38:441–59.
24. Lévy-Bruhl L. La mentalité primitive. Paris: Flammarion; 2010.
25. Pievani T. Evoluti e abbandonati. Torino: Einaudi; 2014.
26. Paul GS. Cross-national correlations of quantifiable societal health with popular religiosity and secularism in the prosperous democracies. J Religion Soc. 2005;7:1–17.
27. Cravioto J, Delecardie ER, Birch HG. Nutrition, growth and neurointegrative development. Pediatrics. 1966;38:319–72.
28. Navarro V, Muntaner C, editors. The financial and economic crises and their impact on health and social well-being. Amityville: Baywood; 2014.
29. Krupeneye C, Kano F, Hirata S, Call J, Tomasello M. Great apes anticipate that other individuals will act according to false beliefs. Science. 2016;354:110–3.
30. Pinker S. The better angels of our nature. New York: Penguin; 2011.
31. Fry DP, Söderberg P. Lethal aggression in mobile forager bands and implications for the origins of war. Science. 2013;341:270–2.
32. Bunge M. Emergence and convergence. Toronto: University of Toronto Press; 2003.

How Is Our Self Related to Its Brain? Neurophilosophical Concepts

Georg Northoff

Abstract The present chapter aims to target yet another central feature of the mind: the self as the subject of all our experience and hence of consciousness. More specifically, the focus is on different concepts of the self and how they are related to recent findings about neural mechanisms related to the self-reference of stimuli. I first introduce different basic concepts of the self as they are currently discussed in philosophy. The first concept of self is the self as mental substance, which was introduced originally by Descartes. This is rejected by current and more empirically oriented concepts of the self where the idea of a mental substance is replaced by assuming specific self-representational capacities. These self-representational capacities represent the body's and brain's physical, neuronal states in a summarized, coordinated, and integrated way. As such, the self-representational concept of the self must be distinguished from the phenomenological concept of self that is supposed to be an integral part of the experience and thus of consciousness. This phenomenal self resurfaces in the current debate as the "minimal self"—a basic sense of self in our experience that is supposed to be closely related to both the brain and body. Current neuroscience investigates the spatial and temporal neural mechanisms underlying those stimuli that are closely related to the self when compared to the stimuli that show no relation or reference to the self. This is described as the self-reference effect. When comparing self- versus non-self-specific stimuli, neural activity in the middle regions of the brain, the so-called cortical midline structures, is increased. Moreover, increased neuronal synchronization in the gamma frequency domain can be observed. The question is how specific these findings are for the concept of self as discussed in philosophy. Neuronal specificity describes the specific and exclusive association of the midline regions with the self. This is not the case since the same regions are also associated with a variety of other functions. This goes along with the quest for the psychological and experimental specificity of psychological functions and experimental paradigms and measures used to test for the self. One may also raise the issue of phenomenal specificity: the concept of phenomenal specificity refers to whether the phenomenal features of the self, that is, mineness and belongingness, are distinguished from other phenomenal features like

G. Northoff (✉)
Mind, Brain Imaging and Neuroethics Research Unit, Institute of Mental Health Research, Royal Ottawa Mental Health Centre, 1145 Carling Avenue, Ottawa, ON, Canada
e-mail: Georg.Northoff@theroyal.ca

© Springer International Publishing Switzerland 2017

A. Ibáñez et al. (eds.), *Neuroscience and Social Science*,
DOI 10.1007/978-3-319-68421-5_19

intentionality or qualia. Finally, one may discuss the question of conceptual specificity that targets the distinction between the concepts of self-reference and self.

Keywords Mental self • Empirical self • Phenomenal self • Minimal self • Cortical midline structures

1 Concept of Self

You read these lines. You are winning a game of tennis, while your girlfriend is watching. You feel pride. Who experiences that pride? You. You are the subject of the experience of boredom. Without you as subject of this experience, you could not experience anything at all, not even boredom. This subject of experience has been described as the "self." Your "self" makes it possible for you to experience things. In other words, it is a necessary condition for experience and thus also for consciousness. It is clear, therefore, that there is much at stake when it comes to the self.

The concept of self has been subject to intense philosophical discussion over the centuries. Different philosophers have suggested different concepts of self. Because of time and space constraints, here we will only focus on those that are relevant in the attempt to map the interface between philosophical and neuroscientific accounts of the self.

There are four main different concepts of self discussed in current philosophy. First is "the mental self," which is based on our thoughts and a specific mental substance. Second is the "empirical self"—this concept of the self represents and reflects the biological processes in one's body and brain. Third is the notion of the "phenomenal self," which gives rise to our experience in consciousness. Our consciousness is accompanied by an awareness of our self, referred to as pre-reflective self-awareness or phenomenal self. Finally, and most recently, philosophers speak of a "minimal self," which emphasizes more on the objective-biological nature of self. This concept of the self is based on our body and its physiological processes. In this chapter, I will discuss each of these different concepts and how they relate to the brain.

Before we do this, I have to shed some light on several related concepts. We experience our self in daily life during, for example, the act of perceiving certain objects, persons, or events in our environment. While making a list of all the things you have to do today, you experience not only the act of thinking and writing but an awareness and experience of your own self. Hence, your self as the very subject of experience seems to be part of that experience. In other words, your self is a content of your consciousness. This is described as self-consciousness. However, there is more to the self than the self itself and our experience of it in self-consciousness. You wake up every morning. Every day. Every week, every year. Your body changes. You become older. You get wrinkled and your hair turns white. Despite all these bodily changes, you nevertheless have the feeling that you are the same self. You still experience your self as being the same self of 20 years ago.

You are one and the same person. There is thus a temporal dimension to your self that seems to be coherent and persistent over time. You and your self are continuous across time. The temporal dimension of your self has consequently been discussed under the umbrella of what is called "personal identity" in philosophy. While our discussion will touch upon the temporal dimension of the self and thus upon personal identity, we will not explicitly discuss it.

In a world of over seven billion people, there are many, many selves: you, your friends, your family, etc. Most interestingly, you can relate to them—you can communicate with other selves and sometimes even feel their emotions as in, for instance, the grief someone might feel when they lose a loved one. Or you might experience pain when your boyfriend's arm is broken. How is this possible? In philosophy, this is called "intersubjectivity." Finally, your self is not isolated from the rest of the world. You can share others' experiences and feel connected to the world. The world and its specific objects, persons, and events have meaning to you—you can relate to it more or less and can appropriate it for your own self. How is such basic integration of your own self within the world possible? And how is that related to your brain and its neuronal mechanisms? That shall be the focus in the following section.

2 Empirical Investigation of Self in Neuroscience

How can we investigate the self? In order to experimentally address the self, we need some quantifiable and objective measures that can be observed from third-person perspective. How can we obtain such measures? Psychologists focusing on memory observed that items related to ourselves were better remembered than those unrelated (see [1]). For example, as a resident of Ottawa, I recall the recent thunderstorm that wiped away several houses locally much better than a person who, perhaps living in Germany, just heard about it in the news.

There is thus superiority in the recollections of those items and stimuli that are related to one's self. This is described as the self-reference effect (SRE). The SRE has been well validated in several psychological studies. Most interestingly, it has been shown to operate in different domains, not only in respect to memory but also in relation to emotions, sensorimotor functions, faces, words, etc. In all these different domains (see below for details), stimuli related to one's own self, known as self-specific stimuli, are recalled much better than those that are unrelated to one's own self, known as non-self-specific stimuli.

How is the SRE possible? Numerous investigations (see, e.g., [2, 3] for summaries) show that the SRE is mediated by different psychological functions. These range from personal memories including autobiographical memories over memories of facts (semantic memories) to those cognitive capacities that allow for self-reflection and self-representation. Hence, the SRE is by itself not a unitary function, but rather a complex multifaceted psychological composite of functions and processes.

How can we link the SRE to the brain? Before the introduction of functional imaging techniques such as fMRI at the beginning of the 1990s, most studies conducted focused on the effect of dysfunction or lesions in specific brain regions caused by brain tumors or stroke. These revealed that lesions in medial temporal regions that are central in memory recall, such as the hippocampus, change and ultimately abolish the SRE effect.

With the introduction of brain imaging techniques such as fMRI, we could then transfer the experimental paradigms of comparing self- and non-self-specific stimuli to the scanner and investigate the underlying brain regions. The basic premise here is that if self-specific stimuli are recalled better than non-self-specific stimuli, they must be processed by the brain in a different way. This might be, for instance, by higher degrees of neural activity and/or different regions. This led to the investigation of numerous experimental designs of SRE-like paradigms in the fMRI scanner. For example, subjects were presented trait adjectives that were either related to themselves (such as, for me, my hometown, Ottawa) or as opposed to (Sydney, an unrelated city for me). In other tests, subjects were presented with images of the own face, and these were compared with faces of other people. Also autobiographical events from the subject's past were compared with those from other people. One's own movements and actions could also be compared with those of other people, implying what is called ownership (e.g., my movements) and agency ('I myself caused that action').

The stimuli belonged to different domains such as memory, faces, emotions, verbal, spatial, motor, or social. Most of the stimuli were presented either visually or auditorily, and the presentation of these stimuli was usually accompanied by an online judgment about whether the stimuli are related and personally meaningful or not to the research subject.

On the whole, we can see that current neuroscience can investigate the self in various experimental ways using mainly functional brain imaging. However, any empirical research relies on certain presuppositions. This also holds true for current neuroscientific research on the self, which aims to reveal the neuronal mechanisms underlying our experience or sense of self. However, before examining the neuroscientific findings, we need to briefly shed some light on the concept of the self and how it has been defined in philosophical discussions.

3 Philosophical Concepts of Self

3.1 Mental Self

What is the self? What must it look like in order to presuppose experience and be the subject of our experience? The self has often been viewed as a specific "thing." Stones are things, and the table on which your laptop stands is a thing. And in the same way, the table makes it possible for the laptop to stand on it, and the self may be a thing that makes experience and consciousness possible. In other words, metaphorically speaking, experience and consciousness stand on the shoulders of the self.

However, another question is whether the self is a thing or, as philosophers such as Rene Descartes suggest, a substance? A substance is a specific entity or material that serves as a basis for something like a self. For instance, the body can be considered a physical substance, while the self can be associated with a mental substance.

Is our self real and thus does it exist? Or is it just an illusion? Let us compare the situation to perception. When we perceive something in our environment, we sometimes perceive it not a real thing but an illusion that in reality does not exist. The question of what exists and is real is what philosophers call a metaphysical question. Earlier philosophers, such as Rene Descartes, assumed that the self is real and exists.

However, Descartes also assumed that the self is different from the body. Hence, self and body exist but differ in their existence and reality. Thus, from this perspective, the self cannot be a physical substance and is a mental substance instead. It is a fcaturc not of thc body but of thc mind.

However, the characterization of the self as a mental entity has been questioned. For example, Scottish philosopher David Hume argued that there is no self as a mental entity. There is only a complex set or "bundle" of perceptions of interrelated events that reflect the world in its entirety. There is no additional self in the world; instead, there is nothing but the events we perceive. Everything else, such as the assumption of a self as mental entity, is an illusion. The self as mental entity and thus as a mental substance does not exist and is therefore not real.

To reject the idea of self as mental substance and to dismiss it as mere illusion are currently popular. One major proponent of this view today is German philosopher Thomas Metzinger [4]. In a nutshell, he argues that through our experience, we develop models of the self, the so-called self-models. These self-models are nothing but information processes in our brain. However, since we do not have direct access to these neuronal processes (e.g., all those processes and activities of the cells, neurons, in the brain), we tend to assume the presence of an entity that must underlie our own self-model. This entity is then characterized as the self (Fig. 1a).

According to Metzinger, the assumption of the self as a mental entity results from an erroneous inference from our experience. We cannot experience the neuronal processes in our brain as such. Nobody has ever experienced their own brain and its neuronal processes. Therefore, the outcome of our brain's neuronal processes, the self, cannot be traced back to its original basis, the brain, in our experience.

Where then does the self come from? We assume that it must be traced back to a special instance different from the brain. This leads us to assume that the mind and the self are mental entities rather than a physical, neuronal entities originating in the brain itself. Metzinger argues that the self as a mental entity simply does not exist. Therefore, Metzinger [4] concludes, selves do not really exist, hence, the title of his book *Being No One*.

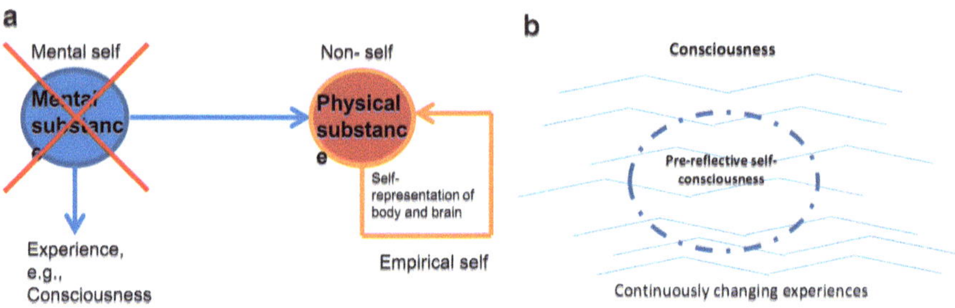

Fig. 1 The figure schematically illustrates different concepts of self, the self as mental substance (**a**) and the phenomenal self (**b**). (**a**) The self is determined as mental substance (left) that is distinguished from the body (and brain) as mere physical substance (right). Thereby the self as mental self controls and directs the body following the earlier French philosopher Descartes. This is denied in current empirical approaches to the self (e.g., vertical red lines). They reject the notion of the self as mental substance and claim that such mental self does not exist. All there is the body as physical substance with the brain allowing for the representation of both body and brain in the brain's neural activity. Such self-representation may then amount to what can be described as empirical self. (**b**) The phenomenal self no longer claims to be outside and prior to any experience. Instead, the phenomenal self is supposed to be "located" or part of the experience itself in the gestalt of pre-reflective self-consciousness. This is indicated by the insertion of the circle within the midst of the experience, e.g., consciousness, itself

3.2 *From the Metaphysical to the Empirical Self*

What is the self if not a mental entity? Current authors, such as Metzinger [4] and Churchland [5], argue that the self as mental substance or entity does not exist. How do we come up with the idea of a self or the self-model as Metzinger calls it? The model of our own self is based on summarizing, integrating, and coordinating all the information from our own body and own brain.

What does such integration look like? Take all that information together, coordinate and integrate it, and then you have a self-model of your own brain and body and their respective processes. In more technical terms, our own brain and body are represented in the neuronal activity of the brain. Such representation of the own brain and body amounts then to a model of your self. The self-model is therefore nothing but an inner model of the integrated and summarized version of your own brain and body's information processing. The self is thus a mere model of one's own body's and brain's processes.

The original mental self, the self as mental substance or entity, is in this line of thinking replaced by a self-model. This implies a shift from a metaphysical discussion of the existence and reality of self to the processes that underlie the representation of body and brain as a self-model. Since this representation is based on the coordination and integration of the various ongoing processes in the brain and body, it is associated with specific higher-order cognitive functions such as working memory, attention, executive function, and memory, among others.

What does this imply for the characterization of the self (presupposing a broader concept of self beyond the self as mental substance)? The self is no longer characterized as a mental substance but as a cognitive function. Methodologically, this implies that the self should be investigated empirically rather than metaphysically.

We therefore need to search for the cognitive processes underlying the special self-representation. The self is consequently no longer an issue of philosophy, but rather one of cognitive psychology and ultimately of cognitive neuroscience. According to this model, the self is no longer a metaphysical matter but a possible subject of empirical investigation.

3.2.1 Phenomenal Self

One of the problems one encounters is that such substance or meta-representation cannot be experienced as such. Nobody ever experienced a mental substance or a meta-representation in consciousness. We are not conscious of any such substance or meta-representation. Therefore, instead of speculating about something that lies beyond the scope of our experience, why not start with experience itself and thus with consciousness? Rather than looking at what lies "outside" our consciousness, like a substance or meta-representation, the self may be found within that very consciousness itself.

However, this localization is denied in phenomenological philosophy precisely because it focuses on consciousness itself and what lies "inside" our experience. More specifically, phenomenological philosophy is interested in investigating the structure and organization of our experience and thus of consciousness. It focuses on how our experience is structured and organized and reveals phenomenal features as we experience them from the first-person perspective.

How does the phenomenal approach determine the self? The concept of "phenomenal self" emphasizes the subjective-experiential nature of self as in the term "phenomenal"—the concept thus refers to what can be described our "sense of self." The subjective-experiential nature of the phenomenal self distinguishes it from a more objective meaning of self as it can be observed—i.e., objective-observational, as it is described by the term "minimal self" (see below).

How is the self related to experience in general? Currently, it is argued that the self is an integral part of experience itself [6]. The self is always present and manifests in the phenomenal features of our experience such as intentionality (e.g., the directedness of our consciousness toward specific contents), qualia (e.g., the qualitative character of our experience, what it is like), etc. Without these features, the self would remain impossible.

Consequently, phenomenological philosophers such as Zahavi [7] consider the self to be an inherent part of consciousness itself. Here, the self is supposed to be always already accompanied by some kind of consciousness of the external world, even if we are not aware of the self being part of that experience. Phenomenological philosophers therefore speak of what they call pre-reflective self-awareness (or pre-reflective self-consciousness).

The concept of pre-reflective self-consciousness contains two main terms, "pre-reflective" and "self-consciousness." "Pre-reflective" means before or prior to reflection and, as indicated by the prefix "pre," means that consciousness in this sense does not yet involve reflection. This is, for instance, well reflected in the recent distinction between report and no-report paradigms in consciousness research [8]. No-report paradigms are those paradigms where the subject does not need to give a response about its awareness of the stimulus—however, the subject may have nevertheless perceived the stimulus in a conscious way without any reflection yet, i.e., in a pre-reflective way. In contrast, report paradigms require the subject to reflect upon the perceived stimulus thus involving reflection. I suggest that we experience our self in a pre-reflective way—it is already there in our experience even if we do not reflect upon. The self is thus pre-reflective. It is simultaneously an inherent part of our experience and thus of our consciousness. The self is consequently no longer outside of our consciousness but an integral part of it, hence the second term, "self-consciousness." Such an approach suggests an intimate and intrinsic link between self and consciousness (Fig. 1b).

Characterizing the self in terms of self-consciousness implies a significant shift. The self is no longer metaphysical as Descartes proposes. Nor is it empirical as advocated by Hume and others such as Metzinger and Churchland. Instead, the self is part of experience and of consciousness itself and can therefore be characterized as the "phenomenal self." Such a phenomenal self is open to systematic investigation of the phenomenal features of our experience, which would complement the metaphysical, empirical, and logical approaches to the self (see Aristegui, this volume).

3.2.2 Minimal Self

How can we describe the pre-reflective self-consciousness in more detail? It is always already there in every experience so that we cannot avoid it or separate it from the experience. The self is always present in our consciousness and thus in our subjective experience. Even if we do not focus on the self as such, we cannot avoid or remove its presence. Hence, the term pre-reflective self-conscious describes an implicit or tacit experience of our self in our consciousness.

Since the self as pre-reflectively experienced is the basis of all phenomenal features of our experience, it must be considered as essential for any subsequent cognitive activity. Such a basic and fundamental self occurs in our experience before any reflection. For instance, when reading the lines of this book, you experience the contents, and in addition, you also experience your self as reading these lines.

Hence, your immediate experience and consciousness come with both the content and your own self, since the experience of such self occurs prior to any reflection and recruitment of higher-order cognitive functions. This is why this concept of self is sort of a minimal version of the self. Current phenomenological philosophers such as Gallagher [9] or Zahavi [7] speak therefore of a "minimal self" when referring to the self as implicitly, tacitly, and immediately experienced in consciousness.

How can we describe the concept of the "minimal self"? The minimal self refers to a basic form of self that is part of any experience. As such, it is not extended across time like it is in the experience of the self as a continuity across time in personal identity. Instead the minimal self describes a basic sense of self at any particular given moment in time but does not yet provide a link between different moments in time and thus continuity across time. Taken in this sense, the minimal sense is not principally different from the "phenomenal self"—it describes the most basic building block of the phenomenal self: the phenomenal self can thus be conceived as the outcome or result of the basic processes provided by the minimal self. The minimal self, as perceived in this sense, is thus much closer to an objective concept of self that provides those mechanisms and processes in the brain which we access by our observation. The concept of minimal self is thus much closer to an objective-observational notion of self than the phenomenal self which is much more subjective-experiential [10, 11]. However, it is clear that there is no contradiction between the objective-observational and subjective-experiential notions of self—rather there is a continuous transition with the minimal self providing the mechanisms and processes that lead to the phenomenal self as outcome.

How can such continuity across time be constituted? Cognitive functions such as memories and autobiographical memories in particular may be central. In this model, the self may become more complex. One might speak of a cognitive, extended, or autobiographical self, as does, for example, Portuguese-American neuroscientist Damasio (see, e.g., [12, 13]).

Another important feature of the minimal self is that although we experience it, we may not be aware of it as such. This means that we might not be able to reflect upon it in order to gain knowledge of it. We are, to put it in technical terms, only pre-reflectively aware of the minimal self. In contrast to such pre-reflective awareness, there is no reflective awareness of the minimal self. How can we become reflectively aware of the minimal self? For that to be possible, the different moments or points in time need to be integrated and, as philosophers say, represented. For such representation to occur, cognitive functions are needed which make it possible to put and link together the different time points.

Finally, the minimal self may also occur prior to verbalization and thus linguistic expression. Rather than being tied to specific linguistic concepts, as is the case with more cognitive concepts of the self, the minimal self must be considered prelinguistic. It is an experience, a sense of self that can barely be put into concepts. We can experience it as self but are not really able to describe these experiences in terms of concepts and thus articulate them in a linguistic way.

Thus, the minimal self is prelinguistic and preconceptual and will therefore, speculatively, not be affected by second-language acquisition. It is the kind of experience, an implicit sense of self, which most likely subjects will take with them as more or less stable when moving to a new country where they have to acquire a new language. However, at the same time, the minimal self provides the essential basis upon which more cognitive forms of self are developed. These are then central and instrumental in providing the ability to learn a second language.

3.2.3 Social Self

How does the self interact with other selves? So far we described the self in an isolated and purely intra-individual way. However, in daily life, the self is not isolated from others but always related to other selves. This is called interindividualism rather than intra-individualism. This raises questions about what is described as the "problem of other minds" or, more generally, questions concerning intersubjectivity. Here we will give a brief description of the problem of intersubjectivity.

How can we make the assumption of attributing mental states and thus self and mind to other people? Philosophy has long relied on what is called the "inference by analogy." What is the "inference by analogy"? "Inference by analogy" goes like this. We observe person A to show the behavior of type X. And we know that in our own case the same behavior X goes along with the mental state type M. Since our own behavior and that of the person A are similar, we assume the other person A to show the same mental state type M we experience when exhibiting behavior X.

What kind of inference do we draw here? There is similar or analogous behavior between ourselves and the other person. In addition, my own behavior is associated with a particular mental state. Since now the other person shows the same behavior, I infer that she also show the same mental state as it is associated with my own behavior. Hence, by indirect inference and analogy via our own case, we claim to obtain knowledge of the other person's mental state. How can we make such inference? We may make it on the basis of our own mental states and their associated behavior. And what we do may also hold true for the other person who in the same way attributes mental states to us by inferring them from the comparison between our behavior and their own mental states.

Why do we make such inferences? Because it seems to be the easiest and best way for us to explain the other people's behavior. The assumption of mental states thus seems to be the best explanation for your behavior. The "inference by analogy" may thus be considered an inference to the best possible explanation.

The inference by analogy describes intersubjectivity in a very cognitive and ultimately linguistic way when attributing mental states and a self to other persons. There might be, however, a deeper level of intersubjectivity. We also feel the other persons' mental states when sharing the emotional pain one's spouse experiences when her father died. Such sharing of feeling is described as empathy and sheds light on a deeper precognitive and preverbal dimension of intersubjectivity. This has been emphasized especially in phenomenological philosophy (see, for instance, Metzinger [4]).

However, both empathy and the attribution of mental states to another person are puzzling: despite the fact that we do not experience the other's mental states and consciousness, we nevertheless either share them (as in empathy) or infer them (as in inference by analogy). We have no direct access to other persons' experience of a self and its mental states in first-person perspective and nevertheless share their mental states and assume that they have a self. How is that possible?

This is where we need to introduce yet another perspective. There is first-person perspective—tied to the self itself and its experience or consciousness of objects, events, or persons in the environment. Then there is the third-person perspective—this perspective allows us to observe the objects, events, or persons in the environment from the outside, rather than from the inside. The picture is not complete.

What is the second-person perspective? The second-person perspective has initially been associated in philosophy with the introspection of one's own mental states. Rather than actually experiencing one's own mental states in first-person perspective, the second-person perspective makes possible to reflect and introspect about one's own mental states. An example of this is when you ask yourself whether the voice you heard was really the voice of your good friend (see also [14]).

The second-person perspective thus allows us to put the contents of our consciousness as experienced in first-person perspective into a wider context, the context of oneself as related to the environment. In other words, the second-person perspective makes it possible to situate and integrate the purely intra-individual self with its first-person perspective into a social context. This transforms the intra-individual self into an interindividual self. Another way of thinking of second-person perspective is to call this concept of the self the "social self."

How can we define the concept of the social self? The concept of the social self describes the linkage and integration of the self into the social context of other selves. This shifts the focus from experience or consciousness in the first-person perspective to the various kinds of interactions between different selves as associated with the second-person perspective. As we already indicated, there may be different kinds of social interactions including affective precognitive and more cognitive ones that involve meta-representation as described above.

4 Neuroscientific Account of the Self

4.1 Spatial Patterns of Neural Activity During Self-Specific Stimuli

How can we relate the various philosophical concepts of the self to the neuroscientific findings of self-reference? Above, we discussed that psychology, and later neuroscience, quantified the self in terms of the self-reference effect (SRE). The SRE describes the different impacts of self-referential and non-self-referential stimuli on psychological (e.g., reaction time, recall, etc.; see above) and neural (e.g.., degree of activity, regions; see below) measures. Below we want to briefly highlight some of the main findings of recent imaging studies on the self-reference effect.

What results did the various imaging studies yield in the fMRI? Two different kinds of regions showed up. First, one could see that the regions specific for the respective domains like emotions or faces were recruited. For instance, there is a region in the back of the brain that processes specifically faces (as distinguished from, say, houses); this is called the fusiform face area. This region is obviously active during the presentation of faces, no matter whether it is one's own face or another person's face.

Importantly, clear differences between self- and non-self-specific stimuli could not be observed in these domain-specific regions in most studies (see [1]).

What about other regions that are not specific to particular domains (also known as domain-independent regions) involved in the neural processing of the self? Meta-analyses of the various studies demonstrated the involvement of a particular set of regions in the middle of the brain. These regions include the perigenual anterior cingulate cortex (PACC), the ventro- and dorsomedial prefrontal cortex (vmPFC, DMPFC), the supragenual anterior cingulate cortex (SACC), the posterior cingulate cortex (PCC), and the precuneus. Since they are all located in the midline of the brain, they have been coined "cortical midline structures" (CMS).

The self-specific stimuli—those that were personally relevant for the subjects—induced higher neural activity in these regions than non-self-specific stimuli or those that remained irrelevant and unrelated to the person. This was observed in the various domains for faces, trait adjectives, movements/actions, memories, and social communication. Therefore, the CMS seem to show a special significance to the self and self-reference.

However, there is also some differentiation within the CMS. The self-specific stimuli may be presented in different ways to the subject in the scanner. If subjects have to make judgments requiring cognitive involvement, the dorsal and posterior regions such as the SACC, DMPFC, and PCC are recruited to a stronger degree. If, in contrast, stimuli are merely perceived without any judgment and thus without any cognitive component, the ventral and anterior regions such as the vmPFC and PACC were highly involved (Fig. 2a, b).

This led to the assumption that the different regions mediate different aspects of self-reference. The ventral and anterior regions, such as the PACC and VMPFC, may be more involved in the representation of the degree of self-reference in the stimulus. However, dorsal regions, such as the SACC and the DMPFC, may be related to monitoring and reflection of the stimulus and its self-reference when we become aware of the stimulus as self-specific.

Finally, the posterior regions, such as the PCC, may be implicated in integrating the stimulus and its degree of self-reference into the autobiographical memory of the respective person. These regions seem to be implicated in the recall and retrieval of especially personally relevant and autobiographical information from the past of that person. Thus, it can be concluded that specific regions in the midline of the brain, the cortical midline structures, seem to be involved in the neural processing of self-reference or attributing personal relevance or self-relevance to stimuli.

4.2 *Temporal Patterns of Neural Activity During Self-Specific Stimuli*

In addition to the spatial patterns of self-reference, its temporal patterns have also been investigated using the EEG. Again self-specific and non-self-specific stimuli have been compared with each other, while the subjects undergo EEG

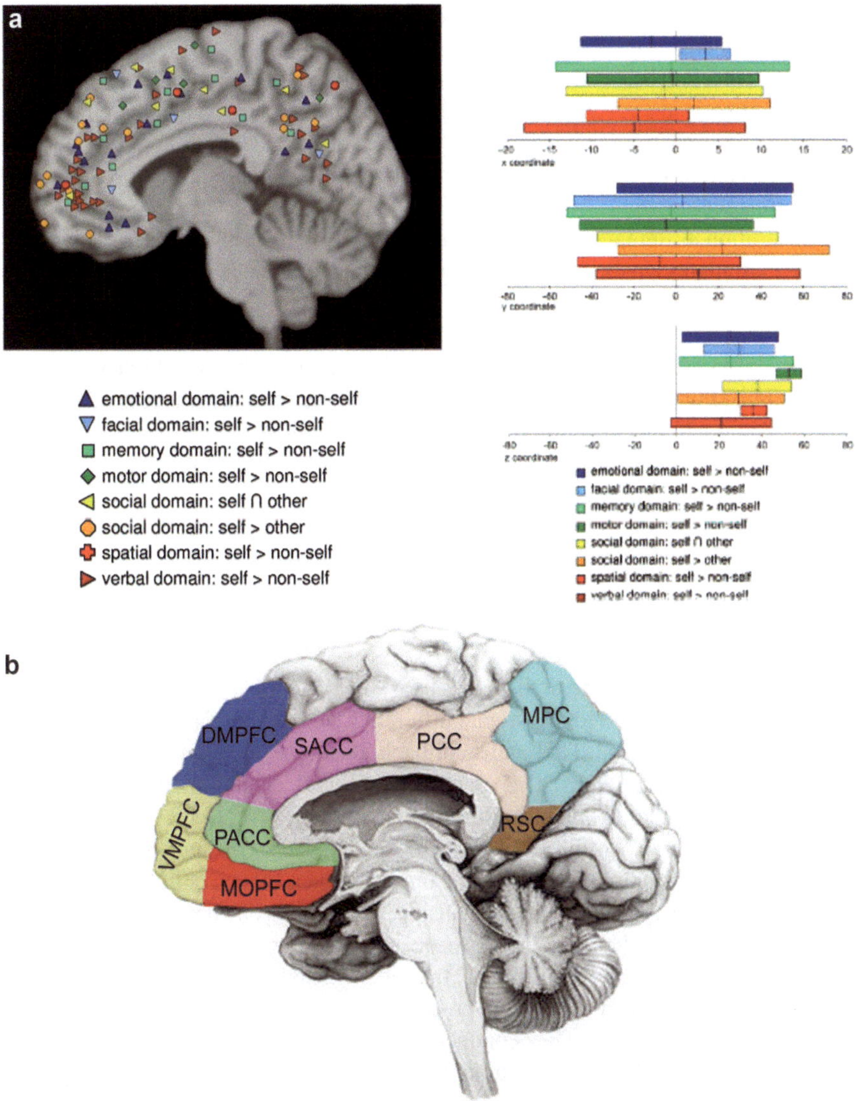

Fig. 2 The figure demonstrates the results of a meta-analysis on imaging studies of self-reference (**a**) and anatomical illustration of the midline regions (**b**). The figure on the left depicts all the imaging studies on the self as plotted in their obtained location on one brain. This includes self-referential stimuli in various domains or functions like memory, social, spatial, etc. as indicated in the lower text with the colors as shown above and on the right. On the right, three different coordinates (x, y, z) are shown that determine the direction (medial-lateral, inferior-superior) of the location in the brain. One can see that all studies locate in the midline regions of the brain (left image) as seen in the x-coordinates that describe the medial-lateral location (right image). The figure shows the anatomical regions in the midline of the brain. *MOPFC* medial orbital prefrontal cortex, *PACC* perigenual anterior cingulate cortex, *vmPFC and DMPFC* ventro- and dorsomedial prefrontal cortex, *SACC* supragenual anterior cingulate cortex, *PCC* posterior cingulate cortex, *MPC* medial parietal cortex, *RSC* retrosplenial cortex

measurement. This revealed early changes during self-specific stimuli at around 100–150 ms after stimulus onset.

More specifically, self-specific stimuli induced different electrical activity changes already at 130–200 ms after their onset when compared to non-self-specific stimuli. This was accompanied by later changes at around 300–500 ms. Hence, the temporal pattern between self- and non-self-specific stimuli shows both early and late differences.

In addition, different frequencies of neural activity were investigated. The neural activity oscillates rhythmically in different frequency ranges in the fluctuations of the neuronal activity.

One frequency often induced by stimuli is gamma frequencies in the range of 30–40 Hz. Interestingly, some EEG (and MEG) studies observed higher power in the gamma range in anterior and posterior midline regions during self-specific stimuli than non-self-specific stimuli. The question though is whether such increase in gamma power is specific to self-specific stimuli since it can also be observed in other functions independent of self-reference (see below).

4.3 Social Patterns of Neural Activity During Self-Reference

How can we investigate the earlier described social nature of the self? Various studies have been conducted to investigate different kinds of interaction between different selves. Pfeiffer et al. [15] and Schilbach et al. [16] distinguish two different methodological approaches. One investigates social cognition, the cognition of mental states in other people, from third-person perspective. Here, social cognition is investigated in an "offline" mode. More recently this "offline" methodological strategy has been complemented by an "online" mode. In the "online mode," social interaction is investigated from the "inside," by taking on the perspective of the interacting selves (rather than the observer's point of view).

Besides conducting several studies, the same group has recently investigated the neural overlap between emotional processing, resting state activity, and social-cognitive processing [16]. They conducted a meta-analysis including imaging studies from all three kinds of investigations, resting state, emotional, and social-cognitive. In a first step, they analyzed the regions implicated in each of the three tasks. This yielded significant recruitment of neural activity in especially the midline regions like the ventro- and dorsomedial prefrontal cortex and the posterior cingulate cortex (bordering to the precuneus). In addition, neural activity in the temporo-parietal junction and the middle temporal gyrus was observed.

In a second step, they overlaid the three tasks, emotional, social-cognitive, and resting state, in order to detect commonly underlying areas. This indeed revealed the midline regions, the dorsomedial prefrontal cortex, and the posterior cingulate cortex, to be commonly shared among emotional and social-cognitive tasks and resting state activity. Based on this neural overlap, the authors concluded that there may be an intrinsically social dimension in our neural activity which might be essential for consciousness of both our own self and other selves. If this is true, it will have radical consequences, not only for the concept of the self but also for consciousness in general.

5 Neurophilosophical Concept of Self

5.1 *Different Forms of Specificity*

So far we have covered philosophical approaches to the concept of the self. We also discussed neuroscientific findings about self-reference. Now, the question is how both philosophical concepts and neuroscientific findings are related to each other. This requires what one may describe as a neurophilosophical discussion. A neurophilosophical discussion directly relates empirical findings in neuroscience to concepts in philosophy.

How are the neuroscientific findings about self-reference related to the philosophical concepts of self? Are the philosophical concepts of self empirically plausible and thus compatible with the neuroscientific findings of self-reference? In order to address these questions, one should start by investigating the degree of specificity for the self of the neuroscientific findings. One can thus speak of neuronal specificity of the cortical midline structure for the self. The concept of neuronal specificity describes the quest for the exclusive association of a particular neuronal measure, like the activity of a certain region or network, with exclusively one specific function.

And they may also be discussed in the context of psychological functions associated with the self and thus psychological specificity. Furthermore, one may question the ability of the experimental designs and measures to really tap into the self. This is called experimental specificity (Fig. 3).

One may also raise the question whether the results really reflect the experiential and thus phenomenal features related to the self. Experience may be, for instance, confounded by features that are not directly related to the own self. One may thus want to speak of phenomenal specificity. Finally, one may want to discuss how the results relate to the different concepts of the self and whether they correspond exclusively to one specific concept. If they do, this would imply conceptual specificity.

5.2 *Neuronal Specificity of Midline Regions*

Let us start with neuronal specificity. The concept of neuronal specificity describes whether the spatial and temporal patterns of neural activity observed in studies about self-specificity are really specific to the self. We roughly distinguished two kinds of different regions, the domain-specific regions and the domain-independent regions.

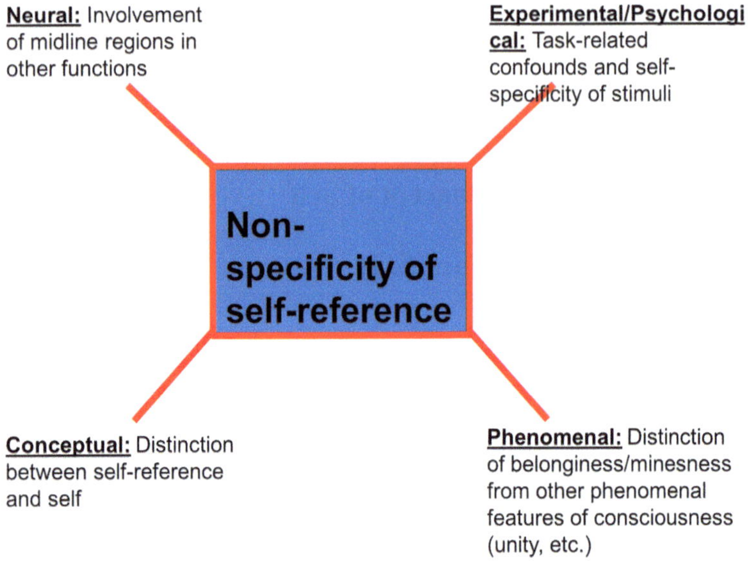

Fig. 3 The figure shows different domains (neural, experimental/psychological, conceptual, phenomenal) where current imaging studies on self-reference suffer from non-specificity. Upper left: there is neural non-specificity because the often observed midline regions are also implicated in functions other than self-reference (as, for instance, in mind-reading, emotion, autobiographical memory, etc.). Upper right: there is experimental/psychological non-specificity because the presentation of self-referential stimuli is often associated with a task like judgment yielding task-related confounds. Moreover, the self-specificity of the stimuli may be confounded by other aspects of the stimuli. Lower left: there is conceptual non-specificity because the studies do not distinguish between self-reference (of tasks and stimuli) and the self itself in their experimental paradigms. They infer from self-reference to the self which though is an inference between two different concepts that are not identical and do not imply each other. Lower right: there is phenomenal non-specificity because the experiential, i.e., phenomenal features characterizing the self, e.g., mineness/belongingness, are not properly distinguished from the ones associated with consciousness in general, e.g., unity, qualia, etc

Domain-specific regions are those that are related to the processing of a content in specific modality (e.g., sensory) or domain (e.g., verbal, sensory, motor). Depending on the stimuli and/or the task, domain-specific regions were activated in the above-described imaging data.

Are these domain-specific regions specific for the self? No, because the imaging data show that the very same regions are also recruited when applying stimuli that are not related to the self at all. For example, you are shown a house in Brazil. For you, a resident of Canada (and has no connection to Brazil), this image has no degree of self-specificity or self-relatedness to you. It nevertheless activates your fusiform face area.

The self-specialist may now want to argue that at least the degree of neural activity in the fusiform face area or other domain-specific areas may be different between self- and non-self-specific stimuli. However, empirical data are not clear. While some studies report some difference, though small in sensory regions, the majority of studies did apparently not observe differences between self- and non-self-specific stimuli in these domain-specific regions. Hence, it seems as if the domain-specific regions like the sensory and motor cortex remain unspecific for the self. This implies neuronal unspecificity.

What about the domain-independent regions like the cortical midline structures? There has been much discussion whether these regions are specific for self-specific stimuli as distinguished from non-self-specific stimuli. Is the self "located" in the midline regions? Initial enthusiasm was in support of the theory that the midline regions are specific for the self; recent investigations implicated the same set of regions in a variety of different functions.

Let me be more specific. Tasks requiring the need to understand other people and their mental states—mind-reading as described in theory of the mind in psychology—strongly recruit the midline regions. Emotional stimuli and emotional tasks also led to strong activation in the midline regions. In addition, various kinds of social tasks that require social exchange and reciprocity also recruit these regions. Finally, daydreaming or mind-wandering and other forms of introspection also recruit these regions.

The involvement of the midline regions in various functions other than self-reference sheds some doubt on the neuronal specificity of the midline structure for the self. Hence, even the domain-independent regions like the midline regions do not seem to show any specificity for the self.

The same diagnosis of neuronal unspecificity also is true of the reported gamma synchronizations. Gamma synchronization is not specific to the self but has been observed in a variety of different functions including sensorimotor, working memory, attention, and episodic memory retrieval. Hence, there is neuronal unspecificity in both a temporal and spatial sense with regard to the self.

5.3 *Psychological and Experimental Specificity*

Most of the fMRI studies above compared self- versus non-self-specific stimuli, such as a grand piano for a professional pianist compared to a saw for a carpenter. In addition to the mere perception, subjects were required to make a judgment after each stimulus, to judge whether it was self- or non-self-specific. This raises questions about what exactly the study is measuring—the perception or the judgment of the stimulus? Is it capturing the effect of the stimulus itself or the task related to that stimulus?

Most likely the results reflect a mixture between stimulus- and task-related effects. This, therefore, casts some doubt on whether the midline regions show psychological specificity for the self. The judgment about self-specificity requires various cognitive functions such as attention, working memory, judgment, and autobiographical memory retrieval. What about when research investigates the self in relation to more basic functions such as movements and actions? Even when subjects perform some motor tasks, we face the same confusion of different functions: the self's components, such as ownership (my own movement), as well as agency (whether I am the agent of the movement), may be confounded by the neural mechanisms underlying the execution of the movement/action by the person.

Such psychological unspecificity highlights the need in neuroscience to specify the experimental design and measures. We need measures that are specific to the self as distinguished from the various associated sensorimotor, affective, and cognitive functions. We also need experimental designs to segregate stimulus-related effects and task-related effects. For example, we might do this by spacing perception and judgment temporally apart from each other.

5.3.1 Self-Specificity and Other Functions

We also need to discuss the relationship between the self and other functions. Recent imaging studies demonstrated strong neural overlap between the self and reward, the self and emotions, and the self and decision-making. For example, when receiving a reward in relation to a specific stimuli, such as money, regions of the reward system like the ventral striatum (VS) and the ventromedial prefrontal cortex (vmPFC) become active [6]. These same regions are also active when the same stimulus is conceived of as self-specific, rather than non-self-specific by the respective subject.

The same effects can be observed in emotions where emotional and self-specific stimuli have been shown to overlap in the anterior midline regions especially. Finally, the same effect can be observed in decision-making: if external cues are provided when making a decision (such as a higher or lower price of the same kind of apples), lateral cortical regions become active. If, in contrast, no such external cues are provided, we need to come up with some internal criterion to guide our decision about which apples to purchase [17]. Such internal criterion can only stem from our self. Studies comparing both kinds of decision-making show predominant involvement of the midline regions in internally guided when compared to externally guided decision-making [17].

Together, this neural overlap between the self and other functions such as reward, emotions, and decision-making raises questions about the relationship between them. Different models could be imagined. Self- and self-specificity could be an independent function just like attention, working memory, emotion, sensorimotor, etc. However, in that case, one would expect specific regions in the brain and specific psychological functions to subserve specifically, and exclusively, self-specificity. However, at this point in time, this cannot be supported empirically.

Finally, one could also suggest that self and self-specificity are basic functions that underlie and provide the basis for all other functions—sensorimotor, affective, cognitive, and social. In this sense, self and self-specificity would occur prior to the recruitment of the other functions. Self-specificity would then always be present, making its involvement and manifestation in the various functions unavoidable. Rather than searching for self-specificity in relation to specific functions, such as language, one would then need to look for more basic functions that must occur prior to sensorimotor, affective, or cognitive functions.

One could, for instance, imagine that the strong involvement of the self in language acquisition requires the recruitment of midline regions. Such involvement of midline structures may be implicitly presupposed in many of the tasks or paradigms described above when presenting self- and non-self-relevant words, known as trait adjectives. While the linguistic tasks themselves seem to involve the lateral cortical regions more, their degree of activity may nevertheless be dependent upon the midline regions and their high resting state activity. Hence, future studies should investigate the relationship between midline regions and the lateral networks implicated in language, which psychologically may correspond to the relationship between self and language.

5.4 *Phenomenal Specificity of Self-Reference*

The assumption of self-specificity brings us back to the concept of the self as a "minimal self" (see above). To recap, the minimal self describes a basic sense of self that occurs immediately and is always already part of our experience of the world. The question now is how the concept of the minimal self is related to the neuroscientific results discussed above. To answer this question, we briefly have to shed light on the experience of the minimal self as manifest in pre-reflective self-consciousness.

Various phenomenal features such as qualia and first-person perspective characterize our consciousness. If the minimal self is part of any experience (rather than being outside of it), the self should be manifested in these phenomenal features, too. What experiential and thus phenomenal features does the self add? One may assume that the self, first and foremost, makes possible the generation of qualia. Without self, there is no point of view and hence no qualitative features in our experience.

Phenomenological philosophers assume that the special contribution of the self consists in what they describe as "belongingness" or "mineness" [7, 9]: the contents of our experience are experienced as belonging to a particular self; they are experienced as mine. For instance, I experience my friend's laptop on which I write for a while as my laptop though I do not own it. This goes along with an experience of a feeling of belongingness, thus being related to myself. However, such experience is not possible for the person sitting beside me who though looking at the same laptop does not experience any relation to the self. Instead, she/he may experience mineness or belongingness of the CD lying beside the laptop because she/he is a composer and it is a CD of her/his work.

This relation to the own self is particularly important when one needs to acquire a second language. The foreign language will appear as totally strange, as having no relation to one's own self and thus no self-relevance will be detected in any of the words. Why? Because none of the new words are yet associated with any experiences in specific contexts and situations.

The words thus do not yet elicit any sense of relation to the self. However, once one immerses oneself more and more into a new culture or learning context and gains new experiences, the novel words will become associated with self-relevance, thus inducing a sense of self. In short, the novel language will increasingly become associated with one's own self and become part of it. It is to be supposed that this self-relevance of language may facilitate the acquisition and learning of the new language.

5.4.1 Phenomenal Specificity and Phenomenal Limits

In order to account for phenomenal specificity, neuroscience needs to demonstrate which neuronal mechanisms underlie the experience of mineness and belongingness. We also need to distinguish those that underlie other phenomenal features of experience, including intentionality, unity, first-person perspective, qualia, and spatiotemporal continuity.

One would therefore require distinct experimental measures and designs for each of these phenomenal features. Only then would we be able to achieve phenomenal specificity and to clearly distinguish the phenomenal or minimal self from phenomenal consciousness. In short, we need to experimentally distinguish between self- and non-self-specific phenomenal measures.

However, the phenomenological philosopher may want to raise the following question: is such phenomenal specificity with the experimental distinction between self and non-self-specific phenomenal measures really possible at all? The minimal self is considered part of the experience and thus of consciousness more generally. Any consciousness of the world goes along with an experience of the self in a pre-reflective way. And the opposite holds true too. Any experience of the self is part of an experience of the world. Both the experience of self and experience of world are thus intrinsically linked.

What does this intrinsic link between the experience of self and the experience of the world imply for the phenomenal specificity of the self? It means that we will remain unable to properly and clearly segregate experimental measures for the minimal self from those of our experience in general. Why? Because these phenomenal features are always already "infected" by the self—they are encoded and ingrained into the self. Hence, the requirement of maximal experimental and phenomenal specificity may have reached its phenomenal limits. If so, we may be forced to acknowledge that there may be principal limitations to what we can and cannot investigate experimentally when it comes to the minimal self.

5.5 *Minimal Self and Body*

What about self and body? We experience our own body as our own body. The body plays a central role in the concept of the minimal self in a dual way. First, because of its basic and minimal character, the minimal self is supposed to be subserved by functions of the brain and body. The sensorimotor functions of the body that link it with the environment are thus supposed to yield and constitute the minimal self. Phenomenologically minded neuroscientists therefore consider the minimal self to be embodied and embedded. This leads us to the characteristic feature of the body, namely, that it can be experienced in consciousness. The body is not only an objective body that can be observed from a third-person perspective. This is the body the neuroscientist and the physician investigate. It can also be experienced from a first-person perspective. This is the body we consciously experience, also known as "lived body."

The lived body is my body as distinguished from others' bodies. Hence, we experience the lived body in relation to our self—in terms of mineness and belongingness. Thus, the experience of the body, the lived body, may be regarded as the first and most fundamental manifestation of the phenomenal or minimal self. Our self in its most basic and minimal form is thus essentially a bodily self.

This relationship to the self is also reflected in what we described earlier as ownership and agency. Ownership describes the fact that I experience my body as my body, rather than some other body. Neuroscientifically, the ownership of the body has been associated with neuronal activity in specific regions of the brain such as the sensory cortex and the parietal cortex. The parietal cortex mediates the spatial position of the body in the world.

Agency is the experience that it is I, rather than some other person, that causes action and movement. I, myself, am the agent of the lines I am currently writing here on my laptop. Neurally, regions such as the premotor cortex and the motor cortex have been associated with agency; these are regions that are implicated in generating movement and action in general.

How is the experience of such bodily self mediated? In determining this, sensorimotor function is considered central, especially the coordination and integration between sensory and motor circuits in the brain. For instance, when generating an action and movement, a copy of such sensorimotor coordination, a so-called efference copy that signals a forward model, is sent to the sensory cortex.

Why? By receiving an efference copy of the intended and to-be-performed action, the sensory cortex can prepare itself to and thus "anticipate" potentially incoming sensory stimuli. It can thus predict more easily the next sensory state on the basis of what the motor cortex is currently doing. Through this process, sensory and motor functions become intrinsically linked together and provide the integration of the body within the environment, known as embeddedness.

5.5.1 Body and Proto-Self

Sensorimotor functions are not only mediated by cortical regions such as the motor and the sensory cortex. In addition, they are already processed in subcortical regions like the periaqueductal gray, the superior and inferior colliculi, and the basal ganglia like the pallidum, the caudate, and the subthalamic nucleus. In addition to the sensorimotor functions, these regions are central in regulating and controlling the vegetative and thus the inner visceral or homeostatic functions of the body. This in turn is central in eliciting emotions.

How are these regions related to the minimal self? Investigations that did not include a strong cognitive or task-related component like a judgment (see above) demonstrated neural activity in these subcortical regions during self-specific stimuli. Because of their involvement in various functions, these regions are definitely not specific for self-referent stimuli. What this shows is that they nevertheless participate in constituting a self, a minimal or phenomenal self.

Current neuroscientists like Jaak Panksepp and Antonio Damasio do, therefore, speak of a bodily self or "proto-self" that occurs prior to the minimal or phenomenal self. They call this the "core or mental self." These subcortical regions seem to coordinate and integrate and map the inputs from the body at each moment in time. This allows to represent the body as one's own body in a most basic way. Panksepp goes so far as to even characterize the term self as "simple ego-type life form," to indicate the basic and most fundamental nature and relevance of the self for the body as a biological organism.

How can we distinguish between the different concepts of self? I suppose that the proto-self is more objective-observational and provides the necessary biological condition of the possible generating a self—these conditions are related to both brain and body. The minimal self constitutes those mechanisms or processes that allow to establish an actual self which is manifest and realized in the phenomenal self. Taken together, this amounts to the following:

1. The proto-self is a necessary condition of possible self, a neural predisposition of self (NPS)—this is the "proto-self."
2. The minimal self is an enabling condition or prerequisite of the actual manifestation of a self, a neural enabling condition of self, i.e., NES—this is the "minimal self."
3. The neural mechanisms that are sufficient for actually realizing the phenomenal self are the neural correlates of self (NCS)—this is the "phenomenal self."

The transition from NPS over NES to NCS marks a transition from objective-observational to the subjective-phenomenal realm.

5.6 Difference Between the Concepts of Self and Self-Reference

The concept of conceptual specificity focuses on the question: do the neuronal findings really reflect the self or some other functions? First and foremost, we must see what exactly is investigated in the imaging studies. Remember that all experimental paradigms are based on the self-reference effect. The self-reference effect assumes the distinct processing of stimuli (e.g., items, objects, persons, or events) that are related to the self when compared to those that are not related to the self.

For instance, a picture of Ottawa has a specific self-reference to me since I live here, while to you, as a resident of Australia, it has no self-reference whatsoever. Due to such difference in self-reference, the picture of Ottawa will be processed differently by your brain and my brain.

The self-reference effect presupposes the distinction between the self and a specific content to which the self may refer or not. What is experimentally investigated in the above-described experimental paradigms is thus not so much the self itself, but rather the degree of reference of a particular content to the self, known as self-reference. Some neuroscientific authors do, therefore, also speak of self-related or self-referential processing that describes the processes that are assumed to constitute the relation of a particular content to the self.

Why is this important? It means that the experimental paradigms do not target and measure the self as such, but rather the degree of the relation of a specific content to the self. For instance, the degree of neural activity in the midline structures reflects the degree of self-reference, rather than the self itself. The same holds true for the gamma oscillations, which at best correspond to the degree of self-reference, rather than to the self itself.

This means that the various empirical findings remain unspecific with regard to the concept of the self itself. They tell us about self-reference as the relationship of particular contents to the self, but not about the self. This means that the empirical findings are conceptually unspecific with regard to the concept of self. How can we resolve this conceptual unspecificity? In order to close the gap between the concepts of self and self-reference, we may need to shift our focus from the self to self-consciousness.

5.6.1 Self as Brain-Based Neurosocial Structure and Organization

What does this imply for the self? Our self may be considered as intrinsically linked to the body. This is called embodied self. Furthermore, since it is based on self-reference, our self may also be intrinsically linked to the environment. This is called the embedded and social self. Our self cannot consequently be regarded as an entity located somewhere in the brain and isolated from both body and environment. Instead our self seems to be intrinsically social, as suggested by the advocates of the concept of a social self (see above).

What does this intrinsically bodily and social nature imply for the conceptual characterization of the self? Our self may be described as structure and organization rather than as an entity—be it mental or physical. Such structure and organization need to develop through childhood and adolescence with persistent changes even throughout adulthood. Despite all the changes, there may also be persistence and continuity across time, which then accounts for what can be described as identity. Identity may describe the persistence and continuity of self over time which, in an exploratory study, has recently been associated with the midline structures and their high intrinsic activity.

We can also see that this concept of self as structure and organization is embodied and embedded. Hence, the virtual structure of the self spans across the brain, body, and environment. At the same, that very same virtual structure is dependent upon the respective environmental context. Freud's characterization of the ego as structure and organization surfaces here in a more specific way as being integrated in body and environment that is embodied and embedded. Put differently, the ego consists in a relation, the one between the brain, body, and environment, and can thus be determined in an intrinsically relational way. Future investigation might link the different features Freud attributed to the ego to the self.

What, however, do we mean by the concepts of structure and organization? The structure must be virtual in that it spans across the physical boundaries of the brain, body, and environment. Does this mean that we have to revert to a mental structure and organization as distinct from the physical structure and organization of the brain? No! The results from neuroscience clearly link the self with neuronal processes related to both intra-individual experiences and interindividual interaction. There is thus a neuronal basis for the distinct aspects of the self within the context of the brain, body, and environment. We therefore reject the mental characterization of the structure and organization that is supposed to define the self.

How can we define the concepts of structure and organization in a more positive way? One way is to characterize structure and organization as social. This distinguishes it from mental or physical features. The social characterization would then be the underlying basis that links and integrates between the purely physical and the purely mental. The self would then be based on the brain but would also extend beyond it to the body and the environment. This means that conceptually, we need to characterize the concept of the self as brain-based rather than brain-reductive (as the proponents of the empirical self tend to do). The brain-based nature of the self also excludes both mind- and consciousness-based approaches to the self.

If the social characterization of the structure and organization as related to the self is indeed basic and fundamental, one would assume that our brain's neural activity is intrinsically neurosocial: the brain cannot avoid including the social environmental context in the encoding of stimuli into its own neural activity. The neural activity is thus by default neurosocial rather than merely neuronal. This is supported by the above-described neural overlap between resting state activity and the neural activity changes during emotional and social-cognitive tasks.

Whether the brain encodes its neural activity in an intrinsically neurosocial way remains unclear at this point. What is clear is that the exact characterization of the brain's neural activity will be essential if we are to develop a truly neurophilosophical, brain-based (rather than brain-reductive) and neurosocial (rather than merely neuronal) concept of the self.

5.6.2 Self, Belief, and Valuation

This chapter has so far focused mainly on the self itself and its nature and structure. In contrast, I left open the capacities of the self. One such capacity of the self is to believe and value. Valuation is a central component in our life. Normative judgment, moral attitudes, and ultimately many decisions we make in daily life rely and are based on valuation and belief. Psychologically, valuation is closely related to reward, namely, to assign reward value to otherwise neutral or valueless stimuli, events, and persons.

Where though does the value come from? It may be closely related to the self that links and relates the seemingly neutral and valueless stimuli to itself and makes possible, thereby the attribution of value to the stimuli. If so one would expect close relationship between self and reward including their underlying neural correlates. We indeed conducted imaging studies that directly compared self and reward. Subjects were presented with stimuli upon which they had either to gamble obtaining reward or to judge their degree of self-relatedness [18].

What are the results? We first analyzed those regions related to the gambling and thus the reward. This yielded, as expected, the ventral tegmental area (VTA), the ventral striatum (VS), and the ventromedial prefrontal cortex (vmPFC) all well known to typically mediate reward. In a second step, we then investigated the neural activity changes in exactly these reward-related regions during the presentation of the same stimuli when they had to be judged with regard to their self-relatedness. Most interestingly, high self-related stimuli induced high activity in all three, VTA, VS, and vmPFC, whereas those stimuli judged by the subjects as low self-related did not induce any activity change in these regions at all. These data illustrate the close relationship between reward and self-relatedness in that the former's regions, VTA, VS, and vmPFC, also mediate self-relatedness.

Are reward and self-relatedness identical? de Greck [19, 20] used the same paradigm with patients exhibiting alcohol and pathological gambling problems. These results show normal reward-related activity in VTA, VS, and vmPFC, whereas neither high nor low self-related stimuli induced activity changes. This suggests that self and reward though being closely related are not identical since otherwise they could not dissociate from each other in their neural activities in VTA, VS, and vmPFC.

What do these data tell us about the relationship between self and valuation? One essential capacity of the self is to give and assign value to otherwise valueless stimuli, etc. Though preliminary, the data suggests that this capacity may be closely related to the reward system and its close relationship with self, the neural overlap between both in typical reward regions like VTA, VS, and vmPFC. However, neural overlap is not to be confused with neural identity between reward and self since for that empirical evidence is not given. Hence, the self seems to utilize reward-related regions though apparently in a slightly different yet unclear way when compared to reward. This is well compatible with the fact that self and valuation are not identical but closely related as we experience almost on a daily basis. Our self is more than just belief and valuation to which the neuroscientist may add that neurally our self is more than just the reward system.

6 Conclusion

The discussion about the self is rather confusing. Various philosophical concepts stand on the one side, while neuroscience provides yet different concepts of self. How are the two sides, neuroscientific and philosophical, related to each other? I here reviewed various philosophical concepts and recent neuroscientific findings. The suggestion is that there is a continuum between objective-observational concepts of self like proto-self and minimal self, as used in neuroscience, and subjective-experiential concepts of self like phenomenal self as suggested in philosophy. I suggest that the proto-self can be regarded as pre-phenomenal [21] and is therefore a neural predisposition of self (NPS). The minimal self concerns the enabling necessary non-sufficient conditions of the actual generation of a self—this amounts to the neural enabling conditions of self (NES). Finally, there is the phenomenal self whose sufficient neural conditions refer to the neural correlates of self (NES). This provides a suitable conceptual framework that allows for both sides, neuroscience and philosophy, to link their respective domains (i.e., empirical and conceptual) in developing an empirically plausible concept of self.

References

1. Northoff G, Heinzel A, de Greck M, Bermpohl F, Dobrowolny H, Panksepp J. Self-referential processing in our brain – a meta-analysis of imaging studies on the self. NeuroImage. 2006; 31(1):440–57.
2. Klein SB. Self, memory, and the self-reference effect: an examination of conceptual and methodological issues. Personal Soc Psychol Rev. 2012;16(3):283–300. https://doi.org/10.1177/1088868311434214.
3. Klein SB, Gangi CE. The multiplicity of self: neuropsychological evidence and its implications for the self as a construct in psychological research. Ann N Y Acad Sci. 2010;1191:1–15. https://doi.org/10.1111/j.1749-6632.2010.05441.x.
4. Metzinger T. Being no one: the self-model theory of subjectivity. Cambridge: MIT Press; 2004.
5. Churchland PS. Self-representation in nervous systems. Science. 2002;296(5566):308–10. https://doi.org/10.1126/science.1070564.
6. Northoff G. Immanuel Kant's mind and the brain's resting state. Trends Cogn Sci. 2012;16(7): 356–9. https://doi.org/10.1016/j.tics.2012.06.001.
7. Zahavi D. Subjectivity and selfhood: investigating the first-person perspective. Cambridge: MIT Press; 2005.
8. Tsuchiya N, Wilke M, Frässle S, Lamme V. No-report paradigms: extracting the true neural correlates of consciousness. Trends Cogn Sci. 2015;19(12):757–70. https://doi.org/10.1016/j.tics.2015.10.002. Epub 2015 Nov 13.
9. Gallagher II. Philosophical conceptions of the self: implications for cognitive science. Trends Cogn Sci. 2000;4(1):14–21.
10. Northoff G. Unlocking the brain. Volume I. Coding. New York: Oxford University Press; 2014.
11. Northoff G. Unlocking the brain. Volume II. Consciousness. New York: Oxford University Press; 2014.
12. Damasio A. Self comes to mind: constructing the conscious mind. New York: Pantheon; 2010.

13. Damasio AR. How the brain creates the mind. Sci Am. 1999;281(6):112–7.
14. Schilbach L, Eickhoff SB, Schultze T, Mojzisch A, Vogeley K. To you I am listening: perceived competence of advisors influences judgment and decision-making via recruitment of the amygdala. Soc Neurosci. 2013;8(3):189–202. https://doi.org/10.1080/17470919.2013.775967.
15. Pfeiffer UJ, Timmermans B, Vogeley K, Frith CD, Schilbach L. Towards a neuroscience of social interaction. Front Hum Neurosci. 2013;7:22. https://doi.org/10.3389/fnhum.2013.00022.
16. Schilbach L, Bzdok D, Timmermans B, Fox PT, Laird AR, Vogeley K, Eickhoff SB. Introspective minds: using ALE meta-analyses to study commonalities in the neural correlates of emotional processing, social & unconstrained cognition. PLoS One. 2012;7(2):e30920. https://doi.org/10.1371/journal.pone.0030920.
17. Nakao T, Ohira H, Northoff G. Distinction between externally vs. internally guided decision-making: operational differences, meta-analytical comparisons and their theoretical implications. Front Neurosci. 2012;6:31. https://doi.org/10.3389/fnins.2012.00031.
18. de Greck M, Rotte M, Paus R, Moritz D, Thiemann R, Proesch U, Bruer U, Moerth S, Tempelmann C, Bogerts B, Northoff G. Is our self based on reward? Self-relatedness recruits neural activity in the reward system. NeuroImage. 2008;39(4):2066–75. https://doi.org/10.1016/j.neuroimage.2007.11.006.
19. de Greck M, Enzi B, Prosch U, Gantman A, Tempelmann C, Northoff G. Decreased neuronal activity in reward circuitry of pathological gamblers during processing of personal relevant stimuli. Hum Brain Mapp. 2010;31(11):1802–12.
20. de Greck M, Supady A, Thiemann R, Tempelmann C, Bogerts B, Forschner L, Ploetz KV, Northoff G. Decreased neural activity in reward circuitry during personal reference in abstinent alcoholics – a fMRI study. Hum Brain Mapp. 2009;30(5):1691–704.
21. Northoff G (2014) Unlocking the brain. Vol II Consciousness. Oxford University Press, oxford, New Yrk, Oxford.

Enaction and Neurophenomenology in Language

Roberto Arístegui

Abstract This chapter situates the conception of language (and communication) in enaction in the context of the research program of the cognitive sciences. It focuses on the formulation of the synthesis of hermeneutics and speech acts and the vision of language according to the metaphor of structural coupling. The exclusion of expressive speech acts in this design is problematized. An examination is offered of the critical steps to the theory of language as a reflection and the linguistic correspondence of cognitivism. We examine the foundations of the proposal in the line of language and social enaction as emergent phenomena which are not reducible to autopoiesis but which constitute a new neurophenomenological position in the pragmatic language dimension. A proposal is made for the integration of hermeneutic phenomenology with genetic and generative phenomenology in social semiotics. The inclusion of expressive speech acts based on the functions of language in the Habermas–Bühler line is also addressed. An opening is proposed of enaction to the expressive dimension of language and meaning holism with the referential use of language.

Keywords Enaction • Neurophenomenology • Performative speech acts • Expressive speech acts • Background • Meaning holism

1 Introduction

In the present chapter, we focus on the research program of the cognitive sciences. This program integrates the fields of neuroscience, psychology, linguistics, artificial intelligence, and philosophy. Its main purpose is to study cognition in an objective and scientific way. In this context, cognitivism has been proposed as the central line of investigation. However, recently different voices, particularly from social neuroscience, have considered this program incomplete, since it has left aside the dimensions of emotion, affect, and motivation. Also, human subjectivity in the study of

R. Arístegui (✉)
Escuela de Psicología, Universidad Adolfo Ibáñez, Santiago, Chile
e-mail: roberto.aristegui@uai.cl

© Springer International Publishing Switzerland 2017

A. Ibáñez et al. (eds.), *Neuroscience and Social Science*,
DOI 10.1007/978-3-319-68421-5_20

the mind has not been addressed, which has opened a growing interest in phenomenology. At the same time, it has been suggested that it is necessary to complement the study of the mind with the contributions of psychology, neuroscience, and biology [1–3].

In the development of the cognitive sciences, four stages are distinguished [4–6]. An early stage is linked to cybernetics. Forged in the late 1940s, this trend laid the basis for establishing models of cognition understood from the metaphor of recursive, goal-oriented, mechanical systems. With the possibility of self-regulation through feedback mechanisms, this conception is still present, with an emphasis on mechanistic schemes.

Next, the approach of cognitivism, which appeared in the 1950s, integrated the dimension of the machine with formally represented mental processes. It is constructed in analogy with a computer program, or software, where the body would correspond to the hardware. The functional model does not consider consciousness or the body in the subjective human dimension.

The next stage, connectionism, emerged in the 1980s and proposed the metaphor of cognition as a neural network, with multiple connections. The strength of the mind varies with the ability to integrate learning rules and with the history of experiences. This model does not consider subjectivity.

Finally, the enaction perspective proposes a metaphor of the mind as a dynamic system embodied in the world. Enaction understands cognition as a temporal phenomenon, as a response to perturbations of a human system. It is not understood as a product of repeated standardized instructions. In addition, it considers that cognitive processes involve the embodiment of sensory motor skills in autonomous individuals.

While it is recognized that the central orientation is given by cognitivism in discussion with connectionism, the fundamentals of both positions are questionable [7–12]. Both the cognitivist and the connectionist models fail to characterize the relation between cognitive processes and the world, understanding mind and world as separate entities. The mind is thus understood in a formal, abstract dimension [2, 3].

The scientific and philosophical study of the phenomena of knowing and consciousness highlights an explanatory void with respect to subjectivity. The enaction project, at the same time, questions the foundations of this predominant orientation, proposing an alternative via phenomenology. It then gives rise to a novel position called neurophenomenology, which is derived from enactive cognitive science. In the context of the neurophenomenology program, enaction and phenomenology share a view of the mind, as it intentionally constitutes objects. In addition, enaction assumes autonomy; and phenomenology is characterized as the main feature of living intentionality [3]. This opens a field of dialogue, where phenomenology provides a philosophical framework for the scientific investigation of consciousness and subjectivity [13–21].

We are interested in deepening the foundations of the alternative to knowledge from enaction. We hold that the emergence of enactive neurophenomenological orientation in the cognitive sciences [4, 22–24] implies the proposition of an epistemological change, in the understanding of experience and cognition. They are both understood, in addition, as phenomena situated in the social world.

In this context, it is particularly relevant to consider the alternative position of enaction in the domain of epistemological assumptions of the theory of knowledge as a reflection, since this conception of knowledge compromises a version of language as a reflection. We refer to the pictorial theory of language. According to the tradition of language as reflex, cognition and language are in a relation of correspondence with reality [25–29]. Enactive cognitive science brings forth a new view that confronts the theory of truth as correspondence implicated by the conception of language as a reflex [4]. Language, understood as enaction, makes a change from the traditional pictorial theory of language that prevails in the cognitive sciences. We propose that it understands the way of being with others in the world through a coordination of action in language modulated by expressiveness.

Up to now, the enactive research program has been largely confined to the main domain of perception. Varela's enaction theory characterizes and describes perception as perceptually guided action. At the same time, there is a parallel development applied in the field of language and communication [30–32], explicitly recognized by Varela [4] as enaction. In this specific context, the proposal of enactive understanding of language [30] is made through hermeneutics [33–35] and speech act theory [36–43] as an alternative to the rationalist orientation in language.

Advancing in the line of enaction, the metaphor of the structural coupling in language is advanced through speech acts as commitments, to make infrequent the disruptions. This enactive orientation in language has led to a prolific development in communicative competence, in the field of philosophy with a hermeneutic perspective of language. In addition, it has integrated areas of management [32, 44, 45], organizational psychology [46–49], and constructivist psychotherapy [50–53].

This raises the notion of enactive cognitive science by bringing to the forefront a research program that integrates the perspective of the first person incarnated with the third person and the position of the second person incarnated in the social relationship. It thus clashes with the traditional, objective, third-person, scientific position that confines the mechanisms underlying the mind and consciousness to the sub-personal, proper to the position of cognitivism. To explain the focus of enaction on embodied language or social practices also makes a difference to the notion of a bridge between the first and third person [4], which is the alternative of enaction in the field of perception research. In other words, we propose that enaction, addressing the dimensions of first, second, and third person in the investigation of subjectivity, including the dimensions of perception and language, reformulates the metatheoretical and methodological field of the theoretical problematic itself of the cognitive sciences, which involves the question of the being of the incarnate consciousness.

By introducing the perspective that values first-person reports, language is introduced, and consequences are followed both in the field of the study of cognition and in some connected disciplines such as phenomenology, psychotherapy, and practice meditation like mindfulness, for example. It is interesting to point out that adding the methodological option in the first and second person involves entering the phenomenon of language as the use of ordinary language, inasmuch as the constitution of shared sense occurs when considering the terms themselves used in communication.

Although established within the approach of enaction, neurophenomenology [4, 20] has insisted that it is important to consider the perspective of the first person versus the traditional approach in the third person in science. At the same time, in order to consider the first person, it becomes essential to analyze first-person reports on a second-person position or perspective [54–58]. These reports, while they can be obtained through objective records, are a way of understanding the language of the first person in the context of an intersubjective relationship. In this context, the neurophenomenological view becomes relevant.

Our perspective is to investigate whether there is a consistent approach in enaction to address the use of language—specifically, in those first-person reports in the context of an interaction with a second person—which raises an interest in knowing how to explain the understanding of language in enaction and in neurophenomenology. At the same time, from a methodological point of view, we propose to examine a contrast between a traditional conception of language as an image (or linguistic correspondence, as a pictorial theory of language formulated in third-person statements) in relation to a conception of language as enaction. This vision integrates the pragmatic dimension of the uses of first and second person in relation.

It seems to us that it is important to examine the way in which enaction in the relationship, in the case of being understood or interpreted in language, includes the use of language in its propositional-performative dimension. This opens the enactive dimension of expressiveness in language. It seems to us that this focus is important insofar as one puts into play the capacity of a coherent approach to enaction, anchored in the emotional experience of the body and language. In this respect, it is especially important to examine the assumptions of the conception of social enaction in language rooted in the synthesis of hermeneutics and speech acts. This leads us to argue that there are two dimensions of language involved in this synthesis: the semantic function of openness in language and the pragmatic function of communication in language [59]. This is more notorious during communicative agreements in speech acts. In particular, in the context of the abovementioned synthesis, consider the language-opening function that is performed by interpreting the meaning of a speech act delivered in the first person.

It is of particular interest to ask about the conception of language that the hermeneutical position implies in the determination of the propositional content of a speech act. At the same time, differentiating propositional content from illocutionary force implies examining the dimension of communication in language. The understanding of first-person reports (centrally in neurophenomenology) from enaction requires a clarification of language philosophy assumptions: from the semantic hermeneutics and the communicational pragmatics to the speech acts.

In this context, this chapter addresses a specific metatheoretical development of enaction. In relation to a conception of enactive language, it includes the holistic perspective on the functions of language (propositional, appealing, and expressive dimension of language) including expressive speech acts. Expressive speech acts have been omitted in the enactive approach of the previously noted language. In order to develop our proposal, we will refer to (1) the enactive orientation, (2) enaction in language, and (3) an analysis of the metatheoric assumptions of enaction in language and in the synthesis of hermeneutics and speech acts.

2 The Enactive Orientation

In the context of the cognitive sciences, Varela [4, 5, 22] proposed the new program as an alternative vision to cognition, understood as representation. Varela is interested in the problem of cognition, which, according to etymology, in its Latin root, refers to knowing by the senses, seeing, knowing, and recognizing [60]. He introduces his vision, alluding to the motion of cognitive sciences, as inheritors of the traditional Greek formulation of the term "epistemology," which refers to the theory of cognition. The central question posed by epistemology is how do you know?

It defines cognitive sciences as the modern scientific analysis of knowledge understood in all its dimensions [60]. Reformulating the epistemological question, ("How do we know?") in the field of cognition, proposes that this question leads to the scientific study of the mind, considered as a valid scientific enterprise [6, 20]. Varela [5] has characterized the stages in the tradition of the cognitive sciences in the following terms:

1. Cybernetics, beginning with the artificial intelligence project (formal logic)
2. The cognitivist (symbolic) position (computation of symbolic representations)
3. The connectionist (sub-symbolic) position (self-organized network interconnections)

Varela [4, 5] stands in opposition to the tradition of representation, and he is dissatisfied with connectionism, understood as formal processing. He challenges three principles underlying this tradition: (1) we inhabit a world defined by particular properties, (2) we capture or retrieve these external properties of the world by symbolically representing them through formal representations, and (3) there is an internal entity, a "subjective us," which accomplishes the above operations.

As an alternative to representation, considered the core of the tradition of cognitive sciences, he proposes enaction. The term enaction means to execute or put into action or perform a performance. It centrally questions the notion of representation that supposes a pre-given world and a pre-given mind. Cognition is rather the putting into action of a world and a mind that arises. It emerges, based on a history of actions carried out in the very act of being-in-the-world.

3 Enaction in Language

In the context of the cognitive sciences, the cognitivist orientation has prevailed. We can recognize that the position of enaction represents an alternative to cognitivism insofar as it integrates experiential, emotional, and bodily factors into scientific study [2, 3, 23]. In the same sense, by including the development of language as enaction, we need to clarify whether the variant of enaction applied to language allows recognizing the presence of the emotional/affective dimension. The Winograd and Flores [30] approach, which intertwines Heidegger's hermeneutics [34] with the theory of speech acts from Searle [38], does not include expressive speech acts. However, a hermeneutic perspective recognizes the background as connected with the sincerity condition of speech acts [34]. However, they do not explicitly address it within a conception of speech acts including the expressive dimension.

At the same time, considering that enaction is characterized precisely by constituting an opening for the study of the emotional (therefore expressive) dimension, the omission of the expressive dimension of language undermines the approach. In that sense, we think that it is necessary to develop relevant concepts and that the discussion in this respect allows opening that dimension to research. We intend to clarify the restriction regarding the study of expressive speech acts [37–39].

To carry out this project of enaction in language, in this first part, we examine the approach of Winograd and Flores [30] in three stages: (1) the critique of enaction to the project of understanding of natural language in cognitivism, (2) the synthesis of hermeneutics and speech acts, and (3) the conversation for action in social organizations.

3.1 Enactive Critique to the Cognitivist Position in Language

In order to examine the enactive conception of language [30], we shall now examine the critical characterization of the tradition of rationalism. The critique of rationalism focuses on correspondence and the model of understanding of language in cognitivism. It is considered that this tradition traces its origins back to antiquity (Plato). Also, it underlies the modern foundations still present in the tradition of the analytical philosophy of the ideal language, which includes authors like Frege [61], Russell [62], Wittgenstein [63], and Carnap [64–67].

These authors initially address a critique of the tradition of rationalism, according to which language is understood as a representation of external reality. This criticism is explicit according to the following formulation:

(a) The main function of sentences of language is to describe the external world.
(b) At the same time, it is assumed that terms in a grammatical construction represent parts of the world or their attributes.
(c) Finally, it is considered that words denote reality.

This conception enters directly into the field of cognitive science through a conception of truth as correspondence-reference and linguistic correspondence. Language is understood as a reflection of reality.

The conception of linguistic correspondence entails [30, 31, 68] that the sentences of ordinary language are translated to a formal language by the application of a system of rules:

(a) There is a system of rules by which ordinary language is translated into a linguistic framework.
(b) There is a system of rules applied in the formal background language, by which correspondences are established between the parts of the sentence and the objects of the world whereby the meaning is established.
(c) There is also a system of rules by which conditions of truth are assigned to the sentences, so correspondence is established.
(d) The structure or standard form of sentences for comprehension corresponds to the indicative sentence.

The enactive conception in language questions the way in which these rules are applied for the understanding of natural language, because it assumes that there is an external reality [30].

Following this characterization of linguistic correspondence, language describes the properties of the objects that exist externally, and the words are understood as they denote these properties. Critically characterizing this tradition, enaction points to the questioning of the propositionally interpreted background. Propositions, assertions, or sentences, according to the last tradition, get their meaning through the propositional linguistic background, and not through an understanding of literal meaning, free of context.

For example, in the semantic version of truth [69], it is understood by correspondence that the sentence "the snow is white" is true *if and only if* the snow is white. This is the standard scheme to establish that something is the case.

Winograd and Flores [30] give examples in which the meaning of the sentence is not the same, depending on the colloquial context of the moment. For example, like this: "…do you mean snow on the mountain?" or "…snow conditions in the freezer?" (p. 55).

In analogy with the formal system of an ideal language, enaction proposes that the understanding of language is characterized in terms of formal representation. The cognitive theory of the mind has its foundations on the computational theory of the mind. In cognitivism, the position of Cartesianism [34] is assumed as a form of understanding, according to which the mental processes correspond to the software, while the body, the physiology, would correspond to the hardware. The manipulation of formal symbols is cognition.

3.2 Language Alternative in Enaction: Synthesis of Hermeneutics and Speech Acts

An alternative vision or a combination of hermeneutics and speech acts is constructed drawing on notions of Heidegger's hermeneutics in Dreyfus [34] version and Austin/Searle's theory of speech acts, in the conception of critical hermeneutics of Habermas's theory of communicative action [41–43].

This position first assumes the vision of Heidegger's hermeneutic phenomenology [34]. Starting from the conception of the being-in-the-world structure, it is possible to differentiate the following modes of being-in-the-world: available-by-hand, unavailable-to-the-hand, present-before-the-eyes, and purely-present-before-the-eyes. These distinctions allow questioning the tradition of the subject-object paradigm and object representation. Heidegger's conception of being-in-the-world [34] allows us to consider cognition not as a position of a subject (subjectivity) versus the object (objectivity) that adopts the attitude of representation (the mirror mind).

It is precisely in this context that the present-before-the-eyes notion (and purely-present-before-the-eyes in the formalization) makes it possible to characterize the idea of linguistic correspondence. When applied to both the understanding of language and the position of a subject before the world, it represents, objectively, as a derived mode of being. The cognitivist position tries to characterize or describe the use of language according to the notion of correspondence-reference and linguistic correspondence, giving rise to a formal representation, as we pointed out previously. However, from the enactive view of language, the representation is not considered the primary. By resorting to the position of the first person, it is also placed in the world, in a background of available-by-hand [30].

Understanding that representation is not primary, the notion of being-in-the-world accounts for how it arises in a world, in which we behave in analogous language to be immersed in the world of action. It is as if one tried to understand the role of a tool in the set of utensils, as taking part of a whole in a background equipment. The basic form in which the world appears does not occur in the representation of an object but in the disposition of utensils. In the background of social practices, meaning arises as part of an already understood social functioning.

When the possibility of accessing the expected background is not available, and there is a flaw in the availability of what the world has been, a breakdown occurs [45] such that talk of unavailability appears. In the same sense, in analogy with the way of understanding language according to this metaphor of the use of utensils, when understanding is available, an availability flows. If the flow is interrupted or a misunderstanding arises, it becomes necessary to restore in practice a background of understanding. A competent speaker who knows the tradition resorts to a background of prior understanding that is not primarily representational or propositional. But instead, it rests on a practice acquired by habit, which has allowed him to have acquired communication skills by repeated use. It is a pre-understanding [70].

When understanding cannot be restored relying the background of the practice following a commonly used interpretation, the enactive view proposes that there would be a breakdown in the background. This situation would give way to the reflective attitude, according to the model of having or putting something opposite, as an object of representation. It is the position called present-before-the-eyes. Criticism of the cognitive tradition is then directed at questioning the belief that primary access to the world would be via representation, what is characterized as the position before the eyes or presence-before-the-eyes. This leads directly to questioning the theory of truth as correspondence and the conception that the basic meaning consists only in designating objects. This critique is faced with the programs of natural language comprehension, based on the notion of linguistic correspondence.

Finally, when the meaning of the context is totally abstracted, by generalizing it to be used in any moment, we are in the purely-present-before-the-eyes dimension. This is where formalization takes place according to logical-linguistic formulations. The formal representation, in the deep structure, as the place of meaning, underlying the surface uses of ordinary language, corresponds to the formal structure that is used to disambiguate the meaning and that ensures the correspondence-reference. It refers to the model of truth conditions for the language expressions that are translated into a linguistic framework. It is linguistic correspondence. In this space, the understanding of language is understood as literal meaning.

In the tradition of hermeneutics, however, the availability of access to a background based on the understanding-interpretation of competent speakers-hearers in a social and historical context is emphasized. In a tradition of using language in a context of use, the notion of background becomes central to the alternative of understanding ordinary language. Following the language conception of Heidegger's [34] position, enaction adopts the notion of a background of shared practices.

3.2.1 Background and Speech Acts

In the context in which the alternative of the synthesis of hermeneutics and speech acts in contrast to the tradition of cognitivism is proposed, hermeneutic understanding of language in a context opens up. Enaction also implies emphasizing the option by ordinary language, the language that is spoken in daily life.

In the line of access to the social context, it adopts the conception of Austin's speech act theory, which introduces a fundamental distinction between constative and performative uses. The former refer to language as a representation of an external reality through representation, truth, and reference. Instead, the latter are used to do things in the world. They correspond to what is done in saying, through the use of first-person, present, indicative language. Thus, Winograd and Flores [30] established the differentiation with respect to the tradition of the use of language as representation.

Moving in that direction, Austin [37] introduced distinctions among what is said (locutionary), what is done in saying (illocutionary), and the effect of what is done when saying (perlocutionary).

Austin's procedure and distinction are opposed to the treatment of language solely in terms of truth conditions [70]. It is not necessary to take language alone in the constative dimension. It is possible, therefore, to leave the itinerary of linguistic correspondence and access the structure of performative action. The so-called felicity conditions [36] account for an appropriate use of language in certain circumstances or conditions of use. For example, coming to tea, after an invitation, is neither true nor false. Rather, they are conditions of compliance following certain invitations, commitment responses, or declarations of intentions. The state of the world is not constituted by the representation of facts in this case but by the declaration of intention.

According to Searle's taxonomy [39], an access to the background of language practices is done through a treatment of Austin's speech act theory. He [38] considers the illocutionary structure of speech acts, which distinguishes, as a central structure, illocutionary force and propositional content (the structure F (p)). Recall that Searle [39], advancing in his conception, rethinks the illocutionary structure of the theory of speech acts. He introduces the condition of input, the condition of sincerity, and the essential condition for each type of speech act. In addition, he characterizes each type of speech act according to its essential condition in the deep structure. This implies that in the analysis of the deep structure of speech acts, logical-linguistic dimension of performativity is crucial. This is manifested in that illocutionary points or types of speech acts are detailed.

In the first development of the enactive theory of language carried out by Winograd and Flores [30], the crossing of hermeneutics with speech acts considers only four illocutionary points or types of speech acts, namely, declarations, commissives, directives, and assertives (expressives are not included). They integrate declarations, directives, and commissives in the background of illocutionary forces of the other speech acts considered in the background. They move the propositional content, in the articulation of the dimension of unavailability, into the breakdown, so that the before-the-eyes dimension of the propositional commonality is present. They classify expressives as speech acts in the dimension of present-before-eyes, considered as the expression of an internal representational state, which does not integrate the illocutionary forces of the background. They also consider that the condition of sincerity is reinterpreted by the dimension of states of mind, in line with Heidegger's hermeneutics.

Table 1 shows the options of the crossing of hermeneutics and signed speech acts [30]. Considering the structure of being-in-the-world, of Heidegger's hermeneutics [34, 71], the types of speech acts are considered in their structure of illocutionary force and illocutionary point.

Table 1 Crossing of hermeneutics and signed speech acts

Available-at-hand	Unavailable-at-hand	Present-before-the-eyes	Purely-present-before-the-eyes
Understanding	*Interpretation*	*Enunciation*	
Declarations			
Directives			
Commissives			
	"Assertives"	Assertives	
		Expressives	

3.3 Application of Enactive Conversation to Social Organization

Winograd and Flores [30] proposed a systemic design in analogy with a closed or an autopoietic hermeneutic system [31, 68, 72], proposing what they call an action conversation applied to the human relationship of organizational systems—thus exemplifying what they call the conversation of action in a social system.

In a development of the proposal, they propose an organizational system as a network of conversations that need to complete or close a conversational cycle [45]. They assume a cybernetic second-order design, that is, of social and nonmechanical systems in conversation. Considering a metaphor of a structural coupling, they realize the commitments in the language allowing to make infrequent breaks. To this end, they propose a system of structural coupling based on the concept of autopoiesis [73], making an extrapolation to the social plane and understanding the social system as a closed system of action conversations [74]. In this conversational system, the conditions of satisfaction of the engagements [41], generated in the conversation of action, according to distinctions of types of speech acts, are fulfilled [30].

This is applied in an organizational system conceived as a network of closed conversations [45]. Interaction is classified according to the following coordination scheme of recursive action in the language background: (1) declarations in correspondence to the strategic apiece, (2) directives in correspondence to the management level, and (3) commissives in correspondence to the operating core.

Assertives are situated on the recursive point, like a way of "self-organization" (instead of "control") in the conversational system. However, in exemplifying what they call an action conversation, they suppress or eliminate expressive speech acts (Fig. 1).

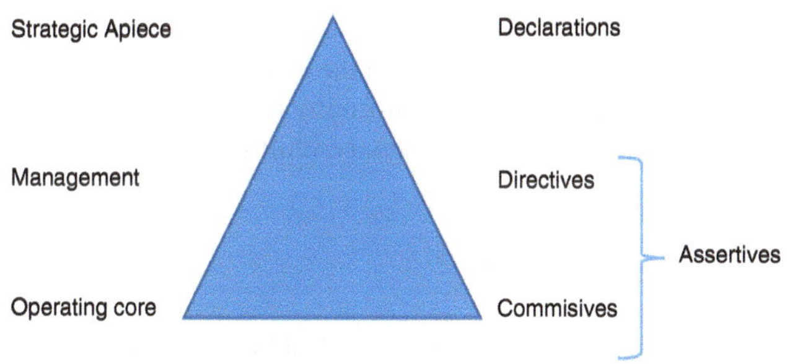

Fig. 1 Structure of an action conversation

4 Analysis of the Metatheoretical Assumptions of Enaction in the Synthesis of Hermeneutics and Speech Acts

In this section, we propose to explain some metatheoretical assumptions of the perspective of enaction in the language that we have presented. According to our view, the metatheoretical assumptions of enaction and neurophenomenology can be clarified by examining some propositions arising in the development of the language shift. To this end, we will develop the following points: (1) the framework of previous analysis following the linguistic turn, (2) an explanation of the assumptions of enaction in language, and (3) the expressiveness and the social application of enaction.

4.1 The Linguistic Turn

The linguistic turn develops in the dimension of analytical philosophy and at the same time in continental philosophy [74–76]. It is proposed to explain the same terms of the problems examined in a way that would not be possible if language is not taken into account. One assumption is that it is possible to absolutely determine the meaning, structure, content, and consistency of a language in a background language. This view of philosophy, as a philosophy of language, is inaugurated by Frege's proposals [61] that question the previous distinctions of subject-object modernity, introducing a new vocabulary with the new distinctions of meaning, as sense and reference.

The reception of Frege's [61] position was carried out both within the analytical philosophy of language and in continental philosophy in a way. In observing the development of language philosophy programs, we can see that the ideal language philosophy addressed problems by focusing on reference, while continental philosophy assumed the dimension of sense. In the context of analytic philosophy, two variants are presented: ideal language philosophy in conjunction with the theory of truth as correspondence and ordinary language philosophy [77].

4.1.1 Ideal Language Philosophy

In the context of the analytical philosophy of ideal language [78], Russell [79] continued to develop a vision that pointed to reference (denotation). In order to establish a philosophical structure in the language and to establish the reference, he addressed the difficulties presented by ordinary language for the formalization and disambiguation of terms. He proposed an underlying deep structure, where it would be possible to establish a logical form, which would allow the exact reference. In this way, the philosophy of ideal language was constituted as a logical-linguistic structure outside ordinary language.

In a next step, Wittgenstein [63] and also Russell [62] provided a theory of truth as correspondence, in which the logical form of the propositions of language allowed to establish a correspondence with the logical structure of the world. An intentional isomorphism between the atomic molecular structure of language (composed of propositions and words) and the atomic molecular structure of the world (composed of facts and combinations of objects) was proposed [63]. It is what is called the pictorial theory of language. In this context, it was considered that the words of the proposition denoted objects of the world.

The development of the Frege [61], Russell [62, 79], and Wittgenstein [63] line in the philosophy of ideal language strongly impressed the members of the Vienna Circle [80, 81], arousing the interest in moving to the domain of science, conceived as an analytical philosophy of science. The conception of referential language was forged in an attempt to promote scientific development.

The doctrine of logical empiricism proposed to consider three types of terms: logical, theoretical, and observational [77]. Then it proposed to translate the theoretical terms into observational terms, using rules of correspondence (operationalization). In this period, Carnap [65] traveled from the syntactic stage to the semantic stage. At first he assimilated the conception of the philosophy of the ideal language and the developments of the theory of truth like correspondence, integrating them in a semantic system which distinguishes among (1) rules of formation (syntactic), (2) rules of transformation (semantic), and [3] rules of truth (to establish the conditions of truth).

Applying the semantic system, it is possible to determine in the language of the background a linguistic framework, introducing a general term and introducing a new type of variables, besides establishing the connectives, the logical apparatus, and the quantifiers. The meaning in a language is established by translating it into the background frame, where meaning and reference of the terms are established.

4.1.2 Ordinary Language Philosophy

The variant of philosophy of ordinary language, opened by the second Wittgenstein [82, 83], questioned the approaches of the philosophy of ideal language. He proposed that the focus of the analysis should not be on formal deep structure but rather on the use of ordinary language, as it is presented in daily conversation. He directly questioned the language-centric game of science considered as a unique game. He advocated the use of language in contexts of use through language games connected with life forms. In a sense, it is considered that this movement of Wittgenstein II inaugurates the pragmatic turn of language [84]. This completes a process inside the linguistic turn, initiated with the syntactic stage, the semantic stage, and now the pragmatic stage.

Within the context of the philosophy of ordinary language, a prominent development is given by Austin's speech act theory [37], which introduces the performative-constative distinction. Constative uses of language that reflect external reality represent the world or the state of the world. In this sphere of uses of language, the theory of truth as correspondence develops. Performative uses, on the other hand, are a form of language use in which "to say is to do." Therefore, they have a constituent role. In the dimension of the constative-performative distinction, Austin differentiates the notions of locutionary acts (what is said), illocutionary acts (what is done by saying), and perlocutionary acts (the effect of the latter). These formulations allow differentiating the dimensions of saying and doing. In this context, Austin [37] distinguishes among five types of speech acts:

(a) Declarations: they establish a correspondence between the propositional content of the speech act and reality.
(b) Directives: they seek to get the hearer to do something.
(c) Commissives: they commit the speaker to some future course of action.
(d) Assertives: they commit the speaker to something being the case.
(e) Expressives: they express a psychological state about a state of affairs.

A next step is given by Searle [38, 85, 86], who developed the illocutionary structure of the theory of speech acts. Recall that Searle [39], advancing in his conception, rethinks the illocutionary structure of the theory of speech acts. He introduces the condition of input, the condition of sincerity, and the essential condition for each type of speech act. He differentiates the illocutionary dimension from the perlocutionary dimension to establish the meaning agreement.

An integrative development in the field of speech act theory is the universal pragmatics of Habermas [41, 42]. It implies that in an ideal speech community, every ideal speaker-listener is able to use all speech acts. At the same time, he includes Searle's distinctions in the illocutionary structure and proposes a theory of commitment in language [41, 74, 86]. It integrates the dimensions of communicative competence according to the functions of language [87]: the representational, appealing, and expressive language ("... who communicates with another, about something"). According to the principle of universal pragmatics, just cited, an ideal speaker-listener is competent to restore understanding in the shared background.

4.1.3 Continental Philosophy

Frege's distinctions [61], sense, and reference, established to overcome the subject-object paradigm, are developed by Husserl's phenomenology [88]. Husserl [89] integrates the notion of meaning in the sphere of intentionality, in relation to the subject, who constitutes the sense as a presentation of meaning in consciousness. However, Husserl's position is linked to the tradition of the intended object. Also, he has proposed immediate access to the world through consciousness. He has centrally considered phenomenological reduction as a methodology to question the presuppositions or prejudices with which we access the world. His famous interweaving of the natural attitude, which gives way to the phenomenological attitude, is recognized as a way to phenomenologically access the meaning in consciousness. It presents centrally the role of reflection through reduction.

Recently, the work of Husserl has been complemented with the publication of unpublished aspects [1–3, 15, 89]. What is called genetic phenomenology is recognized, which happens after transcendental phenomenology. In the stage of genetic phenomenology, Husserl complements his previous approaches by distinguishing against the active synthesis, where the ego is present, and a passive synthesis. It also integrates the pre-reflective dimension, permitting thus to introduce distinctions from emotion and affect. In this area, it poses a way of behaving in the world not guided by directed intentionality to an object. It is the so-called operative intentionality. In a next stage, Husserl integrates the notion of generative phenomenology, where he approaches the world of life and the intergenerational connection in a horizon of sense shared culturally.

4.1.4 Hermeneutic Phenomenology

Gadamer [90] addresses the interaction between the horizon of the text and the interpreter to establish meaning. Interpretation arises as a prejudice or pre-comprehension, anchored in the historicity that refers to tradition in society. Hermeneutics accepts what is called the inevitability of the hermeneutic circle. The meaning of the text is contextual and emerges from the horizon of the interpreter, an horizon that in turn is historical and represents interactions in the language that returns to the pre-understanding. In this line, the question of the role of interpretation in the interaction of the person with the text leads from Gadamer to Heidegger to the understanding of the world as a whole.

Following the lines of meaning, in the tradition of Frege [61] and Husserl [88], Heidegger's conception proposes an entry into language (questioning the preeminence of the referent in the philosophy of ideal language and in the developments of science, supported by logical empiricism in science). He holds a conception of language where meaning is preeminent to the referent. This is how he proposes a conception of language in which understanding-interpretation precedes the utterance [59]. He also systematically opposes the theory of truth as correspondence. He develops a conception of the structure of being-in-the-world according to which the meaning is presented in a background, as availability, as opposed to representing an object as a presence to represent-before-the-eyes. In access to the background, unavailability is faced as an imperative to return to the background.

Heidegger questions the tradition of philosophy and science, anchored in the subject-object paradigm and object representation. The development of hermeneutic philosophy proposes that access to the world is presented or given pre-reflexively, through a state of mind or emotion, as is anguish. We understand the meaning of existence in a pre-theoretical, pre-conceptual way. Heidegger strongly opposes the rationalist tradition and the subject-object distinction, as well as representation and the predominance of reflection, as a way of accessing consciousness to the world. He identifies Husserl with the tradition of rationalistic developments of reflection anchored in Cartesianism [34].

From this hermeneutic phenomenological perspective, the criticism of rationalism arises, which is characterized by a kind of dualism that distinguishes the body from the mind, where the objective or physical reality and the subjective or mental are presented.

Heidegger, from a phenomenological position originated in Husserl, is oriented to the investigation of the comprehension of the being-in-the-world, as a fundamental structure that denies the subject-object separation. Both the interpreter and the interpreted exist in an interdependence. Prejudgment is the condition to access a background, which at the same time allows interpretation. The hermeneutical circle applies as a whole to understanding, preventing all assumptions that can be made explicit. Heidegger reverses the terms of the rationalist tradition oriented to the theory and maintains that our primary access to the world is through a practice with the available-to-hand, at a pre-reflective level. Being situated in the world, acting from a pre-reflective praxis, we are thrown into a pre-conceptual understanding. In close connection with the above, Heidegger questions that the relationship with the world is established from a mental representation in correspondence with the objective world. We act in the world not as a result of a separate theoretical contemplation. The world does not appear before the eyes of an observer who sees it separately as a subject who represents it. The representation is derived.

Consequently, he maintains that meaning is social; language does not arise from the individual mind. Social activity is the basis of intelligibility. Thus, the approach of hermeneutic phenomenology appears showing that there is a transit from the individual mind (which allows explicitly that cognitivism conceives the dependent meaning of the individual mind) toward the social dimension of meaning.

The being-in-the-world orientation, being pre-reflective, also allows us to understand that we act as part of an availability background at hand. In this orientation, it is not presented as the primary to distinguish an object, nor to be a subject that faces the world as an object of representation. When the availability is broken or there is an unavailability to the immediate use at hand, it is presented in sight, before the eyes. For Heidegger, to speak of object and its properties appears in function of an activity.

The understanding of being-in-the-world (*Dasein*, as a mode of being-in-the-world that is an alternative denomination to the traditional subject, presupposing an individual mind) is understood as an understanding of possibilities, not a reality before the eyes. Being thrown or thrown into the world, the *Dasein* understands its possibilities and is projected, in an open state. This opening puts it before a factual situation. This structure of being as a thrown project gives way to a differentiation between understanding and interpreting. To the development of the possibilities involved in the prior understanding, Heidegger calls it "interpretation." The expressive understanding, the interpretation, is conceived having the structure "of something as something." The how is the structure of the expressibility of something as something, which precedes expressiveness in the statement. Heidegger thus proposes the thesis of the derived character of the utterance, inasmuch as all pre-predicative seeing is interpretive understanding. Specifically, he proposes that utterance is a derived mode from interpretation.

4.1.5 Pragmatism and Holism

The tradition of holism and pragmatism arose in parallel with the position of the analytic philosophy of ordinary language. Rorty [27], criticizing the tradition of mirror and mirror mind, has invoked Quine's position [91, 92], epistemological holism, to oppose analyticity (the idea that there are privileged representations). He also refers to Sellars [92], before the so-called Myth of the Given. Finally, he invokes Wittgenstein's position [62, 81, 82], facing reference as a single language game to give the meaning in science.

He develops a critique of the tradition of knowledge as a reflect and mirror mind. The focus of his development is the theory of truth as correspondence, which places in the pictorial tradition of the first Wittgenstein [63] and also in the semantic conception of truth of Tarski [93]. He emphasizes the linguistic turn [94] and the opposition of ideal language philosophy and the ordinary language philosophy.

He opposes epistemological foundationalism, which operates with the position that there is a privileged access to knowledge, to the privileged representation of the world as it is in itself, which today would reside in science. In criticizing correspondence and reference, based on Quine-Davidson's philosophy of language centered on the program of translation-interpretation [94–96], he turns to the understanding-interpretation of meaning according to hermeneutics. He recognizes in the tradition of Heidegger [33], Sartre, and Gadamer [90] a way of understanding that connects with the project of being-in-the-world. His central thesis is to suggest that there is an alternative vocabulary choice, analogous to Sartre's conception of Being and Nothing [97], which includes the central notion of self-choice to characterize the human reality.

4.2 Metatheoretical Assumptions of the Enactive Position in Language

First, the conception of enaction arises in discussion with the correspondence-reference assumptions that compromise the traditional view of the cognitive sciences, both in cognitivism and in connectionism, with the tradition of knowledge as a reflection and with mind as mirror.

In the original approach of enaction, Varela refers to the position of Rorty [27], where the discussion with the paradigm of the image is widely discussed and directed against representation. We could see the enaction approach, aligned with the pragmatic position that questions the theory of truth as correspondence, in the sense that it opposes the tradition of privileged representation in the field of cognitive sciences.

At the same time, Rorty [27] opposes privileged representations, but not representations themselves. In the same sense, recourse to incarnated representations may be a way that is not incompatible with the notion of enaction, if a perspective of holism is adopted to deal with perception and emotion.

Varela argues against the tradition of representation in the context of the world and the mind as something pre-given, which reminds us of Sellars [98, 99]. The questioning of correspondence appears in relation to representing the given. His opposition to foundationalism would lead him to question that there is something there, as data with which to correspond. But centrally, he is treating the tradition of reference to the game of cognitive science. We can say that he is introducing a broader perspective of meaning into a conception of enactive cognitive science. He may be said to attempt to expand the focus of the vocabulary, to include the dimension of meaning in the first person.

In the position of enaction in language versus cognitivism, the argument of linguistic correspondence aims to characterize critically the position of natural language comprehension programs in artificial intelligence (cognitive stage). Under this position, it would be possible to determine the reference in the background linguistic system. In proposing criticism to referential correspondence, enaction is directed against to the idea of determining reference in the background linguistic system. Dissatisfaction with that tradition leads to look for the alternative in hermeneutics and performativity—precisely in the line of meaning understood as not primarily referential.

It should be noted that a first clash in this area was given by Dreyfus's criticism of artificial intelligence attempting to model human intelligence [35]. The argumentation of Dreyfus proposed that the assumptions of artificial intelligence rested on a formulation derived from logical atomism. His alternative was to propose the vision of Heidegger, to be-in-the-world, to show a person situated in the changing world, rather than an abstract intelligence program based on decontextualized rules.

In the same vein, the position of Winograd and Flores [30], questioning the comprehension of language based on a system of linguistic rules, opposes the notion of being-in-the-world, in the domain of being available-to-hand, facing breakdowns of unavailability [100]. The language of the background implies understanding-interpretation according to an incarnated life practice, with others.

Understanding in language does not occur through the abstract representation, in a mirror mind, of a previous world. It is not primarily stated by enunciation. This occurs when a breakdown cannot be restored to the background and the practical attitude is left to give rise to reflection. Then the present-before-the-eyes object appears. The understanding of meaning arises in a world lived pre-reflectively, and not present-before-the-eyes in the reflection.

In the line of Winograd and Flores, Heidegger's hermeneutic phenomenology would allow access to the expressive dimension, also expressive speech acts. According to our view, Dreyfus's critical version, which characterized Husserl as a proto-cognitive, computationalist, precursor of cognitivism [101], influenced the initial version of enaction in language. For Dreyfus [34], expressive speech acts represent an internal representational mental state, which would be externalized as an expression by the means of this kind of speech acts.

According to the perspective discussed in the previous section, they are situated in the present-before-the-eyes dimension, not understanding expressivity that arises in the background of availability. They fail, therefore, in the opposition or duality of mind-world. It seems to us that there lies a possible reason for the exclusion of expressive speech act from the enactive language model of Winograd and Flores [30]. They are in the line of opposition of Heidegger in language (hermeneutics) confronting Husserl's cognitivism (phenomenology).

In the light of the new developments, explicit in relation to genetic phenomenology and generative phenomenology, there appears a dimension of Husserl's phenomenology that is not in opposition to Heidegger's hermeneutic phenomenology [2, 17]. Consequently, the notion of being-in-the-world is similar to the idea of the world of life and the horizon of meaning.

We are interested in establishing similarities between Husserl's horizon of meaning and Heidegger's modes of being-in-the-world. It is especially novel to make a cross between Heidegger's hermeneutic phenomenology and Varela's neurophenomenology as a naturalized version of Husserl's phenomenology. The connection is feasible to be established if we refer to the recent line of open research after the discovery of genetic phenomenology. In this area, we distinguish three key components: nonrepresentational consciousness, intersubjectivity, and passive synthesis.

According to this view, it is possible to make explicit a crossing with the structure of the background of availability and unavailability of hermeneutic phenomenology. Previously, phenomenology was conceived as a reflexive mode derived from a previous pre-reflexive sense. Now, we can situate both developments, hermeneutic phenomenology and genetic phenomenology, articulated from neurophenomenology, with potential for access to the background. From our perspective, we see the access to the background in the enactive dimension of language as a possibility of articulation in the breakdown of unavailability, allowing access to the pre-reflexive dimension of the language.

The opening from the pre-reflective dimension presupposes an access to the world of life, not initially mediated by reduction and reflection. Similarly, the possibility of crossing this unified perspective of phenomenology and hermeneutics with the tradition of language games, the use of language, as language games are connected with life forms, allows a transit with respect to performativity, illocutiveness, and expressiveness. This opens up an integrated consideration of the functions of language, in the path of Habermas [102].

The theory of speech acts, with Habermas [102], has emphasized the triple perspective of language functions of Bühler [87][1] to consider the speech acts. The explicitation is in the following formulation: "I (first person) communicate with someone else (second person) about something (third person)"—what is called expressive function, appealing function, and propositional function, respectively.

Considered holistically, the semantic dimension of each of the functions of language is recognized and has pragmatic implications on the participants. In the

[1] In the context of Bühler's theory of language, the expressive functions are not representational but symbolic. Bühler also categorizes the body under the expressive function of language.

context of the theory of enaction and speech acts reconsidered, we propose that the expressive function of language is presented as a bodily affective unit emerging from the shared background of social life world [40, 86, 103].

The expressive function of language is embodied in gestures and movements. It is synchronized. In the performative language game, illocutive expressivity (first person) is connected with ways of life that include appealing to others (second person). Expressive speech acts, as enaction, are incarnated in the social dimension. This new position has emotional consequences for the participants.

As an example, we can consider a moment of change in a process of psychotherapy, in an episode of change, characterized from the point of view of the use of language, by the use of an expressive speech act in first person, present indicative [54–58]. From our perspective, such use of language is inherently an enaction in language. It is a behavior with itself, expressive that self-relies directly from the pre-reflective level. It arises from a breakdown of unavailability that leads to the search for help. Neurophenomenology does not imply a prior access to the representational consciousness as a condition of the change in the use of language. It is rather a mode of being intersubjectively situated, enacted by one relational self with another.

4.3 Expressiveness and Social Application of Enaction

The antecedents of the position of enaction in the work on autopoiesis [73, 104] make it necessary to take into account what the thesis means. Living systems are conceived as autonomous systems that generate their own way of living. They reproduce their form and generate identity, which does not depend on external inputs. In this sense, the system is determined by its structure. Such a system does not discriminate, in the experience, illusion from perception. At the same time, they produce the components that reproduce their way of life.

The theoretical antecedent of autopoiesis is important epistemologically [4], since the systemic conception is reformulated with respect to the notion of an open system, which is now considered closed. This notion of autopoiesis allows us to question the idea that a system is instructively oriented, determined by the environment. It is questioned that there is an external reference to the system. This consideration in relation to the notion of a closed system allows to establish a simile with a social system in terms of a closed system as a network of closed conversations. These conversations are not understood according to the tradition of the language of correspondence-reference, but in a frame of hermeneutics, they would give place to a system that constructs its own meaning without necessity of considering the world as external, that is, without external reference.

However, Varela [4], in a self-critical development, questions the extrapolation of the idea of autopoiesis beyond the biological systems, at the cellular level in which they were proposed. He appeals to a new proposal, which has to do with the birth of the position of enaction. There is an emergence in a system that accounts for a complex level of organization that is not reduced to neural components. This point is crucial, since the enactive position, although it has an antecedent and an origin in the studies of autopoiesis, surpasses that position considering that it is not possible to extrapolate to a living human system in a social system. It respects the level of organization in which the human phenomenon occurs in the relationship at the cultural level and enaction constitutes a space or a field where a surplus of meaning emerges [105]. He argues that meaning arises in a sense of both the system and the environment, according to a history of structural coupling.

Then we can see in this approach an advance in confrontation to the theory of pictorial language (theory of truth as correspondence and correspondence-reference) in the sense of questioning the position of accessing an exterior of the system. An autonomous system will not be guided from the outside. By analogy, a hermeneutic social system as a closed network of conversations, according to the metaphor of structural coupling, does not recognize an objective exterior.

Varela precisely developed a conception of codetermination analogous to being-in-the-world with others in the language (not the duality of mind-world). Here the concept of structural coupling and emergence is at stake, implying two levels of complexity for an integration of the perspective of enaction in language. Here the concept of a history of structural coupling and emergence is at stake, implying two levels of complexity for an integration of the perspective of enaction and neurophenomenology in language.

In this domain we conceive the functions of language holistically, including the expressive function and, with it, expressive speech acts. This makes possible that emotion, affect, and mood, as enactive phenomena, can be expressed, not translated, in language. An expressive experience could be made explicit in an expressive speech act, as self-reference, not representational. In the context of an illocutive language game, expressive speech acts take place as part of the coordination and "codetermination I-and-Other" [4], p. 251.

Then, a context of cognition understood as generative enaction is also expressive.[2] This is an alternative that opens to the study of enaction in the intersubjective social dimension in psychotherapy, in mindfulness meditation, and in the field of cognitive sciences—neurophenomenology in the line of the new Husserl.

From the point of view of neurophenomenology, there is a connection between the third person and the first person as a bridge in scientific study of consciousness. We have showed the viability of a connection between second person and first person in social application of enaction. It is still necessary to allow the interconnection between the three functions in language in this area of investigation. In this regard, this allows a context of codetermination: the relation between human systems by enaction in language. What appears to us is the alternative of performativity as an enaction: when we use speech acts, the use of language is guided expressively and emotionally from the background of world of life.

[2] In therapeutic conversation, during change episodes, expressive self-references, as performative uses in first person, present indicative, have been observed as constitutive parts of the change moment [105], p. 55–58.

4.3.1 Implications of Structural Coupling and Linguistic Opening of the Social World

The underlying position in the Winograd and Flores approach leads to a close parallel between the notions of structural coupling as a metaphor of the hermeneutic phenomenological approach of language; it is argued that structural coupling would allow an analogy with design, based on commitment in the language to make infrequent recurrent breaks. With the assumption of linguistic openness of the world, according to the structure of being-in-the-world, the availability at hand (Heidegger) precedes the position of the break and the position before the eyes, where it appears what is before the view.

In the same sense, the position of observer (Maturana) that establishes distinctions in language is derived from the previous structural coupling. This conception of language is equivalent to the proposal of an enactive social cognition, in language. Action conversation appears as a network of conversations as conversational design, in which the dimension of social exchange arises as a conversational sequence of roles in different states or stages of the conversation. The question of design becomes crucial to open up possibilities for action.

In this space, it is argued that everything happens in language and that meaning matches in acts of speech with potential to access the background. However, this is where it appears as central to be able to make some distinctions. While the application of an enactive perspective on language is important for social action as a type of conversation or conversation networks, it is worth asking whether this approach reproduces a system of communicative conversation oriented to the understanding between the participants or if it is instrumental, oriented to success. This consideration is important, because it affects the attempt to elucidate the social nature of the enactive approach to language.

Here it seems pertinent to consider the dimensions of language pointed out at the beginning, the semantic openness, and the pragmatic communicative function. The perspective of the linguistic opening of the world, following the assumptions introduced by the enactive position of Winograd and Flores (based on Heidegger), leads to holism and to the thesis that meaning determines the referent. On the other hand, we argue that the discussion of the hermeneutic phenomenological position of Winograd and Flores must take over the position taken on the notion of background. This is because the notion of background (assumed in Heidegger's and Habermas's position) incorporates the notion of holism, which leads to the impossibility of distinguishing between the knowledge of meaning and the knowledge of the world in the realm of the theory of indirect reference.

If we adopt the assumption that meaning determines the referent, in moving toward the notion of background, we do not have the means to distinguish between these types of knowledge. If we consider the model of communicative action that proposes enaction in the language of Winograd and Flores, in the light of the theory of communicative action, we see that it does not meet the criteria of validity of universal pragmatics.

This is equivalent to saying that it does not put communicative competence into action, if it does not meet the spheres of validity. By suppressing the dimension of expressive speech acts, it does not comply with expressive validity. The communication system of action should be able to give the alternative of updating expressive communicative competence, by recognizing that expressive dimension in language. What would allow to mobilize the truth, that is, the speakers take a position in the first person about what is expressed in what is said. However, addressing the dimension of language competence is not enough to establish the meaning.

The design that Winograd and Flores propose for access to the background implies modifying the essential condition of the assertion and proposing a background of illocutionary forces over propositional content. The structure F (P) proves to be a preeminent condition of the illocutionary force over the propositional content. It gives an account of the dimension and function of communication, but it does not address the assumption of indirect reference in the dimension of openness in language. The backtrace of the determination of meaning to the background leads to holism, via the thesis of the preeminence of meaning over the referent. It presupposes that in the treatment of communication via speech acts, it proposes to overcome the hypothesis of literal meaning. The preeminence of the background where meaning determines the referent articulates the knowledge of meaning with the knowledge of the world not allowing differentiation of reference. Consequently, the move to a position such as that sustained by Winograd and Flores, which is based precisely on Heidegger, Gadamer, and Habermas, on a crossing of hermeneutics and speech acts, is based on the idea that meaning is established in commitment through speech acts and that content is articulated in recurrent models of breakdown and potential access to the background. Meaning or everything that exists is established through language, which means that the function of language as opening the world is developed and interpreted considering that language is constitutive of the world.

4.3.2 Enaction in Language, Neurophenomenology, and Meaning Holism

The possibility of establishing social conversation, communication at an intersubjective level, leads us to think that it is necessary to recognize the referential function or dimension of language, as a way to sustain an agreement regarding the dimension in which we communicate. Thus, we can agree intersubjectively about whether we communicate in the realm of norms or the state of the world or an experience. If we reargue that the entry to the agreement is developed through an intermediary (intentional) entity, we immerse ourselves in the problem of not being able to distinguish between the knowledge of the world and meaning. A central function of the criticism to the assumption that the meaning is preeminent to the reference is given by the possibility of generating an intersubjective position in front of an individualist conception, which rests on the idea of multiple individual mental accesses as valid.

Social cognition, understood within the framework of the theory of enaction in language according to Winograd and Flores, impinges on these individualistic assumptions. In this sense, the application of social models according to this scheme of social action does not address the problem of the support agreement based on joint access to a determination of what is meant to guide action.

The enactive conception in the language of Winograd and Flores, the synthesis of hermeneutics and speech acts, which uses the metaphorical framing (as metonymy) of the structural coupling, is subject to this criticism of the supposed individualists. By interpreting the previous understanding, according to the point of view itself, it does not allow access to the dimension of distinction between the internal and external and between what comes from the position itself and what sustains another, crucial to being able to agree intersubjectively about what is said.

The position of the observer who comes late to establish the distinctions of something already played locates the relation of the reference as derived. The previous, available, already elucidated or clarified by the metaphor of the structural coupling, was proposed as the simile to establish the meaning, in the scope of the company and in the social organization.

Taking a step further, the designative, the referential, allows a semiotic approach to learning, as our dealings with the world's images, anchored in the body, as incarnated cognition, open us to establish recursive orders, in communication, as expression of levels in social learning. Self-reference, at the communicative level, allows us the process of intersubjective communication. At the social level, we can raise the orders of learning [106] in a confluence with the Bateson Project in the area of biosemiotics.

If one looks at Bateson from the indirect theory of the referent, the meaning must be placed on the referent as reality of the second order. However, in an enactive perspective, in the world and in addition a direct referent theory, we access meaning recognizing that being able to refer directly to something in the world would allow us to access the distinction of logical types and not only logical levels. Bateson accepts metacommunication by type. Here, Bateson [107] and theorists of meaning holism [59, 95, 96, 108, 109] open possibilities of differentiating the problem of linguistic competence and the extensional dimension, which is addressed in enaction field.

4.3.3 Understanding the Terms in First-Person Reports in Neurophenomenology

In this context, it seems relevant to address the connection with the dimension of the comprehension of the terms used by the participants in first-person reports, as an axis of the proposal of enaction (and the neurophenomenology of Varela).

It seems that the task of understanding the use of first-person language confronts the dimension of translation and interpretation. In this specific context, we refer to the position of Quine and Putnam's semantic holism as they have developed a position that confronts the positions of the intentionalist theory of meaning. Quine has developed holism and indeterminacy, while Putnam agrees with holism and introduces a critical path to intentionalism with a development of direct reference theory.

The position of Quine [92] regarding the indeterminacy of translation is consistent with the statement that there is more than one translation manual, although not logically equivalents. In the context of radical translation, there is no fact of the matter. In the same context, the reference appears to be undetermined behaviorally. Moved to the mother tongue, the inscrutability of the reference is also presented. What makes sense is not to say why objects are the terms of the theory but how they are interpreted or reinterpreted in a background theory, which gives rise to the doctrine of ontological relativity. In a context of translation, at the pragmatic level, we seek an equivalence of meaning, rather than a radical translation.

According to this theoretical context [95], theses of indetermination make it possible to question the assumptions of reference correspondence and linguistic correspondence. In this respect, we agree with the critical perspective of enaction. Moving forward in the semantic dimension of linguistic openness and the pragmatic communicative dimension, from the position of meaning holism, we converge with the position of enaction in language, specifically with the neurophenomenology of Varela, from the delimitation previously pointed out against the assumptions of the synthesis of hermeneutics and speech acts of Winograd and Flores.

Thus, we propose that an alternative is to differentiate the understanding of the descriptions in their designative or referential use regarding the attributive use. From a conception of direct reference theory [108], the designative use of institutional use in a pragmatic context makes it possible to differentiate reference from identification [108]. By understanding a term in an attributive form, it is included in the identification corresponding to a classification that generically preaches belonging to a class of descriptions. In contrast, the referential use allows specifying the referential singularity of a thing, without having to comply or satisfy the belonging to a description. That is, membership (identification) to a core of descriptions implies a property or previous sense, as a way for reference. Whereas, the direct reference proposes an access to the thing, not measured by the fulfillment of the conditions of the description.

This distinction allows an alternative to the thesis that intension determines extension. It is what allows to question the assumption of Heidegger assumed by Winograd and Flores in that the meaning determines the referent. This assumption must be questioned if one wishes to maintain the position of neurophenomenology. We propose that the field in which the central thesis of Varela that affirms the hypothesis of basic work arises: "For a circulation between external and phenomenological analysis: The phenomenological references about the structure of experience and its equivalents in cognitive science are related to one another through mutual restraints" [4], p. 283.

This intends to differentiate the dimension of meaning from the referent, which we have pointed out with the distinction between the attributive uses from the referential use in language. Applied to the understanding of the reports of first person, we propose that enaction in language in the neurophenomenological way allows differentiating those uses and facing the epistemological consequences of non-revisability of the core meaning that comes off to assume the analyticity or the position of an a priori meaning (or the given, as Varela points out).

The social consequences, pointed out in the previous section, lead us to emphasize that in a context of social agreement construction in the conversation, the recursive self-reference dimension in language allows a position of fallibility if one assumes the designative dimension of the terms. The confluence between enaction and Bateson's perspective would allow an approach with direct referential uses within the social field. The pragmatic consequence shows us that it is not necessary to impose a prior vision as a condition of social dialogue in situations of communicational breakdown. Social enactive communication is a choice compatible with communicative self-reference systems.

5 Conclusion

In examining the assumptions of the conception of language in enaction, a critique of the pictorial theory of language and external referentialism appears from the hermeneutical phenomenology of enaction in language. What we realize is the partial nature of the developments in language of cognitivism in cognitive sciences, which assumes the traditional vision of language, resting on representation, in respect to the three functions of language recognized in performative speech act theory.

In the same line of argumentation, the initial perspective on language from the enactive position does not include the expressive function of language, being a partial development. Also, propositional content and referential use are proposed as derived. In the perspective of the crossing of hermeneutics and speech acts, we find a way to integrate hermeneutic phenomenology with genetic phenomenology and generative phenomenology, which allows integrating the dimension of expressive speech acts. From this background, a path of holistic integration of the functions of language in the theory of speech acts, which includes the expressive and the propositional, opens a possibility of a new discussion of social enaction in language.

Besides, the position of enaction, in the path of Varela's neurophenomenology, allows us to propose an alternative to the dimension of the indirect referent that underlies Winograd and Flores's development. We hold that enaction gives an alternative to the linguistic opening of the world, understanding a dimension of pragmatic language as a kind of direct referent not mediated by a prior meaning or concepts or previous social sense.

It seems to us that in this direction, a conception of the direct referent allows us to face the blind road of the conception of indirect reference. In a vision of neurophenomenology in coherence with the theory of direct reference, the language turn can be integrated, without the social consequences of imposing a linguistic opening of the world.

In the same sense, we open a parallel to address enaction and neurophenomenology in the line of understanding emotions and affective states from direct reference, without previously formulated senses—accompanying the emerging. Here is a position of integration between propositional content and the formulation of emotion in language, without reducing it to language.

References

1. Adrián J. La actualidad de la fenomenología husserliana: superación de viejos tópicos y apertura de nuevos campos de exploración. Eidos. 2013;18:12–45.
2. Adrián J. Husserl y la neurofenomenología. Invest Fenomenológicas. 2012;9:173–94.
3. Adrián J. Heidegger y la genealogía de la pregunta por el ser. Herder: Barcelona; 2010.
4. Varela F. El fenómeno de la vida. Dolmen: Santiago; 2002.
5. Varela F. Conocer. Gedisa: Barcelona; 1990.
6. Varela F, Thompson E, Rosch E. De cuerpo presente. Gedisa: Barcelona; 2011.
7. Lakoff G, Johnson M. Philosophy in the flesh: the embodied mind and its challenge to western thought. New York: Basic Books; 1999.
8. Chalmers D. The conscious mind: in search of fundamental theory. Oxford: Oxford University Press; 1996.
9. Dennet D. La conciencia explicada. Paidós: Barcelona; 1995.
10. Chomsky N. Knowledge of language. New York: Praeger; 1986.
11. Fodor J. Representations: philosophical essays on the foundations of cognitive science. Cambridge: Bradford Books, MIT Press; 1981.
12. Fodor J. La modularidad de la mente. Madrid: Morata; 1986.
13. Gallagher S, Zahavi D. La mente fenomenológica. Madrid: Alianza; 2013.
14. Vargas E, Canales-Johnson A, Fuentes C. Francisco's Varela neurophenomenology of time: temporality of consciousness explained? Actas Esp Psiquiatr. 2013;41(4):253–62.
15. Zahavi D. Husserl's noema and the internalist-externalist debate. Inquiry. 2004;47(1):42–66.
16. Zahavi D. Husserl's phenomenology. Stanford: Stanford University Press; 2003.
17. Welton D. The other Husserl: the horizons of transcendental phenomenology. Bloomington: Indiana University Press; 2003.
18. Varela F. La habilidad ética. Debate: Barcelona; 2003.
19. Depraz N, Gallagher S. Phenomenology and the cognitive science: editorial introduction. Phenomenol Cogn Sci. 2002;1:1–6.
20. Varela F, Shear J. The view from within: first-person approaches to the study of consciousness. London: Imprint Academic; 2000.
21. Gendlin E. Experiencing and the creation of meaning, Part IV-B. Evanston: Northwestern University Press; 1997.
22. Varela F. Neurophenomenology: a methodological remedy for the hard problem. J Conscious Stud. 1996;3(4):330–50.
23. Thompson E. Mind in life. Cambridge: Belknap Press of Harvard University Press; 2007.
24. Thompson T, Lutz A, Cosmelli D. Neurophenomenology: an introduction for neurophilosophers. In: Brook A, Akins K, editors. Cognition and the brain: the philosophy and neuroscience movement. New York: Cambridge University Press; 2005. p. 40–97.
25. Rorty R. La filosofía como política cultural. Madrid: Paidós; 2010.
26. Rorty R. El pragmatismo una versión. Barcelona: Ariel; 2000.
27. Rorty R. Philosophy and the mirror of nature. Princeton: Princeton University Press; 1979.
28. Rorty R. Objetividad, relativismo y verdad. Barcelona: Paidós; 1996.
29. Putnam H. Mind language and reality: philosophical papers, vol. II. Cambridge: Cambridge University Press; 1975.
30. Winograd T, Flores F. Understanding computers and cognition: a new foundation for design. Norwood: Ablex Publishing Co; 1987.

31. Winograd T. Language as a cognitive process. Boston: Addison-Wesley Publishing Company, Inc.; 1983.
32. Flores F. Creando organizaciones para el futuro. Santiago: Dolmen; 1994.
33. Heidegger M. Being and time. New York: Harper and Row; 1962.
34. Dreyfus H. Being-in the-world: a commentary on Heidegger's being and time, division I. Cambridge: MIT Press; 1991.
35. Dreyfus H. What computers can't do: a critique of artificial reason. New York: Harper and Row; 1979.
36. Austin JL. Philosophical papers. Oxford: Oxford University Press; 1961.
37. Austin JL. Cómo hacer cosas con palabras. Paidós: Barcelona; 1962.
38. Searle J. Speech acts. Cambridge: Cambridge University Press; 1969.
39. Searle J. Language, mind and knowledge. Minneapolis: University of Minnesota; 1975.
40. Searle J. Expresion and meaning. Cambridge: Cambridge University Press; 1979.
41. Habermas J. Qué significa pragmática universal? In: Habermas J, editor. Teoría de la acción comunicativa: complementos y estudios previos. Madrid: Taurus; 1989.
42. Habermas J. Teoría de la verdad. In: Habermas J, editor. Teoría de la acción comunicativa: Complementos y estudios previos. Madrid: Taurus; 1989.
43. Habermas J. Teoría de la acción comunicativa I: racionalidad de la Acción y racionalización social. Madrid: Taurus; 1989.
44. Flores F. Conversaciones para la acción. Lemoine Editores: Bogotá; 2015.
45. Flores F. Management and communication in the office of the future. San Francisco: Hermenet; 1982.
46. Echeverría R. Ontología del lenguaje. Santiago: J.C. Saez Editor; 2016.
47. Echeverría R. Actos de lenguaje volumen I: la escucha. Santiago: J.C. Saez Editor; 2006.
48. Echeverría R. La empresa emergente. Santiago: Ediciones Granica; 2003.
49. Leiva J. Fundamentación y diseño de un modelo de intervención socioeducativa desde una perspectiva constructivista para su aplicación en organizaciones productivas: Estudio de su aplicación y observación de su impacto en una empresa. PhD thesis. Universidad Ramón Llul; 2008.
50. Guidano V. The self in process: toward a post-rationalistic cognitive therapy. New York: Guilford Press; 1991.
51. Guidano VF. Constructivist psychotherapy: a theoretical framework. In: Neimeyer RA, Mahoney MJ, editors. Constructivism in psychotherapy. Washington, DC: American Psychological Association; 1995. p. 93–108.
52. Maturana H. Desde la biología a la psicología. Viña del Mar: Synthesis; 1993.
53. Neimeyer RA, Mahoney MJ. Constructivism in psychotherapy. Washington, DC: American Psychological Association; 1995.
54. Gaete J, Arístegui R, Krause M. Cuatro prácticas conversacionales para propiciar un cambio de foco terapéutico. Rev Argent Clín Psicol. 2017;26:220.
55. Aristegui R. Construccionismo social y discusión de paradigmas en psicología: Indeterminacion, holismo y juegos de lenguaje vs. la teoria pictórica del lenguaje. Chagrin Falls: Taos Institute; 2015. http://www.taosinstitute.net
56. Arístegui R, Gaete J, Muñoz G, Salazar JI, Vilches O, Krause M, et al. Diálogos y autorreferencia: procesos de cambio en psicoterapia desde la perspectiva de los actos de habla. Rev Latinoam Psicol. 2009;41:277–89.
57. Reyes L, Arístegui R, Krause M, Strasser K, Tomicic A, Valdés N, et al. Language and therapeutic change: a speech acts analysis. Psychother Res. 2008;18:355–62.
58. Arístegui R, De la Parra G, Reyes L, Tomicic A, Ben-Dov P, Dagnino P, et al. Actos de habla en la conversación terapéutica. Ter Psicol. 2004;22(2):131–43.
59. Lafont C. La razón como lenguaje. Madrid: Visor; 1997.
60. Ojeda C. Francisco Varela and the cognitive sciences. Rev Chil Neuropsiquiatr. 2001;39(4):206–95.

61. Frege G. On sense and nominatum. In: Feigh H, Sellars W, editors. Readings in philosophical analysis. New York: Appleton Century Cros; 1949. p. 85–102.
62. Russell B. Logic and knowledge. London: Allen and Unwin; 1956.
63. Wittgenstein L. Tractatus logico-philosophicus. London: Routledge and Kegan Paul; 1961.
64. Carnap R. Introduction to semantics and formalization of logic. Cambridge: Harvard University Press; 1959.
65. Carnap R. Meaning and necessity. New York: Routledge and Kegan Paul Inc.; 1970.
66. Carnap R. La concepción analítica de la filosofía. Madrid: Alianza; 1974.
67. Carnap R. La construcción lógica del mundo. México: UNAM; 1988.
68. Winograd T. What does it means to understand language. Cogn Sci. 1980;4:209–24.
69. Davidson D. Truth and interpretation. Oxford: Oxford University Press; 1984.
70. Lafont C. Gadamer y brandom: sobre la interpretación. Signos Filosóficos. 2010;12(23):99–118.
71. Arístegui R. La conversación para la acción de flores desde el punto de vistade la acción comunicativa de habermas. Psykhe. 2002;11(2):55–70.
72. Maturana H. Biology of language: the epistemology of reality. In: Miller GA, Lennenberg E, editors. Psychology and biology of language and thought: essays in honor of Erich Lenneberg. New York: Academic Press; 1978. p. 27–63.
73. Maturana H, Varela F. Autopoiesis and cognition: the realization and the living. Dordrecht: Reidel Publishing Co.; 1980.
74. Habermas J. Verdad y justificación: cnsayos filosóficos. Madrid: Trotta; 2002.
75. D'Agostini F. Analíticos y continentales. Cátedra: España; 2000.
76. Dummett M. The origins of analytical philosophy. London: Duckworth; 1993.
77. Romanos G. Quine and analytic philosophy. Cambridge: MIT Press; 1983.
78. Martinich AP. A companion to analytic philosophy. Oxford: Blackwell Publishing; 2005.
79. Russell B. Introduction to mathematical philosophy. London: Allen and Unwin; 1920.
80. Putnam H. El positivismo lógico: una mirada desde adentro. Alianza: Barcelona; 2001.
81. Brown H. La nueva filosofía de la ciencia. Madrid: Tecnos; 1984.
82. Wittgenstein L. Philosophical investigations. Oxford: Blackwell; 1953.
83. Wittgenstein L. Los cuadernos azul y marrón. Madrid: Tecnos; 2007.
84. Naishtat F. Una perspectiva pragmática. Buenos Aires: Prometeo; 2005.
85. Searle J. Intentionality. Cambridge: Cambridge University Press; 1983.
86. Searle J. Mind, language and society. New York: Basic Books; 1999.
87. Bühler K. Theory of language: the representational function of language. Amsterdam: John Benjamins; 1934/2011.
88. Husserl E. Investigaciones lógicas. Madrid: Alianza; 1985.
89. San MJ. La nueva imagen de Husserl: lecciones de guanajuato. Madrid: Editorial Trotta; 2015.
90. Gadamer HC. In: Barden C, Cumming J, editors. Truth and method. New York: Seabury Press; 1975.
91. Quine WVO. From a logical point of view. Cambridge: Harvard University Press; 1953.
92. Quine WVO. Ontological relativity and other essays. New York: Columbia University Press; 1969.
93. Tarski A. Logic semantics metamathematics. London: Oxford University Press; 1956.
94. Rorty R. The linguistic turn. Chicago: University of Chicago Press; 1967.
95. Quine WVO. Word and object. Cambridge: MIT Press; 1960.
96. Davidson D. Truth and meaning. Synthese. 1967;17(3):304–23.
97. Sartre J. El ser y la nada. Buenos Aires: Losada; 1966.
98. Sellars W. Empiricism and the philosophy of mind. In: Feigl H, Scriven M, editors. Minnesota studies in the philosophy of science, vol I: the foundations of science and the concepts of psychology and psychoanalysis. Minnesota: University of Minnesota Press; 1956. p. 253–329.
99. Sellars W. Science, perception and reality. New York: Humanities; 1963.
100. Dreyfus H, Dreyfus S. Mind over machine. New York: Mac Millan/The Free Press; 1985.

101. Dreyfus H, Hall H. Introduction. In: Dreyfus HL, Hall H, editors. Husserl, intentionality and cognitive sciences. Cambridge: MIT Press; 1982. p. 1–27.
102. Habermas J. Pensamiento postmetafisico. Madrid: Taurus; 1990.
103. Schutz A, Luckmann T. Las estructuras del mundo de la vida. Buenos Aires: Amorrortu; 2009.
104. Maturana H, Varela F. De máquinas y seres vivos. Santiago: Editorial Universitaria; 1973.
105. Sanhueza J. Modelo de valoración del cambio en intervenciones de consultoría organizacional: Dispositivos técnico-metodológicos de una perspectiva binocular del cambio. PhD thesis. Tilburg University; 2012.
106. Hofffmeyer J, editor. A legacy for living systems: Gregory Bateson as precursor to biosemiotics: 2. Berlin: Springer; 2008.
107. Bateson G. Steps to an ecology of mind. London: Paladin Books; 1973.
108. Putnam H. Realism with a human face. Harvard: Harvard University Press; 1990.
109. Lafont C. Lenguaje y apertura del mundo. El giro linguístico de la hernmenéutica de Heidegger. Idioma: Alianza Universidad; 1997.

A Pluralist Framework for the Philosophy of Social Neuroscience

Sergio Daniel Barberis, M. Itatí Branca, and A. Nicolás Venturelli

Abstract The philosophy of neuroscience has been a dynamic field of research in the philosophy of science since the turn of the century. As a result of this activity, a new mechanistic philosophy has emerged as the dominant approach to explanation and scientific integration in neuroscience. Rather surprisingly, the philosophy of social neuroscience has remained an almost uncharted territory. In this chapter, we advance a pluralistic framework for that field. Our framework seeks to ground the proliferation of modeling approaches, explanatory styles, and integrative trends within social neuroscience. First, we highlight the plurality of modeling approaches pursued by social neuroscientists by reviewing the distinctive features of mechanistic models, dynamical models, computational models, and optimality models. Second, we reject unitary explanatory perspectives and emphasize the plurality of explanatory styles that can emerge from those modeling approaches, considering their contents and vehicles. As regards their content, we present two kinds of information a model may provide, namely, causal/compositional or noncausal/structural information. As regards their vehicles, we examine and illustrate different guiding representational ideals (e.g., precision, generality, and simplicity). Third, we turn to integrative trends in social neuroscience, assessing the prospects of inter-theoretical reduction, mechanistic mosaic unity, and multilevel integrative analysis. We contend that the pluralist framework we develop is an adequate approach to scientific modeling, explanation, and integration in social neuroscience. We additionally address how this pluralistic perspective may shed light on the intersection between

S.D. Barberis (✉)
Universidad de Buenos Aires (UBA), Buenos Aires, Argentina

Agencia Nacional de Promoción Científica y Tecnológica (ANPCyT), Buenos Aires, Argentina
e-mail: sergiobarberis@gmail.com

M. Itatí Branca
Universidad Nacional de Córdoba (UNC), Córdoba, Argentina

Consejo Nacional de Investigaciones Científicas y Técnicas (CONICET), Buenos Aires, Argentina
e-mail: itatibranca@gmail.com

A. Nicolás Venturelli
Instituto de Humanidades (UNC/CONICET), Buenos Aires, Argentina
e-mail: nicolasventurelli@gmail.com

© Springer International Publishing Switzerland 2017

A. Ibáñez et al. (eds.), *Neuroscience and Social Science*,
DOI 10.1007/978-3-319-68421-5_21

the neural and the social realms, in a context of greater interdisciplinary collaboration between neuroscientists and social scientists.

Keywords Social neuroscience • Models • Explanation • Pluralism • Integration

1 Introduction

The development of the philosophy of science from the second half of the twentieth century has been primarily characterized by an increasingly thorough focus on particular disciplinary areas, following the widespread recognition that each scientific area presents different philosophically relevant theoretical and methodological questions (see Bunge, this volume). To offer a panoramic view of recent philosophical work in neuroscience, here we address three prominent issues that are highly relevant for social neuroscience (SN), namely: modeling approaches, scientific explanation, and theoretical integration. Though this selection of topics is admittedly limited, it proves fairly representative of contemporary debates and results.

We will approach the issue of modeling in SN and its relation to the problems of scientific explanation and integration in the field, from the perspective of the working scientist who constructs, revises, and applies models under some specific, concrete objective. This aligns with a growing trend in the philosophy of science aiming to understand the dynamic aspect of scientific knowledge, including the processes underlying the emergence, change, and disappearance of research programs, disciplines, and whole fields, as well as the evolution of scientific instruments, experimental paradigms, models, and theories.

SN emerged only recently, around 1990, as a multilevel approach to the study of the neural bases of social behavior [1]. This approach intended to reject previous cognitive neuroscientific perspectives that primarily focused on the human brain considered in isolation; in this way, most research was overly indifferent to the inherently social nature of human beings, which in turn became the central subject of interest to SN [2, 3]. Since then, SN has experienced significant development, including the establishment of two journals in 2006, *Social Cognitive and Affective Neuroscience* (Oxford University Press) and *Social Neuroscience* (Taylor and Francis), and three societies, the Social and Affective Neuroscience Society (SANS), established in 2008; the European Society for Cognitive and Affective Neuroscience (ESCAN), founded in 2009; and the Society for Social Neuroscience (S4SN), established in 2010. Even though the development of the field has certainly accelerated, SN is still very much in its infancy, full as it is of programmatic questions to be approached and conceptual issues that need to be reviewed.

In what follows, we will focus on a subfield which, though important for SN, is not exclusive to this field. Specifically, we will deal with bottom-up approaches, that is, approaches which take as a general starting point the description of the structural and functional aspects of neuronal mechanisms and neuronal systems, which can in turn be connected with socially relevant psychological phenomena. This is a necessary restriction of our focus, given the methodological distance and differences in scope between SN and other neuroscientific arenas, such as neuroanthropology, neurosociology, or neuroeconomics. Although these share multiple aspects with SN and cognitive SN, they approach research in an inverse direction, hinging on the economic, political, and cultural influences on brain function and development. One common motivation behind the diverse disciplines that adopted a top-down approach has been a growing dissatisfaction with the traditional, non-mentalistic understanding of social phenomena (cf. [5], p. 11) and the realization that neuroscience could provide social science with a rigorous basis for conceptualizing and measuring the mind. Another way of framing this is that while these disciplines, despite their individual differences, tackle the neural dimension of classical social science questions, the subfield of SN we address here applies traditional neuroscientific methods to social phenomena—although we acknowledge SN is by no means restricted to such bottom-up approaches.

The case of SN, as we restrict it here, is peculiar both on account of its complexity (as a hybrid field approaching social phenomenon through neuroscientific methods) and its relative youth within neuroscience. This partially explains why the philosophical reflection specifically directed to problems arising from SN and social cognitive neuroscience is still very incipient. In what follows we thus make an effort to extend some of the more developed themes within the philosophy of neuroscience to enlighten relevant aspects of contemporary social neuroscientific research. This situation makes the advancement of an established philosophy of SN a promising and compelling challenge for the years to come.

2 Models in Social Neuroscience

2.1 Theories and Models in Philosophy of Science

Mainly stemming from the mid-century historical turn led by Thomas Kuhn, scientific models have come to occupy a fundamental and pervasive role in recent philosophy of science. The variety of modeling strategies across disciplines and the varied functions they serve led to the general recognition that models exert major influence in the production of scientific knowledge. This realization contrasts sharply with the merely psychological, pedagogical, or at most heuristic function that logical empiricist philosophers had previously ascribed to models. Relevant as they may have been from a psychological, sociological, or historical point of view, models were relegated to the periphery of the philosophy of science by influential thinkers such as Rudolf Carnap, Carl Hempel, and Karl Popper.

Following Bailer-Jones [4], a scientific model can be seen as an interpretative description of an empirical phenomenon whose primary general function is to facilitate cognitive access to it. This access can be either perceptual or intellectual. In order to grant this kind of otherwise unavailable access, models tend to focus on specific aspects of a phenomenon. This privileged access is achieved, on the one hand, by leaving aside a host of other aspects pertaining to the phenomenon and, on the other, by simplifying or idealizing those aspects considered to be essential for the depiction of the phenomenon or vis-à-vis some specific objective pursued through the modeling effort. Sometimes, modeling may also imply appealing to some unrealistic or fictional assumptions to meet ongoing requirements (which can and cannot be of a representational kind). In this sense, a model is always a partial description of its target phenomenon, under some particular problem context.

The distinction between scientific theories and models involves at least three points of contrast: generality, structure, and function. Although the nature, composition, and rate of change of scientific theories remain a hotly debated topic, here we will conceive them as articulate and wide-ranging constructions that represent and explain general characteristics of a set of phenomena (cf., [5]). A construction of this kind can take on different formats, such as linguistic or mathematical, but, in most cases, it would allow its expression in symbolic notation. From a comparative perspective, a scientific theory is taken to be the most exhaustive and far-reaching presentation of the particular way things are thought to be and function within a certain state of affairs and for a particular scientific community. In this sense, although models can help unleash their representational potential, theories occupy a somewhat distant position with regard to phenomena.

Now, from the point of view of the model's user, a scientific model may also be thought of as a complex tool for thought [6, 7]. As such, it can be directed toward a wide spectrum of related endeavors, such as representing data sets, exploring novel phenomena, orienting or directing experimental design, driving computational simulation efforts, theory application, construction, revision, and so forth. The often-highlighted central position of models, as mediators between theory and phenomena, is inherent to this multiplicity of roles and uses (a multiplicity that is accordingly absent in the case of theory). This also contrasts sharply with the above concept of theory, which can be conceived as a sort of end product of modeling and experimental efforts, though open to revision and adjustment. The model-as-tool notion is thus another important point of departure between both concepts.

One particularly notable aspect of this understanding of scientific models, especially regarding SN as well as other areas of contemporary neuroscience, is the fact that models maintain an important kind of autonomy vis-à-vis theory. This concerns how models are elaborated as well as how they are variously deployed. The philosophical tradition that we are following here (see, e.g., [8]) has emphasized several ways in which this autonomy can be found in the history of science. In particular, the cognitive profit brought about through modeling exceeds by far its representational capacity as derived by theory, and most importantly, it has to be acknowledged even in absence of a firm and fully developed theory within a particular discipline or area of research. This is a situation that fits perfectly with SN as practiced today. In the next subsection, we will consider modeling in neuroscience and particularly in SN, so as to then assess the particular kinds of models that are found in the field.

2.2 *Theoretical Principles and Models in Social Neuroscience*

As already anticipated, models in neuroscience have a preeminent role as well as a specific kind of autonomy regarding theory. The primacy of models in the field partly reflects the fact that theories of brain function are not dominant, as growingly acknowledged within the philosophy of neuroscience. For our purposes here, the main point is not that theories are nowhere to be found in SN (as we will shortly see); rather, we maintain that they do not define a high degree of agreement within relevant communities. In the recent literature, a certain widespread consensus can be found around this idea [9]. In a very early statement, Churchland and Sejnowski [10] defined neuroscience as a data-intensive field while remaining poor with regard to theory. While this premature recognition may not have been cautionary, the years to come have rapidly intensified this particular situation (as we will shortly see, to a degree recently deemed problematic by notable neuroscientists). The growth and sophistication of experimental approaches, greatly fueled by the resonant expansion of different kinds of structural and functional neuroimaging studies, certainly stands as a crucial factor contributing to this trend.

Although a systematic philosophical treatment of the theoretical status in neuroscience is still due, several philosophers have advanced considerations along these lines. The position defended by Valerie Hardcastle is worth considering in some detail. In a series of papers [11, 12], she portrays the theoretical dimension of neuroscience as a collection of loosely related and to some extent autonomous theoretical principles. The main point is that these principles are not (or so far have not been) articulated into a cohesive theory addressing some specific set of phenomena. On the other side, these general principles are used in order to guide experimental research and interpret experimental results. They can be thought of as contributing an interpretative framework that, in a moderate sense, drives research. Inasmuch as this is an accurate picture, theoretical frameworks of this kind may be in part responsible for the fragmentation inherent to almost all fields in cognitive and social neuroscience—and as already pinpointed in the very early moments of SN (e.g., [1]). Some principles that may be mentioned are, for example, the role of functional segregation as an organizing element in the cerebral cortex (e.g., [13]), the assumption that two or more sensory systems are anatomically overlapping (e.g., [14], one case considered by Hardcastle and Stewart), or the increasingly explored idea that neural networks learn statistical regularities from the natural world following Bayesian principles (e.g., [15]).

There have been some attempts to develop general theories in scientific fields that can be considered as part of the constellation of SN. Twenty years before Cacioppo and Berntson's contribution [1], Joseph Bogen and Warren TenHouten coined the term "neurosociology" to refer to "a confluence of neurologic and sociologic observations" and, in particular, to describe a series of studies of sociocultural variations in performance of lateralized cognitive tests ([16], p. 49). From the perspective of neurosociological analysis, the emphasis is on "the social production of thought and the social determination of brain organization and brain function" ([17], p. 10). Crucially, in several works [18, 19], TenHouten has explored a general affect-spectrum theory of emotions. In what follows, we will briefly present it here for illustrative purposes.

The affect-spectrum theory is rooted in Plutchik's [20] psychoevolutionary theory of basic emotions (see TenHouten, this volume). According to Plutchik, there are four fundamental problems of life facing a wide range of species. For each existential problem, there is a negative aspect, or danger, and a positive aspect, or opportunity. The first problem is temporality, which refers to the finite life span of creatures and to the cycle of life, reproduction, and death [19]. The inevitability of separation and loss is definitive of sadness, while the possibility of social integration and support is definitive of joy. The second basic life problem is identity, which concerns membership in social groups. The opposed primary emotions surrounding the notion of identity are acceptance (incorporating) and rejection or disgust (expelling). The third problem is hierarchy, which involves power, authority, status, and prestige. The struggle for dominance defines anger, while the acceptance of lower status defines fear. The fourth problem is territoriality, which includes not only geographical space but commodities and all kinds of symbolic capital [19]. The control of territory defines exploration, while the violation of one's boundaries implies surprise.

TenHouten ([18], p. 55) proposes that each of these four existential problems has evolved into Fiske's [21] four elementary forms of sociality. In this way, the positive pole of Plutchik's temporality can be generalized into what Fiske [21] calls communal sharing, a social relationship of equivalence based on solidarity, unity, and identification with the collectivity, especially with the kinship system. Secondly, Plutchik's identity can be generalized into what Fiske calls equality matching, an egalitarian social relationship between distinct and coequal people in which each person receives roughly an equal share, regardless of the community's needs. Thirdly, hierarchy can be linked to authority ranking, a social relationship in which, according to Fiske, the superiors command and control the production and distribution of goods. Fourthly, since territoriality has been broadened to include all form of possessions, it can be assimilated to Fiske's notion of market pricing, a social relationship based on reciprocal exchanges mediated by values and determined by a market system.

Given these generalizations, Tenhouten [18, 19] defines a *quaternio*, a dynamically related double polarity in which there is an affinity between communal sharing and equality matching, on the one side, and authority ranking and market pricing, on the other. TenHouten's affect-spectrum theory predicts the spectrum of the 36 primary and secondary emotions using a generalization of Plateau's law, to wit: $\Psi_{ij} = kR_j^m iR_j^m j / f\left(d_{ij}\right)$, in which Ψ is the predicted level of the emotion, the Rs are two of the valenced Fiskeian social relations, and f is a function of the distance between the social relations on the quaternion [18]. In this way, the emotional experience is viewed as the product of social relationships: for example, love, which Plutchik defines as joy plus acceptance, is predicted as a product of communal sharing and equality matching. The theory has had certain impact on the sociology of emotions [22]. However, bearing in mind that many other influential theories have been advanced and developed in the field, the mainstream in SN has tended to adopt a bottom-up approach that emphasizes the search for the neural and molecular correlates of emotional states and social relationships (compare, e.g., the bottom-up treatment of love and bondedness reviewed in Sects. 2.3, 3.1, and 4).

Additional examples of general conceptual principles guiding experimental research within the bottom-up approach to SN include views on whether human empathy is to be understood as a cognitive or an emotional process [23] or the extent to which one can define a functionally segregated neural system dedicated to a given social phenomenon as a guide for human brain mapping strategies (see, e.g., [24]). The generality, relevance, and testability of such principles vary greatly, also depending on the line of research and their specific role within it. They nevertheless define the theoretical profile of the field, opening up a sort of theoretical vacuum where models must operate: a mediating role for models which can, on the one hand, help clearly present and interpret experimental data in the light of a given principle or theoretical framework and, on the other hand, help specify the empirical relevance of a given principle or theoretical framework in order to guide or define experimental designs and protocols. As we will shortly appreciate, this middle ground where most modeling work is to be found offers a wide range of modeling strategies to connect theoretical and experimental research as well as a broad repertoire of types of models, which philosophers of neuroscience have identified and characterized.

Some neuroscientific positions must be highlighted which reinforce the picture presented. The contention that the field of neuroscience lacks strong, widely held theoretical constructions has been voiced by several influential neuroscientists. Marder et al. [25] underscore the idiosyncratic role of theoretical models, considering this lack of solid, structuring theories. Stevens (cf. [26], p. 177), in a brief review, goes so far as denying that any theory up to this moment can be considered to have made any fundamental contribution to neurobiology. We have then further reasons to reaffirm the idea that, more than properly neuroscientific theories, theoretical principles are variously deployed to guide the construction of different kinds of models and the development of experiments.

A concomitant fact to be mentioned is the growing trend of model-based cognitive neuroscience [27, 28]. The concurrent use of cognitive modeling to guide and complement different experimental strategies to explore brain functioning (such as electrophysiological and neuroimaging techniques) is a recent attempt to find unifying approaches and to face the dispersion of existing models and the diverse and data-intensive experimental results typical of the field. Although the complementary use of cognitive models and typical neuroscience techniques is not necessarily new, there is a marked and explicit recognition of the need for integrative efforts of this kind (see Sect. 4).

While the first advances toward establishing a bottom-up neuroscientific approach to social phenomena can be traced already to the first half of the 1990s (e.g., [1]), the methodological difficulties in the case of SN were even more comprehensive and more pressing than the ones faced by contemporary cognitive neuroscientists. On the one hand, there were the expectable hurdles accompanying the application of complex areas of research (such as social psychology and social theory) to an already novel group of disciplines. On the other hand, the precise delimitation of target psychological phenomena together with the early realization that the neural systems implied are generally largely distributed entailed additional difficulties for both experimental and theoretical researchers alike.

Now, very specific descriptions are thus applied to the above theoretical anchors, stemming from the different kinds of experimental results obtained. It could be argued that, especially in the case of social cognitive neuroscience, this diversity is somewhat limited by the extended inclination to work with human experimental subjects, in part for obvious reasons concerning the kinds of phenomena under study and in part due to the rapid transformation and increasing availability of neuroimaging technology (and very specially, fMRI). Nevertheless, the distance between theoretical prescriptions and experimental descriptions is still very large, and, as already mentioned, it is within this gap where models come in and are most useful.

In what follows, we will present and analyze the different kinds of models and the associated modeling strategies that have been identified in the philosophical literature. The main aim is to offer a comprehensive picture of the theoretical mosaic which comprises contemporary SN, within the restricted group of bottom-up approaches we are considering. This will offer an outlook of this model-intensive field, inasmuch as it can then be tied to relevant explanatory and integrative efforts. Both of these endeavors will in turn be examined in the two following sections.

2.3 *Kinds of Neuroscientific Models*

Before presenting the main types of models that philosophers of neuroscience have discussed, it can be useful to introduce some standard distinctions commonly used in the neuroscientific literature. Some of the concepts below may overlap with some of the more philosophically oriented categories we will consider, that is, cognitive models, computational models, mechanistic models, and dynamical models. These categories are thoroughly debated in terms of the explanatory and integrative dimensions of neuroscience, in general, and its subfields, in particular. Their neuroscientific counterpart, on the other hand, will provide us with a platform to draw comparisons from and with a more comprehensive picture of modeling in SN.

In the preface of their remarkable 2001 book, Dayan and Abbott present a seemingly exhaustive distinction between descriptive, mechanistic, and interpretative models. While this categorization was proposed in reference to theoretical neuroscience, it can be easily extended to other areas of neuroscience, including SN. Such types of models are presented in terms of the differential questions that drive their construction: what it is that a particular neural system does (descriptive models), how it is that it does it (mechanistic models), and why (interpretative models). This is a very useful and at the same time very broad tripartite distinction that is silent on issues such as the level of description, complexity, theoretical commitment, or explanatory scope of the models.

As Dayan and Abbott (cf., [29], p. 1) state, descriptive models summarize large amounts of experimental data under descriptive purposes. Mechanistic models describe how neural systems operate on the basis of known anatomical and physiological features. Finally, interpretative models focus on the behavioral and cognitive relevance of different aspects of brain function to define the computational principles behind it: the already mentioned efficient coding principle, according to which neural activity is minimized in order to transmit information along a processing stream, is a very general principle that can be used to elaborate specific computational models of brain function.

A related distinction, which goes well beyond the field of neuroscience and has also been thoroughly discussed by general philosophers of science, is the distinction between phenomenological and theoretical models (see, e.g., [30]). First, descriptive models are inherently phenomenological, inasmuch as they aim at representing phenomena—where a phenomenon is a scientifically relevant set of general and relatively stable features of the world. Second, interpretative models are inherently theoretical, positing as they do functional and operational principles that neural systems allegedly embody. Third, mechanistic models are more complex in the sense that they can be partially phenomenological and partially theoretical: to the extent that a model purporting to describe a system's mode of operation incorporates some kind of theoretical entity or hidden mechanism (not an uncommon situation in SN, as well as in other areas of neuroscience), then it exceeds this classical distinction (see Northoff, as well as Aristegui, this volume).[1]

A final related distinction is the one between quantitative and qualitative models. While most properly neuroscientific models are quantitative, or can be precisely expressed through mathematical or computational means, SN has benefited from qualitative models deriving from social psychology and cognitive science. Generally, when a set of phenomena is poorly understood or when its research is still in its infancy, qualitative modeling can be a possible, fruitful starting place. On a similar note, a model's complexity or its level of built-in biological detail can vary widely, according to the level of knowledge achieved on a particular neurobiological structure or neural system and, importantly, on the modeling purposes at hand. As we have already alluded, simplifying assumptions do not always depend on mere lack of knowledge and can instead be deliberately implemented (see, e.g., [32]).

SN, as most other areas of neuroscience, presents a vast range of models, stemming from ideal models designed to contrast intuitions on a conceptual matter (e.g., is empathy a genuine neuroscientific phenomenon, whose neural bases can be identified and described?) to very detailed models of oxytocin's neural pathways implied

[1] It can be pointed out that Craver [31] draws a distinction, not between phenomenological and theoretical models but between phenomenological and explanatory models. Theoretical enrichment, Craver would suggest, isn't necessary nor sufficient for a model to be explanatory. Similarly, a phenomenological model may theoretically enrich the description of the explanandum, as can be the case of LISP-based computational models.

in regulatory behavior related to stress outbursts. What can be called the level of granularity of a given model is certainly a very relevant feature to its assessment and has strong connections to a model's explanatory and integrative power. Below, we consider the different kinds of neuroscientific models that populate the philosophical literature and illustrate them with examples from SN.

At least four general kinds of models have been recently discussed in the philosophy of neuroscience. Although SN models do not figure prominently, the rapid growth of the field during the last decade will most probably be accompanied by an increase in the associated philosophical interest. It is also important to mention that the peculiarity of SN mainly comes from its problem domain, that is, the universe of neural and behavioral phenomena of an inherently social nature, and in this sense the kinds of models generally sought for and developed are on a continuum with other areas of neuroscience directed to cognitive phenomena (with some caveats that we will consider). These are cognitive models, computational models, mechanistic models, and dynamical models. Let us consider them in turn.

Cognitive models, sometimes also called functional models,[2] aim fundamentally at the specification of the operational stages necessary for a given psychological capacity to be carried out. Weiskopf [33] has described these kinds of models in terms of their epistemic aims and the array of techniques adopted to elaborate them. The purpose of cognitive models is to single out the functional properties of the neural system responsible for the psychological capacity under study. In terms of the well-known tripartite distinction between levels of analysis of an information processing system [34], cognitive models work at what Marr called the theory of calculus or computational level: they portray the activity of the system as a projection from one kind of information into another kind, within a series of necessary steps. Models of this kind posit a sequence of representational states and processes, needed for the performance of that particular capacity:

> Specifying such a model involves specifying the set of representations (primitive and complex) that the system can employ, the relevant stock of operations, and the relevant resources available and how they interact with the operations. It also requires showing how they are organized to take the system from its inputs to its outputs in a way that implements the appropriate capacity ([33], p. 323).

Although at first sight one may think these are not properly neuroscientific models, this would be an understatement: within a top-down approach, they can be very important to dismiss idle theoretical avenues and to direct further experimental efforts.

To illustrate this first kind of neuroscientific model, consider an early model of face recognition proposed by Bruce and Young [35]. This, explicitly presented as a functional model, centers on the sort of information (what the authors call "information codes") that has to be generated and accessed in order to recognize a familiar face, on the different stages involved in this process, and their organization. Hinging

[2] It should be noted that Weiskopf [33] understands cognitive models as a subtype of functional models. For reasons of clarity and considering the present context, we preferred to conflate both concepts.

on a host of reaction-time experimental results, data on typical patterns of error, and neuropsychological studies, Bruce and Young's model make clear-cut distinction between information processing operations, such as facial speech analysis and directed visual processing, and functional components of a face recognition system, such as face recognition units and person identity nodes. As is typical in this sort of modeling efforts, they stress the sequential order of relevant operations, claiming, for instance, that visual recognition necessarily precedes access to person knowledge. As we already stated, this kind of modeling work is not at all trivial and, although it may dominate the earliest stages in the study of a given phenomenon or neural system, this need not always be the case, as can be seen in Decety's [36] model of empathy.

Computational models can be likened to Dayan and Abbott's interpretative models as well as understood in terms of Marr's second, algorithmic, level of analysis. Predictably, models of this kind are generally computationally implemented, as this allows for their precise description and their valuable involvement in simulation studies. The growth of computational neuroscience, also due to the increasing level of neurobiological detail built into the models, has led to a proliferation of computational models, also in the field of SN. These models aim at uncovering the computational principles that guide the operation of neural systems, understood as information processing devices. Under this assumption, it is believed that manipulating models implemented in a computer can shed light on neural function, on a theoretical but also on an experimental basis.

In general, what computational models try to specify are the rules that need to be followed in order to produce the specific input-output transformations thought to be necessary for the execution of a given psychological capacity. Part of this endeavor is concerned with defining the computational constraints that govern neural systems, such as defining the computational tractability of an information processing problem or establishing time-related limits to the processing capacity of a given system. Part of the appeal and rationale behind the booming efficient coding research program is precisely a specification of the minimal resources to be employed on different computational operations (see [37] for a careful assessment of the explanatory profile of this sort of minimal models). Clearly, this is partly theoretical work but also a much-needed effort to channel laboratory research by making testable predictions and refining experimental questions.

A case of direct computational interpretation of neural activity can be seen in Behrens et al. [38], a rich review of computational roles attributed to different brain areas thought to be responsible for reward-guided behavior. An interesting example is the case of reinforcement learning algorithms, which state that "future expectations should be updated by the product of the prediction error and the learning rate" ([38], p. 1160). Midbrain dopamine neurons, projecting to the ventral striatum, have been attributed not only the role of predicting expected reward but also that of quantifying the associated deviation in observed reward. Specific model parameters and relative deviations have then been experimentally tested by recording neuronal activity via electrophysiological and neuroimaging methods.

Mechanistic models are the most common in neuroscience, and SN is no exception in this regard. They have recently received an unprecedented degree of attention by philosophers of neuroscience, especially concerning the problem of scientific explanation (see Sect. 3). While they can be understood in terms of Marr's level of implementation, "mechanistic" philosophers have construed this kind of models as part of a whole program of research in the field. For our present purposes, it suffices to say that mechanistic models ideally aim at specifying the set of relevant component parts, features, activities, and organization of the system causally responsible for a given neural or behavioral phenomenon. The identification and specification of a mechanism's structure can be realized on different spatiotemporal levels of the brain's structure, as mechanisms are thought to be hierarchically organized (at least according to the most popular versions such as Carl Craver's or William Bechtel's).

To exemplify, consider available research on oxytocin's role in social phenomena. Oxytocin has been strongly linked to attachment and maternal behavior. Insel and Young [39] review a number of mainly animal studies from molecular, cellular, and systems approaches, which jointly specify oxytocin's contribution to this special kind of selective behavior between a mother and her offspring. The model the authors present follows oxytocin receptors' activity along different pathways and in different cortical and non-cortical brain areas, while also assigning specific functional roles to this activity, both neutrally (such as increasing the activity of noradrenaline cells in the brainstem) and behaviorally (such as decreasing aggressive behavior toward the offspring).

Finally, dynamical models have also been discussed in the philosophical literature. These models focus on the temporal properties of a previously defined system—usually through systems of differential equations—analyzed through mathematical tools derived from general frameworks such as dynamical systems theory and graph theory. Typically, the modeled systems' parameters span the agent's brain and body, as well as relevant features of the environment, meeting a general rejection of the common strategy of partitioning cognitive systems into dedicated components. The research led by Ezequiel Di Paolo on different facets of social behavior is an example of this kind of highly interactive modeling (see, e.g., [40]). In the case of SN, this sort of models is at the moment still in its infancy, of a mostly qualitative nature, and hinging almost exclusively on behavioral parameters. Still, there is a tendency, specially stemming from systems neuroscience to model high-order parameters for large-scale neural systems. How this will unfold for specifically social phenomena will probably be seen in the short term.

3 Explanation in Social Neuroscience

Having reviewed the heterogeneity of modeling practices in SN, we can turn now to the issue of when these models explain. Scientific explanation has been a widely debated subject in the philosophy of neuroscience [33, 37, 41, 42]. In this section, we first introduce some "unitary" perspectives about explanation in neuroscience.

Scientific models can be analyzed considering two main features: (a) their contents or truth-conditions and (b) their vehicles, formats, or representational bearers. Unitary approaches to explanation may hold that explanatory models in neuroscience share the same kind of vehicle, the same kind of content, or both. Examples of unitary approaches are the deductive-nomological model [43] and mechanistic explanation [41, 44, 45]. We argue that these approaches seem to be inappropriate considering the diversity of explanatory practices in SN. Thus, we advance a pluralistic account for model-based explanation in SN. According to explanatory pluralism (EP), models in SN may be explanatory even when they do not exhibit the same kind of representational format nor the same kind of truth-conditions. Explanation in neuroscience, and particularly in SN, requires that modelers evaluate and selectively emphasize different representational ideals to represent different kinds of (causal and/or noncausal) structures in the brain. We think that SN provides an excellent case study for the development of a pluralistic perspective on the explanatory strategies and ideals that partially shape neuroscientific practice.

Concerns about the nature of explanation have a long history in philosophy of science. The first systematic treatment of this subject is Hempel and Oppenheim's classic "Studies in the logic of explanation" [46]. In that paper, they introduce the "deductive-nomological" (DN) model of explanation. The DN model conceives scientific explanation as an inference in which a sentence describing some aspect of an explanandum phenomenon is inferred as a logical consequence from premises describing true laws of nature and information about the antecedent conditions. The key feature of DN explanations is the nomic expectability of the explanandum phenomenon in light of the laws of nature (and the antecedent conditions) described in the explanans.

Several authors have raised serious conceptual concerns about the DN model of explanation. Just to mention some of the main problems, the account does not provide clear criteria to distinguish between true laws and accidental generalizations; it cannot account for the characteristic asymmetry of explanations, and it cannot exclude as non-explanatory inferences based on mere nomic covariations see, [41, 47]. In conjunction with these problems, the DN account does not seem to be representative of the kind of explanations employed in some special, "fragile" sciences, such as biology, neuroscience, or psychology, in which the search for universal laws of nature is at least peripheral. Attending to this feature of special sciences, some authors have claimed that in these disciplines where general laws are scarce and theoretical approaches are not as consolidated as in physics, explanations may adopt a different style.

It has been claimed that explanations in neuroscience and other biological sciences frequently do not address why questions (inquiring on the general conditions that determine the production of the explanandum phenomenon), but rather how questions (concerning the particular way in which the target system, be it cognitive or neuronal, subserves a given higher-level capacity) [41, 42, 48]. In these cases, explanations do not need to exhibit a clear propositional format and may instead involve presenting a scientific model of the underlying local "mechanism" that produces the phenomenon [49].

A scientific model provides a mechanistic explanation of an explanandum phenomenon to the extent that it identifies some aspects of the mechanism responsible for the phenomenon. In particular, a mechanistic model explanation usually involves decomposing the target mechanism into its parts or constituent entities, the activities of those entities, and their organization. This process of decomposition is iterative; thus, the parts identified in a first stage can be further decomposed into subparts. As a result, mechanistic explanations span multiple levels of a mechanism [41, 50]. Finally, this kind of explanation has a local scope, that is to say, mechanistic models are developed for explaining a particular phenomenon and do not extend beyond it. Therefore, the generalizations obtained by this type of explanation are often characterized as limited in scope, mechanistically fragile, and historically contingent ([41], pp. 66–70; [51]).

3.1 The Plurality of Model-Based Explanation in Social Neuroscience

The EP approach we will develop here recognizes both a plurality of representational ideals that may shape explanatory models in neuroscience and a plurality of different kinds of structures (i.e., causal and noncausal) that may be represented by those models. Specifically, we propose that the explanatory heterogeneity of SN can be fruitfully approached by differentiating two main aspects of scientific model explanations: (a) *their content* or truth-conditions, i.e., the kind of structures in the world a model must effectively represent in order to be explanatory, and (b) *their vehicle* or representational format, which may be embodying different representational ideals, like precision or accuracy. Evidently, these two aspects are intimately related in scientific practice. Nevertheless, the claim we want to advance here is that the distinction between them can provide a good framework for analyzing and assessing the explanatory credentials of scientific models in neuroscience.

The *content* of a model explanation is the information it provides about the phenomenon. Depending on the kind and the extent of information it provides, a model may be considered an acceptable explanation. We identify two kinds of content that an explanation may provide about its target system, namely, causal/compositional or noncausal/structural information. On the one hand, scientific models may provide causal explanations by identifying relations of causal dependence, either etiological or constitutive, among the explanandum phenomenon, antecedent conditions, and/or features of the mechanism underlying the phenomenon. This kind of content allows scientists to manipulate and control both the phenomenon and its mechanism in quite precise ways [41, 52]. On the other hand, scientific models may provide noncausal information about the target system. This kind of information includes, for example, the exhibition of counterfactual dependence relations between the design features of the target system and abstract environmental constraints [53]. It could also include purely mathematical relations between empirical phenomena or information about the topological structure of the system. These dependence relations cannot be considered causal, since they are not diachronic nor do they necessarily ground experimental interventions. Furthermore, these relations may not be altered by changing the mechanistic realization of the target system in substantive ways: they are robust [54, 55]. Note that the two kinds of explanatory information a model may provide are perfectly compatible, and both make an important contribution to a thorough understanding of the phenomenon of interest.

Turning now to the *vehicle* of explanation, it may be characterized as the representational bearer of the explanation, that is, the kind of representational structure by which the explanatory information is conveyed, for example, linguistic statements, schematic diagrams, computational simulations, and mathematical equations. These vehicles allow scientists to represent different aspects of the phenomenon of interest and its underlying "mechanism," that is, to represent the intended content of the model. The choice of one representational vehicle over another is guided by several different representational ideals [56], and often modelers are forced to choose a particular vehicle considering the trade-off between different ideals. This is not a novel notion: Levins [57] had already pointed out that modelers often consider the trade-off among at least three representational ideals that cannot be maximized simultaneously: precision, generality, and realism. This trade-off may force some modelers to prioritize the precision and realism of a particular model, for example, in detriment of its generality. Taking into account the differences among the above representational ideals, one of us [58] has advanced a distinction between a mechanistic style, in which modelers tend to privilege structural details and realism, and a functionalist style, in which the ideal of generality is emphasized. The moral is that modelers have to find a preferred balance between the different representational ideals, selecting the most appropriate vehicle for representing the content they are interested in.

Some representational ideals in neuroscience and elsewhere in science are precision, simplicity, and generality. The ideal of precision involves the maximization of the representation's level of detail, either of structural features, component entities and activities, or temporal and spatial features of the system. The ideal of simplicity refers to the search of a model that maximizes the intelligibility of the phenomenon under study and its underpinnings. In many cases, meeting the ideal of simplicity may require scientists to abstract the model from irrelevant details and introduce idealizations. Finally, the ideal of generality refers to the model's ability to be applied across several domains and extrapolated to different target systems. Again, these representational ideals are intimately related to the kind of explanatory information that is conveyed. The analysis we propose might just provide a more complete toolbox for disentangling the varieties of explanation in neuroscience and SN.

With this framework for the analysis of a model's explanatory virtues in place, we now examine some representative cases in SN to exemplify usefulness of this approach. In this direction, we get back to two of the cases presented in the previous section exemplifying different kinds of models: the role of oxytocin in attachment [39] and the mathematical model of reinforcement learning of different patterns of activity related to decision-making processes [38].

Different mechanistic models have addressed the role of oxytocin in attachment [39]. These models have a causal content that pinpoints to a neurobiological mechanism including oxytocin as a major component: the models additionally attempt to determine its activities. The representational structure in these cases often involves diagrams and is guided by the representational ideals of precision and simplicity. The main objective consists in detailing the neural circuits, the different molecular components involved, and their organization related to behavioral expressions of attachment. Here lies the precision ideal displayed by these models. At the same time, the causal structure related to attachment is abstracted from other causal processes and different changes that may be induced in front of different contextual situations. Here we can appreciate the ideal of simplicity followed.

Consider one of the models presented in Insel and Young's review: oxytocin and the bonding behavior that sheep show toward their lambs. The selective and permanent bond appreciated within the 2 h of parturition has been explained by a neurobiological model that posits that:

> Afferent stimulation through the spinal cord from vaginocervical dilation during parturition increases the activity of noradrenaline-containing cells in the brainstem which project to the paraventricular nucleus (PVN) in the hypothalamus as well as to the olfactory bulb. Stimulation of oxytocin cells in the PVN facilitates maternal behaviour through coordinated effects on several regions in which oxytocin increases GABA (γ-aminobutyric acid) and noradrenaline release. Oxytocin in the olfactory bulb and medial preoptic area reduces aggressive or aversive responses to newborn lambs. Oxytocin in the mediobasal hypothalamus inhibits post-partum estrus ([39], p. 2).

This brief extract illustrates how maternal attachment in sheep is explained by a model that identifies different components involved (e.g., oxytocin, noradrenaline), their activities (oxytocin increases GABA release), and their organization.

In another direction, a structural content may be identified in the reinforcement learning model proposed for explaining different social phenomena [38]. In this case, an abstract mathematical structure is employed for expressing the main nuclear organization responsible for different patterns of activity. This model has a mathematical representational bearer (even though it could be represented in computational structures as well), to which two representational ideals may be related: generality and simplicity. Specifically, Behrens et al. [38] show how the simple structure "$Vt + 1 = Vt + atdt$", which includes expectations of future reward ($Vt + 1$), current expectations (Vt), and their discrepancy from the actual outcome that is experienced—the prediction error (dt)—could be related to different patterns of activity observed in decision-making processes. In this case, social phenomena and the activity identified in different brain areas related to them are not explained in terms of precise component activity of neurotransmitters but instead in a more abstract equation that may relate expectancies, previous experience, and reward independently of the specific neurobiological structures that are involved in these functions in different cases. The authors have emphasized that the characteristic abstractness of these formal models makes them suitable for relating information about different neural activities involved in complex social phenomena from different species. In their own terms: "Such a mathematical formalism defines explicit mechanistic hypotheses about internal computations underlying regional brain activity, provides a framework in which to relate different types of activity and understand their contributions to behavior" ([38], p. 1160).

3.2 An Evaluation of the Mechanistic Unitary Approach and Explanatory Pluralism in Social Neuroscience

For some mechanist philosophers, the ideal of mechanistic precision is a universal constraint on the vehicles of explanation (e.g., [45]). In this sense, more detail is always better. This kind of mechanistic approach does not recognize the diversity of ideals that may guide different models nor the trade-off among different representational ideals that is present in many modeling scenarios [33, 58, 59]. Other mechanists endorse [45] the idea that the same target system in neuroscience may be represented by a multiplicity of scientific models, each of them emphasizing a different aspect of the mechanism by selectively emphasizing some representational ideals more than others [60, 61]. However, virtually all mechanist philosophers endorse some kind of unitary approach concerning the content of model-based explanation. According to content unitary perspective, a scientific model provides explanatory information only to the extent that it identifies causal dependence relations underlying the phenomenon of interest [45, 60]. This unitary stance about content implies that cognitive or computational models in cognitive neuroscience, as well as in SN, are just incomplete sketches of mechanisms and that purely dynamical models are mere phenomenal, not explanatory models. We reject content unitary perspectives about explanation in neuroscience and SN.

What is explanatory pluralism? A first claim that should be made is that admitting a plurality of vehicles and contents for model-based explanation in SN should not be equated to the assumption that "anything goes" in explanation or to "the advocacy of retaining all, possibly inconsistent, theories that emerge from a community of investigators" ([62], p. 85). On the contrary, we think that the representational virtues proposed to contribute to a model's explanatory power should be clearly stated. In this sense, a fine balance must be achieved between admitting a plurality of explanatory vehicles and contents and the indistinctive inclusion of any proposed model in the set of explanatory models.

A second issue that we should take into consideration is that the notion of EP has been defined in multiple ways by different authors [33, 37, 62, 63]. To clarify the particular approach we propose here, it is useful to differentiate among three ways in which EP has been defined, to wit: (1) EP about *explanatory levels*, (2) EP about *representational structures*, and (3) EP about *explanatory styles*.

EP about explanatory levels emphasizes the existence of explanations at different levels of entities or size scales, a claim that contrasts with ruthless reductionist perspectives about explanation in neuroscience, like the stance advocated by Bickle [64–68]. The main thesis of EP concerning levels is that in order to explain some phenomenon, entities at different compositional levels or size scales must be relevant. These entities usually are studied from different disciplines or fields, and all these perspectives at different levels of organization should be considered. In addition, it is usually claimed that all perspectives from different levels are complementary to each other and must be ideally integrated. The kind of integration that is expected ranges from complete autonomy to smooth mechanistic integration (see Sect. 4).

EP about *representational structures* admits the possibility and desirability that different scientific representations successfully pick out the same target system, i.e., "the same system in neuroscience can be represented and modelled in a variety of different ways, depending on the particular purposes of the investigation" ([37], p. 148). This conception implies that different representational bearers might be used in perfectly solid explanations of a given phenomenon. Nevertheless, EP about vehicles remains silent about the kind of informational content the different models must convey in order to be explanatory. A philosopher may adopt a unitary stance about the content of explanation, for example, endorsing a causal conception about the contents of explanation and nevertheless admit a plurality of representational structures for representing causes (mathematical equations, computational simulations, visual schemata, etc.).

Finally, EP about *explanatory styles* embraces the idea that different styles of explanation or explanatory virtues should be admitted as providing legitimate explanations [63]. The late Wesley Salmon has suggested this kind of pluralism, when he affirmed that:

> [I]t might be better to list various explanatory virtues that scientific theories might possess, and to evaluate scientific theories in terms of them. Some theories might get high scores on some dimensions, but low scores on others (…) I have been discussing two virtues, one in terms of unification, the other in terms of exposing underlying mechanisms. Perhaps there are others that I have not considered. ([69], p. 20)

Considering that EP about levels or representational structures is not incompatible with unitary accounts about explanatory styles, we consider this third kind of pluralism the most accurate for discriminating between unitary and pluralistic accounts of explanations.

According to our approach, the three kinds of EP are compatible and, in fact, we endorse them all. The idea of a single scientific representation that describes the behavior of the entities that are relevant for a phenomenon at the most fundamental level, that meets all the representational ideals that are appreciated by modelers, and that captures all the causal and noncausal features of the target system is a philosopher's fiction that covers our eyes to the diversity of explanation in neuroscience [70]. Considering model-based explanation in physics, Cartwright [71] has proposed a similar "patchwork" metaphor, according to which different models would be needed to account for the phenomenon under study. In the same direction, Weisberg [56] has highlighted a kind of "idealization of multiple models" which scientists are forced to resort to when dealing with highly complex phenomena. The idea is that there is a variety of explanatory styles in neuroscience and SN, each of them emphasizing different explanatory virtues (in Salmon's sense); as a result, different types of vehicles are used by modelers to and explain to convey causal and noncausal information about different aspects of the target system, thus making very different assumptions about it.

4 The Unity of Social Neuroscience

SN, in the particular strand we are considering here, is an interdisciplinary research program that studies the neurobiological (neuronal, endocrine, and immune) processes that enable social cognition and behavior [72, 73]. The advancement of this scientific field requires the collaboration of researchers from many distinct disciplines, such as cognitive neuroscience, neuropsychology, cognitive science, neuroendocrinology, cellular and molecular neuroscience, social psychology, economics, and political science (see Salles and Evers, this volume). Many neuroscientists and philosophers of neuroscience see theoretical *unity* as a preeminent goal of neuroscience in general and SN in particular [41, 74–76]. How is the unity of SN achieved? In this section, we review three philosophical models of the unity of neuroscience and assess their validity vis-à-vis the modeling and experimental practices aimed at explanation in SN.

The first philosophical model we consider posits that the process of unification proceeds via a kind of reduction in practice [64, 77, 78]. The common experimental technique that grounds this reduction in practice is to intervene causally at lower levels of biological organization (e.g., cellular and molecular levels) in animal models and then to track the specific effects of these interventions on behavior in widely accepted experimental protocols for the target phenomena ([78], p. 230). The empirical success of this reductive experimental technique motivates a "ruthless reductionist" stance, i.e., one according to which, if a class of cognitive phenomena depends upon some molecular mechanisms that can be tracked experimentally, then the research on those molecular mechanisms assumes a kind of methodological priority ([78], p. 232).

Bickle's preferred exemplar of this kind of ruthless reductive unification in social neuroscience is the experimental work on the molecular basis of social recognition memory consolidation in mice [79]. Social recognition memory consists in the ability to remember and recall information tied to particular conspecifics after an initial episode of interaction with them. A standard behavioral protocol aimed to operationalize the concept of social recognition memory is based on Thor and Holloway's [80] idea that, "in the laboratory, social memory can be assessed reliably by measuring the reduction in investigation time of a familiar partner relative to a novel conspecific" ([81], p. 202). Furthermore, social recognition memory is considered to be dependent on the hippocampus, and, as many other forms of hippocampal-dependent long-term memory consolidation, it may be dependent on the activation of cyclic adenosine monophosphate (cAMP) responsive-element binding (CREB) proteins, especially two of its isoforms, α and δ (p. 232). To test this possibility, Kogan et al.

[79] obtained CREB$^{\alpha\delta}$ mutant mice—mice that show no expression of CREB α and δ isoforms—and trained a group of these mutants and a group of wild-type mice in a modified version of Thor and Holloway's [80] behavioral protocol for social recognition. They found that mutant mice CREB$^{\alpha\delta}$ engaged in social investigation (e.g., sniffing) of a given mouse to the same extent after 24 h as they did upon an initial encounter with the same individual. They interpret this finding as implying CREB$^{\alpha\delta}$ mutant mice are impaired in their social recognition abilities and, therefore, that long-term social memory is dependent on CREB function.

The main problem with Bickle's ruthless reductionism is that he seems to think that reduction in practice justifies global reductive claims concerning the molecular basis of some general phenomenon exhibited by organisms in the world (e.g., the molecular basis of social recognition memory *tout court*). However, it is not clear that the particular intervention undertaken by Kogan, Frankland, and Silva directly explains the data observed by another researcher in another laboratory studying the same phenomena but through distinct experimental designs and protocols [82]. What the "intervene molecularly and track behaviorally" technique brings about are "local within-experimental-protocol reductions," and it is not at all clear how these within-lab reductions will converge toward a global reductive claim concerning a general cognitive phenomenon ([82], p. 518). Furthermore, there is the problem of extrapolation. Bickle [78] emphasizes that the same molecular mechanisms for social recognition obtain across a wide variety of different species, from *Drosophila* to *Aplysia*. However, there are species-specific differences that question the generalizability of results obtained in mice to nonhuman primates and human beings [82]. For example, while in most non-primate mammals, social information is encoded via olfactory or pheromonal signals, in human and other primates, individual recognition relies on visual or auditory cues ([77], p. 201). Correspondingly, there are interspecies differences in the brain areas involved in the formation of social recognition memory. These differences cannot be neglected and prevent the sheer elimination of higher-level analyses concerning brain mechanisms that may underlie social cognition and behavior.

The second philosophical model of the unity of neuroscience (and, arguably, SN) incorporates the non-reductive and multilevel character of explanation as a central feature of the account. According to Kaplan and Craver ([45], p. 268), neuroscience is especially interesting to philosophers of science, among other reasons, because it is an interdisciplinary research community that "exemplifies a form of scientific progress in the absence of an overarching paradigm" (cf. [83]). How is this integration possible? Mechanist philosophers claim that the unity of neuroscience is effective when researchers from different scientific fields collaborate to build multilevel mechanistic explanations ([41], p. 18; see also [60, 84]). The product of this collaboration is an "explanatory mosaic" in which distinct scientific models "contribute piecemeal to the construction of a complex and evidentially robust mechanistic explanation" ([41], p. 19). Mechanistic explanations, in this sense, are built from the accumulation of constraints from different fields on the space of possible mechanisms for a given phenomenon. A constraint is a piece of information that shapes the boundaries of the space of possible mechanisms or changes the probability distribution over that space, i.e., the probability that some region of the space describes the actual mechanism. The constraints from different scientific fields are used, like the tiles of a mosaic, to shape the space of possible mechanisms provided by mechanistic research programs.

Embracing mechanistic integration as a working hypothesis, many mechanists accept that modeling strategies from different fields are autonomous to the extent that each of these fields is free to choose which phenomena to explain, which experimental designs to apply, which conceptual resources to adopt, and the precise way in which they are constrained by scientific evidence from adjacent fields [41, 60, 84]. Against Bickle, they claim there is no methodological preeminence of molecular approaches to target phenomena in neuroscience. In fact, the capability of scientific fields to contribute novel constraints to a mechanistic research program demands their relative autonomy: "Because different fields approach problems from different perspectives, using different assumptions and techniques, the evidence they provide makes mechanistic explanations robust" ([41], p. 231). The ideal of a mosaic unity of neuroscience is congenial with Cacioppo and Decety's ([75], p. 166) emphasis on multilevel analysis in SN, that is, the idea that SN "necessitates the integration of multiple levels, and the explication of the mechanisms that link phenomena across these levels."

An example of mechanistic integration in SN comes from research on oxytocin and arginine vasopressin (AVP) as components of the mechanism for pair bonding in monogamous rodents [85–88]. The term "monogamy" refers to a social organization in which each member of a mating pair displays selective affiliation and copulation, nest sharing, and typically biparental care of offspring [87]. Voles provide valuable animal models for comparative studies on the neurobiological mechanisms of pair bonding [89]. Prairie voles (*Microtus ochrogaster*) exhibit a monogamous organization, forming enduring pair bonds following mating. Montane (*Microtus montanus*) and meadow (*Microtus pennsylvanicus*) voles, in contrast, are nonmonogamous species. The experimental protocol that is used in the lab in order to operationalize the concept of pair-bond formation is the partner-preference test. The experimental design includes an apparatus consisting of three chambers connected by tubes. The subject is allowed to move freely throughout the apparatus, while the "partner" and a novel "stranger" are confined to their own chambers. Pair bonding is considered to be present when the subject spends more time with the partner compared to the stranger [87]. The nonapeptides oxytocin and AVP emerged as constitutively relevant components of the mechanism for intense social attachment in voles. While oxytocin seems to be more important in females, AVP is more important in males. Thus, infusion of oxytocin into the cerebral ventricles of female prairie voles facilitates pair bonding, while AVP infusion facilitates pair bonding in male prairie voles. Furthermore, administration of selective oxytocin receptor and AVP receptor 1a (V1aR) antagonists blocks each of these behaviors in females and males, respectively. Considerations from systems neuroscience and evidence from anatomical and pharmacological studies are also relevant to constrain the space of possible pair bonding formation mechanisms. Compared to nonmonogamous species, female prairie voles have higher densities of oxytocin receptors in the prefrontal cortex and nucleus accumbens, while male prairie voles have higher densities of AVP receptors in the ventral pallidum, medial amygdala, and mediodorsal thalamus [88]. These studies indicate that the prefrontal cortex, nucleus accumbens, and ventral pallidum are critical brain regions involved in pair-bond formation. Since these areas are also involved in the mesolimbic dopamine reward system, some researchers have hypothesized that pair bonding may be the result of conditioned reward learning. In this model, "the reinforcing, hedonic properties of mating may become coupled with the olfactory signatures of the mate, resulting in a conditioned partner preference," much in the way drugs of abuse work ([85], p. 1052).

There are two problems affecting the mechanistic ideal of a mosaic unity of (social) neuroscience. According to one criticism, mechanistic integration is too demanding. As mentioned when assessing the ruthless reductive account, within any field in neuroscience (and social neuroscience is not an exception), there is a multiplicity of experimental protocols associated with the "same phenomenon," so it is not at all clear how results obtained from different laboratories, using different experimental protocols, can fit together within a field, before the combined results of that field can be said to set constraints on the space of possible mechanisms for a phenomenon ([82], p. 525). Furthermore, even if a researcher identifies a working part or activity in the mechanism of pair bonding in rodents, it may not be immediately clear that that piece of evidence will constrain the space of possible mechanisms for pair bonding in humans [82]. In this sense, Young and Wang ([85], p. 1052) strongly emphasize that "there are no hard data demonstrating common physiological mechanisms for pair-bond formation in voles and man" and that "the emergence of the neocortex and its ability to modify subcortical function cannot be ignored." The two facts just mentioned are closely related: the multiplicity of experimental protocols concerning a target phenomenon arises in part because the phenomenon itself varies in different species, involving different mechanisms in different species [82].

Moreover, the philosophical issue concerning the level of discontinuity between human and nonhuman minds becomes relevant at this point. Against the dominant tendency in comparative cognitive psychology, Penn, Holyoak, and Povinelli [90] defend the hypothesis that there is a significant functional discontinuity in the degree to which human and nonhuman animals are able to approximate higher-order, abstract, relational capabilities of a physical symbol system. According to their *relational reinterpretation hypothesis* ([90], p. 111), although both humans and nonhumans are capable of learning and acting on the perceptual relations between different aspects of the world, only humans are capable of reinterpreting those relations in a systematic and productive way. For these researchers, the functional discontinuity between human and nonhuman minds pervades nearly every domain of cognition. Particularly, only humans can master general concepts based on structural criteria (beyond any particular source of stimulus control), find systematic analogies between disparate domains, draw logical inferences between higher-order relations, or postulate unobservable mental causes or physical forces as explanations of natural phenomena ([90], p. 110). If they are right and nonhuman minds approximate the capabilities of a physical symbol system to a significantly lesser degree than human minds do, then the prospects of reductionistic or mechanistic integration *across* species are dim.[3]

The second criticism to mechanistic integration points in the opposite direction. Recently, Levy [76] has argued that the mosaic ideal of unity is too minimal, i.e., a version of unity that is "overly modest and for that reason not very attractive." In particular, what the mosaic ideal of unity does not require is the existence of shared theoretical content among the constraints on the space of possible mechanisms for a target phenomenon, that is, "general concepts, principles and explanatory schemas applying across a range of neuroscientific phenomena" ([76], p. 10). Levy compares Craver's "tiles" in the mosaic unity of neuroscience to members of an alliance, i.e., independent states joining efforts. He encourages a stronger, "federal" ideal of unity, in which a set of distinct states are united by general principles. Noticeably, Bickle's ruthless reductive account eschews this problem, since the general principles that unify the different fields of neuroscience are the principles and laws of physics and chemistry that determine molecular and cellular processes within the brain, since "to the extent that we have explained some 'higher level' phenomenon as a sequence (…) of molecular steps, we know that the only way for another 'higher level' process to employ it (…) is via molecular (or lower) mechanisms" ([78], p. 232). The common principles Levy [76] has in mind are not Bickle's physicochemical principles but abstract, recurrent patterns that, according to some recent (and rather speculative) theoretical work in neuroscience, transcend spatial and temporal scales and apply to a range of neural systems. As an example, Levy mentions Sterling and Laughlin's [91] principles of efficient design that apply to the brain as a whole and to different regions at different temporal and spatial scales. One of such principles of neural design is to "minimize wire" (i.e., axon length), which explains, for instance, the placement of ganglia in *Caenorhabditis elegans* nervous system [92] and the organization of neurons in cortical maps in the mammalian visual cortex [93]. Design explanations of this kind allow us to answer why questions such as: Why are neurons in the mammalian visual cortex organized in maps? Or why are neural circuits separate in layers, columns, stripes, or barrels? ([91], p. 446).

There is a very popular research program in SN that aims to provide answers, from the designer perspective to the kind of why questions just mentioned. Given the extraordinary cost of neural material [94], Dunbar [95] asks: why do primates (in particular) have unusually large brains for body size, compared to all other vertebrates? Dunbar's preferred proposal is the social brain hypothesis. According to this hypothesis, large brains are a consequence of natural selection for enhanced social skills, since "an individual's fitness is maximized by how well the group solves the problems that directly affect fitness, and this in turn is a consequence of how well bonded it is (this in turn being a consequence of the individual member's social cognitive skills)" [95]. The social brain hypothesis points to the bondedness of social groups as the intermediate step between brain size and the selective pressures driving brain evolution [96].

---

[3] We thank Warren TenHouten for bringing this issue to our attention.

There are some direct counterexamples to the social brain hypothesis. Lemurs, for example, live in relatively big social groups but have relatively small brains. Some authors have raised deeper concerns about the social brain hypothesis. Cachel ([97], p. 373) contends that if the social brain hypothesis were valid, then we would expect that our closest primate relatives, i.e., the chimpanzee and the bonobo, would exhibit the most complex primate sociality. However, the only truly eusocial nonhuman primates are some New World monkeys, like the tamarins, which exhibit cooperation in the care of their young, reproductive division of labor, and overlap of two or more generations contributing to social life ([98], p. 62), and monkeys are in several respects less intelligent than pongids. Furthermore, according to Cachel, there is a trade-off between social intelligence and natural history intelligence, and only the latter constitutes a principal factor contributing to the formation of a general human-like intelligence. Competitive social behavior is highly demanding in terms of attention and other cognitive resources and also discourages exploration of the natural world. Vervet monkeys, for example, exhibit acute social awareness but are "peculiarly obtuse or stupid about making associations and predictions about the external world" ([97], p. 165). TenHouten states [97]: "Freedom from hypersociality is necessary for the development of complex, symbolic models of the world that can then be subjected to abstract cognition and executive-level decision-making."[4]

In this section, we use the social brain hypothesis merely as an example of an abstract design principle on brain architecture, without endorsing it as a working hypothesis. The rationale behind the social brain hypothesis can be further specified as follows: "Members of social species, by definition, create organizations beyond the individual. These super-organismal structures evolved hand in hand with psychological, neural, hormonal, cellular, and genetic mechanisms to support them" ([75], p. 163). From the standpoint of the social brain hypothesis thus formulated, SN is not a mere alliance of disciplines gathered by the common goal of explaining some target phenomena but represents a broad theoretical paradigm in neuroscience, "a general perspective that underlies a range of theories and methodologies in the field," which presupposes that many central aspects of brain organization and function only make sense in the light of social organization and vice versa ([75], pp. 162–163).

We have reviewed three philosophical accounts of the unity of neuroscience in general and SN in particular. First, SN may become integrated by molecular reductions of social behavior. The challenge for this reductive approach is to account for the existence of a multiplicity of experimental protocols for a given phenomenon, given the different manifestations of that phenomenon across different species. Second, SN may become integrated by the piecemeal accumulation of constraints from autonomous fields on the space of possible mechanisms for the target phenomenon. The challenge for the mechanistic account is twofold. On the one hand, mechanist philosophers have to explain how different results from different laboratories become integrated within a field and how they can be extrapolated from one species to another. On the other hand, the ideal of a mosaic unity may be too minimal, since it does not require the existence of shared theoretical content. In the third place, the

[4] We thank Warren Tenhouten for drawing our attention to these concerns.

unity of SN may be achieved by common general principles and concepts, such as the social brain hypothesis. However, the debate concerning the principles of design and evolution of the social brain is still open. Experimental and modeling practices in SN seem to be quite independent from the development of that theoretical debate. In the absence of general design principles, the multiplicity of experimental protocols becomes the main feature of SN as a laboratory science. A kind of non-reductive pluralism [33, 37, 82, 99] in which that multiplicity is not neglected seems to be the most sensible position concerning the unity of SN at this stage of development.

According to Sullivan ([82], p. 534), there are two fundamental constraints on the experimental process that account for to the multiplicity of experimental protocols in neuroscience: reliability and external validity. These two constraints pull in opposite directions ([82], p. 535). Reliability prescribes simplifying measures in order to keep control in the laboratory and discriminate between competing hypotheses about a laboratory effect. External validity prescribes building into the experimental design as much complexity as possible in order to capture the phenomenon of interest, outside the laboratory. Thus, there is a trade-off between reliability and external validity. This trade-off sheds light at least on some points of intersection of the neural and the social. As emphasized by Callard and Fitzgerald ([100], p. 60), the need for more ecologically valid models (particularly regarding the social environment) in animal research is one of many arenas that would benefit from greater interdisciplinary collaboration between neuroscientists and social scientists.

Consider, for example, the physiological and psychological effects on rodents of laboratory housing conditions [101]. Practically all laboratory-housed rodents live in small "shoe-box" cages which afford little meaningful biological complexity. Physiological and behavioral studies strongly indicate that social isolation is detrimental for rats and mice and that company can be enriching and beneficial. In rodents, usual laboratory conditions may cause impairments in the neural and behavioral development and behavioral stereotypies. Stereotypies are uncommon in free-living wild animals, and they may be caused by the frustration of natural behaviors like finding food or mates, building nests, and avoiding predators. Since animals with stereotypies are poor models of normal behavior, implementing social environmental enrichment is needed in order to regain external validity. In fact, researchers using more naturalistic housing methods have detected deficits in transgenic mice that had been neglected in conventional laboratories [102]. A pluralistic approach predicts that such an increment of external validity will imply an attenuation of experimental reliability and the negotiation of a new equilibrium point between these two constraints of the experimental design.

5 Conclusion

In this chapter we introduced three general philosophical issues stemming from recent and actual research in the field of SN. These issues have become increasingly prominent in the literature and prove highly relevant for the present and near future of SN. In particular, the philosophy of modeling has been intensely debated generally in the philosophy of science. Here, we addressed the philosophical problem of how scientific models and theories relate, while characterizing the different kinds of models and modeling approaches relevant in contemporary SN. Secondly, we presented the issue of scientific explanation, certainly a hot topic in recent philosophy of neuroscience: it can be argued that this problem is responsible for a great deal of the boost philosophy of neuroscience had during the last two decades. The third issue, scientific integration, is, in our opinion, a much pressing topic specifically for SN. The ways different aspects of SN research can be articulated and put into fruitful dialogue, considering specially the characteristic nature of this ambitious neuroscientific approach to social phenomena, are in need of detailed philosophical attention and, we think, will certainly be soon increasingly debated within the philosophical community.

Although we made an effort to present the issues without taking clear-cut sides, we defended a general pluralistic stance toward SN. We started from a resolute recognition of the diversity of modeling approaches today being developed and of the epistemic roles that models can and do play in SN. We then proposed a kind of EP that admits that models tackling different levels, representational bearers, and styles of explanation may be considered legitimately explanatory. In fact, we defended the idea that this plurality is desirable in order to reconstruct the "patchwork" picture of such a complex field as is SN. It is important to highlight, though, that this idea is not equivalent to an "anything goes" principle, and we here suggested a clear framework that might be useful when analyzing the explanatory virtues of different models in SN.

Finally, we reviewed a central question concerning SN: how to best approach the unity of the field. A philosophical account of integration in SN requires an explication of the way in which different empirical results from different laboratories can become integrated within the field and how they can be extrapolated from one model species to another. We have argued that non-reductive pluralism is the most adequate approach to these problems concerning extrapolation and the multiplicity of experimental protocols.

References

1. Cacioppo JT, Berntson GG. Social psychological contributions to the decade of the brain. Doctrine of multilevel analysis. Am Psychol. 1992;47(8):1019–28.
2. Dunbar RIM. Neocortex size and group size in primates: a test of the hypothesis. J Hum Evol. 1995;28(3):287–96.
3. Dunbar RIM. The social brain hypothesis. Evol Anthropol Issues News Rev. 1998;6(5):178–90.
4. Bailer-Jones DM. Scientific models in philosophy of science. Pittsburgh, PA: University of Pittsburgh Press; 2009.
5. Morrison M. Where have all the theories gone? Philos Sci. 2007;74(2):195–228.
6. Cartwright N, Shomar T, Suárez M. The tool box of science: tools for the building of models with a superconductivity example. Poznan Stud Philos Sci Humanit. 1995;44:137–49.

7. Harre R. Cognitive science: a philosophical introduction: SAGE Publications; 2002. p. 344.
8. Morgan MS, Morrison M. Models as mediators: perspectives on natural and social science. Cambridge: Cambridge University Press; 1999. p. 420.
9. Venturelli N. Un abordaje epistemológico de la integración neurocientífica. In: Rodriguez V, Velasco M, editors. Epistemología y prácticas científicas. Córdoba: Editorial Universitaria; 2015. p. 41–71.
10. Churchland PS, Sejnowski TJ. The computational brain. Cambridge, MA: MIT Press; 1994. p. 564.
11. Hardcastle VG, Stewart CM. What do brain data really show? Philos Sci. 2002;69(3):572–82.
12. Hardcastle VG. The theoretical and methodological foundations of cognitive neuroscience. Philos Psychol Cogn Sci. 2007:295–311.
13. Tononi G, Sporns O, Edelman GM. A measure for brain complexity: relating functional segregation and integration in the nervous system. Proc Natl Acad Sci U S A. 1994;91(11):5033–7.
14. Newlands SD, Perachio AA. Compensation of horizontal canal related activity in the medial vestibular nucleus following unilateral labyrinth ablation in the decerebrate gerbil. II. Type II neurons. Exp Brain Res. 1990;82(2):373–83.
15. Lee TS, Mumford D. Hierarchical bayesian inference in the visual cortex. J Opt Soc Am A Opt Image Sci Vis. 2003;20(7):1434–48.
16. Bogen JE, DeZure R, Tenhouten WD, Marsh JF. The other side of the brain. IV. The A-P ratio. Bull Los Angel Neurol Soc. 1972;37(2):49–61.
17. TenHouten WD. Neurosociology. J Soc Evol Syst. 1997;20(1):7–37.
18. TenHouten WD. Explorations in neurosociological theory: from the spectrum of affect to time consciousness. Soc Perspect Emot. 1999;5:41–80.
19. TenHouten WD. A general theory of emotions and social life. New York: Routledge; 2006.
20. Plutchik R. A general psychoevolutionary theory of emotion. In: Emotion: theory, research, and experience, Theories of emotion, vol. 1. New York: Academic; 1980. p. 3–33.
21. Fiske AP. The four elementary forms of sociality: framework for a unified theory of social relations. Psychol Rev. 1992;99(4):689–723.
22. Franks DD. The neuroscience of emotions. In: Stets JE, Turner JH, editors. Handbook of the sociology of emotions, Handbooks of sociology and social research. New York: Springer; 2006. p. 38–62. Available from: http://link.springer.com/chapter/10.1007/978-0-387-30715-2_3.
23. Preston SD, de Waal FBM. Empathy: its ultimate and proximate bases. Behav Brain Sci. 2002;25(1):1–20.
24. Eickhoff SB, Laird AR, Fox PT, Bzdok D, Hensel L. Functional segregation of the human dorsomedial prefrontal cortex. Cereb Cortex. 2016;26(1):304–21.
25. Marder E, Kopell N, Sigvardt K. How computation aids in understanding biological networks. In: PSG S, Grillner S, Selverston AI, Stuart DG, editors. Neurons, networks, and motor behavior. Cambridge, MA: MIT Press; 1997.
26. Stevens CF. Models are common; good theories are scarce. Nat Neurosci. 2000;3:1177.
27. Forstmann BU, Wagenmakers E-J, Eichele T, Brown S, Serences JT. Reciprocal relations between cognitive neuroscience and formal cognitive models: opposites attract? Trends Cogn Sci. 2011;15(6):272–9.
28. Palmeri TJ, Love BC, Turner BM. Model-based cognitive neuroscience. J Math Psychol. 2017;76:59–64.
29. Dayan P, Abbott LF. Theoretical neuroscience: computational and mathematical modeling of neural systems. Cambridge, MA: MIT Press; 2001. p. 576.
30. Frigg R, Hartmann S. Scientific models. In: Sarkar S, Pfeifer J, editors. The philosophy of science: an encyclopedia. New York: Routledge; 2006. p. 740–9.
31. Craver CF. When mechanistic models explain. Synthese. 2006;153(3):355–76.
32. Kronhaus DM, Eglen SJ. The role of simplifying models in neuroscience: modelling structure and function. In: Bio-inspired computing and communication. Berlin, Heidelberg: Springer; 2008. p. 33–44.

33. Weiskopf DA. Models and mechanisms in psychological explanation. Synthese. 2011;183(3):313.
34. Marr D. Vision: a computational investigation into the human representation and processing of visual information. San Francisco: W. H. Freeman and company; 1982. p. 432.
35. Bruce V, Young A. Understanding face recognition. Br J Psychol Lond Engl. 1986;77(Pt 3):305–27.
36. Decety J. A social cognitive neuroscience model of human empathy. In: Harmon-Jones E, Winkielman P, editors. Social neuroscience: integrating biological and psychological explanations of social behavior. New York: Guilford Press; 2007.
37. Chirimuuta M. Minimal models and canonical neural computations: the distinctness of computational explanation in neuroscience. Synthese. 2014;191(2):127–53.
38. Behrens TEJ, Hunt LT, Rushworth MFS. The computation of social behavior. Science. 2009;324(5931):1160–4.
39. Insel TR, Young LJ. The neurobiology of attachment. Nat Rev. Neurosci. 2001;2(2):129–36.
40. Di Paolo E, De Jaegher H. The interactive brain hypothesis. Front Hum Neurosci. 2012;6:163.
41. Craver CF. Explaining the brain: mechanisms and the mosaic unity of neuroscience. Oxford: Oxford University Press; 2007. p. 328.
42. Bechtel W. Mental mechanisms: philosophical perspectives on cognitive neuroscience. New York: Psychology Press; 2008. p. 322.
43. Hempel C. Aspects of scientific explanation and other essays in the philosophy of science. New York: The Free Press; 1965.
44. Wright CD, Bechtel W. Mechanisms and psychological explanation. In: Thagard P, editor. Philosophy of psychology and cognitive science. Amsterdam: Elsevier; 2007.
45. Kaplan DM, Craver CF. The explanatory force of dynamical and mathematical models in neuroscience: a mechanistic perspective. Philos Sci. 2011;78(4):601–27.
46. Hempel CG, Oppenheim P. Studies in the logic of explanation. Philos Sci. 1948;15(2):135–75.
47. Salmon WC. Four decades of scientific explanation. 1st ed. Pittsburgh: University of Pittsburgh Press; 1984. p. 240.
48. Cummins R. "How does it work" versus "what are the laws?": two conceptions of psychological explanation. In: Keil F, Wilson RA, editors. Explanation and cognition. Cambridge, MA: MIT Press; 2000. p. 117–45.
49. Bechtel W, Abrahamsen A. Explanation: a mechanist alternative. Stud Hist Phil Biol Biomed Sci. 2005;36(2):421–41.
50. Craver CF. Levels [Internet]. In: Open MIND. Frankfurt am Main: MIND Group; 2015. [cited 25 Dec 2016]. Available from: http://open-mind.net/papers/levels/getAbstract.
51. Illari PM, Williamson J. Mechanisms are real and local. New York: Oxford University Press; 2011.
52. Woodward J. Making things happen. New York: Oxford University Press; 2003.
53. Wouters AG. Design explanation: determining the constraints on what can be alive. Erkenntnis. 2007;67(1):65–80.
54. Irvine E. Models, robustness, and non-causal explanation: a foray into cognitive science and biology. Synthese. 2014:1–17.
55. Ross LN. Dynamical models and explanation in neuroscience. Philos Sci. 2015;81(1):32–54.
56. Weisberg M. Simulation and similarity: using models to understand the world. New York: Oxford University Press; 2013. p. 211.
57. Levins R. The strategy of model building in population biology. Am Sci. 1966;54(4):421–31.
58. Barberis SD. Functional analyses, mechanistic explanations and explanatory tradeoffs. J Cogn Sci. 2013;14(3):229–51.
59. Nolen S. In defense of dynamical explanation. Philosophical Theses [Internet]. 2013. Available from: http://scholarworks.gsu.edu/philosophy_theses/143.
60. Boone W, Piccinini G. The cognitive neuroscience revolution. Synthese. 2016;193(5):1509–34.
61. Levy A, Bechtel W. Abstraction and the organization of mechanisms. Philos Sci. 2013;80(2):241–61.

62. Mitchell SD. Why integrative pluralism? ECO Spec Double Issue. 2004;6(1–2):81–91.
63. Mantzavinos C. Explanatory pluralism. Cambridge: Cambridge University Press; 2016. p. 237.
64. Bickle J. Reducing mind to molecular pathways: explicating the reductionism implicit in current cellular and molecular neuroscience. Synthese. 2006;151(3):411–34.
65. Abney DH, Dale R, Yoshimi J, Kello CT, Tylén K, Fusaroli R. Joint perceptual decision-making: a case study in explanatory pluralism. Front Psychol. 2014;5:330.
66. Bouwel JV. Pluralists about pluralism? Different versions of explanatory pluralism in psychiatry. In: Galavotti MC, Dieks D, Gonzalez WJ, Hartmann S, Uebel T, Weber M, editors. New directions in the philosophy of science, The philosophy of science in a european perspective. Berlin: Springer; 2014. p. 105–19.
67. Gijsbers V. Explanatory pluralism and the (dis)unity of science: the argument from incompatible counterfactual consequences. Front Psych. 2016;7:32.
68. McCauley RN, Bechtel W. Explanatory pluralism and heuristic identity theory. Theory Psychol. 2001;11(6):736–60.
69. Salmon WC. Scientific explanation: causation and unification. Critica. 1990;22(66):3–23.
70. Venturelli AN. A cautionary contribution to the philosophy of explanation in the cognitive neurosciences. Mind Mach. 2016;26(3):259–85.
71. Cartwright N. The dappled world: a study of the boundaries of science. Cambridge: Cambridge University Press; 1999. p. 264.
72. Cacioppo JT, Berntson GG. Social neuroscience. In: Cacioppo JT, Berntson GG, Adolph R, Carter CS, Davidson RJ, McClintock MK, et al., editors. Foundations in social neuroscience. Cambridge, MA: MIT Press; 2002. p. 3–7.
73. Harmon-Jones E, Winkielman P, editors. A social cognitive neuroscience model of human empathy. New York: Guilford Press; 2007.
74. Bechtel W, Hamilton A. Reduction, integration, and the unity of science: natural, behavioral, and social sciences and the humanities. In: Kuipers T, editor. Philosophy of science: focal issues, Handbook of the philosophy of science, vol. 1. Amsterdam: Elsevier; 2007.
75. Cacioppo JT, Decety J. Social neuroscience: challenges and opportunities in the study of complex behavior. Ann N Y Acad Sci. 2011;1224:162–73.
76. Levy A. The unity of neuroscience: a flat view. Synthese. 2016;193(12):3843–63.
77. Bickle J. Philosophy and neuroscience: a ruthlessly reductive account. Dordrecht: Springer; 2003. p. 235.
78. Bickle J. Ruthless reductionism and social cognition. J Physiol Paris. 2007;101(4–6):230–5.
79. Kogan JH, Frankland PW, Silva AJ. Long-term memory underlying hippocampus-dependent social recognition in mice. Hippocampus. 2000;10(1):47–56.
80. Thor D, Holloway W. Social memory of the male laboratory rat. J Comp Physiol Psychol. 1982;96(6):1000–6.
81. Ferguson JN, Young LJ, Insel TR. The neuroendocrine basis of social recognition. Front Neuroendocrinol. 2002;23(2):200–24.
82. Sullivan JA. The multiplicity of experimental protocols: a challenge to reductionist and non-reductionist models of the unity of neuroscience. Synthese. 2009;167(3):511–39.
83. Kuhn TS. The structure of scientific revolutions. Chicago: University of Chicago Press; 1970. p. 228.
84. Piccinini G, Craver CF. Integrating psychology and neuroscience: functional analyses as mechanism sketches. Synthese. 2011;183(3):283–311.
85. Young LJ, Wang Z. The neurobiology of pair bonding. Nat Neurosci. 2004;7(10):1048–54.
86. Donaldson ZR, Young LJ. Oxytocin, vasopressin, and the neurogenetics of sociality. Science. 2008;322(5903):900–4.
87. Johnson ZV, Young LJ. Neurobiological mechanisms of social attachment and pair bonding. Curr Opin Behav Sci. 2015;3:38–44.
88. Insel TR. The challenge of translation in social neuroscience: a review of oxytocin, vasopressin, and affiliative behavior. Neuron. 2010;65(6):768–79.

89. Carter CS, DeVries AC, Getz LL. Physiological substrates of mammalian monogamy: the prairie vole model. Neurosci Biobehav Rev. 1995;19(2):303–14.
90. Penn DC, Holyoak KJ, Povinelli DJ. Darwin's Mistake: explaining the discontinuity between human and nonhuman minds. Behav Brain Sci. 2008;31(2):109–30.
91. Sterling P, Laughlin S. Principles of neural design. Cambridge, MA: MIT Press; 2015.
92. Cherniak C, Mokhtarzada Z, Rodriguez-Esteban R, Changizi K. Global optimization of cerebral cortex layout. Proc Natl Acad Sci U S A. 2004;101(4):1081–6.
93. Chklovskii DB, Koulakov AA. Maps in the brain: what can we learn from them? Annu Rev. Neurosci. 2004;27:369–92.
94. Aiello LC, Wheeler P. The expensive-tissue hypothesis: the brain and the digestive system in human and primate evolution. Curr Anthropol. 1995;36(2):199–221.
95. Dunbar RIM. Evolutionary basis of the social brain. In: Oxford handbook of social neuroscience. Oxford: Oxford University Press; 2011. p. 28–38.
96. Dunbar RIM, Shultz S. Evolution in the social brain. Science. 2007;317(5843):1344–7.
97. Cachel S. Primate and human evolution. Cambridge, UK: Cambridge University Press; 2006. p. 488.
98. TenHouten WD. Emotion and reason: mind, brain, and the social domains of work and love. New York: Routledge; 2013. p. 279.
99. Mitchell SD, Dietrich MR. Integration without unification: an argument for pluralism in the biological sciences. Am Nat. 2006;168(Suppl 6):S73–9.
100. Callard F, Fitzgerald D. Rethinking interdisciplinarity across the social sciences and neurosciences. Basingstoke, UK: Palgrave Macmillan; 2015. (Wellcome Trust–Funded Monographs and Book Chapters)
101. Balcombe JP. Laboratory environments and rodents' behavioural needs: a review. Lab Anim. 2006;40(3):217–35.
102. Vyssotski AL, Dell'Omo G, Poletaeva II, Vyssotsk DL, Minichiello L, Klein R, et al. Long-term monitoring of hippocampus-dependent behavior in naturalistic settings: mutant mice lacking neurotrophin receptor TrkB in the forebrain show spatial learning but impaired behavioral flexibility. Hippocampus. 2002;12(1):27–38.

Social Neuroscience and Neuroethics: A Fruitful Synergy

Arleen Salles and Kathinka Evers

Abstract Social neuroscience is shedding new light on the relationship between the brain and its environments. In the process, and despite criticism from the social sciences, the field is contributing to the discussion of long-standing controversies concerning, for example, the "nature-nurture" distinction and the relationships between social and neurobiological structures. In this chapter, we argue that in this endeavor social neuroscience would benefit from partnering with neuroethics insofar as their respective areas and methods of explanation are complementary rather than in competition. We provide a richer account of neuroethics than the one given in social neuroscientists' common descriptions of that field and suggest that, when understood in this richer (and in our view more adequate) fashion, neuroethics may open up productive avenues for research and play a key role in allowing us to determine social neuroscience's contribution to unveiling important epistemological as well as ontological notions. Accordingly, social neuroscience and neuroethics may form a constructive partnership.

Keywords Neurobioethics • Empirical neuroethics • Conceptual neuroethics • Social sciences • Social neuroscience • Neuronal epigenesis

1 Social Neuroscience: Expectations and Concerns

The development of social neuroscience is shedding new light on the relationship between the brain and its environments and, notably, its social interactions. This field grew out of the attempt to incorporate the methods of neuroscience into social psychology, and it is devoted to understanding the neural, hormonal, and genetic

A. Salles (✉)
Centre for Research Ethics and Bioethics, Uppsala University, Uppsala, Sweden

Centro de Investigaciones Filosóficas, Buenos Aires, Argentina
e-mail: arleen.salles@crb.uu.se

K. Evers
Centre for Research Ethics and Bioethics, Uppsala University, Uppsala, Sweden
e-mail: Kathinka.Evers@crb.uu.se

© Springer International Publishing Switzerland 2017

A. Ibáñez et al. (eds.), *Neuroscience and Social Science*,
DOI 10.1007/978-3-319-68421-5_22

underpinnings of social structures, emotional processes, behaviors, and the reciprocal impact of social and neural processes [1]. Because it considers that scientific epistemological reductionism falls short of comprehensive explanations of social phenomena, social neuroscience calls for "increasing the scope of analysis to include contributions of factors from both social and neuroscientific perspectives" [2, 3]. Indeed, social neuroscience recognizes that the brain is in constant interaction with its social and cultural environments. Thus, it uses a multilevel approach, one that combines social and biological theories and methods (from molar to micro-levels) and that is organized in terms of a number of grounding, heuristic principles (see Barbeis, this volume). It has been argued that the field can contribute to long-standing controversies about human behavior (moral and otherwise), human identity, and human societies and inform ways to recognize and deal with socially problematic behaviors [4].

Because of its goals to bridge what—by a rather traditionalist view reminiscent of the sociobiological conflicts of the 1970s [5]—is taken to be an "impassable abyss between the social and the biological" [6, 7], this field may appear well positioned to avoid some of the pitfalls of neuroscientific research in general. Indeed, criticism of how neuroscientific research is typically carried out is not new [8], and it mainly revolves around the idea that the neurological turn taking place is too immersed in an unwarranted sense of objectivity and neutrality (based on the alleged legitimacy of brain facts separate from cultural and social environments) and that it is metaphysically and epistemologically reductionist, i.e., that the mind can be replaced by the brain and philosophy of mind by neuroscience [9]. In contrast, social neuroscientists have the conviction that social aspects shape the brain and behaviors and try to open their field to the arguments and evidence provided by the social sciences (see Haye, this volume).

And yet, there is no consensus on the role that the human sciences play and ought to play in the field [4, 5]. Furthermore, it has been suggested that social neuroscience and its allegedly novel findings rest on unexamined and often questionable social and cultural assumptions, hopes, and priorities [10], for example, the dichotomy between nature and culture [11], poor conceptions of the social context it intends to include, or biased understandings of what counts as natural and what does not [12] (see Cornejo, this volume).

Over the past few years, the social sciences and the humanities have become increasingly interested in the multiple issues raised by neuroscience in general and social neuroscience in particular. Within the humanities, this interest is evident in the emergence of neuroethics, a young field that is an interface between the empirical brain sciences, philosophy of mind, moral philosophy, ethics, and psychology. In general terms, we can say that neuroethics is particularly concerned with the numerous questions that arise when scientific findings about the brain are carried into philosophical analyses, medical and legal practice, and health and social policy.

A number of sociologists have focused on neuroscience as well, often taking the field to task [13]. In fact, several social scientists have recently developed what is known as "critical neuroscience," a multidisciplinary approach that calls for more refined and self-reflexive neuroscientific research and practice in general. It argues that there are many reasons, such as the specific attention that social neuroscience gives to sociocultural contexts, its aim to understand human social capacities, its reliance on neuroimaging—a method that some argue promotes a problematic type of reductionism [8]—and the impact of potential applications of its findings that make it important not only to situate neuroscientific knowledge within webs of social and cultural meaning but also to identify tacit operative frameworks, to create spaces of self-reflection, and to open up channels for dialogue within diverse fields [14–16]. In this perspective, historical, social, anthropological, ethnographic, and conceptual analyses play key roles in addressing a number of issues; for example, how is context incorporated in social neuroscience studies and why? What social aspects are considered relevant and on what grounds? What are the social factors that shape what counts as a satisfactory explanation of the brain and why? What are the philosophical and methodological commitments of the field [14, 16]?

Several sociologists who criticize the interpretation and use of social neuroscientific findings believe that philosophy can contribute to a better understanding of the relevant research. Indeed, they claim that scientists should be introduced to a range of critical tools including "basic philosophical theory and conceptual analysis" [11]. However, they tend to sideline neuroethics. As we explain below, among other concerns, some see neuroethics as "in the thrall of neuroscience" and question the extent to which it is able to critically assess neuroscientific research and findings in general [17].

While some of these concerns might be justified, it should be noted, first, that they do not concern neuroethics as such but only specific versions of it and second, that the social sciences are not immune to similar criticism. Indeed, some sociologists call for a more self-critical attitude within their own field and urge their discipline to examine the meaning of its reluctance to accept neuroethics and to recognize that such reluctance creates higher expectations about the role of the social sciences themselves [18]. Recent objections to the critical neuroscience field, in particular, range from the charge that it is too critical and ultimately unproductive to objections about its preserving problematic dichotomies between the social and the biological and giving too much priority to the social thus unequivocally positing a different kind of reductionism [19].

Social neuroscience attempts to offer a novel approach and thus contribute to the discussion of a number of topics typically broached by the humanities. In the process, it raises a number of concerns informally debated by social neuroscientists themselves (indeed, the title of this volume suggest that this is a topic of no indifference to social neuroscientists) and formally debated by both neuroethicists and sociologists. In this chapter, we propose an approach to examine the philosophical and social issues raised by social neuroscience that can avoid the potential pitfalls of both the social sciences and neuroethics as commonly understood. For this, rather than rejecting neuroethics, we suggest a more foundational understanding of the field, one that puts special emphasis on critical reflection and conceptual examination and that can be taken as the basis upon which to fully address the ontological, epistemological, and ethical impact of social neuroscience in particular [20]. We believe that when adequately grounded, neuroethics may open up productive avenues for research and play a key role in allowing us to determine social neuroscience's contribution to unveil important epistemological as well as ontological notions and to address long-standing controversies. In short, we suggest that a conceptually based neuroethics might become a key partner with social neuroscience.

2 Neuroethics: Solution or Problem?

In the last decades, neuroethics has focused on the social, ethical, and legal issues raised by neuroscientific advances. However, within the social science community (and sometimes even within the bioethics community), there appears to be some agreement that neuroethics' adequacy in addressing many (if not all) of the relevant issues can be questioned. Among the concerns we find the following:

1. Neuroethics either mirrors or buys into scientific expectations, and, therefore, it does not offer much from a critical perspective [18, 21]. In short, it tends to be too solicitous to neuroscience and its ideology [13, 17, 22, 23].
2. Because of the above, neuroethics tends to overstate neuroscientific findings and their impact, guarding neuroscience from societal criticisms and legitimizing the neurodisciplines [13]. Thus, it is in an unlikely position to guard and monitor developments within neuroscience [18, 24].
3. Neuroethics covers the same topics that neuroscience covers and shares the same basic assumptions, questions, and arguments without forging ahead [13].
4. Neuroethics is too speculative and often limited to the discussion of hypothetical scenarios and uncritically accepted scientific futures [17, 25]. If it is to be useful, it should be more straightforwardly clinical.
5. Neuroethics is saddled with mainstream bioethical reasoning [18, 24].
6. Neuroethics fundamental questions are not different from bioethics fundamental questions [26, 27], and in its eagerness to hide the crucial similarities between neuroethics and bioethics, neuroethicists unjustifiably hype the brain and its importance in determining human uniqueness.

Broadly, we can say that the first three concerns revolve around the idea that the boundaries between neuroethics and neuroscience are not clear enough and that neuroethics, considered to be largely internal to neuroscience itself, is overly optimistic about and accepting of neuroscience's explanatory power. The fourth concern revolves around the idea that as currently practiced, neuroethics is, to a great extent, practically irrelevant. The last two state that neuroethics is pretending to be different when it is not, and it does so by controversially affording primacy to the brain—see also [28] for a review of relevant issues.

While they might appear unfair to many neuroethicists, not all of these concerns are necessarily unjustified. Since its appearance as an organized field, neuroethics has taken at least three different forms according to the methodological approaches used and the topics addressed. Some of the criticisms are appropriate in some of these forms or versions. However, in what follows, we argue that they are not justified against all versions, so those who raise them have to be careful not to open themselves to the charge of creating a "straw man" or at least of providing unjustified generalizations.

3 Three Methodological Approaches

In previous work, we made a distinction between three different, albeit not fully independent and often overlapping, versions of neuroethics according to their *methodological* approach. We called them (a) neurobioethics, (b) empirical neuroethics, and (c) conceptual neuroethics [20].

Arguably, neuroethics' most common version is as neurobioethics, which uses methods and deals with the kinds of topics found in what Adina Roskies famously labeled the "ethics of neuroscience" [29]. While some neuroethicists use the term "neurobioethics" to refer to neuroethics in general, on the grounds that this term "grounds an understanding and use of neuroscience to the methodology of ethics and the interdisciplinarity and practicality of bioethics" [30], we believe that the different versions of neuroethics show different methodological approaches and, therefore, not all neuroethics are "bioethical in its methods and general field of vision" [30].

Neurobioethics applies ethical theory and reasoning to a wide range of issues related to neuroscience, from those raised by neuroscientific research, e.g., informed consent, the handling of incidental findings, and privacy, and those with clinical relevance, i.e., raised by treatment of pathologies of the brain, to public communication, media representation, and cultural and societal understanding of neuroscience's impact, including policy considerations regarding some of the potential uses of neurotechnology (e.g., cognitive and moral neuro-enhancement and "mind reading") (see Mantilla, Di Marco, and Golombek, this volume). In practice, this is the version of neuroethics reasoning predominant in healthcare, in regulatory contexts, and in the neuroscientific research setting. In theory, this neurobioethical approach is either discussed or exemplified in some of the written work by a number of well-known neuroethicists, including Joseph Fins [25, 31], Martha Farah [32–34], Judy Illes [35], Neil Levy [36], and Walter Glannon [37, 38].

Neurobioethics generally mirrors bioethical methodology and goals. The bioethical methodology is typically grounded in applied philosophy and moral theory and is usually based on the use and application of moral norms and principles (see Rosas, this volume). Even when bioethics draws from a number of other different methodological approaches, such as casuistry and pragmatism, in general, an important component of the bioethicist toolkit is traditional moral philosophy.

Something similar happens in neurobioethics. More or less sophisticated philosophical argumentation is key when assessing the relevant neuroethical issues. But, fundamentally, neurobioethical reasoning has a very specific practical goal—to solve concrete ethical issues, for example, whether newly developed research methods or specific brain interventions are safe enough (where the moral imperative "do no harm" becomes key), whether the use of fMRI to monitor and examine brain activity might threaten subjects' privacy (where the moral imperative to respect people's autonomy appears to play a bigger role), or how to provide ethical care to neurological patients.

A second neuroethical approach can be called "empirical neuroethics" [39] (see Northoff, this volume).[1] It uses neuroscientific data, specifically the relationship of the structures and different cognitive and affective processes in the brain, to inform theoretical issues (e.g., how to understand moral reasoning or how to understand informed consent and moral judgment) and practical issues (such as who can give truly informed consent or which beings can be considered moral agents). To this extent, empirical neuroethics overlaps with social neuroscience and comprises what Roskies calls "the neuroscience of ethics." Although it focuses on theoretical issues, its basis is still fundamentally empirical: it takes as a starting point the view that "adequate" and in some cases "sufficient" knowledge about human beings (e.g., who they are, how they think and judge morally, and how they act) can be achieved by looking at the empirical data on the workings of the nervous system and the brain. This line of reasoning can be found in different degrees in some of the work by Patricia Churchland [40], Michael Gazzaniga [41], Neil Levy [42], and Joshua Greene [43]. From a methodological perspective, it tends to afford priority to the scientific method. A main characteristic of the approach is that while neuroscientific findings are used to discuss a number of ethical and ontological questions, there is not much emphasis on translational concerns, that is, on the issue of whether and how neuroscientific findings can be so used [39, 44].[2]

In contrast to the above, "conceptual neuroethics" is specifically concerned with how neuroscientific knowledge is constructed and why or how observations about the brain can be relevant to philosophical, social, and ethical concerns. In doing so, conceptual neuroethics focuses on one of the main challenges posed by neuroscientific research, the plausibility and legitimacy of translation from laboratory to clinical and social applications (see Roussos, this volume).

There are different conceptual approaches. One proposes that neuroethics can be understood as a type of metaethics, clarifying the operative underlying presuppositions and commitments involved in moral deliberation and moral action [45]. James Giordano and colleagues sometimes go further, suggesting that neuroethics also needs to address "whether, how and why neuroscience can and/or should be employed in various endeavors and circumstances" [30] and calling for the need to "sustain the epistemic probity of neuroscience-inclusive validity of the tools and techniques utilized to develop neuroscientific (and neuroethical) theories" [46]. The authors advance (even if somewhat timidly) that attention should be paid to how neuroscientific concepts are shaped in order to determine the usefulness of neuro-

[1] Note that our understanding of the term "empirical neuroethics" is influenced by Northoff's view [39] and thus not equivalent to Judy Illes' use of the term [71].

[2] A mix of both neurobioethics and empirical neuroethics articles is usually found in classic neuroethics texts. See, for example, [68–70].

science in addressing fundamental human issues, even though, in general, they appear to take the legitimacy of neuroscientific interpretations of results for granted.

A different conceptual approach was advanced by one of us (KE) – "fundamental neuroethics" [5]. Fundamental neuroethics refers to basic research that combines philosophical and scientific, theoretical and empirical perspectives. This approach acknowledges the potential impact that knowledge of the brain's structural and functional architecture and its evolution can have on our understanding of personal identity, consciousness, and intentionality, including the development of moral thought and judgment. However, not only does it focus on the extent to which neuroethics can clarify presuppositions and limit the claims made by philosophers on issues such as how to understand moral deliberation; it also underscores the extent to which the same is needed for neuroscience. Thus, it takes the question of how natural science can deepen our understanding of thought, including moral thought and judgment, to be a key topic of discussion.

Several neuroethicists have noted the challenge typically presented by interpretation of neuroscientific data [34, 47, 48]. From a methodological perspective, fundamental neuroethics argues that a philosophical/conceptual level of interpretation of the scientific evidence allows for an integrated picture of a legitimate connection between scientific evidence and philosophical concepts and issues, without assuming that the empirical and the conceptual correspond one to one. Furthermore, it also plays a key role in stifling unrealistic expectations regarding neuroscientific advances [5, 49]. It does not deny the value of the empirical methodology and of scientific interpretation; however, it considers that while providing important information, empirical considerations by themselves are insufficient to generate an adequate conceptual understanding of data, including data about the brain. In this sense, philosophical analysis is not intended to replace but to complement scientific interpretation of theories and data, fostering understanding of the meaning and use of the main scientific concepts. This, in turn, entails that the neuroethical discourse does not have to be limited by previously constructed neuroscientific interpretations.

Fundamental neuroethics has clear affinities with a third conceptual approach, known as "theoretical neuroethics," which specifically addresses the methodological and conceptual aspects of the link between neuroscientific research and ethics [9, 39].

Conceptual neuroethic approaches generally consider that the integration of scientific and philosophical methodologies is necessary because how significant the role of neuroscience in addressing fundamental philosophical issues is depends on how the relevant philosophical and neuroscientific concepts are interpreted. This does not entail endorsing the already mentioned approach espoused by critical neuroscience, which in its urge to avoid giving primacy to science ends up giving primacy to the sociocultural, possibly unintentionally making the kind of mistake that it accuses social neuroscience of making. Nor does a conceptual approach render neurobioethics and empirical bioethics superfluous. Both neurobioethics and empirical neuroethics raise and address important questions about neuroscience and neuroscientific findings. But a conceptual approach purports to offer more: specifically, the theoretical foundations needed to examine either the neurobiological basis of moral reasoning or the applied ethical problems raised by neuroscientific practice [5, 49].

4 Revisiting Concerns

We noted earlier that three major types of concerns have been presented regarding the nature of neuroethics and its role in addressing the issues raised by the neurosciences: (1) that the boundaries between neuroscience and neuroethics are problematically blurred, (2) that the field is too speculative, and (3) that the field is not different from bioethics and (3i) that if there is such a difference, this is because the unjustified priority that neuroethics gives to the brain. In what follows, we argue that while the first two versions of neuroethics might be vulnerable to some of these criticisms, a third conceptual version of the field is not.

4.1 Lack of Boundaries Between Neuroscience and Neuroethics

Without denying the important practical role played by neurobioethics, it is true that historically, in this version the philosophical analysis has typically remained within the discursive limits previously established by neuroscientists. As a consequence, neurobioethics has tended to shy away from critical examination or serious questioning of the relevant philosophical and neuroscientific discourses, their operating assumptions, and underlying ideologies.

This can be illustrated by the ongoing and fashionable discussion on the moral permissibility of brain optimization or neuro-enhancement, which is fraught with a variety of problematic framing assumptions regarding the utility and value of scientific evidence. Relevant articles devote considerable space to discussing potential consequences of the practice, but there is little room for ambiguity or even any questioning about what the science says, how to interpret what it says, and why what it says is relevant [50]. In this sense, neurobioethics appears to justify the simplistic view that some social scientists have of the relationship between neuroscience and neuroethics.

A similar issue becomes evident in the case of empirical neuroethics that deploys neuroscience to demonstrate the biological basis of a number of phenomena and experiences. It often suggests that unveiling the behavioral correlate of a phenomenon entails a full explanation of its existence and claims that neuroscientific results can illuminate fundamental philosophical questions, justify a change in some beliefs we hold about basic concepts such as autonomy and personhood [51, 52], drastically alter our understanding of moral agency, and possibly refine and enhance the moral tools ethicists use [29, 51, 53]. However, it does not provide a careful examination of the explanatory power of neuroscience. Again, this lack of focus on the issue of how and why makes this version of neuroethics vulnerable to the first set of concerns. Indeed, for empirical neuroethics, the critical examination of one aspect of the moral reality (neuroscience) remains external to the neuroethical task. Instead, it relies on the assumption that ethics deals with a set of answers provided by neuroscientists, without wondering about those answers, how they are produced, and how to understand the concepts used.

The lack of attention given to the issue of why and how is not a minor concern. It suggests a lack of critical attitude toward the potential links between biological explanations, culture, and sociality. A number of assumptions about brain facts, their value, and their normative weight underlie the claim that neuroscientific findings will lead us to revise particular metaphysical and ethical notions [39, 45]. But there is no explanation of how to interpret those suggested brain facts, and unless one supposes that brain facts and normative concepts correspond one to one, it is not clear why such facts are so significant. It seems, then, that empirical neuroethics tends to take as a given both that science is an objective source of knowledge and that the relation between science and philosophical or moral notions is unproblematic. Thus, it does not account for one of the most challenging tasks for neuroethics: the determination of how biological data can have either explanatory or normative relevance. This is the reason why, indirectly, empirical neuroethics justifies the first set of concerns regarding the lack of boundaries between neuroscience and neuroethics, and unintentionally, it may actually prop up certain reductionistic understandings.

Note, however, that this is not a problem for conceptual versions of neuroethics. Instead of being uncritical cheerleaders of neuroscience, conceptual approaches offer a more nuanced perspective. In particular, conceptual approaches are accepting and appreciative of the value and potential contributions of social neuroscience while recognizing that assessing such contributions requires more than uncritical acceptance of common beliefs about the interpretation, value, and applicability of neuroscientific findings.

Consider the conceptual approach we favor, fundamental neuroethics. For this approach, the link between descriptive considerations derived from observations about the brain and their impact on normative considerations is not self-evident: "brain facts" upon which a number of possible applications are predicated are not uncontroversial. Scientific research poses conceptual issues, and understanding them requires conceptual interpretation. Thus, fundamental neuroethics calls for developing a methodological modus operandi for fruitfully linking scientific and philosophical interpretations without necessarily giving primacy to either neuroscience or philosophy [20]. Insofar as it does, it is neither internal to neuroscience nor opposed to it: its task is to accurately assess both philosophical and neuroscientific assumptions, traditions, and practices in order to achieve a better understanding of the relevant issues.

4.2 Neuroethics Is Too Speculative

It is indeed the case that some neurobioethicists have tended to focus on highly speculative issues, discussing futuristic scenarios where people take a pill to become "more moral" and where airport security screening can detect "terrorists." However, as a general concern, this is often blown out of proportion. In fact, the existence of works on issues raised by imaginary scenarios should not obscure the fact that most neuroethical discussions (whatever form they take) are not highly speculative and a healthy percentage of such discussions focus specifically on clinically relevant and/ or socially and legally urgent issues.

4.3 No Significant Difference Between Neuroethics and Bioethics

While it should be evident by now that this objection is simply false with respect to empirical and conceptual neuroethics, it is, to a certain extent, a fair point if focused on what we have called neurobioethics. For not only does neurobioethics use the bioethical methodology, it is also primarily concerned with the kind of applied topics typical of bioethical reasoning and practice. Furthermore, neurobioethicists often claim that their field's uniqueness is related to the fact that it focuses on the brain and its crucial role. This has usually led then to the additional critique that by affording a special status to the brain, neuroethics in general shows a very reductionist understanding of human beings.

Empirical neuroethics is sometimes vulnerable to this last critique as well, to the extent that it appears to be unconcerned by an important fact, namely, that the relevance of neuroscience to understanding thought and judgment and its usefulness in assessing the moral implications of neuroscientific research depend on both how to understand morality and which theoretical model of the brain is used. When they do not provide nor do they explain key notions, neurobioethics and empirical neuroethics can be accused of holding a naïve view of the explanatory power of neuroscience.

Conceptual approaches are more aware of the need to examine the relevant notions, not just moral but scientific as well. To illustrate, fundamental neuroethics provides an empirically based, philosophical account of the brain and how its functional and structural architecture grounds human thought and behavior [20]. Specifically, it offers a rich theoretical model of the brain, informed materialism, that is scientifically, ontologically, and ethically relevant to providing an adequate theoretical framework for addressing issues such as the development of moral thought or the human tendency to build normative systems.

In neuroscience, the concept of informed materialism is used to oppose both dualism, which posits the existence of a mind or soul independent of the material body, and naïve reductionism, which excludes the subjective perspective from scientific study [54]. Informed materialism rejects the view of the brain as a rigid, automatic device whose operations are determined. Instead, it emphasizes the brain's dynamic, plastic, projective, and evaluative nature while underscoring that it is fundamentally constrained by values and emotions.

The human brain is an intrinsically active and motivated neural system genetically predisposed to explore the world and to classify what it finds. Because of how our brains acquire knowledge of ourselves and the world, according to the informed materialism view, an adequate understanding of our subjective experience must take into account self-reflective information, physiological observations, and physical measurements [5, 54, 55]. Moreover, informed materialism underscores the plasticity of the brain and its epigenetic development in response to learning and experience. Genetic control over the brain's development is important but by no means absolute: the epigenetic model of neuronal development postulates that the connections between neurons are not prespecified in the genes but that learning and experience influence the brain's development within the boundaries of a "genetic envelope" [56] (for further details, see [5, 49]). Endorsing such a dynamic model of the brain evolving in biological-sociocultural symbiosis [5] allows fundamental neuroethics to avoid the criticism that it is reducing human beings to their biology, narrowly understood.

5 Social Neuroscience and Fundamental Neuroethics

We have argued here for the need for a more conceptual neuroethical approach. We have also suggested a specific conceptual framework, fundamental neuroethics, which we think can give social neuroscience the kind of methodological novelty that may promote a more productive space of shared collaboration. Fundamental neuroethics does not privilege any particular mode of explanation, coming from natural or social science, nor is it biased in its critique. Instead, it remains constructively critically alert. But in order to attain this balance, substantial scientific ground and conceptual clarity are needed, and this can only be achieved by joining scientific and philosophical interpretations [57]. In short, fundamental neuroethics adds conceptual interpretation to the empirical field of social neuroscience and is instrumental in clarifying a number of scientific, social, and ethical issues.

However, until now this discussion has been kept at an abstract level. A concrete illustration of the suggested complementarity of social neuroscience and fundamental neuroethics would be useful. One possible illustration would depict how fundamental neuroethics complements social neuroscience in addressing perennial philosophical-cum-scientific issues, such as the "nature-nurture" distinction which has bearings on, notably, the understanding of human identity, learning processes, and of the possibility of moral change and the acquisition of moral rules. In this final section, we provide a brief discussion of recent studies of neuronal epigenesis and the impact of culture on brain architecture that we hope will illustrate how fundamental neuroethics and social neuroscience may form a fruitful synergy.

During the past decade, advances in social neuroscience have been considered key in understanding moral compliance and the possibility of moral change. However, at times this has happened at the expense of accuracy and conceptual clarity: many thinkers, including philosophers, have tried to bridge the gap between neuroscientific results and broad-brush explanations of morality, overlooking translational concerns in order to support research into practices such as moral neuro-enhancement—see, for example, [58, 59]. As we noted when discussing both neurobioethics and empirical neuroethics, a careful evaluation of the explanatory power of the neuroscientific evidence, its implications, and limits is sometimes wanting.

Fundamental neuroethics contributes a type of analysis that intends to address this problem. It appreciates the importance of the empirical research and of the scientific methodology while taking a critical stance and carefully considering (a) the precise nature of the scientific evidence in question, (b) the limits of suggested interpretations, and (c) the potential implications of alternative theoretical frameworks.

First, neuroscientific evidence shows that the neuronal organization of our adult brain develops in the course of a 25-year-long period following birth, during which and, to a lesser extent, after which it is subject to cultural influence both at the individual and social group level and across generations [60–62]. Throughout this exceptionally long period of postnatal development, an intense synaptogenesis steadily occurs in the human cerebral cortex, which persists, yet to a smaller extent, in the adult. The adult human brain builds up from a complex intertwining of cultural circuits progressively laid down during development within the framework of a human-specific genetic envelope. There is no compelling evidence that culturally acquired phenotypes will sooner or later be genetically transmitted. What the evidence shows is that they have to be learned by each generation, by children from adults, and epigenetically transmitted from generation to generation, beginning in the mother's womb and up until the adulthood. As a consequence of the steady interaction with the physical, social, and cultural environment, an active epigenetic selection of neuronal networks results in the internalization, in particular, of the cultural and ethical rules prevalent in the social community to which the child and her/his family belongs.

Second, that cultural imprints have a physical reality in the human brain shows, in the first place, that no plausible understanding of the brain can be achieved by assuming that it is somewhat independent from the experiences that shape it. Secondly, it makes evident that knowledge of the brain is highly relevant to understanding social structures, including social and other, e.g., moral, norms as well as how and why people comply with or disrespect those norms. Thirdly, it shows that the discussion over what is normatively "acceptable" or deemed "moral" and what is not requires a normative conceptual examination that goes beyond neuroscientific findings. The limitation of individual disciplines, notably of social neuroscience, social science, and philosophy by themselves, and the consequent necessity for them to collaborate in achieving a deeper and more multifarious understanding of the symbiosis of the brain and its natural and sociocultural contexts must be acknowledged when trying to unveil humans, their behavior (including moral behavior), and their world.

Finally, on the basis of the above, one can legitimately conclude that synaptic epigenesis theories of cultural and social imprinting on our brain architecture (which differ from less discriminative epigenetic modifications of nuclear chromatin) [63, 64] suggest that rather than artificially manipulating individual brains (as proponents of neuro-enhancement suggest), we could potentially be epigenetically proactive (a concept introduced by Evers [5, 65]) and adapt our social structures, in both the short and the long term, to benefit, influence, and constructively interact with the ever-developing neuronal architecture of our brains [5, 65, 66]. We should, however, note that this is a limited claim regarding the potential contribution of social neuroscience. It does not entail that it can dramatically alter our understanding of *moral* change but rather of *change*. Nor does it suggest that social neuroscience can tell us what moral action is or what moral change specifically entails. Whether change qualifies as moral or not and what kind of actions can be deemed morally desirable and thus, if promoted, will morally enhance society depends on a different kind of analysis that goes beyond what neuroscience, including social neuroscience, can tell us [67].

6 Conclusion

Social neuroscience is shedding new light on the relationship between the brain and its environments. In the process, and despite some criticisms from the social sciences, the field is attempting to contribute to the discussion of long-standing controversies concerning, for example, the "nature-nurture" distinction. In this article, we argued that social neuroscience might benefit from partnering with neuroethics insofar as their respective areas and methods of explanation are complementary rather than in competition. In arguing for this point, we provided a richer account of what neuroethics can do than the one given in social neuroscientists' descriptions of that field. For many critics of neuroethics, this field is no more than a kind of pretentious bioethics that unjustifiably gives primacy to the brain. In contrast, and driven by our weariness with a simplistic understanding of the field of neuroethics, we suggested a conceptual understanding that puts special emphasis on critical reflection and conceptual examination in order to fully address the ontological, epistemological, social, and ethical impact of neuroscience in general and of the potential contributions of social neuroscience in particular.

Acknowledgments We wish to thank our colleagues at the Centre for Research Ethics and Bioethics and especially Michele Farisco, for their constructive comments and suggestions. This project has received funding from the European Union's Horizon 2020 research and innovation-program under grant agreement 720270 (HBP SGA1).

Conflict of Interest The authors declare that they have no conflict of interests.

References

1. Decety JP, Keenan J. Social neuroscience: a new journal. Soc Neurosci. 2006;1(1):1–4.
2. Cacciopo J, Berntson G, Decety J. Social neuroscience and its relationship to social psychology. Soc Cogn. 2010;28(6):675–85.

3. Decety CJ. Social neuroscience: challenges and opportunities in the study of complex behavior. Ann N Y Acad Sci. 2011;1224:162–73.
4. Rose N, Abi-Rached J. Neuro: the new brain sciences and the management of the mind. Princeton: Princeton University Press; 2013.
5. Evers K. Quand la matière s'éveille. Paris: Éditions Odile Jacob; 2009.
6. Cacciopo J, Berntson G. Social neuroscience. In: Cacciopo J, Berntson G, Adolphs R, et al., editors. Foundations in social neuroscience. Cambridge, MA: MIT Press; 2002. p. 3–11.
7. Cacciopo J, Berntson G, Adolphs R, et al. Foundations in social neuroscience. Cambridge, MA: MIT Press; 2002.
8. Rose S. The need for a critical neuroscience. In: Choudhury S, Slaby J, editors. Critical neuroscience. A handbook of the social and cultural contexts of neuroscience. Chichester: Wiley; 2012. p. 53–66.
9. Northoff G. Minding the brain: a guide to philosophy and neuroscience. Croydon: Palgrave Macmillan; 2014.
10. Cromby J. Integrating social science with neuroscience: potentials and problems. BioSocieties. 2007;2(2):149–69.
11. Choudhury S, Nagel S, Slaby J. Critical neuroscience: linking neuroscience and society through clinical practice. BioSocieties. 2009;4:61–77.
12. Slaby J, Choudhury S. Proposal for a critical neuroscience. In: Choudhury S, Slaby J, editors. Critical neuroscience: a handbook of the social and cultural cotnexts of neuroscience. Chichester: Wiley; 2012. p. 29–50.
13. Vidal F. Brainhood, anthropological figure of modernity. Hist Human Sci. 2009;22(1):5–36.
14. Choudhury S, Slaby J. Introduction critical neuroscience-between lifeworld and laboratory. In: Choudhury S, Slaby J, editors. Critical neuroscience. A handbook of the social and cultural contexts of neuroscience. Chichester: Wiley; 2012.
15. Kirmayer L. The future of critical neuroscience. In: Choudhury S, Slaby J, editors. Critical neuroscience. A handbook of the social and cultural contexts of neuroscience. Chichester: Wiley; 2012. p. 367–83.
16. Slaby J. Steps towards a critical neuroscience. Phenomenol Cogn Sci. 2010;9:397–416.
17. Conrad E, De Vries R. Field of dreams: a social history of neuroethics. In: Advances in medical sociology, vol. 13. Bingley: Emerald; 2011.
18. Brosnan C. The sociology of neuroethics: expectation, discourses and the rise of a new discipline. Sociol Compass. 2011;5(4):287–97.
19. Fitzgerald D, Callard F. Social science and neuroscience beyond interdisciplinarity: experimental entanglements. Theory Cult Soc. 2015;32(1):3–32.
20. Evers K, Salles A, Farisco M. Theoretical framing of neuroethics: the need for a conceptual approach. In: Racine E, Aspler J, editors. Debates about neuroethics: perspectives on its development, focus and future. Cham: Springer International Publishing; 2017.
21. Ashcroft R. Ethics of neuroscience or neuroscience of ethics? Lancet Neurol. 2006;5:211.
22. Pickersgill M. The co-production of science, ethics, and emotion. Sci Technol Hum Values. 2012;37(6):579–603.
23. De Vries R. Framing neuroethics: a sociological assessment of the neuroethical imagination. Am J Bioeth. 2005;5(2):25–7.
24. De Vries R. Who will guard the guardians of neuroscience? Firing the neuroethical imagination. EMBO Rep. 2007;8(Special Issue):S65–9.
25. Fins J. A leg to stand on: Sir William Osler and Wilder Penfield's "neuroethics". Am J Bioeth. 2008;8:37–46.
26. Moreno J. Neuroethics: a agenda for neuroscience and society. Nat Rev Neurosci. 2003;4:149.
27. Parens E, Johnston J. Does it make sense to speak of neuroethics? EMBO Rep. 2007;8:S61–4.
28. Racine E. Pragmatic neuroethics. Cambridge, MA: MIT Press; 2010.
29. Roskies A. Neuroethics for the new millenium. Neuron. 2002;35:21–3.
30. Giordano J. Neuroethics: interacting "traditions" as a viable meta-ethics. Am J Bioeth Neurosci. 2011;2(2):17–9.

31. Fins J. Rights come to mind: brain injury, ethics, and the struggle for consciousness. New York: Cambridge University Press; 2015.
32. Farah M. The ethical, legal and societal impact of neuroscience. Annu Rev Psychol. 2012;63:571–91.
33. Farah M, Hutchinson B, Phelps E, Wagner A. Functional MRI-based lie detection: scientific and societal challenges. Nat Rev Neurosci. 2014;15:123.
34. Farah M, Smith M, Gawuga C, Lindsell D, Foster D. Brain imaging and brain privacy: a realistic concern? J Cogn Neurosci. 2009;21(1):119–27.
35. Farah M, Illes J, Cook Degan R, Gardner H, Kandel E, King P, et al. Neurocognitive enhandement: what can we do and what should we do? Nat Rev Neurosci. 2004;5:421.
36. Levy N. Neuroethics: ethics and the sciences of the mind. Philos Compass. 2009;4(1):69–81.
37. Glannon W. Neuroethics. Bioethics. 2006;20(1):37–52.
38. Glannon W. Brain, body and mind: neuroethics with a human face. New York: Oxford University Press; 2011.
39. Northoff G. What is neuroethics? Empirical and theoretical neuroethics. Curr Opin Psychiatry. 2009;22:565–9.
40. Churchland P. Braintrust: what neuroscience tells us about morality. Princeton: Princeton University Press; 2011.
41. Gazzaniga M. The ethical brain. Chicago: University of Chicago Press; 2005.
42. Levy N. Neuroethics: a new way of doing ethics. Am J Bioeth Neurosci. 2011;2(2):3–9.
43. Greene J. Moral tribes: emotion, reason, and the gap between us and them. London: Atlantic Books; 2015.
44. Wagner NF, Northoff G. A fallacious jar? The peculiar relation between descriptive premises and normative conclusions in neuroethics. Theor Med Bioeth. 2015;36:215–35.
45. Shook J, Giordano J. A principled and cosmopolitan neuroethics: considerations for international relevance. Philos Ethics Humanit Med. 2014;9(1):1.
46. Avram J, Giordano J. Neuroethics: some things old, some things new, some things borrowed and to do. Am J Bioeth Neurosci. 2014;5(4):23–5.
47. Racine E, Bar Ilan O, Illes J. fMRI in the public eye. Nat Rev Neurosci. 2005;6:159.
48. Illes J, Racine E. Imaging or imagining? A neuroethics challenge informed by genetics. In: Glannon W, editor. Defining right and wrong in brain science. New York: Dana Press; 2007.
49. Evers K. Toward a philosophy for neuroethics. EMBO Rep. 2007;8(special issue on neuroscience and society):S48–51.
50. Melo Martin I, Salles A. Moral bioenhancement: much ado about nothing? Bioethics. 2015;29(4):223–32.
51. Levy N. Introducing neuroethics. Neuroethics. 2008;1(1):1–8.
52. Farah MJ, Helberlein AS. Personhood and neuroscience: naturalizing or nihilating? Am J Bioeth. 2007;7(1):37–48.
53. Greene J. The secret joke of Kant's soul. In: Sinnott-Armstrong W, editor. Moral psychology, vol. 3. Cambridge, MA: MIT Press; 2008. p. 35–80.
54. Changeux JP. The physiology of truth. Cambridge, MA: Belknap Press; 2004.
55. Dehaene S, Sergent C, Changeux JP. A neuronal network model linking subjective reports and objective physiological data during conscious perception. Proc Natl Acad Sci U S A. 2003;100:8520–5.
56. Changeux JP, Courrege P, Danchin A. A theory of the epigenesis of neuronal networks by selective stabilization of synapses. Proc Natl Acad Sci U S A. 1973;70:2974–8.
57. Evers K. Neuroethics: a philosophical challenge. Am J Bioeth. 2005;5(2):31–3.
58. Persson I, Savulescu J. The perils of cognitive enhancement and the urgent imperative tp enhance the moral character of humanity. J Appl Philos. 2008;25:162–77.
59. Douglas T. Moral enhancement. J Appl Philos. 2008;25:228–45.
60. Lagercrantz H, Hanson MA, Ment L, Peebles D. The newborn brain: neuroscience and clinical applications. 2nd ed. New York: Cambridge University Press; 2010.
61. Lagercrantz H. I barnets hjarna. Stockholm: Bonnier Fakta; 2005.

62. Collin G, van den Heuvel MP. The ontogeny of the human connectome: development and dynamic changes of brain connectivity across the life span. Neuroscientist. 2013;19(6):616–28.
63. Changeux JP. Neuronal man. New York: Fayard-Pantheon Books; 1983–1985.
64. Kitayama S, Uskul AK. Culture, mind and the brain: current evidence and future directions. Annu Rev Psychol. 2011;62:419–49.
65. Evers K. Can we be epigenetically proactive? In: Metzinger T, Windt JM, editors. Open mind: philosophy and the mind sciences in the 21st century. Cambridge, MA: MIT Press; 2016. p. 497–518.
66. Evers K, Changeux JP. Proactive epigenesis and ethical innovation: a neuronal hypothesis for the genesis of ethical rules. EMBO Rep. 2016;17:1361–4.
67. Salles A. Proactive epigenesis and ethics. EMBO Rep. 2017. 10.15252/embr.201744697. Published online 25.07.2017.
68. Farah M. Neuroethics: an introduction with readings. Cambridge: Cambridge University Press; 2010.
69. Illes J. Neuroethics: defining the issues in theory, practice, and policy. New York: Oxford University Press; 2006.
70. Glannon W. Bioethics and the brain. New York: Oxford University Press; 2006.
71. Illes J. Empirical neuroethics. EMBO Rep. 2007;8(Special Issue):S57–60.

MIX
Papier aus verantwortungsvollen Quellen
Paper from responsible sources
FSC® C105338

If you have any concerns about our products,
you can contact us on
ProductSafety@springernature.com

In case Publisher is established outside the EU,
the EU authorized representative is:
Springer Nature Customer Service Center GmbH
Europaplatz 3, 69115 Heidelberg, Germany

Printed by Libri Plureos GmbH
in Hamburg, Germany